P9-AGV-211

Dictionary of Acronyms and Technical Abbreviations

Springer
London
Berlin
Heidelberg
New York
Barcelona
Hong Kong
Milan
Paris
Singapore
Tokyo

Jakob Vlietstra

Dictionary of Acronyms and Technical Abbreviations

For Information and Communication Technologies and Related Areas

Second Edition

Springer

Jakob Vlietstra

ISBN 1-85233-397-9 2nd edition Springer-Verlag London Berlin Heidelberg
ISBN 3-540-76152-7 1st edition Springer-Verlag Berlin Heidelberg New York

British Library Cataloguing in Publication Data
Vlietstra, J. (Jakob), 1932–
Dictionary of acronyms and technical abbreviations: for information and communication technologies and related areas—2nd ed.
1. Technology—Abbreviations—Dictionaries
I. Title
601.4′8
ISBN 1852333979

Library of Congress Cataloging-in-Publication Data
Vlietstra, J.
Dictionary of acronyms and technical abbreviations: for information and communication technologies and related areas/Jakob Vlietstra.—2nd ed.
p.cm.
ISBN 1-85233-397-9 (alk. paper)
1. Telecommunication—Dictionaries. 2. Telecommunication—Acronyms.
3. Telecommunication—Abbreviations. 4. Information technology—Acronyms.
5. Information technology—Abbreviations. I. Title.

TK5102.V55 2001
621.382′01′48—dc21 00-052664

Typeset by Florence Production, Stoodleigh, Devon
Printed and bound at the Cromwell Press, Trowbridge, Wiltshire
69/3830-543210 Printed on acid-free paper SPIN 10767303

Introduction

This second edition contains a collection of technical abbreviations, acronyms, and identifiers (in short 'terms') that are used in information and communication technologies and other related areas. They have become part of the 'normal' vocabulary in many industries, institutes, organizations and universities. Too often they are used without mentioning what they stand for. The main area covered by this dictionary is Information and Communication Technology (ICT). This includes computer and communication hardware and software, communication networks, the Internet and the World Wide Web, and automatic control. Other areas covered are ICT-related techniques,solutions, products, processes and activities.

The dictionary also contains symbolic names of organizations and institutions directly connected to the subjects listed above, as well as the abbreviated names of conferences, symposia, workshops where the mentioned subject areas are treated. In some cases the standard two-letter country codes are listed between parenthesis at the end of every explained term and the country where the term originated.

This is a reference book that is important for all practitioners and users in the areas mentioned above. Technical publications often omit the meaning of terms and confront the reader with jargon too often difficult to understand.

Readers can use this book as a complete reference guide without having to guess what the letters of the terms stand for. This new edition contains close to 35.000 terms. Approximately ten thousand new items have been added. Obsolete and less relevant terms have been deleted.

To save space certain terms have been combined into one. In many cases terms may refer to more than one meaning. If one of the meanings of a specific letter in a term identifies different words, use is made of the "or"-symbol (|) to indicate that two terms then have been combined into one. For example:

CA Channel Adapter|Attachment
refers to
Channel Adapter or
Channel Attachment

and

CADS Computer Aided Digitizing|Drafting System
refers to
Computer Aided Digitizing System or
Computer Aided Drafting System

This representation has been limited to one "or"-symbol per listed term only. This has been done to avoid definitions that would become too complicated.

Most of the terms are listed in capital letters. Exceptions are those that reflect a certain type of measurement like "ips" (inches per second). A limited amount of terms contain small letters or a combination of small and capital letters but only in situations where the originators of these terms specifically intended their use. However, some terms originally consisting of small letters only appear often in the literature with capital letters. Some terms contain letters that refer to other terms. The meaning of these referred terms are listed between parenthesis in situations where they have more than one definition.

Finally, one should note that the letter X in a term may have meanings that don't start with an X or don't contain this letter at all. Examples are: extreme, extension, index, cross, connection etc.

The sorting has been carried out using the terms as the first key and the definitions as the second key. Special characters (space, comma, slash, quotation marks etc.) have been ignored. Numbers have precedence over letters in the sorting process. No distinction has been made between small and capital letters.

France/California, Summer 2000.

A

a	ab(solute)
A	Ampere
a	anode
a	area
a	atto
AA	Absolute Address(ing)
AA	Analytic Application
AA	Application Architecture
AA	Application Association
AA	Arithmetic Average
AA	Audible Alarm
AA	Auto Answer
AAAI	American Association for Artificial Intelligence
AAB	All-to-All Broadcast
AAC	Advanced Airborne Computer
AAC	Automatic Aperture Control
AAC	Autonomous Activation Condition
AACE	Association for the Advancement of Computing in Education
AACR	Anglo-American Cataloguing Rules
AACS	Advanced Automatic Compilation System
AAD	Analog Alignment Diskette
AAD	Analog Analog Digital
AAD	Automatic Answering Device
AADC	All Application Digital Computer
AADS	Automatic Application Development System
AADSF	Advanced Automatic Directional Solidification Furnace
AAE	Advanced Computational Aerodynamics
AAE	Applied Automated Engineering
AAE	Architecture & Architectural Engineering
AAE	Audible Alarm Equipment
AAEE	American Association of Electrical Engineers
AAES	American Association (of) Engineering Societies
AAGR	Average Annual Growth Rate
AAI	Administrative Authority Identifier
AAID	Alliance Against Internet Defamation
AAIMS	An Analytical Information Management System
AAL	Absolute Assembly Language
AAL	Asynchronous transfer mode Adaption Layer
AAL	ATM Adaptation Layer (ATM = Asynchronous Transfer Mode)
AAL	Automated Assembly Line
AALx	ATM Adaptation Layer protocol x (x = 1, 2, 3, 4, or 5; ATM = Asynchronous Transfer Mode)
AAM	Application Activity Model
AAMAS	Autonomous Agents and Multi-Agent Systems
AAMOF	As A Matter Of Fact

AAMSI	American Association for Medical Systems Informatics
AAO	Authorized Acquisition Objective
AAOS	Automatic Assembly Ordering System
AAP	Analyst Assistance Program
AAP	Applications Access Point
AAP	Associative Array Processor
AAP	Attached Application Processor
AAR	A-Address Register
AAR	Associative Array Register
AAR	Automatic Alternate Routing
AARN	Australian Academic Research Network
AARNET	Australian Academic and Research Network
AARP	AppleTalk Address Resolution Protocol
AARS	Automatic Address Recognition Subsystem
AARTC	Algorithms and Architectures for Real-Time Control (conference)
AAS	Advanced Automated|Automation System
AAS	All-to-All Scatter
AAS	Automatic Addressing System
AASP	ASCII Asynchronous Support Package
AAT	Amos Audio Tool
AAT	Arbitrated Access Timer
AAT	Automatic Assembly Terminal
AAT	Average Access Time
AAU	Address Arithmetic Unit
AAUI	Apple Attachment Unit Interface
AAVD	Automatic Alternate Voice Data
AB	Abort (session)
AB	Address Bus
AB	Alternating Burst
AB	Answer Back
AB	Arithmetic Bus
AB	Asymmetric Balance
aba	automatic bass (compensation or control)
ABA	Automatic Blood Analysis
ABAC	Association of Business and Administrative Computing (US)
ABACUS	Agents and Brokers Automated Computer Users System
ABATS	Automatic Bit Access Test System
ABB	Array of Building Blocks
ABBET	A Broad-Base Environment (for) Test(ing)
ABBR	Abbreviation
ABC	Airborne Computer
ABC	Approach By Concept
ABC	Association for Business Communication
ABC	Atanasoff-Berry Computer
ABC	Automatic Bandwidth Control
ABC	Automatic Binary Computer
ABC	Automatic Brightness Control

ABCS	Automatic Base Communication Systems
ABD	Abbreviated Dialing
ABD	Automatic Block Diagramming
ABDL	Automatic Binary Data Link
ABE	Arithmetic Building Element
ABEL	Advanced Boolean Expression Language
abend	abnormal end
ABF	Application By Forms
ABFD	Affordable Basic Floppy Disk
ABGR	Alpha Blue Green Red
ABI	Application Binary Interface
ABIC	Adaptive Bi-level Image Compression
ABIOS	Advanced Basic Input/Output System
ABIST	Automatic Built-In Self-Test
ABL	Accepted Batch Listing
ABL	Architectural Block (diagram) Language
ABL	Atlas Basic Language
ABL	Automatic Bootstrap Loader
ABLB	Alternate Binaural Loudness Balance
ABLE	Activity Balance Line Evaluation
ABM	Activity Based Management
ABM	Asynchronous Balanced Mode
ABME	Asynchronous Balanced Mode Extended
ABMPS	Automated Business Mail Processing System
ABMS	Automated Batch Manufacturing System
ABN	AT&T Business Network
ABN	Australian Bibliographic Network
ABOL	Adviser Business Oriented Language
ABP	Actual Block Processor
ABP	Alternate Bipolar
ABPS	Automated Bill Payment System
ABR	Automatic Baud Rate
ABR	Automatic Bit Rate (detection)
ABR	Available Bit Rate
ABRS	Automated Book Request System
ABS	Absolute
ABS	Average Busy Season
ABSBH	Average Busy Season Busy Hour
ABSLDR	Absolute Loader
ABT	Abort
ABT	Abort Timer
ABT	Abstract Windowing Toolkit
ABT	Advanced Board Technology
ABT	Answer Back Tone
ABTC	Adaptive Block Truncation Coding
ABTS	ASCII Block Terminal Services
ABVS	Audit Bureau of Verification Services
ABW	Advise By Wire

AC	Accept Command|Control
AC	Accumulator
AC	Acoustic Coupler
AC	Activity|Adaptive Control
AC	Actual Count
AC	Address Check|Counter
AC	Alteration|Alternate Cancellation
AC	Alternating Current
AC	Analog Computer
AC	Answer Center
AC	Aperture Card
AC	Application Context
AC	Association Control
AC	Author's Correction
AC	Automatic Calling|Computer
AC	Automatic Configuration|Control
ACA	Adjacent Channel Attenuation
ACA	Advanced Calculator Attachment
ACA	Advanced Computing Attachment
ACA	Asynchronous Communications Adapter
ACA	Automatic Circuit Analyzer|Assurance
ACAM	Augmented Content-Addressed Memory
ACAMPS	Automated Communications And Message Processing System
ACAP	Advanced Computer for Array Processing
ACAP	Application Configuration Access Protocol
ACAPE	Aerospace Computer Automatic Program Evaluator
ACARS	Aircraft Communication Addressing and Reporting System
ACARS	ARINC Communications Addressing and Reporting System (ARINC = Aeronautical Radio Incorporated)
ACASP	Adaptive systems in Control And Signal Processing (international symposium)
ACASS	Automated Credit Authorization Sub-System
ACAT	Acquisition Category
ACATS	Acquisition Category System
ACB	Access|Adapter Control Block
ACB	Allocate|Application Control Block
ACB	Asynchronous Communications Base
ACBGEN	Application Control Block Generation
ACBH	Average Consistent Busy Hour
ACBR	Accumulator Buffer Register
acc	acceleration
ACC	Accumulator
ACC	Accuracy Control Character
ACC	Active Congestion Control
ACC	Adaptive Control Constraint
ACC	Additive Card Code
ACC	Advanced Concepts Center
ACC	Application Control Code

ACC	Asynchronous Communications Control
ACC	Audio Communications Controller
ACC	Automatic Call-back Calling
ACC	Automatic Chrominance Control
ACC	Automatic Closed-loop Control
acc	automatic colour compensation
ACC	Automatic Colour Control
ACC	Automatic Congestion|Console Control
ACCA	Asynchronous Communications Control Attachment
ACCAP	Autocoder to Cobol Conversion Aid Program
ACCAT	Advanced Command and Control Architectural Testbed
ACCEL	Automated Circuit Card Etching Layout
ACCESS	Access Characteristics Estimation System
ACCESS	Advanced Customer Connection Evolution Systems
ACCESS	Argonne Code Center Exchange and Storage System
ACCESS	Automatic Checking and Control for Electrical Systems Support
ACCESS	Automatic Computer Controlled Electronic Scanning System
ACCH	Associated Control Channel
ACCI	Australian Computer and Communications Institute
ACCLAIM	Advanced Concurrent Constraint Languages And Implementation (EU)
ACCLAIM	Automated Circuit Card Layout And Implementation
ACCOLC	Access Overload Class
ACCORD	Analog Computer Check-Out Routine Digitally
ACCPAC	A Complete and Comprehensive Program for Accounting Control
ACCS	Access
ACCS	Automated Calling Card Service
ACCS	Automated Cross-Sectional Scanning
ACCT	Access Time
ACCT	Account
ACCW	Alternating Current Carrier Wave
ACCY	Accessory
ACCY	Accuracy
ACD	Adaptive Call Distributor
ACD	Advanced Circuit Design
ACD	All-Channel Decoder
ACD	Automatic Call Distribution|Distributor
ACD	Automatic Circuit Design
ACDA	Automatic Call Disposition Analyzer
AC/DC	Alternating Current/Direct Current
ACDF	Access Control Decision Function
ACDI	Asynchronous Communication(s) Device Interface
ACDMS	Automated Control of a Document Management System
ACE	Adaptive Communication Environment
ACE	Adaptive Computer Experiment
ACE	Adaptive Critic(al) Element
ACE	Advanced Composition Explorer (US)

ACE	Advanced Computing Environment(s)
ACE	Adverse Channel Enhancement(s)
ACE	Analytic Computer Equipment
ACE	Asynchronous Communication Element
ACE	Asynchronous Communications Exchange
ACE	Atmospheric Collection Equipment
ACE	Automated Computing Engine
ACE	Automatic Calibration and Equalization
ACE	Automatic Calling|Checkout Equipment
ACE	Automatic Computing Engine|Equipment
ACE	Automatic Control Engineering
ACE	Automatic Cross-connection Equipment
ACEA	Association of Computing in Engineering and Architecture (US)
ACeDB	A Caenorhabditis elegans Data Base
ACEF	Access Control Enforcement Function
ACELP	Algebraic Code-Excited Linear Prediction
ACES	Adaptively Controlled Explicit Simulation
ACES	Applied Computational Electromagnetics Society
ACES	Autodesk's Comprehensive Education Solution
ACES	Automated Code Evaluation System
ACET	Advisory Committee on Electronics and Telecommunications
ACF	Access Control Facility|Field
ACF	Advanced Communication Facility|Function
ACF	Auto-Correlation Function
ACF	Availability Control File
ACF/NCP	Advanced Communications Function/Network Control Program
ACF/TCAM	Advanced Communications Function/Telecommunications Access Method
ACF/VTAM	Advanced Communications Function/Virtual Telecommunications Access Method
ACF/VTAME	Advanced Communications Function/Virtual Telecommunications Access Method Entry
ACG	Advanced Computing Group
ACG	All-Conditions Gear
ACG	Automatic Code Generator
ach	attempts per circuit per hour
ACH	Automated Clearing House
ACH	Automatic Cartridge Handler
ACHEFT	Automated Clearing House Electronic Funds Transfer
ACI	Access Control Information
ACI	Adjacent Channel Interface|Interference
ACI	Alteration Cancellation Installation
ACI	Alternating Current Inputs
ACIA	Asynchronous Communications Interface Adapter
ACIAS	Automated Calibration Interval Analysis System
ACIB	Arithmetic Checker In Bus

ACIC	Aeronautical Charting and Information Center
ACID	Atomicity, Consistency, Isolation, and Durability
ACID	Automatic Cross-referencing and Indexing Document generator
ACIDIC	Associative, Commutative, Identities, Distributive, Inverses, Closure
ACIS	Academic Information System
ACIS	Alan, Charles, Ian, Spatial (Alan Greyer, Charles Lang, Ian Braid, Spatial Technologies)
ACIS	American Committee for Interoperable Systems
ACK	Acknowledge (character)
ACK	Amsterdam Compiler Kit
ACK0	Acknowledgment, even positive
ACK1	Acknowledgment, odd positive
ACKI	Acknowledgment Input
ACKO	Acknowledgment Output
ACL	Accelerator Control Listing
ACL	Access Competency level
ACL	Access Control List
ACL	Advanced Complementary (metal-oxide silicon) Logic
ACL	Agent Communication Language
ACL	Application Control Language
ACL	Audio Communication Line
ACL	Audit Command Language
ACLIS	Australian Council on Library and Information Services
ACLR	Access Control Logging and Reporting
ACLS	Automatic Carrier Landing System
ACM	Access Control Machine|Matrix
ACM	Active Contour Model
ACM	Address Complete Message
ACM	Alterable Control Memory
ACM	Alternating Current Modulation
ACM	Association for Computing Machinery
ACM	Associative Communications Multiplexer
ACM	Asynchronous Communication control Module
ACM	Audio Compression Manager
ACM	Authorized Controlled Material
ACM	Automatic Clutter Mapper
ACME	Automatic Correction of Multiple Errors
ACMS	Advance Configuration Management System
ACMS	Application Control (and) Management System
ACNS	Academic Computing and Network Services
ACNS	Advanced Communications Network Service
ACO	Adaptive Control Optimization
ACO	Administrative Contract Officer
ACO	Alternating Current Outputs
ACO	Automatic Call Origination
ACO	Average Calculation Operation
ACOB	Analog Circuit Observer Block

ACOB	Arithmetic Checker Out Bus
ACOE	Automatic Check-Out Equipment
ACOF	Attendant Control Of Facilities
ACOL	Application Control Language
ACOL	Atlantic Canada On-Line
ACORD	Automatic Component Ordering
ACORN	Advanced Collaborative Open Resource Network
ACORN	Associative Content Retrieval Network
ACORN	Automatic Coder Report Narrative
ACOS	Application Control Operating System
ACOUSTINT	Acoustical Intelligence
ACP	Access Control Points
ACP	Advanced Circuits Program
ACP	Advanced Computational Processor
ACP	Allied Communications Procedures|Publication
ACP	Ancillary Control Processor
ACP	Arithmetic and Control Processor
ACP	Auxiliary Control Process
ACP	Azimuth Change Pulse
ACPA	Association of Computer Programmers and Analysts (US)
ACPI	Advanced Configuration and Power Interface
ACPM	Association Control Protocol Machine
acpt	accept
ACPX	Advanced Circuit Packaging Extended
ACQM	Automatic Circuit Quality Monitoring
ACQUILEX	Acquisition of Lexical knowledge (EU)
ACR	Abandon Call and Retry
ACR	Access Control Register
ACR	Achromatic Colour Removal|Replacement
ACR	Add Character Register
ACR	Address Control Register
ACR	Allowed Cell Rate
ACR	Alternate CPU Recovery (CPU = Central Processing Unit)
ACR	Automatic Carriage Return
ACRD	Anticipated Card Release Date
ACRE	APAR Control Remote Entry (APAR = Automatic Programming And Recording)
ACRE	Automatic Call Recording Equipment
ACRITH	Accuracy Arithmetic
ACRONYM	Abbreviated Coded Rendition Of Name Yielding Meaning
ACROSS	Automated Cargo Release and Operations Service System
ACS	Academic Computing Services
ACS	Access
ACS	Access Control Set
ACS	Access Control Store|System
ACS	Adjacent Channel Selectivity
ACS	Advanced Communications Service|System
ACS	Alternate Channel Suppression

ACS	Alternating Current Synchronous
ACS	Analog Computer|Control System
ACS	Application Control System
ACS	Application Customer Service
ACS	Assembly Control System
ACS	Associated Computer System
ACS	Asynchronous Communication(s) Server
ACS	Attitude Control System
ACS	Australian Computer Society
ACS	Automated Commercial System
ACS	Automatic Call Selector
ACS	Automatic Checkout System
ACS	Automatic Container Stacking crane
ACS	Automatic Control System
ACS	Auxiliary Core Storage
ACSAP	Automated Cross-Section Analysis Program
ACSES	Automated Computer Science Education System
ACSL	Advanced Continuous Simulation Language
ACSMS	Automated Computer Security Management System
ACSNET	Academic Computing Services Network
ACSNET	Australian Computer Science Network
ACSS	Applied Conceptual Sub-Schema
ACSTI	Advisory Committee for Scientific and Technical Information
ACSU	Advanced (t-1) Channel Service Unit
ACSYS	Accounting Computer System
ACT	Abend
ACT	Active Control Technology
ACT	Actual
ACT	Actuarial (programming language)
ACT	Actuator
ACT	Advanced Computer Techniques
ACT	Algebraic Compiler and Translator
ACT	Alternate Current Trigger
ACT	Analogical Circuit Technique
ACT	Audio Conferencing Terminal
ACT	Automated Contingency Translator
ACT	Automatic Code Translation
ACTA	Automatic Computerized Transverse Axial
ACTAS	Alliance of Computer-based Telephony Application Suppliers
ACTD	Advanced Concept Technology Demonstrations
ACTE	Approvals Committee for Terminal Equipment
ACTEL	Alternating Current Thin-film Electro-Luminescence
ACTGA	Attendant Control of Trunk Group Access
ACTI	Activate Invites
ACTLU	Activate Logical Unit
ACTOR	Interactive object Oriented programming language
ACTPU	Activate Physical Unit
ACTRAN	Autocoder-to-Cobol Translation

ACTS	Advanced Channel Testing System
ACTS	Advanced Communications Technologies and Services (EU)
ACTS	Advanced Communications Technology Satellite
ACTS	Application Control and Teleprocessing System
ACTS	Architecture Computerized Test System
ACTS	Automated Computer Time Service
ACTSU	Association of Computer Time-Sharing Users (US)
ACTT	Advanced Communication and Timekeeping Technology
ACTYS	Automatic Computerized Type-Setting
ACU	Address|Alarm Control Unit
ACU	Arithmetic Control Unit
ACU	Association of Computer Users (US)
ACU	Autocall Unit
ACU	Automatic Calling Unit
ACUA	Automatic Calling Unit Adapter
ACUG	Association of Computer User Groups (US)
ACUTE	Accountants Computer Users Technical Exchange
ACV	Access|Address Control Vector
ACV	Address Control Vector
ACVS	Automatic Computer Voltage Stabilizer
ACW	Access Control Word
ACW	Alternating Continuous Waves
ACWC	Analog Carrier Wave Computer
ACX	Active Control Experts
ACY	All Cycles
AD	Activity Discard
AD	Addendum
AD	Adder
AD	Address
A-D	Advance-Decline (line)
A/D	Analog/Digital
AD	Andorra
AD	Application Development|Documentation
AD	Application Dialogue
AD	Architectural Design
AD	Area Discriminator
ADA	Action Data Automation
ADA	Activity Discard Acknowledgment
ADA	Administrative Data Acquisition
ADA	Advanced Design Aids
ADA	Analog Differential Analyzer
ADA	Analog-Digital-Analog
ADA	Automatic Data Acquisitions (programming language named after Augusta, Lady Ada Lovelace)
ADAAC	Automatic Data Acquisition And Computer
ADABAS	Adaptable Data Base System
ADAC	Analog-Digital Attitude Converter
ADAC	Analog to Digital/digital to Analog Converter

ADAC	Automatic Design Automatic Correction
ADAC	Automatic Detection And Correction
ADAC	Automatic Diagnosis Automatic Correction
ADAC	Automatic Direct Analog Computer
ADACS	Automated Data Acquisition and Control System
ADAEX	Automatic Data Acquisition and Executive
ADAIC	ADA (language) Information Clearinghouse
ADAL	Action Data Automation Language
ADALINE	Adaptive Linear Elements
ADALINE	Adaptive Linear Neurons
ADAM	Adaptive Dynamic Analysis and Maintenance
ADAM	Advanced Data Access Method
ADAM	Advanced Data Management (system)
ADAM	Advanced Detection Availability Manager
ADAM	Architecture for Design, Archiving and Manufacturing
ADAM	Automatic Design Archiving Management system
ADAM	Automatic Direct Access Management
ADAMS	Automatic Dynamic Analysis of Mechanical Systems
ADAPS	Automated Design And Packaging Service
ADAPSO	Association of Data Processing Service Organizations (US)
ADAPT	Advanced Data Adapter/Processor Tester
ADAPT	Automated Data Analysis and Planning Technique
ADAPT	Automatic Density Analysis Profile Technique
ADAPTICOM	Adaptive Communications
ADAPTS	Analog/Digital/Analog Process and Test System
ADAS	Adaptive Digital Acquisition Sampling
ADAS	Airborne Data Acquisition System
ADAS	Automatic Disk Allocation System
ADAS	Auxiliary Data Annotation Set
ADAT	Automatic Data Accumulation and Transfer
ADAU	Auxiliary Data Acquisition Unit
ADB	Active Data Base
ADB	A Debugger
ADB	Agreement Data Base
ADB	Apple Desktop Bus
ADBS	Advanced Data Base System
ADBT	Access Decision Binding Time
ADC	Action Delay Character
ADC	Actual Device Code
ADC	Adaptive Data Compression
ADC	Add with Carry
ADC	Airborne Digital Computer
ADC	Air Data Computer
ADC	Analog Derived Clock
ADC	Analog-Digital Control
ADC	Analog-to-Digital Converter
ADC	Analysis Date Concentrator
ADC	Asynchronous Digital Combiners

ADC	Automatic Device Control
ADC	Automatic Digital Computer
ADCAD	Airways Data Collection And Distribution
ADCAPT	AC/DC Chip Automatic Production Tester
ADCAT	Automatic Data Correction And Transfer
ADCC	Advanced Data Communication Control
ADCC	Analog Direct Current Computer
ADCC	Asynchronous Data Communications Channel|Controller
ADCCP	Advanced Data Communication(s) Control Procedures|Protocol
ADCHEM	Advanced control in Chemical processes (international symposium)
ADCI	Automatic Display Call Indicator
ADCIS	Association for the Development of Computer-based Instruction Systems
ADCOM	Association of Data Center Owners and Managers (US)
ADCON	Address Constant
ADCON	Analog-to-Digital Converter
ADCP	Advanced Data Communication Protocol
ADCR	All Digital Control Runs
ADCS	Automatic Digital Circuit Switch
ADCU	Association of Data Communications Users (US)
AD/Cycle	Application Development Cycle
ADD	Addition
ADD	Address
ADD	Advanced (optical) Disk
ADD	Algebraic Decision Diagram
ADD	Analog-Digital-Digital
ADD	Automatic Document Detection|Distribution
ADDAC	Analog Data Distributor And Computer
ADDACE	Analog-to-Digital, Digital-to-Analog Conversion Equipment
ADDAM	Adaptive Dynamic Decision Aiding Methodology
ADDAR	Automatic Digital Data Acquisition and Recording
ADDAS	Automatic Digital Data Assembly System
ADDD	A Depository of Development Documents
ADDDS	Automatic Direct Distance Dialing System
ADDF	Address Field
ADDL	Additional
ADDM	Addendum
ADDM	Automated Drafting and Digitizing Machine
ADDMD	Administrative Data and Directory Management Domain
ADDR	Adder
ADDR	Address Register
ADDS	Advanced Data Display System
ADDSB	Address Disable
ADDSRTS	Automated Digitized Document Storage, Retrieval and Transmission System
ADE	ADA Development Environment (ADA = programming language ADA)

ADE	Advanced Data Entry
ADE	Application Development Environment
ADE	AutoCad Data Extension
ADE	Automated Design (and) Engineering
ADE	Automated Design Equipment
ADE	Automatic Data Exchange
ADE	Automatic Draughting Equipment
ADEE	Automated Design & Engineering for Electronics (conference)
ADEM	Automatic Data Equalized Modem
ADEPT	Analog Data Extractor & Plotting Table
ADEPT	Automatic Dynamic Evaluation by Programming Testing
ADES	Automated Data Entry System
ADES	Automatic Digital Encoding System
ADEV	Appletalk Device
ADEW	Advanced Discrete Electrical Wiring
ADEW	Andrew Development Environment Workbench
ADEX	Advanced Data Entry Executive
ADF	Access control Decision Function
adf	adapter description file (file extension indicator)
ADF	Adaptive Development Framework|Facility
ADF	Advanced Disk File
ADF	Application Development Facility
ADF	Automatic Direction Finder
ADF	Automatic Document Feed(er)
ADGA	Acyclic Directed Graph with Attributes
ADHA	Analog Data Handling Assembly
ADI	Access control Decision Information
ADI	Alternate Digital Inversion
ADI	Alternating Direction Implicit|Iterative
ADI	Application Data Interface
ADI	Application Directory
ADI	Autocad Device Interface (driver)
ADI	Automatic Direction Indicator
ADIO	Analog Digital Input/Output
ADIOS	Automatic Diagnosis|Digital Input/Output System
ADIP	Alternating Direction Implicit Procedure
ADIP	Automated Data Interchange systems Panel
ADIS	A Data Interchange System
ADIS	Administrative Information System
ADIS	Automatic Data Interchange System
ADIT	Analog-to-Digital Integrating Translator
ADIT	Automatic Detection and Integrated Tracking
ADJ	Adjustable
ADL	Acoustic Delay Line
ADL	A Data Language
ADL	Address Data Latch
ADL	Advances in Digital Libraries (conference)
ADL	Application program interface Definition Language

ADL	Architecture Description Language
ADL	Asymmetrical Digital subscriber Line
ADL	Automated Display|Draughting Language
ADL	Automatic Data Link|Logger
ADLC	Adaptive Lossless Data Compression
ADLC	Advanced Data Link Controller
ADLC	Asynchronous Data Link Control
ADLIB	Algorithmic Description Language for the Internal Behaviour of systems
ADLIPS	Automatic Data Link Plotting System
ADLM	Account Data List Management
AdLM	Autodesk License Manager
ADLS	Airborne Data Link System
ADLT	Airborne Data Link Terminal
ADM	Activity Data Method
ADM	Adaptive Database Manager
ADM	Adaptive Delta Modulation
ADM	Advanced Development Model(s)
ADM	Application Description Manual
ADM	Application Development and Maintenance
ADM	Architecture Development Method
ADM	Asynchronous Disconnected Mode
ADMA	Advanced Direct Memory Access
ADMACS	Apple Document Management And Control System
ADMD	Administrative Management Domain (X.400)
ADMINID	Administrative Identification
ADMIRE	Automated Diagnostic Maintenance Information Retrieval
ADMIRES	Automated Diagnostic Maintenance Information Retrieval System
ADMIS	Automated Data Management Information System
ADMM	Application Data Mobilizer and Manager
ADMS	Advanced Data Management System
ADMS	Automated Document Management System
ADMS	Automatic Digital Message Switch(ing)
ADMSC	Automatic Digital Message Switching Center(s)
ADMU	Add-Drop Multiplexing Unit
ADN	Abbreviated Dialing Number
adn	add in utility (file extension indicator)
ADN	Autodesk Developers Network
ADO	Active|ActiveX Data Object
ADO	Active Desktop Objects
ADO	Advanced Development Objective
ADONIS	Automatic Digital On-line Instrumentation System
ADOPT	Approach to Distributed On-line Processing Transaction
ADOS	Advanced Diskette Operating System
ADOT	Automatic Digital Optical Tracker
ADP	Acoustic Data Processor
ADP	Administrative Data Processing

ADP	Advanced Data Processing
ADP	Application Design Package
ADP	Association of Database Producers (GB)
ADP	Assumed Decimal Point
ADP	Automated|Automatic Data Processing
ADP	Automated Deferment Process
ADP	Automatic Duplicator Printer
ADPACS	Automated Data Processing And Communications Service
ADPC	Automatic Data Processing Center
ADPCM	Adaptive Delta Pulse Code Modulation
ADPCM	Adaptive Differential Pulse Code Modulation
ADPD	Automatic Display Power Down
ADPE	Automatic Data Processing Equipment
ADPES	Automatic Data Processing Equipment and Software
ADPM	Automatic Data Processing Machine
ADPREP	Automatic Data Processing Resource Estimating Procedures
ADPS	Automatic Data Processing System
adpt	adapter
ADP/TM	Automatic Data Processing/Telecommunications Management
ADR	Adder
ADR	Address
ADR	Alternate Data Retry
ADR	Analog-Digital Recorder
ADR	Application Definition Record
ADR	Automatic Data Retrieval
ADR	Automatic Dialogue Replacement
ADR	Automatic Digital Relay
ADRA	Automatic Dynamic Response Analyzer
ADRAC	Automatic Digital Recording And Control
ADRAM	Alien Device Recognition Access Mechanism
ADRCHK	Address Check
ADREN	Address Enable
ADRG	Anti-reflective coating Digitized Raster Graphics
ADRI	Add-Delete Rework Instruction
ADRM	Analog/Digital Re-Master
ADRMPS	Auto-Dialed Recorded Message Players
ADROIT	Automated Data Retrieval and Operations Involving Time series
ADRS	Adaptive Data Reporting System
adrs	address
ADRS	A Departmental Reporting System
ADRS	Analog-to-Digital Recording System
ADRS	Asset Depreciation Range System
ADRT	Analog Data Recording Transcriber
ADS	Accurately Defined System
ADS	Activity Data Sheet
ADS	Additional Device Support
ADS	Advanced Data System

ADS–ADTS

ADS	Advanced Digital|Debugging System
ADS	AION Development System (AION = Automatic Input/Output for Network environments)
ADS	Analog Digital Subsystem
ADS	Analog-to-Digital Sensing
ADS	Application Development Solutions|System
ADS	Astrophysics Data System
ADS	Autocad Development System
ADS	Autographed Document Signed
ADS	Automated Design|Diagnostic System
ADS	Automatic Distribution System
ADS	Auxiliary Data System
ADSC	Address Status Changed
ADSC	Automatic Data Service Center
ADSE	Alternative Delivery Schedule Evaluator
ADSEL	Address Selective
ADSG	Alternative Delivery Schedule Generator
ADSI	Analog Digital Subscriber Interface
ADSI	Analog Display Services Interface
ADSL	Asymmetric(al) Digital Subscriber Line|Loop
ADSL	Asynchronous Digital Subscriber Loop
ADSM	Adstar Distributed Storage Manager
ADSM	Advanced Distributed Storage Manager
ADSOC	Administrative Support Operation Center
ADSOL	Analysis of Dynamical Systems On-Line
ADSP	Advanced Digital Signal Processor
ADSP	AppleTalk Data Stream Protocol
ADSP	Automatic Data Set Protection
ADSS	Advanced Software System
ADSTAR	Advanced Storage And Retrieval
ADSTAR	Automatic Document Storage And Retrieval
ADSU	ATM Data Service Unit (ATM = Asynchronous Transfer Mode)
ADSUP	Automated Data Systems Uniform Practices
ADT	Abstract Data Type
ADT	Active Disk Table
ADT	Advanced DRAM Technology
ADT	Application Data Types
ADT	Application Development Tool
ADT	Asynchronous Data Transceiver|Transfer
ADT	Attribute Distributed Tree
ADT	Automatic Data Translator
ADT	Automatic Detection and Tracking
ADT	Autonomous Data Transfer
ADTD	Anticipated Data Transmission Date
ADTD	Anticipated Design Transmit Date
ADTD	Association of Data Terminal Distributors (US)
ADTR	Average Date Transfer Rate
ADTS	Amplitude Degradation Test System

ADTS	Automated Data and Telecommunications Service
ADU	Automatic Data Unit
ADU	Automatic Dialing Unit
ADV	Advanced
ADV	Array Dope Vector
ADV	Automatic Design Verification
ADVE	Analytical Deviations in the Vertical Error
ADVFS	Advanced File System
ADVICE	Aid in Design Verification for Integrated Circuit Engineering
ADX	Asymmetric|Automatic Data Exchange
ADX	Automatic Digital Exchange
AE	Above or Equal
AE	Acoustic Emission
AE	Activity End
AE	Application Engineering|Execution
AE	Application Entity|Environment
AE	Approximately Equal
AE	Arithmetic Element
AE	Automatic Exchange
AE	United Arab Emirates
AEA	Activity End Acknowledgment
AEB	Analog Expansion Bus
AEB	Automated Estimating & Buying
AEC	Active Elementary Components
AEC	Adaptive Echo Cancellation
AEC	Advance Exit Common
AEC	Architecture, Engineering and Construction
AEC	Automatic Editing|Energy Control
AEC	Automatic Error Correction
AECT	Association for Educational Communications and Technology
AECU	Arithmetic Element Control Unit
AED	Advanced Engineering Data
AED	Association of Equipment Distributors (US)
AED	Automated Engineering Design
AED	Automatic Error Detection
AEDA	An Experimental Design Approach
AEDF	Advanced Engineering Data Form
AEDPS	Automated Engineering Documentation Preparation System
AEDS	Advanced Electronic Distribution System
AEDS	Advanced Engineering Data Set
AEDS	Association for Educational Data Systems (US)
AEF	Access control Enforcement Function
AEF	Address Extension Facility
AEG	Active Element Group
AEGIS	An Existing General Information System
AEI	Application Entity Information
AEI	ATM Electrical Interface (ATM = Asynchronous Transfer Mode)
AEIMP	Apple Event Inter process Messaging Protocol

AEIMS	Administrative Engineering Information Management System
AEL	Active Edge List
AEL	Audit Entry Language
AEP	AppleTalk Echo Protocol
AEP	Application Environment Profile
AER	Auditory Evoked Response
AERIS	Automatic Electronic Ranging Information System
AES	Advanced Encryption Standard
AES	Application Environment Specification
AES	Audio Engineering Society (US)
AES	Authentication and Encryption Service
AES	Automatic Extraction System
AESC	Automatic Electronic Switching Center
AESOP	An Evolutionary System for On-line Planning
AESOP	Angewandte EDV-Systems fuer Optimierung gmbh (DE) (EDV = Elektronische Daten Verarbeitung)
AESOP	Automated Engineering and Scientific Optimization Programming
AESS	Aerospace and Electronics Systems Society
AET	All Events Trace
AET	Application Entity Title
AET	Automatic Exchange Tester
AEVS	Automatic Electronic Voice Switch
AF	Address Family (sockets)
AF	Address Field
AF	Advanced Function
AF	Afghanistan
AF	Arithmetic Flag
AF	Aspect Factor
AF	Audio Frequency
AF	Auxiliary carry Flag
AFA	Abstract Family of Acceptors
AFA	Accelerated File Access
AFACTS	Automatic Facilities Test System
AFADS	Automatic Force Adjustment Data System
AFAIC	As Far As I'm Concerned
AFAICS	As Far As I Can See
AFAICT	As Far As I Can Tell
AFAIK	As Far As I Know
AFAIR	As Far As I Remember
AFC	Advanced Fiber Communication(s)
AFC	Automatic Fare Collection
AFC	Automatic Field|Flight Control
AFC	Automatic Filing Cabinet(s)
AFC	Automatic Font Change
AFC	Automatic Frequency Control
AFCAD	Automatic File Control And Documentation
AFCE	Automatic Flight Control Equipment

AFCET	Association Française pour la Cybernétique Economique et Technique
AFCS	Adaptive Flight Control Subsystem|System
AFCS	Air Flight Control System
AFCS	Attitude Flight Control System
AFCS	Automatic Fare Collection System
AFD	Adaptive Forward Differing
AFD	Automatic File Distribution
AFE	Apple File Exchange
AFE	Application Functional Entity
AFEI	Association For Enterprise Integration
AFES	Automatic Feature Extraction System
AFF	A Flip-Flop
AFG	Analog Function Generator
AFG	Analytic|Automatic Function Generator
AFG	Arbitrary Function Generator
AFI	AppleTalk Filing Interface
AFI	Authority and Format Identifier
AFI	Automatic Fault Isolation
AFII	Association for Font Information Interchange
AFIRM	Automated Fingerprint Image Reporting and Matching (system)
AFIS	Automated Financial Information System
AFIS	Automated Fingerprint Identification System
AFIS	Automatic Flight Information|Inspection System
AFK	Away From Keyboard
AFL	Abstract Family of Languages
AFL	Active Face List
AFM	Abrasive Flow Machining
afm	Adobe font metrics (file extension indicator)
AFM	Application Functions Module
AFMR	Anti Ferro-Magnetic Resonance
AFN	Absolute Frame Number
AFN	Available For Net
AFNOR/CEF	Association Française de Normalisation/Comité Electronique Française
AFO	Addressed|Advanced File Organization
AFO	Application Functional Operation
AFOS	Advanced Field Operating System
AFOS	Automated Field Operations and Services
AFOS	Automation of Field Operations and Services
AFP	Advanced Flexible Processor
AFP	Advanced Flowcharting Package
AFP	Advanced Function Presentation|Printing
AFP	AppleTalk Filing Protocol
AFP	Automatic Floating Point
AFP	Automatic Flowcharting Package|Program
AFPA	Automatic Flow Process Analysis

AFPC	Automatic Frequency/Phase Control
AFR	Advanced Fault Resolution
AFR	Application Function Routine
AFR	Arithmetic Factor Register
AFR	Automatic Field|Format Recognition
AFR	Automatic Frequency Control
AFRC	Automatic Frequency Ration Controller
AFRP	ARCNET Fragmentation Protocol
AFS	Andrew File System
AFSK	Audio Frequency Shift Keying
AFSM	Asynchronous Finite State Model
AFT	Active File Table
AFT	Analog Facility Terminal
AFT	Application File Transfer
AFT	Automated Fund(s) Transfer(s)
AFT	Automatic Fine Tuning
AFT	Available For Test
AFTAD	Analysis-Forecast Transport And Diffusion
AFTEL	Association Française de Télématique
AFTI	Automated Filing of Tariffs and Information system
AFTP	Anonymous File Transfer Protocol
AFTS	Automated Funds Transfers System
AFU	Autonomous Functional Unit
AFUTT	Association Française des Utilisateurs du Téléphone et des Télécommunications
AFUU	Association Française des Utilisateurs d'Unix
AFW	Auxiliary Function Word
AG	Address Generator
AG	Antigua and Barbuda
AG	Application Generation
AG	Association Graph
AG	Attribute Grammar
Ag	Silver
AGA	Advanced Graphic Adapter
AGBH	Average Group Busy Hour
AGC	Automatic Gain|Generation Control
AGCA	Automatic Ground Controlled Approach
AGCL	Automatic Ground Controlled Landing
AGCR	Automatic Gain Control Range
AGCS	Aktien-Gesellschaft Communication Systems (DE)
AGCS	Automatic Ground Control System
AGCSC	Automatic Ground Control System Computer
AGDTE	Aging Date
AGE	A Generalized Expert system
AGE	Automatic Guidance Electronics
AGEC	Automatic Ground Environment Computer
AGFS	Aviation Gridded Forecast System
AGI	Association for Geographic Information

AGL	Above Ground Level
AGL	Apple's Graphic Library
AGL	Atelier de Génie Logiciel (FR)
AGP	Accelerator|Advanced Graphics Port
AGRAS	Anti-Glare anti-Reflective Anti-Static
AGS	Asynchronous Gateway Server
AGT	Arithmetic Greater Than
AGU	Address Generation Unit
AGV	Automated Guided Vehicle
AGVS	Automated Guided Vehicle System
AH	Acceptor Handshake
ah	ampere-hour
AHAC	Automatic High-Accuracy Comparator
AHB	Area Hierarchy Builder
AHC	Association for History and Computing
AHCS	Advanced Hybrid Computing System
AHDL	Analog Hardware Description Language
AHP	Analytic(al) Hierarchy Process
AHR	Add Half-word Register
AHR	Address Hold Register
AHT	Average Holding Times
A&I	Abstracting & Indexing
AI	Adobe Illustrator
AI	Analog Input
AI	And Inverter
AI	Anguilla
AI	Application Identifier
AI	Artificial Intelligence
AI	Authentication Information
AI	Automatic Identification
AIA	Accident/Incident Analysis
AIA	Aerospace Industries Association (US)
AIA	Applications Integration Architecture
AIA	Automation I (one) Association
AIB	Analog Input Base
AIB	Asynchronous Interface Board
AIC	AIXwindows Interface Composer
AIC	Application Interpreted Construct
AIC	Automatic Intercept Center
AICC	Automatic Intercept Communications Controller
AICCP	Association of the Institute for Certification of Computer Professionals
AICK	Apple Internet Connection Kit
AICS	Automated Industrial Control System
AID	Advanced Information Development
AID	Algebraic Interpretative Dialogue
AID	Analog Interface Device
AID	Attention Identifier

AID	Automated Industrial Drilling
AID	Automatic Information Distribution
AID	Automatic Interaction Detection|Detector
AID	Automatic Interactive Design
AIDA	Artificial Intelligence Development Approach
AIDAPS	Automatic Inspection, Diagnostic And Prognostic System
AIDAS	Advanced Instrumentation and Data Analysis System
AIDDE	Ames' Interactive Dynamic Display Editor
AIDE	Accountability In Data Entry
AIDE	Automated Integrated Design and Engineering
AIDE	Automatic In-line Device Evaluator
AIDES	Automated Image Data Extraction System
AIDE/TPS	Advanced Interactive Data Entry/Transaction Processing System
AIDS	Acoustic Intelligence Data System
AIDS	Advanced Impact Drilling System
AIDS	Advanced Interactive Debugging|Display System
AIDS	Advanced Interconnection Development System
AIDS	Aircraft Integrated Data System
AIDS	All-purpose Interactive Debugging System
AIDS	Automated and Integrated Design System
AIDS	Automatic Illustrated Documentation System
AIDS	Automatic Integrated Data System
AIDS	Automatic Interactive Debugging System
AIDS	Automatic Inventory Dispatching System
AIEE	American Institute (of) Electrical Engineers
AIET	Average Instruction Execution Time
AIF	Addressless Instruction Format
AIF	Audio Interchange Format
AIFF	Amiga Image File Format
AIFF	Audio Identification Friend or Foe
AIFF	Audio Interchange File Format
AIFS	Automatic Invoice Forwarding System
AIFU	Automated Instruction Fetch Unit
AIG	Address Indicating Group
AIG	Applications Inter-operability Guidelines
AII	Advanced Internet Integration
AIIM	Association for Information and Image Management (US)
AIIP	Association (of) Independent Information Professionals
AIIT	Asynchronous Interrupt Interpretation Table
AIL	Arithmetic Input Left
AIL	Array Interconnection Logic
AILAS	Automatic Instrument Landing Approach System
AILT	After Image Log Tape
AIM	Access Isolation Mechanism
AIM	Advanced Informatics in Medicine
AIM	Alarm Inhibit Signal
AIM	Analog Intensity Modulation
AIM	Appended Intelligent Module

AIM	Application and Integration Middleware
AIM	Application Interpreted Model
AIM	Asset Inventory Management
AIM	Association for Interactive Media
AIM	Associative Index Method|Model
AIM	Asynchronous Intelligent subscriber Module
AIM	Asynchronous Interface Module
AIM	Automated Inventory Management
AIM	Automatic Identification Manufacturers
AIMDM	Artificial Intelligence in Medicine & medical Decision Making (joint European conference)
AIMIS	Advanced Integrated Modular Instrumentation System
AIMS	Advanced Information Memos
AIMS	Advanced Inventory Management System
AIMS	Auto-adaptive Inventory Management System
AIMS	Auto-Indexing Mass Storage
AIMS	Automated Industrial Management System
AIMS	Automated Integrated Manufacturing System
AIN	Advanced Intelligent Network
AIOD	Automatic Identification of Outward Dialling
AIP	Alphanumeric Impact Printer
AIP	Automated Imagery Processing
AIP	Automated Improvement Process
AIPL	Alternative Initial Program Load
AIPLA	American Intellectual Property Law Association
AIPR	Average Input Pulse Rate
AIPS	Astronomical Image Processing System
aips	average instructions per second
AIPU	Associative Information Processing Unit
AIR	Acoustic Intercept Receiver
AIR	Alternative Internet Resource
AIR	Apple Internet Router
AIR	Arithmetic Input Right
AIR	Association for Interactive Research
AIR	Attention Interrupt Request
AIR	Automated Interrogation Routine
AIRES	Automated Information Resource System
AIRS	Alliance of Information and Referral Services
AIRS	Artificial Intelligence Research Support
AIRS	Automatic Image Retrieval System
AIRTC	Artificial Intelligent based on Real-Time Control (international symposium)
AIS	Action Item System
AIS	Administrative Information System
AIS	Advanced Information System(s)
AIS	Advanced Instructional System
AIS	Alarm Indication Signal
AIS	Analog Input Section|System

AIS	Application Installation Service
AIS	Application Interface Specification
AIS	Automated Information System
AIS	Automatic Imaging Systems
AIS	Automatic Intercept System
AISB	Artificial Intelligence and the Simulation of the Brain
AISB	As I Said Before
AISC	Assessment and Information Services Center
AISC	Association of Independent Software Companies (US)
AISI	As I See It
AISP	Association of Information Systems Professionals (US)
AISS	Automated Information Service System
AISSP	Automated Information Systems Security Program
AIT	Advanced Information Technology
AIT	Advanced Intelligent Tape
AIT	And Inverter Terminate
AITO	Association Internationale pour les Technologies Objets
AIU	Abstract Information Unit
AIUI	As I Understand It
AIV	Aviation Impact Variable
AIX	Advanced Interactive Executive
AJ	Anti-Jam
AJAR	Automatic Junction Analyzer and Recorder
AJD	Anti-Jam Display
AJG	Automatic Job stream Generator
AJR	Automatic Job Recovery
AK	Acknowledge
AK	Alaska (US)
AKA	Also Known As
AKCL	Austin Kyoto Common Lisp
AKM	Automatic Key Management
AKO	A Kind Of
AKR	Address Key Register
AL	Alabama (US)
AL	Albania
Al	Aluminum
AL	Application Layer
AL	Arm Language
AL	Artificial|Assembly Language
AL	Asynchronous Line
ALA	All Letters Answered
ALABOL	Algorithmic And Business Oriented Language
ALADIN	Algebraic Automated Digital Iterative Network
ALAP	AppleTalk Link Access Protocol
ALAP	AppleTalk-LocalTalk link Access Protocol
ALAP	Arctalk Link Access Protocol
ALAP	Associative Linear Array Processor
ALARM	Automatic Log And Restart Mechanism

ALAS	Array Logic And Storage
ALB	Arithmetic and Logic Box
ALB	Assumed Leading Bit
ALBO	Automatic Line Build-Out
ALC	Adaptive Learning Control
ALC	Adaptive Logic Circuit
ALC	Airline Link Control
ALC	Analog Leased Circuit
ALC	Arithmetic and Logic Circuits
ALC	Assembler Language Code
ALC	Automatic Level|Load Control
ALC	Automatic Library Call
ALC	Automatic Light Compensation
ALCAT	Allocator
ALCOM	Algebraic Compiler|Computer
ALCOM	Algorithms and Complexity (EU)
ALCOM-IT	Algorithms and Complexity in Information Technology (EU)
ALCU	Arithmetic Logic and Control Unit
ALCU	Asynchronous Line Control Unit
ALD	Advanced Logic Design
ALD	Asynchronous Limited Distance
ALD	Automated Logic Design|Diagram
ALDEP	Automated Layout Design Program
ALDP	Automatic Language Data Processing
ALDPS	Automatic Language Data Processing System
ALDS	Analysis of Large Data Sets
ALDS	Automatic Literature Distribution System
ALE	Address Latch Enable
ALE	Analog Local Exchange
ALE	Application Link Enabling
ALE	Application Linking and Embedding
ALE	Automatic Line Equalization
ALEC	Algebraic Exponents and Coefficients
ALEC	Analysis of Linear Electronic Circuits
ALEC	Authorized Linux Education Center
ALEG	Address Lifetime Extension Group
ALEP	Advanced Language Engineering Platform
ALERT	Architecture to Logic Equation Realization Technique
ALERT	Automated Linguistic Extraction and Retrieval Technique
ALEX	Automatic Login Executor
ALF	Application Library File
ALF	Automatic Line Feed
ALFA	Automated Loading of Features and Assemblies
ALFA	Automatic Ledger Feed Attachment
ALFA	Automatic Line Fault Analysis
ALFC	Automatic Load Frequency Control
ALFE	Analog Line Front End
ALG	Algebraic

ALG	Asynchronous Line Group (interface)
ALGOL	Algorithmic Language
ALI	Automatic Location Information
ALIA	Australian Library and Information Association
ALIBI	Adaptive Location of Inter networked Bases of Information
ALICE	Application Language Interface Conversion and Extension
ALIS	Advanced Life Information System
ALIT	Automatic Line Insulation Test
ALIWEB	Archie Like Indexing in the Web
ALK	Automated Lamellar Keratoplasty
ALL	Address Locator Logic
ALL	Application Load List
ALLPS	Agreement for Local Licensed Program Support
ALM	Alarm Master
ALM	AppleWare Loadable Module
ALM	Approximate Linear Model
ALM	Asynchronous Line Module|Multiplexer
ALMS	A Library Management System
ALMS	Automated Logic Mapping|Matrix System
ALN	Asynchronous Learning Network
ALN	Attribute Level Number
ALNA	Automatic Logic Network Analysis
ALNICO	Aluminum Nickel Cobalt
ALO	At Least Once
ALOHA	American Limited Online Hours Act
ALOHANET	Aloha Network
ALP	Abstract Local Primitive
ALP	Agreement for Licensed Programs
ALP	Allocation and Loading Program
ALP	Arithmetic and Logic Processor
ALP	Assembly Language Program
ALP	Asynchronous Line Pairs
ALP	Automated Learning Process
ALPS	Advanced Linear Programming System
ALPS	A Language for Process Specification
ALPS	Associative Logic Parallel|Processor System
ALRS	Arithmetic Logic Register Stack
ALRU	Automatic Line Record Update
ALS	Advanced Low-power Schottky
ALS	Application Layer Structure
ALS	Arithmetic Logic Section
ALS	Automated List Service
ALSC	Automated Level and Slope Control
ALSI	Associazione nazionale dei Laureati in Scienze dell'informazione e Informatica (IT)
ALSPEC	Automated Laser Seeker Performance Evaluation System
ALT	Algebraic Language Translator
ALT.	Alternate lifestyle

ALT Alternate (mode)
ALTAC Algebraic Translator And Compiler
ALTAC Algebraic Translator Assembler Compiler
ALTAPE Automatic Line Tracing And Processing Equipment
ALTARE Automatic Logic Testing And Recording Equipment
ALTINSAR Alternate Instruction Address Register
ALTRAN Algebra Translator
ALTRAN Algorithmic Language Translator
ALU Advanced Logical Unit
ALU Application Layer User
ALU Arithmetic and Logic(al) Unit
ALU Asynchronous Line Unit
AM Access Manager|Method
AM Accounting Machine
AM Active Monitor
AM Address(ing) Mode
AM Address Mark(er)|Modifier
AM Administrative Module
AM Alignment Matrix
AM Amplitude Modulation
AM Armenia
AM Asset Management
AM Associative Memory
AM Asynchronous Modem
AM Automated Mapping
AMA Asset Management Account
AMA Associative Memory Address|Array
AMA Asynchronous Multiplexer Adapter
AMA Automatic Message Accounting
AMACS Automatic Message Accounting Collection System
AMA/MTR Automatic Message Accounting – Magnetic Tape Recording
AMANDA Advanced Maryland Automatic Network Disk Archiver
AMANDA Automatic Measurement And Analysis
AMANDDA Automated Messaging And Directory Assistance
AMAP Advanced Multiprogramming Analysis Procedure
AMAR Analog Multiplexer Address Register
AMARC Automatic Message Accounting Recording Center
AMASE Automatic Message Accounting Standard Entry
AMAT Advanced Maintenance Ability Test
AMAT Automatic Message Accounting Transmitter
AMATIST Analogue and Mixed-signal Advanced Test for Improving System-level Testability (EU)
AMATPS Automatic Message Accounting Tele-Processing System
AMAVU Advanced Modular Audio Visual Unit
AMBAS Adapted Method Base Shell
AMBIT Algebraic Manipulation By Identity Translation
AMC Advanced Memory Chip
AMC Advanced Multimedia Concepts

AMC	Automatic Message Counting
AMC	Automatic Modulation Control
AMC	Autonomous Multiplexer Channel
AMCAP	Advanced Microwave Circuit Analysis Program
AMCAT	Addressograph Multigraph Computer Access Terminal
AMCS	Adaptive Micro-programmed Control System
AMD	Active Matrix Display
AMD	Advanced Maintenance Development
AMD	Advanced Memory Development
AMD	Aggregate, Management and Detailed
AMD	Alpha Mosaic Display
AMD	Associative Memory Data
AMDA	Advanced Maintenance Development Activity
AMDF	Absolute Magnitude Difference Function
AMDS	Advanced Microcomputer Development System
AME	Advanced Modelling Extension
AME	Amplitude Modulation Equivalent
AME	Asynchronous Modem Eliminator
AME	Autocad Modelling Extension
AME	Automatic Microfiche Editor
AME	Automatic Monitoring Equipment
AME	Average Magnitude of Error
AMEDA	Automatic Microscope Electronic Data Accumulator
AMEDS	Automated Measurement Evaluator and Director System
AMERITECH	American Information Technologies
AMES	Automatic Message Entry System
AMFIS	Automatic Microfilm Information System
AM/FM	Automated Mapping/Facilities Management
AMH	Application Message Handler|Handling
AMH	Automated Material Handling
AMH	Automated Medical History
AMHS	Automated Message Handling System
AMI	Access Method Interface
AMI	Alternat(iv)e Mark Inversion
AMIA	American Medical Informatics Associations
AMIAP	Associate Member (of the) Institution (of) Analysts (and) Programmers
AMII	Agile Manufacturing Information Infrastructure
AMINET	Amiga Network
AMIP	Atmospheric Model Inter comparison Project
AMIS	Acquisition Management Information System
AMIS	Agri-business Management Information System
AMIS	Audio Message|Messaging Interchange Standard
AMIS	Automated Management Information System
AMIS	Automatic Measurement and Inspection System
AML	Advanced Mathematical Library
AML	A Manufacturing Language
AML	Amplitude Modulated Link

AML	Application Macro|Module Library
AMLC	Asynchronous Multi-Line Controller
AMLCD	Active Matrix Liquid Crystal Display
AMM	Additional Memory Module
AMM	Agent Management Module
AMM	Analog Monitor Module
AMMA	Advanced Memory Management Architecture
AMMA	Automated Media Management system
AMME	Automated Multi-Media Exchange
AMMINET	Automated Mortgage Management Information Network
AMMS	Automated Multi-Media Switch
AMNIPS	Adaptive Man-machine Non-arithmetic(al) Information Processing System
AMNL	Amplitude-Modulation Noise Level
AMODE	Addressing Mode
AMODEUS	Assaying Means Of Design Expression for Users and Systems (EU)
AMOR	Associative Memory with Ordered Retrieval
AMOS	Adjustable Multiclass Organizing System
AMOS	Associative Memory Organizing System
AMOS	Automated Meteorological Observing System
AMOSS	Adaptive Mission Oriented Software System
amp	ampere
AMP	Amplitude
AMP	Automated Maintenance Program
AMP	Automated Manufacturing Planning
AMPC	Automatic Message Processing Center
AMPCR	Alternate Micro-Program Count Register
AMPERES	Automatic Maintenance Performance and Engineering Reliability Evaluation System
amp-hr	ampere hour
AMPL	Amplifier
AMPL-A	A Modelling Programming Language for mathematical programming
AMPM	Auto/Manual Probe Multiplexer
AMPP	Advanced Micro-Programmable Processor
AMPR	Application Module Processing Routine
AMPS	Advanced Mobile Phone Service
AMPS	Automatic Message Processing System
AMPS	Automatic Music Program Search
AM/PSK	Amplitude Modulation with Phase Shift Keying
AMPT	Automatic Module Production Tester
AMR	Acoustic Magnetic Resonance
AMR	Anisotropic Magneto-Resistance
AMR	Arithmetic Mask Register
AMR	Automated Management Report(s)
AMR	Automatic Message Registration|Routing
AMS	Access Method Services

AMS	Advanced Management System
AMS	American Mathematical Society
AMS	Analog Master Slice
AMS	Andrews Message System
AMS	Application Management Specification|System
AMS	Asymmetric Multiprocessing System
AMS	Automated Maintenance System
AMSC	Automatic Message Switching Center
AMSCO	Access Method Services Cryptographic Option
AMSDL	Acquisition Management System and Data requirements control List
AMSEC	Analytical Method for System Evaluation and Control
AMSP	Array Machine Simulation Program
AMSR	Automated Microform Storage and Retrieval
AMST	Automated Maintenance Support Tool
AMSU	Advanced Microwave Sounding Unit
AMT	Active Memory Technology
AMT	Address Mapping Table
AMT	Amount
AMT	Autodin Multimedia Terminal
AMT	Automated Microfiche Terminal
AMTD	Automatic Magnetic Tape Dissemination
AMTOR	Amateur-radio Transmit Or Receive
AMTRAN	Automatic Mathematical Translator
AMTS	Architecture of a Modular Test System
AMU	Analog Multiplier Unit
AMU	Association of Minicomputer Users (US)
AMU	Auxiliary Memory Unit
AMUSING	Algorithms, Models, User and Service Interfaces for Geography (EU)
AMV	Analog Measured Value
AMVFT	Amplitude Modulated Voice Frequency Telegraph
AMX	Advanced Multitasking Executive
AN	Access Node
AN	Analyzer
AN	Anode
AN	Associated Number
AN	Netherlands Antilles
ANA	Article Number Association
ANA	Automatic Number Analysis|Announcement
ANAC	Automatic Number Announcement Circuit
ANACOM	Analog Computer
ANACONDA	Analytical Control and Data
ANALIT	Analysis of Automatic Line Insulation Tests
ANAPAC	Analysis Package
ANATRAN	Analog Translator
ANBACIS	Automated Nuclear, Biological, And Chemical Information System

ANBH Average Network Busy Hour
ANC Active Noise Control|Canceller
ANC All Numbers Calling
ANC American Network Communications
ANCARA Advanced Networked Cities And Regions Association (US)
ANCHOR Alpha-Numeric Character generator
ANCHORS AT&T Network Component Handling and Outage Reporting System
ANCOVA Analysis of Co-Variance
ANCS American Numerical Control Society
ANCS-II Automated Nautical Charting System II
AND Alpha-Numerical Display
AND Automatic Network Dialling
ANDF Architectural Neutral Distributed Format
ANDMS Advanced Network Design and Management System
ANDVT Advanced Narrow-band Digital Voice Terminal
ANEBA Alphanumeric Electronic Bit Analyzer
ANEC Analysis Evaluation and Computation
ANETTE Academic Network for Technology Transfer in Europe
ANF AppleTalk Networking Forum
ANF Automatic Number Forwarding
ANFSCD And Now For Something Completely Different
ANG Angle
ANGS Angstrom
ANI Animated (cursor)
ANI Automatic Number Identification
ANIF Automatic Number Identification Failure
ANIRC Annual National Information Retrieval Colloquium
ANL Automatic New Line
ANL Automatic Noise Limiting
ANLP Alphanumeric Logic Package
ANM Advanced Network Management
ANMP Account Network Management Program
ANN Annotations
ANN Artificial Neural Network
ANN Auditing Network Needs
ANOCOVA Analysis Of Covariance
ANOM Analysis Of Means
ANON-FTP Anonymous File Transfer Protocol
ANOVA Analysis Of Variance
ANR Alphanumeric Replacement
ANRF Area Normalization with Response Factors
ANRS Automatic Noise Reduction System
ANS American National Standard
ANS Answer
ANS Autonomous Navigation Simulation
ANSA Advance Networked System Architecture

ANSC-X3 American National Standards Committee for computers and information processing
ANSC-X4 American National Standards Committee for office machines and supplies
ANSI American National Standards Institute
ANSIM Analog Simulator
ANSORC Analog System for Optical Reading of Characters
ANSTI African Network of Scientific and Technological Institutions
ANSWER Algorithm for Non-Synchronized Wave form Error Reduction
ANSYS Analysis System
ant antenna
ANTC Advanced Networking Test Center
ANTIOPE L'Acquisition Numérique et Télévisualisation d'Images Organisée en Pages d'Ecriture (FR)
ANTS Active Node Transfer System
ANX Automotive Network Exchange
ANZLIC Australia New Zealand Land Information Council
AO Abstract Object
AO Account Of
AO Analog Output
AO Angola
AO Automated Office|Operator
AO Automated Operator
AOA Accessible Operand Affiliation
AOA Activity-On-Arrow
AOA Application Oriented Algorithm
AOC Advice Of Charge
AOC All Ones Counter
AOC Application Oriented Chip
AOC Array Out Counter
AOC Automatic Output Check|Control
AOC-D Advice Of Charge – During call
AOC-E Advice Of Charge – End of call
AOCE Apple Open Collaborative Environment
AOCL Average Outgoing Count Limit
AOCR Advanced Optical Character Reader|Recognition
AOC-S Advice Of Charge – at call Setup
AOCS Altitude and Orbit Control System
AOCU Arithmetic|Associative Output Control Unit
AOD Arithmetic Output Data
AOE Acceptable Operator Exposure level
AOE Application Operating Environment
AOE Auditing Order Error
AOF Advanced Operating Facilities
AOG And/Or Graph
AOH Add-On Header
AOI Acousto-Optical Imaging
AOI And-Or Inverter

AOI	Automated Operator Interface
AOI	Automatic Optical Inspection
AOIPS	Atmospheric and Oceanographic Information Processing System
AOL	Absent Over Leave
AOL	America On Line
AOL	Application Oriented Language
AOM	Acousto-Optical Modulator
AOM	Application OSI Management (OSI = Open Systems Interconnect)
AOP	Annual Operating Plan
AOP	Arithmetic Operation
AOPI	And-Or Power Inverter
AOPX	And-Or Power Extender
AOR	Activity-On-Row
AOR	Add One to the Right
AOR	Alphanumeric Optical Reader
AOR	Application Owning Region
AORS	Abnormal Occurrence Reporting System
AOS	Academic Operating System
AOS	Acquisition Of Signal
AOS	Add Or Subtract
AOS	Advanced Operating System
AOS	Algebraic Operating System
AOS	Alternate Operator Service
AOSIP	Airline Open Systems Interconnection Profile
AOSP	Automatic Operating and Scheduling Program
AOSS	Auxiliary Operator Service System
AOT	Analog Output Timer
AOT	Average Operation Time
AOU	Arithmetic|Associative Output Unit
AOV	Analysis Of Variance
AOV	Attribute Object Value
AOX	And-Or Extender
AP	Adjunct Processor
AP	Advances Placement
AP	Allocation Point
AP	All Points
AP	Alternate Print
A&P	Analysis & Prediction (program)
AP	Application Package|Profile
AP	Application Process(or)
AP	Application Program|Protocol
AP	A Pulse
AP	Argument Pointer
AP	Arithmetic|Array Processor
AP	Associative|Attached Processor
AP	Audio Processing

AP203	Application Protocol 203
APA	All Points Addressable
APA	Application Portability Architecture
APA	Arithmetic Processing Accelerator
APACHE	Analog Programming And Checking
APADI	Adaptive-Potential Alternating-Direction Implicit
APAL	Array Processor Assembly Language
APAM	Array Processor Access Method
APAN	Asian Pacific Advanced Network
APAR	Authorized Program Analysis Report
APAR	Automatic Programming And Recording
APAREL	A Parse Request Language
APAREN	Address Parity Enable
APAS	Adaptable Programmable Assembly System
APAT	Adaptive Programmable Automatic Tester
APATS	Automatic Programmer And Test System
APAVC	Almost Periodic Amplitude Variation Coding
APB	All Points Bulletin
APB	Application Program Block
APC	Adaptive Predictive Coding
APC	Administrative Processing Center
APC	All-Purpose Computer
APC	Area Positive Control
APC	Association for Progressive Communications (US)
APC	Associative Processor Control
APC	Asynchronous Procedure Call
APC	Automatic Peripheral|Process Control
APC	Automatic Phase|Picture Control
APC	Automation Potential Control
APC	Available Page Count
APCBE	Associated Page Control Block Entry
APCC	Association of Professional Computer Consultants (US)
APCHE	Automatic Programmed Checkout Equipment
APCI	Application-layer Protocol Control Information
APCM	Adaptive Pulse Code Modulator
APCM	Authorized Protective Connecting Module
APCON	Approach Control
APCS	Associative Processor Computer System
APCS	Attitude and Pointing Control System
APCUG	Association of Personal Computer User Groups (US)
APD	Advanced Process Development
APD	Amplitude Probability Distribution(s)
APD	Approach Progress Display
APD	Automated Payment and Deposit
APD	Auto Power Down
APD	Average Pulse Duration
APDA	Apple Programmers and Developers Association
APDE	Application Protocol Development Environment

APDL	Algorithmic Processor Description Language
APDM	Associative Push Down Memory
APDN	Ameritech Public Data Network
APDU	Application Protocol Data Unit
APDU	Association of Public Data Users
APE	All-Pass Element
APE	Application Engineering
APE	Application Program Evaluation
APE	Automatic Positioning Equipment
APEM	Automated Production Equipment Model
APET	Application Program Evaluator Tool
APEX	Advanced Project for European information Exchanges
APEX	Amplitude plus Phase Extraction
APEX	Application Program Exchange
APEX	Assembler and Process Executive
APEX	Automated Planning and Execution control system
APEX	Automatic Production of Executable programs
APF	Advanced Printer Function
APF	Advanced Program Feature
APF	All Pins Fail
APF	Application Processing Function(s)
APF	Authorized Program Facility
APG	Application Program Generator
APG	Automated Process Generator
API	And Power Inverter
API	Application Process Innovation
API	Application Program(ming) Interface
API	Automatic Priority|Program Interrupt
APIA	Application Program Interface Association (US)
APIC	Advanced Programmable Interrupt Controller
APICS	American Production and Inventory Control Society
APIF	Automated Processing Information File
APIS	Advanced Passenger Information System
APIS	Array Processing Instruction Set
APL	Absolute Program Loader
APL	Advanced Programming Language
APL	Algorithmic Programming Language
APL	A Programming Language
APL	Automatic Program Load(ing)
APL	Average Picture Level
APLD	Automatic Program Locate Device
APLDI	A Programming Language Data Interface
APLL	Analog|Automatic Phase-Locked Loop
APLS	Automated Process Line System
APLSV	A Programming Language (APL) Shared Variable
APLT	Advanced Private Line Termination
APM	Advanced Power Management
APM	Advanced Program Monitor

APM	Amplitude and Phase Modulation
APM	Analog Panel Meter
APM	Application Performance Management
APM	Application Program Maintenance
APM	A Priori Model
APM	Automated Production Management
APM	Automatic Plant Manager
APM	Automatic Predictive Maintenance
APM	Auxiliary Processing Machine
APMS	Automatic Performance Management System
APN	Asia Pacific Network
APNSS	Analog Private Network Signalling System
APO	Advanced Planning and Optimization
APO	Advanced Programming Option
APO	Automatic Power Off
APO	AWIPS Program Office
APOLLO	Article Procurement with On-Line Local Ordering
APOTA	Automatic Positioning Telemetering Antenna
APP	Advanced Procurement Plan
app	application (file extension indicator)
APP	Application Portability Profile
APP	Associative Parallel Processor
APP	Automatic Program Proving
APP	Automatic Proposal Preparation
APPA	Artificial Pilot Phased Array
APPARC	Applications that are Performance-critical in Parallel Architectures (EU)
APPC	Advanced Peer-to-Peer Communications
APPC	Advanced Program-to-Program Communications
APPC/PC	Advanced Program-to-Program Communications/Personal Computers
APPI	Advanced Peer-to-Peer Inter networking
APPI	Advanced Planning Procurement Information
APPI	Advanced Program-to-Program Inter networking
APPLE	Analog Phased Processing Loop Equipment
APPLE	Associative Processing Programming Language
APPN	Advanced Peer-to-Peer Networking
APPO	Advanced Product Planning Operation
approx	approximately
APPU	Application Program Preparation Utility
APQ	Analog Process Quantity
APR	Active Page Register
APR	Alternate Path Retry
APR	Automated Punch Requisition
APR	Automatic Parts|Picture Replacement
APR	Automatic Production Recording
APR	Automatic Programming and Recording
APR	Automatic Purchase Request

APRICOT	Automatic Printed circuit board Routing with Intermediate Control Of the Tracking
APRIL	Accounts Payable, Receivable, Inventory Library
APRIL	Associative Processor with Integrated Logic
APRIL	Automatically Programmed Remote Indication Logging
APRM	Application Programmers Reference Manual
APROC	Adaptive statistical Processor
APRP	Adaptive Pattern Recognition Processing
APRS	Automatic Position Reference System
APRST	Averaged Probability Ratio Sequential Test
APS	Accessible Program Sequence
APS	Advanced Planning and Scheduling (system)
APS	Alphanumeric Photocomposition System
APS	Alternate Pocket Select
APS	Antenna Pointing Subsystem
APS	Application Processing|Production Services
APS	Application Programming System(s)
APS	Application Protocol Suite
APS	Array Processing Subroutine
APS	Assembly Programming System
APS	Asynchronous Protocol Specification
APS	Automated Patent System
APS	Automatic Protection Switch
APS	Automatic Publication Service
APS	Auxiliary Power Supply
APSE	ADA Programming Support Environment
APSL	Access Path Specification Language
APSM	Auxiliary Power Supply Module
APSP	Array Processor Subroutine Package
APSS	Automated Program Support|Search System
APT	Address Pass Through
APT	Advanced Parallel Technology
APT	Application Programming Tools|Testing
APT	Assigned Probability Technique
APT	Automatically Programmed Tools
APT	Automatic Picture Transmission
APT	Automatic Position Telemetering
APTE	Automatic Production Test Equipment
APTF	Automated Program Testing Facility
APTI	Automatic Print Transfer Instrument
APTIF	Average Process Time Inverted File
APTR	Association Printer
APTS	Application Program Test System
APTS	Automatic Picture Transmission Subsystem
APU	Analog|Analytic Processing Unit
APU	Arithmetic|Asynchronous Processing Unit
APU	Auxiliary Power Unit
AQ	Accumulator-Quotient register

AQ	Antarctica
AQD	Analysis of Quantitative Data
AQE	Allocated Queue Element
AQUARIUS	A Query And Retrieval Interactive Utility System
AQuIS	Achieving Quality In Software (international conference)
AR	Acceptance Review
AR	Accumulator Register
AR	Add|Address Register
AR	Alarm Report
AR	Amplitude Ratio
A&R	Analysts & Researchers
AR	Application Residence
AR	Argentina
AR	Arithmetic Register
AR	Arkansas (US)
AR	Aspect Ratio
AR	As Required
AR	Associative Register
AR	Attributes for representing Relationships
AR	Augmented Reality
AR	Automatic Register|Reverse
ARA	Apple(Talk) Remote Access
ARA	Attribute Registration Authority
ARA	Automatic Route Advance
ARAB	A Register A Bit
ARAC	Array Reduction Analysis Circuit
ARACOR	Advanced Research and Applications Corporation
ARAL	Automatic Record Analysis Language
ARAMIS	Automation Robotics And Machine Intelligence System
ARAP	AppleTalk Remote Access Protocol
ARAR	Applicable or Relevant and Appropriate Requirement
ARAT	Automatic Random Access Transport
ARB	Active Request Block
ARB	Address Reorder Buffer
ARB	Automatic Ring-Back
ARBB	A Register B Bit
ARBY	A Rule Based expert system of Yale (University)
arc	archive (file extension indicator)
ARC	Attached Resource Computer
ARC	Audio Response Controller
ARC	Automated Radio-Theodolite
ARC	Automatically Repaired Computer
ARC	Automatic Recovery Computer
ARC	Automatic Relay Calculator
ARC	Automatic Revenue Collection
ARC	Average Response Computer
ARCA	Advanced RISC Computing Architecture (RISC = Reduced Instruction Set Computer)

ARCADE	Advanced Realism Computer Aided Design Environment
ARCADE	Automatic Radar Control And Data Equipment
ARCAIC	Archives and Record Cataloguing And Indexing by Computer
ARCAS	Automatic Radar Chain Acquisition System
ARCB	A Register C Bit
ARCH	Articulated Computer Hierarchy
ARCNET	Attached Resource Computer Network
ARCS	Accounts Receivable Control System
ARCS	Advanced Reconfigurable Computer System
ARCS	Advanced RISC Computing Specification (RISC = Reduced Instruction Set Computer)
ARCS	Automated Reproduction and Collating System
ARCS	Automated Revenue Collection System
ARCS	Automated Ring Code System
ARD	Answering, Recording and Dialling
ARD	Application Remote Database
ARD	Asset Record Date
ARD	Automatic Release Data
ARDA	Analog Recording Dynamic Analyzers
ARDI	Analysis, Requirements determination, Design and development, and Implementation and evaluation
ARDIS	Advanced national Radio Data and Information Service
ARDS	Advanced Remote Display System
ARE	All Routes Explorer
AREG	A Register
ARELEM	Arithmetic Element
ARENTO	Arab Republic of Egypt National Telecommunications Organization
ARF	Alarm Reporting Function
ARF	Automatic Report Failure
ARF	Auto Renumbering Facility
ARFT	Adjusted Remaining Flow Time
arg	argument
ARGO	A Really Good Open system interconnection
ARGS	Advanced Raster Graphics System
ARGUS	Automatic Routine Generating and Updating System
ARI	Add Register Immediate instruction
ARI	Address Recognized Indicator bit
ARI	Application Reference Index
ARI	Automated Readability Index
ARIES	ATM Research and Industrial Enterprise Studies (ATM = Asynchronous Transfer Mode)
ARIES	Automated Reliability Interactive Estimation System
ARIMA	Auto-Regressive Integrated Moving Average
ARIS	Activity Reporting Information System
ARIS	Architecture of Integrated (information) Systems
ARIS	Automated Reactor Inspection System
ARIST	Annual Review of Information Science and Technology

ARITH	Arithmetic
ARJE	Advanced Remote Job Entry
ARL	Acceptable Reliability Level
ARL	Access Rights List
ARL	Adjusted Ring Length
ARL	Asset Reuse Library
ARL	Attendant Release Loop
ARL	Average Run Length
ARLL	Advanced Run Length Limited
ARM	Acorn|Advanced RISC Machine (RISC = Reduced Instruction Set Computer)
ARM	Application Reference Model
ARM	Asynchronous Response Mode
ARM	Automated Routing Management
ARMA	Association of Records Managers and Administrators (US)
ARMA	Auto-Regressive Moving Average
ARMAN	Artificial Methods Analyst
ARMM	Automated Retroactive Minimal Moderation
ARMM	Automatic Reliability Mathematical Model
ARMMS	Automated Reliability and Maintainability Measurement System
ARMS	Automatic Resistance Measuring System
ARNES	Academic and Research Network of Slovenia
ARO	After Receipt of Order
AROM	Alterable|Associative Read Only Memory
AROS	Alterable Read-only Operating System
ARP	Address Resolution Protocol
ARP	Advance Replacement Parts
ARP	Analogous Random Process
ARP	Asset Record Performance
ARPA	Advanced Research Projects Agency
ARPANET	Advanced Research Projects Agency Network
ARPL	Adjust Requested Privilege Level
ARPS	Advanced Real-time Processing System
ARPS	Advanced Regional Prediction System
ARQ	Answer/Return Query
ARQ	Automatic Recovery Quotient
ARQ	Automatic Repeat Request
ARR	Address Recall Register
ARR	After Re-Record (signal)
ARR	Automatic Repeat Request
ARRAPS	Automatic Repetitive Reset And Program Start
ARROW	Anti-Resonant Reflecting Optical Wave guide
ARS	Active Ready Signal
ARS	Additional Read Station
ARS	Advanced Record System
ARS	Alternate Route Selection
ARS	Audio Response System
ARS	Automatic Regulating System

ARS	Automatic Route Selection|Setting
ARSB	Automated Repair Service Bureau
ARSTEC	Adaptive Random Search Technique
ART	Actual Retention Time
ART	Adaptive Resonance Theory (network)
ART	Advanced Robot Technology
ART	Airborne Radiation Thermometer
ART	Asynchronous Remote Takeover
ART	Authorization and Resource Table
ART	Automated Reasoning Tool
ART	Automated Request Transmission
ART	Automatic Reporting Telephone
ART	Automatic Revision Tracking
ART	Average Run Time
ARTA	Apple Real Time Architecture
ARTCC	Air Route Traffic Control Center
ARTEMIS	Advanced Relay Technology Mission
ARTI	Advanced Rotocraft Technology Integration
ARTILECT	Artificial Intellect (nth generation computer)
ARTRIM	Automatic Resistor Trimmer
ARTS	Asynchronous Remote Takeover Server
ARTS	Audio Response Time-shared System
ARTS	Automated Radar Terminal System
ARTS/DB	Analysis of Real-Time Systems/Data Base oriented
ARTT	Asynchronous Remote Takeover Terminal
ARTTIS	Advanced Real-Time Total Information System
ARU	Address Recognition Unit
ARU	Application Resource Unit
ARU	Automatic Response Unit
ARU	Auxiliary Read-out Unit
AR-WIRE	Address-Read Wire
ARWM	A Register Word Mark
ARX	Autocad Runtime Extension
AS	Accept Session
AS	Activity Start
AS	Adaptive Search|Source
AS	Address Space
AS	Advanced System
AS	American Samoa
AS	Application|Automation System
AS	Autonomous System
AS	Auxiliary Storage
AS3AP	ANSI SQL Standard Scalable And Portable
AS/400	Application System/400
ASA	Agents Systems & Applications (international symposium)
ASA	American Statistical Association
ASA	As Soon As
ASA	Asynchronous Adapter

ASA	Asynchronous/Synchronous Adapter
ASA	Automatic Spectrum Analyzer
ASAI	Adjunct Switch Application Interface
ASAP	Advanced Scientific Array Processor
ASAP	Advanced Symbolic Artwork Preparation
ASAP	Analog System Assembly Pack
ASAP	A Simple Agent Platform
ASAP	As Soon As Possible
ASAP	Australian Science Archives Project
ASAP	Automated Statistical Analysis Program
ASAP	Automatic Spooling with Asynchronous Processing
ASAP	Automatic Switching And Processing
ASARS	Advanced Synthetic Aperture Radar System
ASB	A-Synchronous Balanced mode
ASC	Accredited Standards Committee
ASC	Adaptive Speed Control
ASC	Additional Sense Code
ASC	Advanced Space Computer
ASC	Advanced Storage Control
asc	ascii text (file extension indicator)
ASC	Associative Structure Computer
ASC	Asynchronous Sequential Circuit
ASC	Authorized Support Center
ASC	Automatic Selectivity|Sequence Control
ASC	Automatic Service|Switching Center
ASC	Automatic Switching Control(led)
ASC	Automatic Synchronized Control
ASC	Automatic System Controller
ASC	Auto Search Code
ASC	Auxiliary Station Control
ASCA	Application System Control and Auditability
ASCA	Automatic Subject Citation Alert
ASCB	Address Space Control Block
ASCC	Automatic Sequence-Controlled Calculator
ASCENT	Assembly System for Central processor
ASCI	Accelerated Strategic Computing Initiative
ASCII	American Standard Code for Information Interchange
ASCOM	Association of Telecommunication services (US)
ASCON	Automated Switched Communications Network
ASCOT	Advanced Storage Control Test
ASCP	Automatic System Checkout Program
ASCQ	Additional Sense Code Qualifier
ASCS	American Standard Character Set
ASCU	Association of Small Computer Users (US)
ASCUE	Association of Small Computer Users in Education (US)
ASD	Adaptive Software Development
ASD	Adjustable Speed Drives
ASD	Advanced System(s) Development

ASD	Air Situation Display
ASD	Application Systems Department
ASDAR	Aircraft to Satellite Date Relay
ASDI	Automatic Selective Dissemination of Information
ASDL	Abstract-type and Schema-Definition Language
ASDL	Advanced Systems Development Laboratory
ASDL	Asynchronous Digital Subscriber Loop
ASDM	Automated Systems Design Methodology
ASDR	Airborne Sample Data Reduction
ASDSP	Application-Specific Digital Signal Processor
ASDT	Application Software Data Techniques
ASDU	Application layer Service Data Unit
ASE	Adaptive Server Enterprise
ASE	Advanced Software Environment
ASE	Application Service Element|Entity
ASE	Application Support Environment
ASE	Associative Search Element
ASE	Automatic Stabilization Equipment
ASES	Automated Software Evaluation System
ASET	Applied Science and Engineering Technology
ASETS	Automatic Storage Evaluator Test System
ASF	Accounting Structure File
ASF	Active Segment Field
ASF	Active Streaming Format
ASF	Automatic Sheet Feeder
ASFTS	Auxiliary Systems Functional Test Standards
ASI	Adapter Support Interface
ASI	Alternate Space Inversion
ASI	American Statistical Index
ASI	Application Service Interface
ASI	Automated System Implementation|Installation
ASI	Automated System Initialization
ASI	Automation in the Steel Industry (conference)
ASIC	All-purpose Symbolic Instruction Code
ASIC	Application Specific Integrated Circuit
ASIC	Application System Information Center
ASID	Address Space Identifier
ASIDIC	Association of Information and Dissemination Center (US)
ASIDP	American Society of Information and Data Processing
ASII	American Science Information Institute
ASIO	Asynchronous Start Input/Output
ASIP	Application-Specific Instruction (set) Processor
ASIS	Abort Sensing and Implementation System
ASIS	Alpha Search Inquiry System
ASIS	American Society for Information Science
ASIS	Application Software Installation Server (EU)
ASISNET	American Society (for) Industrial Security Network
ASIT	Adaptable Surface Interface Terminal

ASIT	Advanced Security and Identification Technology
ASK	Amplitude Shift Keying
ASKS	Automatic Station Keeping System
ASL	Adaptive Speed Levelling
ASL	Advanced Simulation Language
ASL	Approve Source List
ASL	Arithmetic Shift Left
ASL	Automatic Stereo level
ASLA	Application Service Level Agreement
ASLD	Advanced Solid Logic Dense
ASLT	Advanced Solid Logic Technology
ASM	Access State Matrix
ASM	Advanced Solid Modeller
ASM	Algorithmic State Machine|Matrix
ASM	Alternate Stimulus Mode (protocol)
ASM	Analog Storage Module
asm	assembler source language (file extension indicator)
ASM	Association for Systems Management
ASM	Asynchronous Sequential Machine
ASM	Asynchronous Subscriber Module
ASM	Auxiliary Storage Manager|Medium
ASM/GEN	Assembler Generating system
ASMP	American Society of Media Photographers
ASMP	Asymmetric Multi-Processing
ASMS	Atmospheric Sensing and Maintenance System
ASN	Automatic Selector Network
ASN	Autonomous System Number
ASN	Average Sample Number
ASO	Application Service Object
ASOS	Automated Surface Observation System
ASP	Abstract Service Primitive
ASP	Acoustic Signal Processor
asp	active server pages (file extension indicator)
ASP	Advanced Support Processor
ASP	Aggregated Switch Procurement
ASP	AppleTalk Session Protocol
ASP	Application Service Provider
ASP	Association of Shareware Professionals (US)
ASP	Association Storing Processor
ASP	Attached Support Processor
ASP	Automated Spooling Priority
ASPA	Automated System Performance Analysis
ASP-DAC	Asia Pacific Design Automation Conference
ASPEN	Automatic System for Performance Evaluation of the Network
ASPEX	Automated Surface Perspectives
ASPI	Advanced SCSI Programming Interface
ASPI	Asynchronous Synchronous Programmable Interface
ASPL	Auto-Sketch Programming Language

ASPLS	Any Special Symbol
ASPP	Application-Specific Programmable Processor
ASPRIN	Application Systems and Programs Reference Information Network
ASPS	Advanced Signal Processing System
ASQ	Analytic Solution to Queues
ASQ	Any Sequence Queue
ASR	Accumulator Shift Right
ASR	Active Status Register
ASR	Address Shift|Start Register
ASR	Address Space Register
ASR	Advanced Systems Research
ASR	Analog Shift Register
ASR	Arithmetic Shift Right
ASR	Assigned Slot Release
ASR	Automatic Send/Receive
ASR	Automatic Speech Recognition
ASR	Automatic Storage/Retrieval
ASR	Automatic System Recovery
ASR	Auxiliary storage Save/Restore
ASR0	Arithmetic Shift Right (toward) zero
ASRA	Automatic Stereophonic Recording Amplifier
ASRL	Average Sample Run Length
ASRS	Automatic Storage and Retrieval System
ASS	Address Selection Switch
ASSB	Associative Buffer
ASSET	Advanced Systems Synthesis and Evaluation Technique
ASSET	Asset Source for Software Engineering Technology
ASSET	Automated System for the Support of Engineering Tests
ASSGN	Assign
ASSIST	Atomic-Scale control of Surfaces and Interfaces in Silicon Technology (EU)
ASSM	Assembler
ASSM	Associative Memory
ASSN	Association
ASSORT	Automatic System for Selection Of Receiver and Transmitter
ASSP	Application-Specific Signal Processor
ASSP	Application-Specific Standard Part
ASSP	Approved Species-Specific Protocol
ASSR	Automated Systems Service Request
ASST	Azienda di Stato per i Servizi Telefonici (IT)
AST	Abstract Syntax Tree
AST	Accessible Subsidiary Tasks
AST	Accurate Screen Technology
AST	Address Synchronizing Track
AST	Add-Subtract Time
AST	Applied System Technology
AST	Asterisk

AST	Asynchronous System Trap
AST	Asynchronous to Synchronous Transmission (adapter)
AST	Attribute Syntax Tree
AST	Automatic Scan Tracking
AST	Automatic Summary Table
ASTA	Association of Short-circuit Testing Authorities
ASTA	Automatic System Trouble Analysis
ASTAP	Advanced Statistical Analysis Program
ASTI	Automated System for Transport Intelligence
ASTOR	Address Storage Register
ASTOR	Associative Storage
ASTRA	Advanced Structural Analyzer
ASTRA	A Skew-To-Retiming Algorithm
ASTRA	Automatic Scheduling and Time-dependent|Time-integrated Resource Allocation
ASTRAIL	Analog Schematic Translator to Algebraic Language
ASTROS	Advanced Systematic Techniques for Reliable Operational Software
ASTRP	Application System Trouble Reporting Procedure
ASTUTE	Association of System 2000 Users for Technical Exchange
ASU	Add/Subtract Unit
AS/U	Advanced Server for Unix
ASU	Auxiliary Storage Unit
ASUG	America's SAP Users' Group (SAP = Systems, Applications and Products)
ASV	Automatic Self Verification
ASVIP	American Standard Vocabulary for Information Processing
ASVS	Automatic Signature Verification System
ASW	Applications Software
ASYLCU	Asynchronous Line Control Unit
ASYNC	Asynchronous (transmission)
ASYTRAN	Asynchronous Transmission
AT	Absolute Time
AT	Access Tandem
AT	Address Translator
AT	Advanced Technology
AT	After Total
AT	Ampere Turn
AT	Anomalous Transmission
AT	Appropriate Technology
A&T	Assemble & Test
AT	Attention
AT	Austria
AT	Automatic Test|Transmission
AT	Automatic Threshold|Ticketing
ATA	Advanced Technology Attachment
ATA	ARCNET Trade Association
ATA	Asynchronous Terminal Adapter

ATA	At Attachment
ATA	Automatic Trouble Analysis
ATA-2	Advanced Transfer Adapter 2
ATAE	Associated Telephone Answering Exchanges
ATAP	Automated Time and Attendance Procedures
ATAPI	Advanced Technology Attachment Packet Interface
ATARS	Automated Travel Agents Reservation System
ATARS	Automatic Traffic Advisory and Resolution Service
ATAS	Analog Test Access System
ATAS	Automated Telephone Answering System
ATB	Address Trace Block
ATB	All Trunks Busy
ATBM	Average Time Between Maintenance
ATC	Advanced Technology Components
ATC	Allocate To Component
ATC	Asynchronous Terminal Controller
ATC	Authorized Training Center
ATC	Automated Transmission Control
ATC	Automatic Temperature|Train Control
ATC	Automatic Tool Changer
ATC	Automatic Transmission Control
ATC	Automatic Tray Changer
ATC	Automatic Tuning Control
ATC	Available Transmission Capacity
ATCA	Automatic Transistor Card Analyzer
ATCCIS	Army Tactical Command and Control Information Systems
ATCD	Automatic Telephone Call Distribution
ATD	Actual To Date
ATD	Association of Telecommunications Dealers (US)
ATD	Asynchronous Time Division
ATDE	Advanced Technology Demonstrator Engine
ATDG	Automated Test Data Generator
ATDM	Asynchronous Time Division Multiplexing
ATDNet	Advanced Technology Demonstration Network
ATDP	Attention Dial Pulse
ATDS	Auxiliary Track Data Storage
ATDT	Attention Dial Tone
ATDU	Alphanumeric Terminal Display Unit
ATE	Altitude Transmitting Equipment
ATE	Asynchronous Terminal Emulation
ATE	Automatic Test Equipment
ATEA	Automatic Test Equipment Association
ATEC	Automated Technical Control
ATEC	Automatic Test Equipment Complex
ATEMIS	Automatic Traffic Engineering and Management Information Systems
ATEX	Automatic Test Equipment conference and Exhibition

ATF	Actuating Transfer Function
ATF	After The Fact
ATF	Automatic Text Formatter
ATFC	Automatic Traffic Flow Control
ATFG	Analog Test Function Generator
AtFS	Attributed File System
ATG	Acoustically Coupled Tone Generator
ATG	Automatic Test Generation|Generator
ATGF	Automatic Test Generation Facility
ATH	Abbreviated Trouble History
ATH	Attention Hang-up
ATI	Advanced Technical Information
ATI	Automatic Test Inhibit
ATI	Automatic Track Initiation
ATI	Auxiliary Tape Input
ATI	Average Total Inspection
ATIM	Advanced Technology Insertion Module
ATIS	Advanced Traveller Information Systems
ATIS	Allocated Tele-Information Service
ATIS	Atherton Technology tool Integration System
ATIS	Automatic Terminal Information Service
ATIS	Automatic Transmitter Identification System
ATK	Andrew Toolkit
ATL	Active Task List
ATL	Active Template Library
ATL	Analog Threshold Logic
ATL	Applications Terminal Language
ATLAS	Abbreviated Test Language System
ATLAS	Automated Testing and Load Analysis System
ATLAS	Automatic Tabulating, Listing And Sorting system
ATLAS	Automatic Tape Load Audit System
ATLAS	Autonomous Temperature Line Acquisition System
ATM	Abstract Test Method
ATM	Adobe Typeface Manager
ATM	Asynchronous Time Multiplexing
ATM	Asynchronous Transfer Mode
atm	atmosphere
ATM	Automated Task Module
ATM	Automatic Teller Machine
ATM	Automatic Test Mode
ATM	Auxiliary Tape Memory
ATMAC	Advanced Technology Microelectronic Array Computer
ATMARS	Automatic Transmission Measuring And Recording System
ATM-DXI	Asynchronous Transfer Mode – Data Exchange Interface
ATME	Automatic Transmission Measuring Equipment
ATMS	Advanced Terminal Management System
ATMS	Advanced Text Management System

ATMS	Advanced Traffic Management System
ATMS	Assumption-based Truth Maintenance System
ATMS	Automated Trunk Measurement System
ATMS	Automatic Transmission Measuring System
ATMSS	Automatic Telegraph Message Switching System
ATN	Attention
ATN	Augmented Transition Network
ATN	Automated Test Network
ATO	Automatic Train Operation
AtoD	Analog to Digital
ATOLS	Automatic Testing On-Line System
ATOM	Automatic Testing, Operating, and Maintenance
ATOMS	Automated Technical Order Maintenance Sequence(s)
ATOT	After Total
ATP	Advanced Technology Program
ATP	Advanced Terminal Processor
ATP	Advanced Testing Procedures
ATP	All Tests Pass
ATP	AppleTalk Transaction Protocol
ATP	Application Transaction Processing
ATP	Automatic Test Procedure
ATP	Available To Promise
ATPG	Automatic Test Pattern Generation
ATPS	AppleTalk Print(ing) Services
ATQ	AppleTalk Transition Queue
ATR	Accumulator Track Read
ATR	Alternate Trunk Routing
ATR	Anti-Transmit-Receive
ATR	Attribute
ATR	Automated Target Recognition
ATR	Automatic Terminal Recognition
ATR	Average Transmission Rate
ATRAC	Angle Tracking Computer
ATRC	Advanced Television Research Consortium (Philips, Thomson, NBC)
ATRS	Advanced Technology Reader/Sorter
ATRS	Automated Trouble Reporting System
ATRVAL	Attribute Value
ATS	Abstract Test Suite
ATS	Acoustical Testing System
ATS	Administrative Terminal System
ATS	Advanced Terminal|Text System
ATS	Analytic Trouble Shooting
ATS	Apple Terminal Services
ATS	Applications Technology Satellite
ATS	Automated Test|Text System
ATS	Automatic Train Stop
ATS	Automatic Transfer Service

ATSC	Advanced Television Systems Committee
ATSEC	Advanced Test generation and testable design methodology for Sequential Circuits (EU)
ATSG	Automatic Transmission Service Group
ATSP	Absolute Text and Storage Protect
ATSS	Automatic Telecommunications Switching System
ATSS	Automatic Test Support Systems
ATSS-D	Automatic Telecommunications Switching System – Digital
ATSU	Association of Time-Sharing Users (US)
ATT	Address Translation Table
AT&T	American Telephone & Telegraph company
ATT	Applied Transmission Technologies
ATT	Attachment
ATT	Attribute
ATT	Audit Trail Tape
ATTC	Automatic Transmission Test and Control circuit
ATTCOM	American Telephone & Telegraph Communications
ATTD	Avalanche Transit Time Diode
ATTGIS	AT&T/Global Information Solutions
ATTIS	AT&T Information Systems
ATTN	Attention
ATTR	Attribute
ATTS	Automatic Tape Time Selector
ATTWS	AT&T Wireless Services
ATU	Analog Trunk Unit
ATU	Analysis and Transformation Unit
ATV	Account Transfer Voucher
ATW	Accumulator Track Write
a-type	abstract type
AU	Access|Adder Unit
AU	Adaptive|Administrative Unit
AU	Additional Use
AU	Angstrom Unit
AU	Arithmetic|Astronomical Unit
AU	Australia
AU	Autocoder
Au	Gold
AUA	Additional Use Authorization
AUC	Additive|Alternate Unit Codes
AUC	Authentication Certificate
AUD	Asynchronous Unit Delay
AUD	Audible
AUDDIT	Automatic Dynamic Digital Test system
AUDICS	Auditable Internal Control Systems
AUDIT	Automated Data Input Terminal
AUDIT	Automatic Unattended Detection Inspection Transmitter
AUDIX	Audio Information Exchange
AUDREY	Automatic Digit Recognizer

AUGRAI	Association d'Usagers du Groupe de Recherche en Automatisation Intégrée (methodology)
AUI	Adaptable User Interface
AUI	Apple attachment Unit Interface
AUI	Attached|Attachment Unit Interface
AUI	Attachment Universal Interface
AUI	Autonomous|Auxiliary Unit Interface
AUP	Acceptable Use Policy
AUP	Appropriate Use Policies
AUR	Automatic Rerouting
AURP	AppleTalk Update Routing Protocol
AURS	Advanced Unit Record Systems
AURS	Alone Unit Record System
AUS	Additional Use
AUSDEC	Australasian Spatial Data Exchange Center
AUSDEC	Australasian STEP Data Exchange Center (STEP = Standard for The Exchange of Product model data)
AUSTIN	Automatic Substrate Tinner In-line
AUT	Advanced User Terminal
AUT	Automated Unit Test
AUTEL	Asociación Espanola de Usuarios de Telecommunicaciones (ES)
AUTO	Automatic
AutoCAD	Automated Computer Aided Design
AUTOCON	Automatic Contour(ing)
AUTODIN	Automated Digital Information Network
AUTODIN	Automatic Digital Network
AUTODOC	Automated Documentation
AUTOEXEC	Automatic Execution
AUTOFACT	Automated Factory
AUTOFACT	Automated integrated Factory of tomorrow (conference and exhibition)
Auto ID	Automatic Identification
AUTOMAD	Automatic Adaption Data
AUTOMAST	Automatic Mathematical Analysis and Symbolic Translation
AUTOMAT	Automatic Methods And Times
AUTOMEX	Automatic Message Exchange
AUTONET	Automatic Network
AUTOPIC	Automatic Personnel Identification Code
AUTOPIT	Automatic Programming Including Technology
AUTOPSY	Automatic Operating System
AUTOSATE	Automated data Systems Analysis Technique
AUTOSEVCOM	Automatic Secure Voice Communications
AUTOSEVOCOM	Automatic Secure Voice Communications (system)
AUTOSTART	Automatic Start
AUTOSTP	Automatic Stop
AutoSTRUCT	Automated Structures
AUTOTRAN	Automatic Translation
AUTOVON	Automatic Voice Network

AUTRAN	Automatic Translation
AUTRAN	Automatic Utility Translator
AUTRAX	Automatic Traffic Recording and Analysis Complex
AUUG	Australian Unix Users Group
AUX	Apple Unix
AUX	Auxiliary
aux	auxiliary file (file extension indicator)
AUXF	Auxiliary Frame
AUXR	Auxiliary Register
AV	Analysis of Variance
AV	Array/Vector
AV	Attribute Value
AV	Audio/Video
AV	Audio/Visual
AV	Available
AV	Average
AVA	Absolute Virtual Address
AVA	Attribute Value Assertion
AVA	Audio Visual Aids
AVA	Audio Visual Authoring (language)
AVA	Audio Visual software fair organizing Association (international)
AVC	Absolute Value Computer
AVC	Audio Visual Connection
AVC	Automatic Voltage|Volume Control
AVCS	Advanced Vidicon Camera System
AVD	Alternate Voice Data
AVDS	Automated Vehicle Diagnostic System
AVE	Automatic Volume Expansion
AVG	Average
AVHRR	Advanced Very High Resolution Radiometer
AVI	Audio Video Interleave
AVI	Audio Visual Information|Interleave(d)
AVI	Automated Visual Inspection
AVI	Automatic Vehicle Identification
avigation	avionics and navigation
AVIP	Association of Viewdata Information Providers
AVIS	Automated Visual Inspection System
AVISAM	Average Index(ed) Sequential Access Method
AVK	Audio Video Kernel
AVL	Adelson-Velskii and Landis (balanced binary search tree)
AVM	Analysis Virtual Machine
AVM	Automatic Voltage Margin
AVN	Audio-Visual Network
AVN	Automated Voice Network
AVNL	Automatic Video Noise Limiting
AVO	Advanced Video Option
AVO	Amperes-Volts-Ohms

AVOCON	Automated Vocabulary Control
AVOS	Acoustic Valve Operating System
AVOS	Advisor Virtual memory Operating System
AVP	Architectural Verification Program
AVP	Attached Virtual Processor
AVPD	Anti-Virus Product Developers
AVPM	Attribute Value Pair Model
AVR	Automatic Voice|Volume Recognition
AVR	Automatic Voltage Regulation
AVS	Application Visualization System
AVS	Automated Verification System
AVS	Automatic Version Synchronization
AVS	Automatic Volume Switching
AVSS	Audio-Video Support System
AVSSCS	Audio-Visual Service-Specific Convergence Sub layer
AVT	Address Vector Table
AVT	Application Virtual Terminal
AVT	Applied Voice Technology
AVT	Attribute Value Time
AVTA	Automatic Vocal Translation Analyzer
AVVC	Altitude Vertical Velocity Computer
AVVID	Architecture for Voice, Video and Integrated Data
AW	Aruba
AWARDS	Automated Weather Acquisition and Retrieval Data System
AWB	Analyst Workbench
AWC	Association for Women in Computing
AWC	Automatic Work Changer
AWE	Asymptotic Wave form Evaluation
AWESPICE	Asymptotic Wave form Evaluation in SPICE
AWF	Acceptable Workload Factor
AWGN	Additive White Gaussian Noise
AWIP	Automatic Wave Information Processor
AWIPS	Advanced Weather Information Processing System
A-WIRE	Address Wire
AWIS	Automatic Wave Information System
AWK	Aho, Weinberger, Kernigham (pattern scanning language)
AWN	Active Web Networks
AWOL	Absent Without Leave
AWOS	Automated Weather Observing System
AWS	Active Work Space
AWS	Advanced Workstations and Systems
AWS	Automatic Warning System
AWT	Abstract Window(ing) Toolkit
AWT	Allocate Warning System
AW-WIRE	Address-Write Wire
AX	And(-operation) Extender
AX	Architecture Extended
AX	Automatic Transmission

aXe	an Executable (text)editor
AXP	Advanced Architecture Processor
AXP	Associative Cross point Processor
AYOR	At Your Own Risk
AZ	Arizona (US)
AZ	Azerbaijan
AZAS	Adjustable Zero Adjustable Span
AZ-EL	Azimuth-Elevation
AZM	Azimuth

B

b	base (of a transistor)
b	bass
B	Battery
B	Bean unit
B	Bell (character)
B	Bias
B	Bottom
B	Branch (instruction)
B	Bus
B	Byte(s)
b	susceptance
B2B	Business-To-Business
B2Bi	Business-To-Business integration
B2C	Business-To-Consumer
B2C	Business-To-Customer
B2X	Binary To Hexadecimal
B6ZS	Bipolar with 6 Zero Substitution
B8ZS	Bipolar with 8 Zero Substitution
BA	Base
BA	Battery
BA	Black Anodize
BAB	Byte Address Buffer
BA/BASIC	Business Applications in BASIC
BABAUD	Bits At BAUD (= Baudot code)
BABCOM	Babylon 5 Communication system
BAC	Basic Access Control
BAC	Binary Asymmetric Channel
BAC	Bistatic Alerting and Cueing
BAC	Bus Adapter Control
BAC	Business Advisory Council
BACE	Basic Automatic Checkout Equipment
BACPAC	Berkeley Advanced Chip Performance Calculator
BACS	Bank Automated Clearing System
BACT	Best Available Control Technology
BAD	Bit Anomaly Detector
BADC	Binary Asymmetric Dependent Channel

BAEQS	Bid-Asked Electronic Quotation System
BAGNET	Bay Area Gigabit Network
BAIC	Binary Asymmetric Independent Channel
BAIS	Bulletin Articles Information Subsystem
BAK	Back At Keyboard
bak	backup (file extension indicator)
BAK	Binary Adaptation Kit
BAL	Balance(d)
BAL	Basic Assembler Language
BAL	Branch And Link (instruction)
BAL	Business Application Language
BALLOTS	Bibliographic Automation of a Large Library making use Of a Time-sharing System
BALM	Block And List Manipulator
BALR	Branch And Link Register (instruction)
BALT	Balance Test
BALUN	Balanced/Unbalanced
BAM	Basic Access Method
BAM	Bi-directional Associative Memory
BAM	Block Allocation|Availability Map
BAM	Boolean Approximation Method
BAM	Business Analysis Model
BAMAF	Bellcore Asynchronous Multiplexer Adapter Format
BAN	Best Asymptotically Normal
BAO	Binary Arithmetic Operation
BAP	Band Amplitude Product
BAP	Basic Assembler Program
BAP	Bus Available Pulse
BAP	Byte Address Buffer
BAPI	Bridge Application Program Interface
BAPI	Business Application Programming Interface
BAPT	Basic Applications Programmer Training
BAPTA	Bearing And Power Transfer Assembly
BAR	Barrier
BAR	Base Address Register|Relocation
BAR	Bookkeeping Address Register
BAR	Buffer|Byte Address Register
bar	generic file name extension (file extension indicator)
BARD	Bodleian Access to Remote Databases
BAREG	Buffer A Register
BARMINT	Basic Research for Microsystems Integration (EU)
BARON	Business/Accounts Reporting Operating Network
BARRNet	Bay Area Regional Research Network
BARTS	Bell Atlantic Regional Time-Sharing
BAS	Basic Activity Subset
bas	basic language (file extension indicator)
BAS	Bell Audit Systems
BAS	Block Arrow Structure

BAS	Block Automation System
BASBOL	Basic Business Oriented Language
BASE	Basic Semantic Element
basecom	base communications
BASEPTA	Base Application et Standards d'Echange pour le Poste de Travail de l'Automaticien (FR)
BASH	Born Again Shell
BASIC	Beginners All-purpose Symbolic Instruction Code
BASM	Built-in Assembler
BASP	Backspace (character)
BASYS	Balanced Automated Systems
BASYS	Balanced Automation Systems (international conference)
BAT	Base Adjustment Test
bat	batch processing (file extension indicator)
BAT	Battery
BAT	Biomechanical Automated Technical
BAT	Block Access Table
BAT	Block Address Translation
BATEA	Best Available Technology Economically Achievable
BATS	Basic Additional Teleprocessing Support
BATSE	Burst And Transient Source Experiment
baud	signalling elements per second
BAW	Bulk Acoustic Wave
BAWR	Bulk Acoustic Wave Resonator
BB	Band Block
BB	Barbados
BB	Baseband Breadboard
BB	Begin Bracket
BB	Branch on Bit (instruction)
BB	Breadboard
BB	Broadband
BB	Building Block
BB	Bulletin Board
BB	Bus Bar
BB	Busy Bit
BBA	Blackboard Architecture
BBAC	Bus-to-Bus Access Circuit
BBC	Black Box Checking
BBC	British Broadcasting Corporation
BBC	Broadband Coaxial (cable)
BBC	Broadband Control
BBC	Byte Bus Channel
BBD	Bucket Brigade Device
BBDB	Big Brother Data Base
BBH	Bouncing Busy Hour
BBL	Basic Business Language
BBL	Be Back Later
bbl	bibliography (file for TeX etc.) (file extension indicator)

BBL	Branch Back and Load
BBM	Break Before Make
BBP	Base-Band Processor
BBP	Binary Back-off Procedure
BBPN	Breitband Pilot Netz (CH)
BBRC	Biochemical and Biophysical Research Communications
BBREG	Buffer B Register
BBROYGBVGW	Black, Brown, Red, Orange, Yellow, Green, Blue, Violet, Gray, White
BBS	Baseband Signal
BBS	Be Back Soon
BBS	Bulletin Board System
BBS	Business Batch System
BBSP	Building Block Signal Processor
BBSR	Bucket Brigade Shift Register
BBT	Base Band Transmission
BBT	Bucket Brigade Technique
BBU	Battery Backup
BBU	Bit Buffer Unit
BBX	Boundary Box
BBXRT	Broad-Band X-Ray Telescope
BC	Basic Control (mode)
BC	Binary Code
BC	Blind Copy
BC	Block Check
BC	Branch on Condition
BC	Broadcast
BC	Broadcast Control
BC	Bulk Core
BC	Business Computer
BC	Business Control function
BC	Byte Counter
BCA	Basic Channel Adapter
BCA	Binary Coded Address
BCA	Bisynchronous Communications Adapter
BCA	Bit Count Appendage
BCAM	Basic Communication Access Method
BCB	Bit Control Block
BCB	Block Control Byte
BCB	Broadcast Band
BCB	Buffer Control Block
BCC	Binary Character Code
BCC	Binary Coded Character
BCC	Block Cancel|Center Character
BCC	Block Check Code
BCC	Block Check(ing)|Control Character
BCC	Blocked Call Cleared
BCC	Body-Centered Cubic

BCC	Broadcasting Control Center
BCC	Business Computer Center
BCD	Binary Coded Decimal (notation)
BCD	Blocked Call(s) Delayed
BCD/B	Binary Coded Decimal/Binary
BCDIC	Binary Coded Decimal Interchange Code
BCDNAF	Binary Coded Decimal Non-Adjacent Form
BCDS	Board Command Data Sort
BCDS	Board Command De-stack Select (program)
BCE	Basic Comparison Element
BCE	Before Common Era
BCEF	Block Correction Efficiency Factor
BCFSK	Binary Coded Frequency Shift Keying
bch	bits per circuit per hour
BCH	Block Control Header
BCH	Blocked Calls Held
BCH	Bose-Chaudhuri-Hocquenghem code
BCHNR	Batch Number
BCHP	Board Command History Purge (program)
BCI	Basic Command Interpreter
BCI	Begin Chain Indicator
BCI	Broadcast Interference
BCIA	Bounded Carry Inspection Adder
BCIU	Bus Control Interface Unit
BCJS	Buffer Control Junction Switch
BCK	Block
BCL	Base Coupled Logic
BCL	Basic|Batch Command Language
BCL	Branch on C Latch (instruction)
BCL	Broadcast Listener
BCM	Basic Control Memory|Mode
BCM	Binary Choice Model
BCM	Bound Control Module
BCN	Backbone Concentrator
BCN	Beacon
BC-Net	Business Consultancy Network
BCNU	Be Seeing You
BCO	Binary Coded Octal
BCP	Batch Communications Program
BCP	Binary Communications Protocol
BCP	Bit Control Panel
BCP	Bulk Copy Program
BCP	Business Communications Project
BCP	Byte Control(led) Protocol(s)
BCP	Byte Count Protocol
BCPL	Basic Computer Programming Language
BCPL	Bootstrap Combined Programming Language
BCR	Blocked Calls Released

BCR	Byte Count Register
BCRK	Bell Crank
BCROS	Balanced Capacitor Read-Only Storage
BCRT	Bright Cathode Ray Tube
BCS	Bar Code Scanner|Sorter
BCS	Bardeen-Cooper-Schrieffer
BCS	Basic Clock Signal
BCS	Basic Combined Subset
BCS	Basic Communications Support
BCS	Basic Computer|Control System
BCS	Batch Change Supplement
BCS	Batch Communication Subsystem
BCS	Binary Compatibility Standard
BCS	Block Check Sequence
BCS	Board Command Sort (program)
BCS	Boeing Computer Services
BCS	British Computer Society
BCS	Business Communications Systems
BCS	Business Computer System
BCST	Broadcast
BCT	Between Commands Testing
BCT	Branch on Count
BCU	Basic Counter Unit
BCU	Block|Buffer Control Unit
BCU	Bus Control Unit
BCUA	Business Computers Users Association (US)
BCW	Buffer Control Word
BCW	Burst Code Word
BD	Balanced Dimension(s)
BD	Bangladesh
BD	Binary Decoder
BD	Binary Digit
BD	Blank Display|Draft
BD	Block Diagram
BD	Bright Dip
BDA	Batch Data Acquisition
BDA	BIOS Data Area (BIOS = Basic Input/Output System)
BDAM	Basic Direct Access|Address Method
BDBE	Basic Data Base Environment
BDC	Binary Decimal Counter
BDC	Binary Differential Computer
BDCB	Buffered Data and Control Bus
BDD	Binary Decision Diagram
BDD	Binary to Decimal Decoder
BDD	Business Development Directives
BDDN	Broadband Data Dissemination Network
BDE	Basic|Batch Data Exchange
BDE	Borland Database Engine

BDES	Batch Data Exchange Services
BDF	Basic Display File
BDF	Bi-directional Distribution Function
BDF	Binary Distribution Format
BDF	Bitmap Description|Distribution Format
BDF	Business Design Facility
BDFS	Basic Display Filing System
BDI	Batch Data Interchange
BDI	Binary Discrete Input
BDIHIST	Board Interface History
BDK	Bean Development Kit
BDL	Basic Design Language
BDL	Build Definition Language
BDL	Building Description Language
BDLC	B-synchronous Data Link Communication
BDM	Basic Data Management
BDM	Binary Decision Machine
BDN	Base Distinguished Name
BDN	Bell Data Network
BDO	Buffered Data Output
BDOS	Basic Disk Operating System
BDP	Basic Data Processing
BDP	Binary Decision Program
BDP	Business Data Processing
BDPSC	Basic Data Processing Systems Center
BDR	Backup Designated Router
BDR	Basic Data Records
BDR	Bi-Duplexed Redundancy
BDR	Bus Device Request
BDS	Block Data Set
BDS	Building Distribution System
BDS	Bulk Data Switching
BDS	Business Definition System
BDT	Billing Data Transmitter
BDT	Box Diagnostic Time
BDT	Bulk Data Transfer
BDTS	Buffered Data Transmission Simulator
BDU	Basic Device|Display Unit
BDW	Block Description Word
BDX	Bar Double-X
BDY	Boundary
BE	Back End
BE	Band Elimination
BE	Beam Expander
BE	Belgium
BE	Below or Equal
BE	Bose Einstein
BE	Bus Enable

BEAMOS	Beam Addressed Metal Oxide Semiconductor
BEAMS	Base Engineer Automated Management System
BEAR	Bulk Easy Access Readout
BEAV	Binary Editor And Viewer
BEB	Beam Entry Buffer
BEB	Binary Exponential Back-off
BEC	Backward Error Correction
BEC	Belgisch Elektrotechnisch Comité (BE)
BEC	Boolean Equivalence Checker
BEC	Bose-Einstein Condensation
BECAUSE	Benchmark of Concurrent Architectures for their Use in Scientific Engineering
BECC	Binary Error Correcting Code
BECN	Backward Explicit Congestion Notification
BECS	Basic Error Control System
BEDC	Binary Error Detecting Code
BEDIS	Booktrade Electronic Data Interchange Standards
BEDO	Burst Extended Data Out
BEEC	Binary Error Erasure Channel
BEEF	Business and Engineering Enriched Fortran
BEEP	Building Energy Estimating Program
BEF	Band Elimination Filter
BEF	Best Excitatory Frequency
BEG	Begin(ning)
BEGNR	Beginning Number
BEGOREC	Beginning Of Record
BEGSR	Begin Subroutine
BEITA	Business Equipment and Information Technology Association
BEL	Bell (character)
BELLCORE	Bell Communications Research
BEM	Basic Editor Monitor
BEM	Boundary Element Method
BEMA	Business Equipment Manufacturers' Association (US)
BEN	Bending
BEN	Bus Enable
BENET	Boundary Element resources Network
BEP	Beam Entry Processor
BEP	Bit Error Probability
BEP	Burst Error Processor
BER	Basic Encoding Rules
BER	Bit Error Rate
BERM	Bit Error Rate Monitor
BERT	Bit Error Rate Test(er)
BES	Basic Executive System
BEST	Basic Emitter Self-aligned Technology
BEST	Basic Executive Scheduler and Timekeeper
BEST	Boundary Element Software Technology

BEST	Business EDP System Technique (EDP = Electronic Data Processing)
BET	Balanced Emitter-Transistor
BET	Between
BET	Block Equivalence Tape
BETRS	Basic Exchange Telecommunications Radio Service
BEU	Basic Encoding Unit
BEV	Billion Electron Volts
BEX	Broadband Exchange
BEZS	Bandwidth Efficient Zero Suppression
B/F	Background/Foreground
BF	Bad Flag
BF	Basic Frequency
BF	Blocking Factor
BF	Bold Face
BF	Bridge Function
BF	Bridging Fault
BF	Brought Forward
BF	Burkina Faso
BF	Framing Bit
BFAP	Binary Fault Analysis Program
BFAS	Basic File Access System
BFD	Basic Floppy Disk
BFD	Best Fit Decreasing
BFD	Binary File Descriptor
BFF	Binary File Format
BFF	Buffered Flip-Flop
BFI	Batch Freedom Input
BFIC	Binary Fault Isolation Chart
BFICC	British Facsimile Industry Consultative Committee
BFL	Back Focal Length
BFL	Buffered FET Logic
BFN	Beam Forming Network
BFN	Bye For Now
BFO	Beat Frequency Oscillator
BFOS	Balanced File Organization Scheme
BFPDDA	Binary Floating Point Digital Differential Analyzer
BFPR	Binary Floating Point Resistor
BFR	Buffer
BFR	Business Functional Requirement
BFRST	Buffer Start (address)
BFS	Breadth-First Search
BFS	Business File System
BFT	Binary File Transfer|Transmission
BFT	Boundary Function Table
BFT	Bulk Function Transfer
BG	Background
BG	Bulgaria

BGA	Ball Grid Array
BGE	Branch if Greater or Equal
bgi	Borland graphic interface (file extension indicator)
BGNG	Beginning
BGP	Border Gateway Protocol
BGS	Background Storage
BGS	Basic Graphics System
BGS	Bell Global Solutions
BGT	Branch if Greater Than
BGU	Business Graphics Utility
BH	Bahrain
BH	Block Handler
BHC	Block Hardware Code
BHC	Busy Hour Call(s)
BHCA	Busy Hour Call Attempts
BHI	Branch if Higher
BHIS	Branch if Higher or Same
BHL	Busy Hour Load
BHM	Basic Health Management (international)
BHR	Block Handling Routine
BHS	Block Handling Set
BHSI	Beyond (very) High Speed Integration
BI	Backplane Interconnect
BI	Batch Input
B/I	Batch Interactive
BI	Binary Input
BI	Buffer Index
BI	Burundi
BI	Bus Interconnect|Interface
BIA	Burned-In Address
BIAS	Broader Industry Automation System
BIB	Balanced Incomplete Block
bib	bibliography (file extension indicator)
BIB	Bus Interface Board
BIBD	Balanced Incomplete Block Design
BIBO	Bounded-Input, Bounded-Output
BIC	Block Ignore Character
BIC	Branch Information Center
BIC	Buffer Interlace Controller
BIC	Byte Input Control
BICEPS	Basic Industrial Control Engineering Programming System
B-ICI	Broadband Inter-Carrier Interface
BICMOS	Bipolar Complementary Metal-Oxide Semiconductor
BID	Binary Instruction Deck
BID	Business Information Directory
BIDEC	Binary to Decimal
BIDEC	Binary to Decimal Converter
BIDI	Bi-Directional

BIDS	Baggage Information Display System
BIDS	Bath Information and Data Services
BIDS	Broadband Integrated Distributed Star
BIE	Boundary Integral Equation
BIF	Best Inhibitory Frequency
BIF	Binary Information File
BIF	Bulletin Information Function call
BIFET	Bi-metallic Inductive Field-Effect Transistor
BIFF	Binary Interchange File Format
BIFFET	Bipolar combined with junction Field Effect Transistor
BIFS	Bell Information Flow System
BILBO	Built-In Logic Block Observation
BILE	Balanced Inductor Logical Element
BILT	Before Image Log Tape
BIM	Beginning of Information Marker
BIM	Bit Image Memory
BIM	Bus Interface Module
BIMOS	Bipolar Metal-Oxide Semiconductor
BIMS	Business Information Management System
BIN	Basic Information Network
bin	binary (file extension indicator)
BINAC	Binary Automatic Computer
BINAS	Bio-safety Information Network and Advisory Services
BINCL	Binary Cell
BIND	Berkeley Internet Name Domain
BINDIG	Binary Digit
BINHEX	Binary-Hexadecimal
BINNOSYS	Binary Number System
BINOP	Binary Operation
BINPT	Binary Point
BINTR	Binary Trigger
BINVAR	Binary Variable
BIO	Buffered Input/Output
BIOD	Bell Integrated Optical Device
BIOI	Block Input/Output Input
BIO-L	Bi-phased Level
BIO-M	Bi-phased Mark
BIOO	Block Input/Output Output
BIOR	Business Input/Output Rerun
BIOS	Basic Input/Output System
BIOS	Bioregional On-line Information Service
BIO-S	Bi-phased Space
BIOSIS	Bio-Science Information Service
BIP	Binary Image Processor
BIP	Bit Interleave Parity
BIP	Break-In Point
BIP	Bulk Information Processing
BIP	Business Improvement Program

BIPCO	Built-In-Place Components
BIPLEX	Binary Pattern Laser|Logic Extraction
bips	billion instructions per second
BI-QUIN	Bi-Quinary
BIR	Binary Incremental Representation
BIR	Bus Interface Register
BIRDIE	Battery Integration + Radar Display Equipment
BIRS	Basic Indexing and Retrieval System
BIRS	Basic Information Retrieval System
BIS	Basic Instruction Set
BIS	Bootstrap Initiating Switch
BIS	Boundary Intermediate System
BIS	Brain Information Service
BIS	Bureau of Information Science
BIS	Business Information System
BISAD	Business Information Systems Analysis and Design
BISAM	Basic Indexed Sequential Access Method
BISDN	Basic Integrated Services Digital Network
B-ISDN	Broadband Integrated Services Digital Network
BISP	Business Information System Program
BIST	Built-In Self-Test
Bisync	Binary synchronous
BISYNC	Binary Synchronous Communications
BIT	Basic Interconnection Test
bit	binary digit
BIT	Built-In Test|Tracing
BitBlt	Bit Block Transfer
BITE	Built-In Test Equipment
BITN	Bilateral Iterative Network
BITNIC	Bitnet Network Information Center
BITS	Biotechnology Information Toolkit Software
bit/s	bits per second
BITS	Building Integrated Timing System
BIU	Basic Information Unit
BIU	Buffer Image Unit
BIU	Bus Interface Unit
BIVAR	Bivariant
BIX	Binary|Byte Information Exchange
BIZ	Business
BIZMAC	Business Machine Computer
BJ	Benin
BJC	Bubble Jet Colour (printer)
BJF	Batch Job Foreground (mode)
BJT	Bipolar Junction Transistor
BJT	Bubble Jet Technology
BK	Backup
bk!	backup (file extension indicator)
BK	Break(-in)

BKERT	Block Error Rate Tester
BKSP	Backspace
BKUP	Backup
BKWD	Backward
BL	Backlit
BL	Basic Language
BL	Bit Length
BL	Blank(ing)
BL	Block Length
BL	Bus Link
BLA	Binary Logical Association
BLADE	Basic Level Automation of Data through Electronics
BLADES	Bell Laboratories Automatic Design System
BLAISE	British Library Automated Information Service
BLAS	Basic Linear Algebra Subroutines
BLAST	Basic Local Alignment Search Tool
BLAST	Blocked Asynchronous/Synchronous Transmission
BLC	Board Level Computer
BLC	Boundary Layer Control
BLC	Buffer Location Counter
BLCS	Board Line Command Select
bld	basic load graphics (file extension indicator)
BLD	Binary Load Dump
BLDL	Build List
BLDS	Busy Line/Don't Answer
BLE	Branch if Less or Equal
BLEC	Building Local Exchange Carrier
BLER	Block Error Rate
BLERT	Block Error Rate Test(er)
BLES	Buffer Layer Engineering in Semiconductors (EU)
BLF	Branch Loss Factor
BLF	Bubble Lattice File
BLF	Busy Lamp|Line Field
BLIF	Berkeley Logic Interchange Format
BLIM	Buffer location counter Limit
BLIMP	Boundary Layer Integral Matrix Program
BLINC	Battery-powered Light Intra-plant Communication
BLINK	Backward Linkage
BLIP	Base Language Input Program
BLIP	Bit Line Inductance Program
BLIPS	Benthic Layer Interactive Profiling System
BLIS	Bell Laboratories Interpretive System
BLIS	Brunel Library and Information Service
BLISS	Basic Language for the Implementation of System Software
BLISS	Basic List-oriented Information Structured System
BLK	Block
BLKCNT	Block Count
BLKR	Block Rule

BLL	Base Locator Linkage
BLL	Below Lower Limit
BLLE	Balanced Line Logical Element
BLM	Ball Limiting Memory
BLM	Basic Language Machine
BLM	Bus and Lining Module
BLMC	Buried Logic Macro-Cell
BLMPX	Block Multiplexer
BLMUX	Block Multiplexer
BLN	Bell Labs Network
BLNK	Blank
BLN/T	Bell Labs Network for terminal Traffic
BLNT	Broadband Local Network Technology
BLOB	Binary Large Object
BLOS	Branch if Lower Or Same
BLP	Bar Line Printer
BLP	Buffer Load Point
BLP	Bypass Label Processing
BLS	Buffered Line Selector
BLT	Block Transfer
BLU	Basic Length|Link Unit
BLU	Bus Link Unit
BLUE	Best Linear Unbiased Estimate
BLV	Busy Line Verification
BM	Base Machine
BM	Basic Materials
BM	Bermuda
BM	Breakdown Maintenance
BM	Bubble Memory
BM	Buffer Mark|Module
BM	Business Machine
BM	Byte Machine|Mask
BMA	Broadcast Multiple Access
BMAS	Bill of Material Automated System
BMAS	Business Management Accounting System
BMC	Basic Mode Control
BMC	Block Multiplex(er) Channel
BMC	Bubble Memory Controller
BMC	Bulk Media Conversion
BMC	Burst|Byte Multiplex(er) Channel
BMCV	Biologically Motivated Computer Vision (international workshop)
BMD	Baseband Modulator/Demodulator
BMD	Benchmark Monitor Display system
BMD	Bubble Memory Device
BMDP	Biomedical Data Processing
BME	Boundary Merging and Embedding
BMFT	Bundes Ministerium Für Technologie (DE)

BMI	Branch if Minus
BMIC	Bus Master Interface Controller
BMICS	Bill of Material Inventory Control System
BMIDE	Bus Managed Integrated Drive Electronics
BMIS	Blank Management Information System
BML	Basic Machine Language
BML	Bean Markup Language
BMLC	Basic Mode Link Control
BMLC	Basic Multi-Line Controller
BMLD	Binaural Masking Level Differences
BMM	Basic Modular Memory
BMMRG	Board Multiply Merge (program)
BMO	Basic Machine Operation(s)
BMOM	Base Maintenance and Operational Model
BMOS	Bytex Matrix Operating System
BMP	Basic Mapping Support
BMP	Batch Message Partition
BMP	Batch Message Processing|Program
BMP	Benchmark Plan
BMP	Benchmark Program
BMP	Bill of Materials Processor
bmp	bitmap graphics (file extension indicator)
BMP	Bit Mapped Pattern
BMP	Bitmap Picture
BMPX	Block Multiplexer
BMS	Basic Mapping Support
BMT	Basic Machine Time
BMT	Beam Management Terminal
BMT	Bipolar Memory Technology
BMT	Business Management Team
BMU	Basic Measurement Unit
BMVC	Biologically Motivated Computer Vision (workshop)
BN	Binary|Block Number
BN	Brunei Darussalam
BNC	Baby 'N' Connector
BNC	Bayonet Navy|Nut Connector
BNC	Bayonet-Neill-Concelnan
BNC	Branch Network Controller
BNC	British National Connector
BNCC	Base Network Control Center
BNDC	Bulk Negative Differential Conductivity
bndg	binding
BNE	Branch if Not Equal
BNET	Berkeley Network(ing)
BNF	Backus Normal Form
BNG	Branch No Group
BNN	Boundary Network Node
BNPF	Beginning, Negative, Positive, Finish

BNR	Bell-Northern Research (CA)
BNRCVUUCP	Batch News Receive Via Unix to Unix Copy Program
BNS	Backbone Network Service
BNS	Binary Number System
BNS	Business Network Service
BNT	Broadband Network Termination
B-NT1	B-ISDN Network Termination 1 (ISDN = Integrated Services Data Network)
BNU	Basic Network(ing) Utilities
BNZ	Byte Not Zero
BO	Beat Oscillator
BO	Binary Output
BO	Blocking factor Output
BO	Bolivia
BO	Break Out
BO	Bus|Byte Out
BOA	Basic Object Adapter
BOA	Binary-translation Optimized Architecture
BOB	Best Of Both
BoB	Break-out Box
BOC	Back Office Crunch
BOC	Basic Operator Console
BOC	Bell Operating Company
BOC	Block Oriented Computer
BOC	Branch On Condition
BOC	Breach Of Contract
BOC	Bus Out Check
BOC	Byte Output Control
BOCA	Board Of Customer Advisors
BOCA	Borland Object Component Architecture
BOCS	Berard Object (and) Class Specifier
BOD	Bandwidth On Demand
BOE	Beginning Of Extent
BOF	Beginning Of File
BOF	Bill Of Forms
BOF	Birds Of a Feather
BOFADS	Business Office Forms Administration Data System
BOGO	Buy One Get One
BOH	Balance On Hand
BOI	Basic Operators Interface
BOI	Beginning Of Information
BOI	Branch Output Interrupt
BOIS	Branch Office Information Sources|System
BOL	Business On-Line
BOL	Business Oriented Language
BOLD	Bibliographic On-Line Display
BOLD	Bit-Oriented Line Discipline
BOLT	Beam Of Light Transistor

BOM	Basic Operating Memory|Monitor
BOM	Beginning Of Message
BOMAG	Bill Of Material Analyzer and Generator
BOMP	Bill Of Material Processor
BOMS	Base Operations Maintenance Simulator
BONDING	Bandwidth On Demand Interoperability Group
BOOL	Boolean
BOOTP	Bootstrap Protocol
BOP	Basic Operator Panel
BOP	Binary Output Program
BOP	Bit Oriented Protocol
BOP	Break Off Portion
BOPA	Basic Operating Programming Aid
bops	billion operations per second
BOR	Basic Output Report
BOR	Beginning Of Record
BOR	Bus Out Register
BORAM	Block Organized Random Access Memory
BORIS	Batch Oriented Reporting Information System
BORSHT	Battery, Over voltage, Ring, Supervision, Hybrid Test
BOS	Balance Of System
BOS	Basic|Batch Operating System
BOS	Begin Of String
BOS	Bit Organized Storage
BOS	Business Operating System
BOSR	Base Of Stack Register
BOSS	Basic Operating Software System
BOSS	Business Object Server Solution
BOSS	Business On-line Scheduling System
BOSS	Business Oriented Software System
BOT	Beginning Of Table
BOT	Beginning Of Tape|Track
BOT	Boolean Operation Table
BOT	Bottom
BOT	Build, Operate, Transfer
BOUT	Bus Out
BP	Back Propagation
BP	Band Pass
BP	Base Pointer
BP	Batch Processing
BP	Block Packaging
BP	Blueprint
BP	B Pulse
BP	Buffered Printing
BP	Business Process
BP	Bypass
BPA	Bandpass Amplifier
BPAM	Basic Partitioned Access Method

BPB	BIOS Parameter Block (BIOS = Basic Input/Output System)
BPC	Basic Peripheral Channel
BPC	Block Parity Check
BPDU	Bridge Protocol Data Unit
BPE	Basic Programming Extensions
BPE	Beacon Processing Equipment
BPEW	Board Pre-Edit Wires
BPF	Bandpass Filter
BPFS	Bit Parallel Fault Simulation
BPI	Baseline Privacy Interface
bpi	bits per inch
BPI	Block Protection Information
Bpi	Bytes per inch
BPICS	British Production and Inventory Control Society
BPKFL	Block Packaging Flag
BPL	Binary Program Loader
BPL	Branch if Plus
BPL	Business Planning|Programming Language
BPM	Batch Processing Mode|Monitor
bpmm	bits per millimeter
BPNRZ	Bipolar Non-Return-to-Zero
BPOC	Bell Point Of Contact
BPOS	Block print Position
bpp	bits per pixel
BPP	Brokered Private Peering
BPPF	Base Program Preparation Facility
BPR	Business Process Redesign|Re-engineering
BPR	Bypass Ratio
BPRZ	Bipolar Return-to-Zero
BPS	Basic Programming Support|System
BPS	Bill Processor System
BPS	Binary Program Space
bps	bits per second
BPS	Block Parity System
BPS	Business Planning System
Bps	Bytes per second
BPSA	Batch Process Scheduling Algorithm
bpsi	bits per square inch
BPSK	Binary Phase Shift Keying
BPSS	Basic Packet Switching Services
BPSS	Bell Packet Switching System
BPT	Basic Programming Training
BPT	Bisynchronous Pass-Through
BPU	Basic Processing Unit
BPU	Branch Protection Unit
BPV	Bipolar Violation
BPX	Buffer Prefix
BQS	Basic Query System

BR	Bad|Base Register
BR	Blocks Received
BR	Branch
BR	Branch Register
BR	Brazil
BR	Break
BR	Break|Bus Request
BR	Buffer Register
BRA	Baseline Risk Assessment
BRA	Basic Rate Access
BRA	Basic Record Audit
BRAB	B Register A Bit
BRADS	Business Report Application Development System
BRAID	Bi-directional Reference Array, Internally Derived
BRAP	Broadcast Recognition with Alternating Priorities
BRB	Be Right Back
BRBB	B Register B Bit
BRC	Bit Reversion Circuit
BRC	Block Representation Code
BRC	Blocks Received Correctly
BRC	Bounded Right Context
BRC	Branch Conditional
BRCB	B Register C Bit
BRCD	Binary information with a Residue Check Digest
BRDF	Bi-directional Reflectance Distribution Function
BREAD	Basic Routine for Enquiries And Data
BREG	B Register
BREGAB	B Register A Bit
BREGBB	B Register B Bit
BREGCB	B Register C Bit
BREGWDMK	B Register Word Mark
B-rep	Boundary representation
BRF	Band Reject Filter
BRG	Baud Rate Generator
BRGC	Binary Reflected Gray Code
BRI	Basic Rate Interface
BRI	Basic Rate ISDN (= Integrated Services Digital Network)
BRI	Brain Response Interface
BRIEF	Basic Reconfigurable Interactive Editing Facility
BRITE/EURAM	Basic Research in Industrial Technologies for Europe/European Research on Advanced Materials
BRK	Break
BRKPT	Breakpoint
BRKT	Bracket
BRLC	Branch on Load Card
BRM	Basic Reference Model
BRM	Basic Remote Module
BRM	Binary Rate Multiplier

BRM	Bit Rate Multiplier
BRMN	Branch on Minus
BRNZ	Branch on Non-Zero
BROADCAST	Basic Research On Advanced Distributed Computing: from Algorithms to Systems (and) Technologies (EU)
BROM	Bipolar Read Only Memory
BROS	Basic Read-Only Storage
brouter	bridge-router
BROV	Branch Overflow
BROWSER	Browsing On-line With Selective Retrieval
BRP	Business Resumption Plan
BRR	Before Record (signal)
BRS	Basic Record Structure
BRS	Break Request Signal
BRT	Byte Result Trigger
BRTM	Basic Real-Time Monitor
BRU	Basic Resolution Unit
BRU	Branch Unconditional
BRUCE	Buffer Register Under Computer Edit
BRUIN	Brown University Interpreter
BRULE	Block Rule
BRWM	B Register Word Mark
BRZ	Branch on Zero
B&S	American wire gauge
BS	Back Space (character)
BS	Back Spread
BS	Bahamas
BS	Banded Signalling|Setting
BS	Base Station
BS	Beam Splitter
BS	Binary Subtract
BS	Bit Space
b/s	bits per second
BS	Block Scale|Signal
BS	Bureau of Standards
B/s	Bytes per second
BSA	Binary Synchronous Adapter
BSA	Business Software Alliance
BSAL	Block Structured Assembly Language
BSAM	Basic Sequential Access Method
BSAW	Band Sawing
BSB	Back Space Block
BSBG	Burst and Sync Bit Generator
BSBH	Busy Season Busy Hour
BSC	Back Space Character
BSC	Base Station Controller
BSC	Base Stock Control
BSC	Basic Systems Center

BSC	Binary Symmetric Channel
BSC	Binary Synchronous Communication|Control
bsc	boyan script (file extension indicator)
BSC	Branch or Skip on Condition
BSC	Business Service Center
BSC	Bus State Controller
BSCA	Binary Synchronous Communication Adapter
BSCC	Binary Synchronous Communications Controller
BSCEL	Binary Synchronous Communication Equivalence Link
BSCM	Binary Synchronous Communications Macro
BSCM	Bi-Synchronous Communications Module
BSCS	Bachelor of Science in Computer Science
BSCS	Binary Synchronous Communication System
BSC/SS	Binary Synchronous Communications/Start Stop
BSD	Berkeley Software Design
BSD	Berkeley Software|Source Distribution
BSD	Berkeley Standard Distribution
BSD	Bulk Storage Device
BSDC	Binary-Symmetric Dependent Channel
BSDL	Boundary Scan Descriptor Language
BSDP	Bell System Data Processing
BSDS	Berkeley Software Distributions
BSE	Basic Service Element
BSE	Basic Support Equipment
BSE	Basic System Extension
BSELCH	Buffered Selector Channel
BSF	Back Space File
BSF	Band Stop Filter
BSF	Bit Scan Forward
BSFA	Band-Selectable Fourier Analysis
BSI/BEC	British Standard Institute/British Electrotechnical Committee
BSIC	Binary-Symmetric Independent Channel
BSIE	Banking Systems Information Exchange
BSKT	Basket
BSL	Basic Systems Language
BSL	Bit Serial Link
BSLC	Bi-Synchronous data Link Communication
BSM	Basic Storage Module
BSM	Basic Systems Memory
BSM	Batch Scan|Spool Monitor
BSM	Bound Storage Manager
BSM	Bus Server Module
BSMPY	Board Sort Multiply (program)
BSMTP	Batched Simple Mail Transfer Protocol
BSMTP	Batch Simple Message Transfer Protocol
BSN	Broadband Service Node
BSOD	Blue Screen Of Death
BSOR	Block Successive Over Relaxation

BSP	Binary Space Partition (tree)
BSP	Business Service Provider
BSP	Business Systems Planning
BSPE	Board Sort Pre-Edit (program)
BSPG	Business Process Simulation Game
B-spline	Basic spline
BSPS	Basic Semantics (for) Process Simulation
BSQI	Basic Schedule of Quantified Items
BSR	Back Space Record
BSR	Bit Scan Reverse
BSR	Bit Serial Recording
BSR	Buffered Send/Receive
BSRAM	Burst Static Random Access Memory
BSRF	Basic System Reference Frequency
BSRFS	Bell System Reference Frequency Standard
BSRN	Baseline Surface Radiation Network (UN)
BSS	Basic Synchronized Subset
BSS	Block Started by Symbol
BSS	Bulk Storage System
BSSC	Basic System Support Center
BSSDF	Bi-directional Surface-Scattering Distribution Function
BST	Basic Services Terminal
BST	Beam Switching Tube
BST	Binary Synchronous Transmission
BST	But Seriously Though
BSTAT	Basic Status (register)
BSU	Basic Spatial Unit
BSU	Behavioral Science Unit
BSU	Business System Unit
BSY	Binary Synchronous
BSY	Busy
BSYNC	Binary Synchronous Communications (protocol)
BT	Before Total
BT	Bhutan
BT	Bias Temperature
BT	Binary Trigger
BT	Bit Test
BT	British Telecom
BT	Burst Trapping
B-TA	B-ISDN Terminal Adapter (ISDN = Integrated Services Data Network)
BTAC	Branch Target Address Cache
BTAIM	Be That As It May
BTAM	Basic Telecommunications Access Method
BTAM	Basic Terminal|Transmission Access Method
BTB	Branch Target Buffer
BTC	Batch Terminal Controller
BTC	Bistable Trigger Circuit

BTC	Bit Test and Complement
BTC	Block Transfer Controller
BTCU	Buffered Transmission Control Unit
BTDF	Bi-directional Transmission Distribution Function
BTDL	Basic Transient Diode Logic
BTDT	Been There, Done That
BTE	Bench Test Equipment
BTE	Bi-directional Transceiver Element
BTE	Broadband Terminal Equipment
BTE	Business Terminal Equipment
B-TE1	B-ISDN Terminal Equipment 1 (ISDN = Integrated Services Data Network)
BTF	Batch Transaction|Transmission File
BTF	Bulk Transfer Facility
BTL	Backplane Transceiver Logic
BTL	Board Test Language
BTL	Bridge Tied Load
BTL	Business Translation Language
BTLZ	British Telecom Lempel Ziv
BTM	Basic Transport Mechanism
BTM	Batch Time-sharing|Transaction Monitor
BTMA	Busy Tone Multiple Access
BTNS	Basic Terminal Network Support
BTNW	Board Tester Not-Wired (program)
BTO	Bachman Turner Overdrive
BTOA	Binary To ASCII
BTP	Batch Transfer Program
BTP	Build To Plan
BTPON	Business Telephony on Passive Optical Network
BTR	Behind Tape Reader (system)
BTR	Bit Test and Reset
BTR	Bit Transfer Rate
BTRON	Business TRON (= The Real-time Operating system Nucleus)
BTRY	Battery
BTS	Base Transceiver Station
BTS	Basic Tape System
BTS	Batch Terminal Simulation
BTS	Bit Test and Set
BTS	Board Tracking System
BTS	Burster, Trimmer, Stacker
BTSS	Basic Time-Sharing System
BTSS	Braille Time-Sharing System
BTST	Bootstrap
BTT	Basic Time Track
BTT	Blank Transmission Test
BTU	Basic Transmission Unit
BTW	Board Tester Wired (program)
BTW	By The Way

BTX	Bildschirmtext (DE)
BTYPE	Block Type
BU	Basic|Branch Unit
BU	Bottom Up
BU	Business Unit
BUBL	Bulletin Board for Libraries
BUC	Bus Control
BUCH	Board Update Command History (program)
BUDTIF	Block Up and Down Two Interval Forced
BUE	Built-Up-Edge
BUF	Buffer
BUFR	Binary Universal Form for Representation (of meteorological data)
BUFSTOR	Buffer Storage
BUG	Basic Update Generator
BUIC	Back-Up Interceptor Control
BUILD	Base for Uniform Language Definition
BUMP	Back-Up Module Program
BUMP	Bottom Up Modular Programming
BUP	Bottom Up Programming
BUR	Backup Register
BURS	Basic Unformatted Read System
BUS	Basic Utility System
BUS	Business
BUSEN	Bus Enable
BUSREQ	Bus Request
BUT	Board Under Test
BV	Bouvet Island
BV	Breakdown Voltage
BV	Bulk Volume
BVP	Boundary Value Problem
BW	Band-Width
B/W	Black and White
BW	Botswana
BW	Business Wire
BWA	Backward Wave Amplifier
BWAR	Board Wire Assignment and Routing
BWC	Band-Width Compression
BWC	Buffer Word Counter
BWD	Basic Word Data
BWG	Birmingham Wire Gauge
BWH	Basic Warehouse
BWM	Block Write Mode
BWO	Backward Wave Oscillator
BWPA	Backward Wave Power Amplifier
BWT	Backward Wave Tube
BWTS	Band-Width Test Set
BX	Armoured and insulated cable

BX	Base Exchange|Indexed
BX	Bit X
BXH	Branch on Index High
BXLE	Branch on Index Low or Equal
BY	Backlog Yield
BY	Belarus
BYMUX	Byte Multiplexer
BYP	By-Pass (character)
BYTECY	Byte Cycle
BZ	Belize
BZL	Branch on Z Latch

C

c	candle
C	Capacitor (symbol)
C	Carbon
C	Carry (bit)
C	Celsius
C	C language
C	Clock
C	Code
C	Collector
C	Common
C	Copper
C	Coulomb
c	C source code (file extension indicator)
C	Cylinder
c	speed of light in vacuum
C2C	C To C
C2D	Character To Decimal
C2F	Character To Floating point
C2PC	Command & Control Personal Computer
C2X	Character To Hexadecimal
C3	Command, Control and Communications
C3DS	Control & Coordination of Complex Distributed Services
C3I	Command, Control, Communications and Intelligence
C3PO	Custom Third Party Objects
C4	CAD, CAM, CAE, CIM (CAD = Computer Aided Design; CAM = Computer Aided Manufacturing; CAE = Computer Aided Engineering; CIM = Computer Integrated Manufacturing)
C4	Controlled Collapse Chip Connection
C4I	Command, Control, Communications, Computers and Intelligence
CA	Cable
CA	California (US)
CA	Canada
CA	Cancel

C&A	Certification & Accreditation
CA	Change Accumulation
CA	Channel Adapter|Attachment
C&A	Classification & Audit
CA	Communications Adapter|Attachments
CA	Computer Aided
CA	Computer Animation (international conference)
CA	Connecting Arrangement
CA	Continue Any
CA	Control Area
CA	Coverage Analysis
CAA	Chinese Association of Automation
CAA	Computer Aided Accounting
CAA	Computer Aided Artwork|Assembly
CAA	Computer Assisted Arts
CAAD	Computer Aided Architectural Design
CAAIS	Computer Assisted Action Information System
CAAN	Common Administration Architecture for windows NT (= New Technology)
CAAS	Computer Aided Approach Sequencing
CAATS	Canadian Automated Air Traffic System
CAB	Cabinet
CAB	Channel Address Buffer|Bus
CAB	Compare And Branch
CAB	Computer Address Bus
CABD	Computer Aided Building Design
CABR	Central Address Buffer Register
CABS	Carrier Access Billing System
CABS	Computer Aided Batch Scheduling
CAC	Cache Array Controller
CAC	Call A Computer
CAC	Carrier Access Code
CAC	Circuit Administration Center
CAC	Computer Acceleration Control
CAC	Computer Advisory Committee
CAC	Computer Aided Classification
CAC	Computer Aided Conferencing|Correspondence
CAC	Computer Assisted Cartography
CAC	Continuous Action Controller
CACA	Computer Aided Circuit Analysis
CACD	Compromise Approach to Compiler Design
CACD	Computer Aided Circuit Design
CAChe	Computer Aided Chemistry
CACHKR	Channel Adapter Check Register
CACM	Communications of the Association for Computing Machinery
CA-CO	Cancelling Code
CACS	Center for Advanced Computer Studies
CACS	Computer Assisted Communication System

CACS	Content Addressable Computing System
CACS	Customer Administration Communication System
CACSD	Computer Aided Control System Design
CAD	Cartridge Activated Device
CAD	Channel Adapter
CA/D	Character Assemble/Disassemble
CAD	Character Assembler and Distributor
CAD	Clear & Add
CAD	Computer Aided Design
CAD	Computer Aided Detection|Diagnosis
CAD	Computer Aided Dispatch(ing)
CAD	Computer Assisted|Automated Design
CAD	Control Alternate Delete
CAD	Control And Display (unit)
CADA	Computer Aided Data Analysis
CADA	Computer Assisted Distribution and Assignment
CADAE	Computer Aided Design And Engineering
CADAM	Computer Aided Design And Manufacturing
CADAM	Computer graphics Augmented Design And Manufacturing system
CADAPSO	Canadian Association of Data Processing Service Organizations
CADAR	Computer Aided Design, Analysis and Reliability
CADAS	Computerized Automatic Data Acquisition System
CADAVRS	Computer Assisted Dial Access Video Retrieval System
CADB	Channel Adapter Data Buffer
CAD/CAM	Computer Aided Design/Computer Aided Manufacturing
CADCOM	Computer Aided Design for Communications
CADD	Computer Aided Design and Drafting
CADD	Computer Aided Detector Design
CADDA	Combined Analog Digital Differential Analyzer
CADDETC	Computer Aided Design Data Exchange Technical Center
CADDIA	Cooperation in the Automation of Data Documentation for Import/export and Agriculture (EU)
CADE	Client/server Application Development Environment
CADE	Computer Aided Design and Engineering
CADE	Computer Aided Design Evaluation
CADE	Computer Aided Design of Electronics
CADE	Computer Assisted Data Entry|Evaluation
CADES	Computer Aided Design for Electronic Systems
CADES	Computer Aided Development and Evaluation System
CADET	Computer Aided Design Experimental Translator
CADEX	Computer Aided Design geometry data Exchange
CADfC	Computer Aided Design for Construction
CADFISS	Computer And Data Flow Integrated Sub-System
CAD*I	Computer Aided Design Integration
CADIC	Computer Aided Design of Integrated Circuits
CADICS	Computer Aided Design of Industrial Cabling Systems
CADIS	Computer Aided Design Interactive System

CADIS	Computer Aided Design of Information Systems
CADLIC	Computer Aided Design of Linear Integrated Circuits
CADMAT	Computer Aided Design Manufacture And Testing
CADO	Computer Aided Document Origination
CADOCR	Computer Aided Design of Optical Character Recognition
CADOS	Computational Aerodynamic Design by Optimization System
CADP	Computer-Aided Dimension Planning
CADPIN	Customs Automated Data Processing Intelligence Network
CADPRO	Communications And Data Processing Operation
CADR	Computer Aided Drafting
CADS	Computer Aided Design System
CADS	Computer Aided Digitizing|Drafting System
CADS	Computer Analysis and Design System
CADS	Contents Addressable Store
CADSS	Combined Analog/Digital Systems Simulator
CADSYS	Computer Aided Design System
CADTES	Computer Aided Design and Test
CADV	Combined Alternate Data/Voice
CADVANCE	CAD Advanced (CAD = Computer Aided Design)
CAE	Client Application Enablement|Enabler
CAE	Common Application(s) Environment
CAE	Compare Alpha Equal
CAE	Components Applications Engineering
CAE	Computer Aided|Assisted Education
CAE	Computer Aided|Assisted Engineering
CAE	Computer Assisted Entry|Estimating
CAEDS	Computer Aided Engineering Design System
CAEN	Computer Aided Engineering Network
CAF	Communication Application Factor
CAF	Computer Assisted Fraud
CAF	Conversion Assist Feature
CAF	Current Approved File
CAFC	Computer Automated Frequency Control
CAFE	Computer Aided Film Editor
CAFM	Computer Aided Facility Management
CAFRS	Client Accounting and Financial Reporting System
CAFS	Content Addressable File Store
CAG	Column Address Generator
CAG	Computer Assisted Graphics|Guidance
CAG	Cooperative Automation Group
CAGD	Computer Aided Geometric Design
CAGE	Computer Aided Graphic Expression
CAG-mix	Commitment Article Group mix
CAI	Call Assembly Index
CAI	Channel Available Interrupt
CAI	Common Air Interface
CAI	Computer Administered Instruction
CAI	Computer Aided|Assisted Instruction

CAI	Computer Aided Inspection
CAI	Computer Augmented Instruction
CAI	Conditional Assembly Instruction
CAIC	Computer Assisted Indexing and Classification
CAID	Computer Aided Industrial Design
CAIN	Computer Aided Indexing
CAIN	Computerized Aids Information Network
CAINS	Computer Aided Instruction System
CAINT	Computer Assisted Interrogation
CAIO	Computer Analog Input/Output
CAIP	Computer Analysis of Images and Patterns (international conference)
CAIRS	Computer Assisted Information Retrieval System
CAIS	Canadian Association for Information Services
CAIS	Common APSE Interface Specification
CAIS	Computer Aided Insurance System
CAiSE	Conference on Advanced information Systems Engineering
CAIVR	Computer Assisted Instruction with Voice Response
CAK	Command Acknowledge
CAKE	Categorization-based Knowledge Engineering
CAKE	Computer Assisted Key Entry
CAL	Calendar
cal	calibrate
CAL	Checking Automation Languages
CAL	Common Assembly Language
CAL	Computer Aided|Assisted Learning
CAL	Computer Aided Logistics
CAL	Conversational Algebraic Language
CALA	Computer Aided Loads Analysis
CALAS	Calendering And Scheduling (system)
CALB	Computer Aided Line Balancing
CALB	Computer-Assembly Line Balancing
CALC	Calculator
CALC	Customer Access Line Charge
CALFIN	California Fisheries Information Network
CALL	Computer-Assisted Language Learning
CALM	COBOL Automatic Language Modifier
CALM	Common Assembly Language for Microprocessors
CALM	Computer Aided Load Manifesting
CALRS	Centralized Automated Loop Reporting System
CALS	Commerce At Light Speed
CALS	Computer aided Acquisition and Logistics Support
CALS	Continuous Acquisition and Lifecycle Support
CALSSP	Common Assembly Language Scientific Subroutine Package
CAM	Calculated Access Method
CAM	Channel Access Method
CAM	Chinese Access Method
CAM	Clear Add Magnitude

CAM	Common Access Method
CAM	Communications Access Manager|Method
CAM	Computer Aided|Assisted Manufacturing
CAM	Computer Aided Management
CAM	Computer Aided Modelling
CAM	Computing Accounting|Attachment Machine
CAM	Content Addressable|Addressed Memory
CAMA	Centralized Automatic Message Accounting
CAMA	Computer Assisted Maintenance
CAMA	Control and Automation Manufacturers Association
CAMAC	Channel Allocation Monitor And Control
CAMAC	Computer Aided Measurement And Control
CAMAC	Computer Automated Measurement And Control
CAMA-ONI	Centralized Automatic Message Accounting – Operator Number Identification
CAM-BIT	Coaxial Anisotropic Magnetic films with Barrier-reduced Time-constant
CAMCOS	Computer Assisted Maintenance planning and Control System
CAMD	Computer Aided Mechanical Design
CAME	Computer Assisted Medicine
CAMELOT	Computerization And Mechanization of Local Office Tasks
CAMEO	Computer Aided Management of Emergency Operations
CAMEO	Content Auditing for Microsoft Exchange Organizations
CAMIS	Center (for) Advanced Medical Informatics (at) Stanford
CAMM	Computer Aided Machining and Manufacturing
CAMM	Computer Assisted Material Management
CAMP	Central|Common Access Monitor Program
CAMP	Communications Administrative Message Program
CAMP	Compiler for Automatic Machine Programming
CAMP	Computer Assisted Menu Planning
CAMP	Controls And Monitoring Processor
CAMP	Current Amplifier
CAMPRAD	Computer Assisted Message Preparation Relay And Distribution
CAMPS	Computer Assisted Message Processing System
CAMPUG	Capital Apple Mac Performa User Group
CAMR	Channel Adapter Mode Register
CAMS	Computer Activity Monitor System
CAMS	Computer Aided Manufacturing Systems
CAMS	Computer Automated Mailing System
CAMTech	Center for Advanced Media Technology (SG)
CAN	Campus Area Network
CAN	Cancel (character)
CAN	Controller Area Network
CANDE	Command And Edit(ing) (language)
CANDE	Culvert Analysis and Design
CANDO	Computer Analysis of Networks with Design Orientation
CANTRAN	Cancel Transmission

CAO	Central Applications Office
CAO	Continuous Analog Output
CAOS	Completely Automatic Operational System
CAOS	Computer Applications Operating System
CAOS	Computer Augmented Oscilloscope System
CAP	Cable Access Point
CAP	Capacitor|Capacity
CAP	Capital
CAP	Capture
CAP	Carrierless Amplitude and Phase(-modulation)
CAP	Central Access Point
CAP	Columbia AppleTalk Package
CAP	Communication Application Platform
CAP	Competitive Access Provider
CAP	Computation Acceleration Processor
CAP	Computer Aided Planning
CAP	Computer Aided Programming|Publishing
CAP	Conflict Analysis Program
CAP	Consumer Analysts and Programmers
CAP	Cryotron Associative Processor
CAPABLE	Controls And Panel Arrangements By Logical Evaluation
CAPARS	Computer Aided Placement And Routing System
CAPC	Computer Aided Production Control
CAPD	Computing to Assist Persons with Disabilities
CAPE	Communications Automatic Processing Equipment
CAPE	Computer Aided Planning and Estimating
CAPE	Computer Aided Product(ion) Engineering
CAPE	Computer Applications in Product(ion) (and) Engineering
CAPER	Computer Aided Pattern Evaluation and Recognition
CAPER	Core Analysis and Program Evaluation Reorder
CAPERTSIM	Computer Assisted Program Evaluation Review Technique Simulation
CAPICS	Canadian Production and Inventory Control Society
CAPM	Computer Aided Production Management
CAPMEC	Computer Application for Planning Manufacturing Engineering Costs
CAPOSS	Capacity Planning and Operation(s) Sequencing System
CAPP	Computer Aided Part Planning
CAPP	Computer Aided Process|Production Planning
CAPP	Content(s) Addressable Parallel Processor
CAPR	Catalog of Programs
CAPRI	Card And Printer Remote Interface
CAPRI	Computer Automated Procurement payment and Receipt of material Inclusive
CAPRI	Computerized Analysis for Programming Investments
CAPS	Cassette Programming System
CAPS	Computer Aided Packaging System
CAPS	Computer Aided Part Selection

CAPS Computer Aided Planning System
CAPS Computer Aided Problem Solving
CAPS Computer Aided Project Study
CAPS Computer Application Service
CAPSE Computer Aided Parallel Software Engineering
CaPSL Canon Printing System Language
CAPTAIN Character And Pattern Telephone Access Information Network
CAPTAINS Character And Pattern Telephone Access Information Network System
CAPUR Computer Assisted Programming User Remote
CAPY Capacity
CAQ Computer Aided Quality
CAQA Computer Aided Quality Assurance
CAR Card Adapter Register
CAR Carry Register
CAR Central|Channel Address Register
CAR Check Authorization Record
CAR Computer Access & Retrieval
CAR Computer Aided Retrieval|Robotics
CAR Console Address Register
CAR Contents of the Address part of the Register in LISP (= List Programming language)
CAR Current Address Register
CARAD Computer Aided Reliability And Design
CARCAS Computer aided Archiving and Change Accounting System
CARD Carrier Detector
CARD Compact Automatic Retrieval Display
CARDA Computer Aided Reliability Data Analysis
CARDS Central Archive for Reusable Defense Software
CARDS Computer Aided Reliability Data System
CARHSPD Carry High Speed
CARL Commercial Analysis Reference Library
CARL Concordia's Automated Response Line
C-ARMS Commercial Accounts Routing and Management System
CAROT Centralized Automatic Reporting On Trunks
CARP Computed Air Release Point
CARPS Calculus Rate Problem Solver
CARR Carriage
CARR Carrier
CARR Carry
CARR Combination Acknowledgment/Rescheduling Request
CARRET Carriage Return
CARS Computer Aided Routing System
CARS Computer Assisted Radiology & Surgery (international congress and exhibition)
CARS Computer Audit Retrieval System
CARS Computerized Audit and Reporting System
CARSIMS Carry Simultaneous

CART	Central Automatic Reliability Tester
CAS	Cable Audit System
CAS	Cartridge Access Station
CAS	Centralized Attendant Service|System
CAS	Chain Acquisition System
CAS	Channel|Circuit Associated Signalling
CAS	Circuits And Systems
CAS	Collision Avoidance Systems
CAS	Column Address Select|Strobe
CAS	Communicating Application Specification|Standard
CAS	Communications Application Specification
CAS	Computer Aided Styling
CAS	Computer Automation System
CAS	Computerized Auto dial System
CAS	Contents Addressable Search|Storage
CAS	Controlled Access System
CASA	Centre for Advanced Spatial Analysis
CASA/SME	Computer and Automated Systems Association of the Society of Manufacturing Engineers
CASC	Commercial Analysis Support Center
CASCADE	Centralized Administrative Systems Control And Design
CAS/CPA	Computer Accounting System/Computer Performance Analysis
CASD	Computer Aided Software Development
CASD	Computer Aided System Design
CASE	Common Application Service Element
CASE	Computer Aided Software|System Engineering
CASE	Computer Aided Structural Engineering
CASE	Computer Aided System(s) Evaluation
CASH	Computer Aided Stock Holdings
CASH	Computer Aided System Hardware
CASL	Control Automation System Logic(s)
CASL	Crosstalk Application Scripting Language
CASMAC	Core Australian Specification for Management and Administrative Computing
CASMIT	Control Automation System Manufacturing Interface Tape
CASNET	Casual Associative Network
CASO	Computer Assisted System Operation
CASOE	Computer Accounting System for Office Expenditure
CASPAR	Cambridge Analog Simulator for Predicting Atomic Reactions
CASS	Coding Accuracy Support System
CASS	Common Address Space Section
CASS	Computer Aided Schematic Symbol (editor)
CASS	Computer Assisted Search Service
CASS	Computer Automatic Scheduling System
CASS	Conditional Access Sub-System
CASSCF	Complete Active-Space Self-Consistent Field
CASSIS	Classified And Search Support Information System
CASSM	Context Addressed Segment Sequential Memory

CAST	Computer Aided Software Testing
CAST	Computer Aided Storage and Transportation
CAST	Computer Applications Systems Technology
CAST	Computer Assisted Scanning Techniques
CAST	Cooperation in Applied Science and Technology
CASTER	Computer Aided System for Total Effort Reduction
CASTER	Conversational And Service Terminal
CASTR	Channel Adapter Status Register
cat	catalog (file extension indicator)
CAT	Category
CAT	Common Authentication Technology
CAT	Communications Authority of Thailand
CAT	Computer Aided Testing|Tomography
CAT	Computer Aided Transcription|Translation
CAT	Computer Aided Typesetting
CAT	Computer Architecture Team
CAT	Computer Assisted Teleconferencing|Telephony
CAT	Computer Assisted Testing|Typesetter
CAT	Computerized Axial Tomography
CAT	Concatenate
CAT	Craft Access Terminal
CAT	Customer Acceptance Test
CATC	Computer Aided Tolerance Chart|Control (system)
CATC	Computer Aided Traffic Control
CATC	Computer Assisted Test Construction
CATCH	Computer Aided Tracking and Characterization of Homicides
CATE	Computer Aided Test Engineering
CATE	Computer Aided Time Efficiency
CATE	Computer Automated Translation and Editing
CATH	Cathode
CATIA	Computer graphics Aided Three-dimensional Interactive Application
CATIS	Common Applications and Tools Integration Services
CATLAS	Centralized Automatic Trouble Locating and Analysis System
CATLINE	Cataloguing on-Line
CATS	Calculative And Typing Selectric
CATS	Centralized Automatic Test System
CATS	Computer Aided Trouble Shooting
CATS	Computer Analysis of Tape Systems
CATS	Computer Assisted Training System
CATSCAN	Computerized Axial Tomography Scan
CATT	Computer Aided Technologies for Thailand (conference and exhibition)
CATT	Conveyorized Automatic Tube Tester
CATV	Cable Antenna Television
CATV	Cable Telecommunications and Video
CATV	Cable Television
CATV	Community Antenna Television

CATVA	Computer Assisted Total Value Assessment
CAU	Connection Arrangement Unit
CAU	Controlled Access Unit
CAU	Crypto Auxiliary Unit
CAUCE	Coalition Against Unsolicited Commercial E-mail
CAU/LAM	Controlled Access Unit/Lobe Attachment Module
CAVD	Computer Aided VLSI Design
CAVE	Cave Automated Virtual Environment
CAVE	Computer Animated|Assisted Virtual Environment
CAVE	Computer Assisted Visual Education
CAVE	Computer Automated Vacuum Equipment|Evaporator
CAVE	Computerized Automatic Virtual Environment
CAVERN	Computer Assisted Virtual Environment Research Network
CAW	Channel Address Word
CAW	Computer Aided Writing
CAX	Community Automatic Exchange
CAX	Computer Aided Mix
cb	C beautifier
Cb	Centibel(s)
CB	Check Bit|Burst
CB	Circuit Breaker
CB	Citizen's Band
CB	Cobol
CB	Collector Base
CB	Column Binary
CB	Command Bus
CB	Communications Buffer
CB	Condition Bit
CB	Contact Breaker
CB	Control Block
CBA	Computer Based Automation
CBA	Concrete Block Association
CBBS	Computerized Bulletin Board Service
CBC	Chain Block Controller
CBC	Cipher Block Chain(ing)
CBC	Computer Based Conferencing|Concentrator
CBC	Consecutive Blank Column
CBCR	Channel Byte Count Register
CBCT	Customer Bank Communication Terminal
CBD	Component Based Development
CBDP	Communications Based Data Processing
CBDS	Circuit Board Design System
CBDS	Constraint Based Diagnostic System
CBDT	Card Board Description Tape
CBE	Computer Based Education
CBED	Convergent Beam Electron Diffraction
CBEMA	Canadian Business Equipment Manufacturers Association
CBEMA	Computer and Business Equipment Manufacturers Association

CBFM	Constant Bandwidth Frequency Modulation
CBI	Charles Babbage Institute
CBI	Common Bus Interface
CBI	Complementary Binary Input
CBI	Compound Batch Identification
CBI	Computer Based Instruction
CBIE	Computer Based Information Exchange
CBIPO	Custom-Built Installation Process Offering
CBIS	Computer Based Information System
CBL	Character Base Line
cbl	COBOL source code (file extension indicator)
CBL	Common Base Language
CBL	Computer Based Learning
CBL	Convective Boundary Layer
CBM	Commodore Business Machines
CBMIS	Comprehensive Budget and Management Information System
CBMIS	Computer Based Management Information System
CBMS	Computer Based Mail(ing) System
CBMS	Computer Based Medical Systems
CBMS	Computer Based Message Service|System
CBO	Continuous Bit stream Oriented
CBODR	Channel Bus Out Diagnostic Register
CBP	Coded Block Pattern
CBPDO	Custom-Built Product Delivery Offering
CBPU	Central Branch Processing Unit
CBQ	Class-Based Queuing
CBR	Carson Bandwidth Rule
CBR	Case-Based Reasoning
CBR	Computer Based Reference
CBR	Constant Bit Rate
CBR/L	Case Based Reasoning and Learning
CBS	Communication Based System
CBT	Computer Based Terminal
CBT	Computer Based Testing|Training
CBT	Core Based Tree
CBT	Core Block Table
CBW	Convert Byte to Word
CBX	Computer controlled Branch Exchange
CBX	Computerized Branch|Business Exchange
CC	Cable Connector
CC	Carbon Copy
CC	C Compiler
CC	Cell Control
CC	Center Conductor
CC	Central Computer|Control
CC	Chain Command
CC	Channel Control(ler)
CC	Charge Coupled

CC	Circuit Card
CC	Clearing Center
CC	Closed Circuit
CC	Cluster Controller
CC	Cocos (Keeling) Islands
CC	Colour Compensation (filter)
CC	Colour Control
CC	Command Chain(ing)
C/C	Command/Control
CC	Common Control
CC	Communication(s) Channel|Computer
CC	Communication(s) Control(ler)
CC	Compute Clock
CC	Computer Center
CC	Computerized Conferencing
CC	Computing Center
CC	Condition Code
CC	Conformance Class
CC	Connect(ion) Confirm
CC	Control Check|Cycle
CC	Control Computer|Counter
CC	Country Code
cc	cubic centimeters
CC	Cursor Control (character)
CCA	Carrier Controlled Approach
CCA	Cascade(d) Correlation Algorithm
CCA	Central Computer Agency
CCA	Channel to Channel Adapter
CCA	Common Communication Adapter
CCA	Common Cryptographic Architecture
CCA	Communications Channel Adapter
CC&A	Computer Control & Auditing
CCA	Conceptual Communications Area
CCA	Constant-Current Anemometer
CCA	Country Customized Application
CCA	Cross-Call Access
CCA	Current Controlled Amplifier
CCAA	Center for Computer Aided Analysis
CCAID	Charge Coupled Area Imaging Device
CCAM	Conversational Communication Access Method
CCAP	Communications Control(led) Application Program
CCAP	Computer Controlled Application Program
CCAR	Component Commitment Analysis Report
CCAS	Communication Control Aid System
CCB	Channel Command|Control Block
CCB	Character|Command Control Block
CCB	Communications Control Batches|Block
CCB	Configuration Control Board

CCB	Convertible Circuit Breaker
CCBS	Clear Channel Broadcasting System
CCC	Central Communications Controller
CCC	Central Computer Center
CCC	Ceramic Chip Carriers
CCC	Channel Control Check
CCC	Channel to Channel Connection
CCC	Charge Carrier Concentration
CCC	Clear Channel Capability
CCC	Command, Control, and Communications
CCC	Command Control Center|Character
CCC	Communication Control Character|Console
CCC	Computer Communication Console
CCC	Computer Control Center
CCC	Concourse Computer Center
CCC	Copy Control Character
CCC	Cryogenic Current Comparator
CCC	Custom Chip Checking
CCC	Cycle Control Counter
CCCB	Completion Code Control Block
CCCC	Colour Calibration Communication and Control
CCCCM	Command, Control, and Communications Counter Measures
CCCE	Computer Controlled Checkout Equipment
CCCI	Command, Control, Communications, and Intelligence
CCCL	Complementary Constant-Current Logic
CCD	Central Control Desk
CCD	Centralized Call Dispatch
CCD	Charge-Coupled Device
CCD	Closed-Circuit Device
CCD	Common Commencement Date
CCD	Component Circuit Diagram(s)
CCD	Computer-Controlled Display
CCD	Custom Chip Design
CCDA	Cable|Circuit Card(s) Design Automation
CCDC	Computer Communication in Developing Countries
CCDM	Charge-Coupled Device Memory
CCDN	Corporate Consolidated Data Network
CCDS	Control Command Data Status register
C/CDSB	Command/Control Disable
CCE	Channel Control Error
CCE	Communication Control Equipment
CCEB	Combined Communications Electronics Board
CCETT	Centre Commun d'Études de Télévision et de Télécommunications (FR)
CCF	Cache Controller Feature
CCF	Central Computing Facility
CCF	Communications Control Field
CCF	Complex Coherence Function

CCF	Compressed Citation File
CCF	Controller Configuration Facility
CCF	Corporate Central File
CCF	Customer Calling Features
CCFL	Cold Cathode Fluorescent Lamp
CCFT	Cold Cathode Fluorescent Tube
CCG	Cache Conflict Graph
CCG	Centro de Computaçao Gráfica (PT)
CCH	Channel Check Handler
cch	connections per circuit per hour
cch	hundred-circuit-hours
CCHS	Cylinder-Cylinder-Head Sector
CCI	Common Carrier|Client Interface
CCI	Common Content Inspection
CCI	Computer Carrier Interrupt
CCI	Computer Controlled Instruction
CCIA	Computer and Communication Industry Association
CCIC	Control Card Installation Card
CCIP	Command and Control Information Processing
CCIP	Continuously Computed Impact Point
CCIR	Comité Consultatif Internationale des Radiocommunications
CCIRN	Coordinating Committee for Intercontinental Research Networks
CCIS	Command Control Information System
CCIS	Common Channel Interface|Interoffice Signalling
CCIS	Computer Controlled Interconnect System
CCITT	Comité Consultatif International du Telephone et Telegraphique
CCITT-2	Baudot character code for telegraphy (5 bits)
CCITT-5	Character code equal to ASCII code (7 bits)
CCIU	Command Channel Interface Unit
CCL	Command Control Language
CCL	Common Command|Control Language
CCL	Communications Control Language
CCL	Composite Cell Logic
CCL	Concise Command Language
CCL	Connection Control Language
CCL	Console Command Language
CCL	Customized Card List
CCLCD	Customized Card List Control Date
CCLSS	Customer Code Line Specification Sheet
CCM	Central Customer Manager
CCM	Communications Control (and) Management
CCM	Communications Control Mode|Module
CCM	Constant Current Modulation
CCM	Control Computer Module
CCM	Copper Circuitized Module
CCM	Correlation Coefficient Method
CCM	Counter Counter Measure(s)

CCMF	Computer Controlled Manufacturing Facility
CCMP	Communications Control (and) Management Processor
CCMS	Computer Center Management System
CCMT	Computer Controlled Machine Tool
CCMU	Computer Controlled Multiplexer Unit
CCN	Cluster Control Node
CCN	Common Carrier Network
CCN	Contact Center Network
CCN	Customer Control Node
CCNC	Common Channel Network Controller|Center
CCNC	Computer Controlled Network Center
CCO	Computer Controlled Operation
CCO	Current Controlled Oscillator
CCOMP	COPICS Communications Oriented Manufacturing Plan
CCOO	Comprehensive Cost Of Ownership
CCP	Certificate in Computer Programming
CCP	Certified Computer Programmer
CCP	Command Control Program
CCP	Communication Control Package|Program
CCP	Communication Control Panel|Processor
CCP	Compression Control Protocol
CCP	Computer Circuit Protector
CCP	Conditional Command Processor
CCP	Configuration Control Program
CCP	Console Command Processor
CCP	Coordinated Commentary Programming
CCPC	Control Cable Parallel Connector
CCPDS	Command & Control Processing Display System
CCPF	Common Customer Profile (system)
CCPG	Central Clock Pulse Generator
CCPS	Constant Current Pulse Source
CCPT	Computer Controlled Positioning Table
CCPT	Controller Creation Parameter Table
CCR	Central Control Room
CCR	Channel Control Reconfiguration
CCR	Clock Count Register
CCR	Commitment, Concurrency, and Recovery
CCR	Compare Character Registration instruction
CCR	Computer Character Recognition
CCR	Computer Controlled Retrieval
CCR	Condition Code Register
CCR	Configuration Control Register
CCR	Console Card Reader
CCR	Covenants, Conditions, and Restrictions
CCR	Customer Controlled Reconfiguration
CCR	Customized Communications Routine
CCROM	Card Capacitor Read-Only Memory
CCROS	Card Capacitor|Capacity Read-Only Storage

CCRP Continuously Computed Release Point
CCRS Canada Centre for Remote Sensing
CCRSE Commitment, Concurrency, and Recovery Service Element
CCS Call Confirmation Signal
CC&S Central Computer & Sequencer
CCS Centum (hundreds) Call Seconds
CCS Coincident Current Selection
CCS Collective Call Sign
CCS Colour Calibration System
CCS Column Code Suppression
CCS Command & Communication System
CCS Common Channel Signalling
CCS Common Command Set
CCS Common Communications Services|Support
CCS Communications Control System
CCS Computer Center Services
CCS Computer Control System
CCS Continuous Commercial Service
CCSA Common Control Switching Arrangement
CCSA Customer Controlled Switching Arrangement
CCSB Computer and Communications Standards Board
CCSD Cellular Circuit-Switched Data
CCSDS Consultative Committee for Space Data Systems
CCSE Corporate Communications Switching Equipment
CCSP Communications Concentrator Software Package
CCSPS Central Computer Support Programming System
CCSS Commodity Command Standard System
CCSS Country Currency Sub-System
CCST Center for Computer Sciences and Technology
CCT Carriage Control Tape
CCT Center for Computer Technology
CCT Central Computer and Telecommunications agency (GB)
CCT Central Control Terminal
CCT Charge Control Tape
CCT Circuit
CCT Computer Compatible Tape
CCT Computer Concepts Trainer
CCT Computer Controlled Typesetting
CCT Current Coincidence Technique
CCTA Central Computer and Telecommunications Agency (GB)
CCTAC Computer Communications Trouble Analysis Center
CCTN Correction
CCTS Comité de Coordination des Télécommunications par Satellite
CCTS Configuration Control Test System
CCTV Closed-circuit Cable Television
CCTV Closed-Circuit Television
CCTV Computer Controlled Television
CCTV/LSD Closed Circuit Television/Large Screen Display

CCU	Camera|Camera Control Unit
CCU	COLT Computer Unit (COLT = Central Office Line Tester)
CCU	Command Chain Unit
CCU	Communication(s)|Computer Control Unit
CCU	Correlation Control Unit
CCUF	Components Control Usage File
CCUF	Consolidated Component Usage File
CCV	Common Control Vector
CCVS	COBOL Compiler Validation System
CCW	Channel Command|Control Word
CCW	Counter Clockwise
CCY	Card Cycle(s)
Cd	Cadmium
CD	Call Deflection
cd	candela
CD	Capability Data
CD	Card
CD	Carrier Detect
CD	Chain Data
CD	Change Directory
CD	Check Digit
CD	Class Directory
CD	Clock Driver
CD	Colour Display
CD	Column Distributor
CD	Common Data
CD	Compact Disc
CD	Component Demand
CD	Computer Drum(s)
CD	Contents Directory
C/D	Control Data
C/D	Control/Display
CD	Core Driver
CD	Count Data
CD	Current Data|Density
CD	Cycle Delay
CDA	Call Data Accumulator
CDA	Capability Data Acknowledgment
CDA	Cascade Design Automation
CDA	Central Data Acquisition
CDA	Chain Data Address
CDA	Class Directory Area
CDA	Command and Data Acquisition
CDA	Communications Decency Act (US)
CDA	Composite Design Aid
CDA	Compound Document Architecture
CDAC	Communications Dual Access Controller
CDAR	Customer Dialled Account Recording

CDB	Command Descriptor Block
CDB	Common Data Bus
CDB	Current Data Bit
CDB	Customer Distributed Bulgiest
CDC	Call Direct(ing)|Directly Code
CDC	Channel Data Check
CDC	Character-Deletion Character
CDC	Code Directing Character
CDC	Computer Display Channel
CDC	Constant Duty Cycle
CDC	Coordination Division Code
CDC	Count Down Clock
CDCCP	Control Data Communications Control Procedure
CD-CHRDY	Card Channel Ready
CDCS	Central Data Collection System
CDDA	CD(-ROM) Direct Access (CD = Compact Disc; ROM = Read Only Memory)
CD-DA	Compact Disc – Digital Audio
CDDI	Copper Distributed Data Interface
CD-DIS	Compact Disc Development Information System
CDE	C Development Environment
CDE	Channel Data Error
CDE	Common Desktop Environment
CD-E	Compact Disc – Erasable
CDE	Complex Data Entry
CDE	Contents Directory Entry
CDE	Control and Display Equipment
CDEC	Central Data Conversion Equipment
CDES	Conversational Data Entry System
CDEV	Control panel Device
CDEX	Check stub Data Exchange
CDF	Channel Definition Format
CDF	Combined Distribution Frame
cdf	comma delimited format (file extension indicator)
CDF	Common Data Format
CDF	Component Data File
CDF	Component Description File
CDF	Consolidated Design File
CDF	Context-Dependent File
CDF	Contiguous Disk File
CDFG	Control and Data Flow Graph
CDFS	Compact Disc File System
CD+G	Compact Disc plus Graphics
CDG	Control Dependence Graph
CDH	Command and Data Handling
CDHS	Comprehensive Data Handling System
CDI	Circle Digit Identification
CDI	Collector Diffused Isolation

CD-I	Compact Disc – Interactive
CDIF	CASE Data Interchange Format (CASE = Computer Aided Software Engineering)
CDIF	Common Data Interchange Format
CDIP	Ceramic Dual In-line Package
CDIS	Conceptual Design Information Server
CDK	Channel Data Check
CDL	Computer Description|Design Language
CDL	Computer Development Laboratory
CDL	Computer Driving License
CDL	Custom Dynamic Logic
CDLC	Cellular Data Link Control
CDLIS	Commercial Drivers' License Information System
CDM	CATIA Data Manager
CDM	Code|Colour Division Multiplexing
CDM	Computer Description Manual
CDM	Conditional Delay Matrix
CDM	Construction Database Model
CDMA	Cartridge Direct Memory Access
CDMA	Code Division Multiple Access
CDMF	Commercial Data Masking Facility
CDML	Claris Dynamic Markup Language
CDMM	Component Demand Maintenance Module
CD-MO	Compact Disc – Magneto Optical
CDMR	Conventional Double Magnetic Resonance
CDMS	Commercial Data Management System
CDMS	Communications Data Management System
CDN	Coded Decimal Notation
CDN	Corporate Data Network
CDOCS	Classified Document Control System
CDOP	Component Description Operation Procedure
CDOS	Concurrent Disc Operating System
CDP	Centralized Data Processing
CDP	Certificate in Data Processing
CDP	Certified|Checkout Data Processor
CDP	Communication(s)|Configuration Data Processor
CDP	Conventional Data Processing
CDPA	Certified Data Processing Auditor
CDPD	Cellular Digital Packet Data
CDPF	Composed Document Printing Facility
CD/PL	Component Demand/Plug List
CDPP	Centralized Data Processing Program
CDPR	Customer Dial Pulse Receiver
CD-PROM	Compact Disc Programmable Read Only Memory
CDPS	Computing and Data Processing Society
CDR	Call Detail Record(ing)
CDR	Call Dial Rerouting
CDR	Channel Data Register

CDR	Common Data Representation
CD-R	Compact Disc – Recordable
CDR	Contents of the Decrement part of the Register
CD-RDx	Compact Disc – Read-only exchange (standard)
CDRL	Contract Data Requirement(s) List
CD-ROM	Compact Disc Read Only Memory
CD-ROM_XA	Compact Disc ROM Extended Architecture
CDRS	Conceptual Design and Rendering Software|System
CD-RTOS	Compact Disc – Real Time Operating System
CD-RW	Compact Disc – Rewrite
CDS	Call Dispatch System
CDS	Case Data System
CDS	Catalogued Data Set
CDS	Central Dynamic System
CDS	Chip Design System
CD-S	Compact Disc – Single
CDS	Comprehensive Display System
CDS	Conceptual Data Store
CDS	Configuration Data Set
CDS	Consumer Data Services|System
CDS	Control Data Set
CDS	Control Display System
CDS	Current Directory Structure
CDSS	Clinical Decision Support System
CDSS	Customer Digital Switching System
CDT	Cable Data Tape
CDT	Call Data Transmitter
CDT	Cambridge Display Technology
CDT	Command Definition Table
CDT	Connectionless Data Transmission
CDT	Control Data Terminal
CDT	Corel Draw Template
CDTI	Cockpit Display (of) Traffic Information
CDTL	Common Data Translation Language
CDTP	Colour Diffusion Transfer Paper
CDTV	Commodore Dynamic Total Vision
CDU	Cartridge Disk Unit
CDU	Central Display Unit
CDU	Control and Display Unit
CDU	Coolant Distribution Unit
CDV	Cell Delay Variation
CD-V	Compact Disc – Video
CDW	Computer Distributed Warehouse
CD-WO	Compact Disc – Write Once
CD-WORM	Compact Disc – Write Once/Read Many
cdx	compound index (file extension indicator)
CD-XA	Compact Disc – Extended Architecture
CE	Cache Enable

CE	Card Error
CE	Ceramic
CE	Channel Enable|End
CE	Channel End
CE	Chip Enable
CE	Communications and Electronics
CE	Communications Entity
CE	Concurrent Engineering
CE	Convert Enable
CE	Correctable Error
CE	Critical Examination
CE	Customer Engineer
CEA	Column Exits Addresses
CEA	Communications Electronics Agency
CEASE	Cost Engineering Automated System Estimate
CEBAF	Continuous Electron Beam Accelerator Facility
CeBIT	Centrum für Büro- und Informations- und Telekommunikationstechnologien (DE)
CEBO	Customer Engineering Branch Office
CEBPS	Customer Engineering Basic Parts System
CEBus	Consumer Electronics Bus
CEC	Central Electronic Complex
CEC	Character-Erase Character
CEC	Chip Electronic Commerce
CEC	Code Extension Character
CEC	Commission of the European Communities
CEC	Competitive Equipment Characteristics
CEC	Computers, Electronics and Control symposium
CECS	Customer Engineering Communication(s) System
CECS	Customer Engineering Contract Services
CECUA	Confederation of European Computer Users Association
CED	Computer Entry Device
CEDA	Communications Equipment Distributors Association (US)
CEDAC	Computer Energy Distribution and Automated Control
CEDAR	Central European (environmental) Data Request facility
CEDAR	Computer-aided Environmental Design Analysis and Realization
CEDDA	Center for Experiment Design and Data Analysis (US)
CEDEX	Courrier d'Entreprise Distribution Exceptionelle (FR)
CEDL	Center for Excellence in Distance Learning
CEDR	Comprehensive Epidemiological Data Resource
CEECS	Computer Environment Energy Control System
CEF	Cable Entrance Facility
CEFACT	Center for E-business Facilitation and Trade
CEFIC	Conseil Européen des Fédérations de l'Industriel Chimique
CEG	Continuous Edge Graphics
CEGL	Cause and Effect Graph Language
CEH	Common Error Handler

CEH	Customer Engineering Hours
CEI	Certified Enterprise Integrator
CEI	Chip Enable Input
CEI	Communications Electronics Instructions
CEI	Comparably Efficient Interconnection
CEI	Conducted Electromagnetic Interference
CEI	Connection Endpoint Identifier
CEI	Contract End Item
CEIM	Customer Engineering Installation Manual
CEIR	Customer Engineering Instruction Reference
CEISD	Customer Engineering Instruction System Diagram
CELEX	European Community Legal Expert database
CELP	Code Excited Linear Predictive coding
CEM	Central Enhancement and Maintenance
CEM	Customer Engineering Manual
CEMAS	Cell Management System
CEMAST	Control of Engineering Material, Acquisition, Storage and Transport
CEMDM	Customer Engineering Maintenance Diagram Manual
CEMF	Counter Electro-Motive Force
CEMI	Customer Engineering Manual of Instruction
CEMISS	Customer Engineering Management Information Support System
CEMM	Customer Engineering Machine|Maintenance Manual
CEMON	Customer Engineering Monitor
CEMS	Central Electronic Management System
CEMS	Constituent Electronic Mail System
CEN	Central
CEN-CENELEC	Comité Européen de Normalisation – Comité Européen de Normalisation Electrotechnique
CENEL	Comité Européen de coordination des Normes Electriques
CENOUT	Central Output
CENT	Carry Entry
CENTREX	Central Exchange
CEO	Chip Enable Output
CEO	Comprehensive Electronics Office
CEOP	Conditional End Of Page
CEP	Channel Event Processor
CEP	Circular Error Possibility
CEP	Civil Engineering Package
CEP	Column Exit Pulse
CEP	Computer Entry Punch
CEP	Connection End-Point
CEP	Continuous Estimation Program
CEPA	Civil Engineering Program Application
CEPA	Society for the advancement of computers in Engineering, Planning and Architecture
CEPI	Connection End-Point Identifier

CEPICOS	Customer Engineering Parts Inventory Control On-line System
CEPICS	Customer Engineering Parts Inventory Control System
CEPIS	Council of European Professional Informatics Societies
CEPPA	Customer Engineering Product Performance Analysis
CEPS	Colour Electronics Prepress System
CEPS	Customer Engineering Parts System
CEPT	Comité Européen des administrations des Postes et des Télécommunications
CEPURS	Customer Engineering Parts Usage Reporting System
CER	Cell Error Ratio
CER	Ceramic
CERC	Computer Entry and Read-out Control
CERC	Concurrent Engineering Research Center
CERDIP	Ceramic Dual In-line Package
CERE	Computer Entry and Read-out Equipment
CERI	Center for Environmental Research Information
CERM	Customer Engineering Reference Manual
CERMET	Ceramic Metallic
CERN	Centre European de Recherches Nucleaires (EU)
CERS	Customer Engineering Reporting System
CERST	Customer Engineering Replenishment Strategy Tool
CERT	Certified
CERT	Character Error Rate Test(er)
CERT	Computer Emergency Response Team
CERT	Constant Extension Rate Test
CES	Circuit Emulation Service
CESD	Composite External Symbol Dictionary
CET	Computer Enhanced Telephony
CET	Console Electric Typewriter
CETIA	Control, Electronics, Telecommunications, Instrument Automation
CETO	Customer Engineering Technical Operations
CEU	Channel Extension Unit
CEU	Communications Expansion Unit
CEVI	Common Equipment Voltage Indicator
CEX	Carry Exit
CF	Carried|Carry Forward
CF	Cathode Follower
CF	Central African Republic
CF	Central File
CF	Centrifugal Focusing
CF	Certainty Factor
CF	Chain Flag
CF	Clock Frequency
CF	Cluster Feature
CF	Command Function
CF	Communications Facility
CF	Context Free

CF	Control Flag
CF	Conversion Facility|Factor
CF	Count Forward
CFA	Carrier Frequency Amplifier
CFA	Common Function Area
CFA	Company|Component Flow Analysis
CFA	Computer Family Architecture
CFAC	Call Forwarding All Calls
CFAR	Constant False Alarm Rate
CFB	Call Forwarding Busy
CFB	Cipher Feedback
CFB	Colour Frame Buffer
CFB	Configurable Function Block
CFC	Channel Flow Control
CFC	Continuous Flow Centrifuge
CFC	Control Field Chart
CFCU	Common File Control Unit
CFD	Communications Facilities Determination
CFD	Computational Fluid Dynamics
CFD	Control Flow Diagram
CFE	Command Facility Extension
CFE	Complement Field End
CFF	Critical Flicker Frequency
cfg	configuration (file extension indicator)
CFG	Context Free Grammar
CFG	Control Flow Graph
CFI	CAD Framework Initiative (CAD = Computer Aided Design)
CFIA	Component Failure Impact Analysis
CFIM	Compatibility Feature Initialize Mode
CFL	Context Free Language
CFLC	Compatibility Feature Load Constants
CFLV	Compatibility Feature Load Variables
CFM	Central Facility Maintenance
CFM	Circuit Feasibility Model
CFM	Code Fragment Manager
CFM	Computer Feasibility Model
CFM	Confirmation
CFMS	Chained File Management System
CFMS	Compatibility Feature Mode Set
CFMU	Central Flow Management Unit
CFNR	Call Forwarding No Reply
CFP	Call For Papers
CFP	Computers, Freedom, and Privacy
CFP	Concept Formulation Package
CFP	Creation Facilities Program
CFPIM	Certified Fellow in Production and Inventory Management
CFR	Code of Federal Regulations
CFR	Computerized Facial Recognition

CFRONT	C++ Front processor
CFS	Combined File Search
CFS	Common File System
CFS	Continuous Forms Stacker
CFSC	Compatibility Feature Store Constants
CFSS	Combined File Search System
CFSV	Compatibility Feature Store Variables
CFTG	Context Free Transduction Grammar
CFU	Call Forwarding Unconditional
CG	Center of Gravity
CG	Character Generator
CG	Cloud to Ground
CG	Colour Graphics
CG	Command Generator
CG	Computer Graphics
CG	Congo
CG	Contracted Ground
CG	Course Generator
CGA	Colour Graphics Adapter
CGB	Convert Gray to Binary
CGCT	Compagnie Générale de Constructions Téléphoniques (FR)
CGDA	Computer Game Developers Association
CGDC	Computer Game Developers Conference
CGE	Common Graphics Environment
CG-hi	Channel Grant high – first priority
CGI	Common Gateway Interface
CGI	Computer Graphics Interface
CGIM	Computer Graphics & Imaging (international conference)
CGL	Computer Generated Letter
CG-lo	Channel Grant low – third priority
cgm	computer graphics metafile (file extension indicator)
CG-med	Channel Grant medium – second priority
CGMID	Character Generation Module Identifier
CGMIF	Computer Graphics Metafile Interchange Format
CGN	Concentrator Group Number
CGO	Continue to Go On
CGP	Colour Graphics Printer
CGPC	Cellular General Purpose Computer
CGRP	Compound file element Group chunk
cgs	centimetre-gramme-second
CGS	Cycle Group Signal
CGSA	Cellular Geographic Serving Area
CGSA	Computer Graphics Structural Analysis
CGSE	Counter Group Selector Entry
CGSEX	Counter Group Selector Exit
CGT	Computer Graphics Technology
CGU	Character Generator Utility

CGW	Color Graphics Workstation
CH	Change
CH	Channel (hold)
CH	Character
CH	Clearing House
CH	Control Heading
CH	Switzerland
CHA	Closed Hard Access
CHAI	Cisco Hosted Applications Initiative
CHAID	Chi-squared Automatic Interaction Detector
CHAM	Chained Access Method
CHAMP	Character Manipulation Procedures
CHAMP	Communications Handler for Automatic Multiple Programs
CHAN	Channel
CHAOS	Chicago university Asynchronous Operating System
CHAP	Challenge Authentication Protocol
CHAP	Challenge Handshake Authentication Protocol
CHAP	Channel Processor
CHAR	Character
CHARGEN	Character Generator
CHARM	Comprehensive Human Animation Resource Model (EU)
CHAT	Cheap Access Terminal
CHAT	Computer Hypertext Access Technology
CHAT	Conversational Hypertext Access Technology
CHBIT	Check Bit
CHC	Channel Control
CHCK	Channel Check
CHCP	Change Code Page
CHCU	Channel Control Unit
CHCV	Channel Control Vector
CHDIR	Change Directory
CHDL	Computer Hardware Definition|Description Language
CHDS	Chip Hierarchical Design System
CHDSt	Chip Hierarchical Design Standard
CHE	Channel End
CHE	Character Early
CHE	Chip Enable
CHECS	Check Handling Executive Control System
ChemMIST	Chemical Management Information System Tool
CHEOPS	Chemical Engineering Optimization System
CHES	Cryptographic Hardware & Embedded Systems (workshop)
CHF	Critical Heat Flux
CHFN	Change Finger (Unix)
CHG	Change
CHGRP	Change Group
CHI	Communication Hardware Interface
CHI	Computer Human Interface
CHIF	Channel Interface

CHIGFET	Complementary Heterostructure Insulated Gate Field Effect Transistor
CHIL	Current Hogging Injection Logic
CHILD	Cognitive Hybrid Intelligent Learning Device
CHIMES	Chip Manufacturing Expert System
CHIN	Characteristic Information
CHIN	Community Health Information Network
CHIO	Channel Input/Output
CHIP	Coded Hexadecimal Interpretive Programming
CHIPS	Computer-Heterogeneous Information Processing Systems (EU)
CHK	Check
chk	check disk (file extension indicator)
CHKDSK	Check Disk
CHKNO	Check Number
CHKPROB	Check Problem
CHKPT	Checkpoint
CHKR	Checker
CHL	Character Late
CHMOD	Change Mode
CHN	Chaining (check)
CHNLENT	Channel Entry
CHOL	Common High Order Language
CHOOSE	Swiss (CH) special interest group for Object-Oriented Systems and Environments
CHOP	Change in Operational Control
CHOWN	Change Owner
CHP	Channel Pointer|Processor
CHP	Chapter
CHPM	Change Priority Mask
CHR	Channel Hardware Reconfiguration
CHR	Channel Register
CHR	Character
CHR	Collision Handling Routine
CHR	Compare Halfword Register instruction
CHRP	Common Hardware Reference Platform
CHS	Character Sense
CHS	Cylinder(s) Head(s) Sector
CHT	Call Holding Time
CHT	Chart
CHT	Collection, Holding, and Transfer
CHT	Cylinder Head Temperature
CI	Call Indicator
C/I	Carrier-to-Interference ratio
CI	Carry In
CI	Cartridge Image
CI	Cluster Interconnect
CI	Colour Index
CI	Communications|Component Interface

CI	Computer Industry
CI	Computer Interconnect(ion)
CI	Configuration Item
CI	Connect Indication
CI	Control Interval
CI	Copy Inhibit
CI	Cote d'Ivoire
CI	Couple Input
CI	Cubic Inches
CI	Current awareness Information
Ci	input capacitance
CIA	C Information Abstractor
CIA	Collection Interface Agents
CIA	Computer Industry Association
CIA	Computerized Image Analysis
CIA	Current Instruction Address
CIAC	Ceramics Information Analysis Center
CIAC	Computer Incident Advisory Capability
CIB	Channel Interface Base
CIB	Command Input Buffer
CIB	Console Interface Board
CIB	Core Image Buffer
CIBER	Cellular Inter-carrier Billing Exchange Roamer
CIC	Carrier Identification Code
CIC	Central Information Center
CIC	Computer Integrated Construction
CIC	Control Inquiry Card
CIC	Coordination and Information Center
CIC	Corporate Information Center
CIC	Customer Initiated Call
CIC	Custom Integrated Circuit
CICA	Construction Industry Computing Association (US)
CICERO	Control Information system Concepts based on Encapsulated Real-time Objects
CICI	Confederation of Information Communication Industries
CICLOP	Customizable Inference and Concept Language for Object Processing
CICP	Communication Interrupt Control Program
CICS	Customer Information Control System
CICS/ISC	Customer Information Control System/Inter Systems Communications
CICSPARS	Customer Information Control System Performance Analysis Reporting System
CICS/VS	Customer Information Control System/Virtual Storage
CID	Charge Injection Device
CID	Charge-injection Imaging Device
CID	Collection of Indexable Data
CID	Communication Identifier

CID	Compatibility Initialization Deck
CID	Component Identification (number)
CID	Computer Initialization Deck
CID	Computer Integrated Design
CID	Configuration, Installation, and Distribution
CID	Connection Identifier
CIDA	Channel Indirect Data Addressing
CIDAS	Compusult Integrated Data Access System
CIDAS	Conversational Iterative Digital/Analog Simulator
CIDF	Control Interval Definition Field
CIDH	Common Import Data Handler
CIDIN	Common Icao Data Interchange Network
CIDR	Classless Inter-Domain Routing
CIDS	Chemical Information and Data System
CIE	Comité Internationale de l'Eclairage
Cie	Compagnie (FR)
CIE	Computer Integrated Enterprise
CIE	Computer Interrupt Equipment
CIE	Customer Initiated Entry
CIEE	Computer Integrated Electronic Engineering
CIEEM	Center of Integrated Electronics and Electronics Manufacturing
CIELAB	Commission International l'Eclairage + Lightness + A*, B* axis labels
CIELUV	Commission International l'Eclairage + Lightness + U*, V* axis labels
CIESIN	Consortium for International Earth Science Information Network
CIF	Caltech Intermediate Form
CIF	Cells In Frames
CIF	Central Information File
CIF	Common Information Facilities
CIF	Common Interchange|Intermediate Format
CIF	Computer Integrated Factory
CIF	Crystallographic Information File
CIF	Customer Information Feed|File
CIFA	Communications Interface
CIFAM	Communications Interface Module
CIFAX	Ciphony/Facsimile
CIFS	Common Internet Filing System
CIG	Commercial Internet (exchange) Group
CIGOS	Canadian Interest Group on Open Systems
CIH	City Information Highway
CII	Call Identity Index
CIIR	Center (for) Intelligent Information Retrieval
CIKM	Conference on Information and Knowledge Management
CIL	Circuit verification environment Interval Language
CIL	Computer Integrated Logistics
CIL	Computer Integration Laboratories

CIL	Computer Interpreter Language
CIL	Core Image Library
CIL	Current Injection Logic
CIL	CVE's Interval Language (CVE = Circuit Verification Environment)
CILE	Call Information Logging Equipment
CILS	Communications Information Library Sources
CIM	Central Inventory Management
CIM	Common Information Model
CIM	Communications Interface Monitor
CIM	Compuserve Information Manager
CIM	Computer Input Microfilm
CIM	Computer Integrated Manufacturing
CIM	Console Interface Module
CIM	Corporate Information Management
CIMA	Computational Intelligence Methods & Applications (international conference)
CIMAP	Circuit Installation Maintenance Access Package
CIMC	Communications Intelligent Matrix Control
CIM&CDF	Computer Integrated Manufacturing & Communications and Data Facility
CIME	Computer Integrated Manufacturing and Engineering
CIMIA	Computer Integrated Manufacturing and Industrial Automation
CIMITI	Center for Information Management and Information Technology Innovation
CIMMOD	Computer Integrated Manufacturing Modeling
CIMOSA	Computer Integrated Manufacturing Open System Architecture
CIMP	Computer Integrated Manufacturing Process
CIMS	Computer Installation Management System
CIMS	Computer Integrated Manufacturing|Measurement System
CIMS	Countermeasures Internal Management System
CIMS-ERC	Computer Integrated Manufacturing Systems – Engineering Research Center
CIMT	Computer Integrated Manufacturing Technology
CIN	Change Identification Number
CIN	Corporate Information Network
CINEMA	Configurable Integrated Multimedia Architecture
CINV	Control Interval
CIO	Carrier Injection Oscillator
CIO	Central Input/Output (multiplex)
CIO	Chief Information Officer
CIO	Control Input/Output
CIOC	Central Input/Output Control
CIOCS	Communications Input/Output Control System
CIOS	Cisco's ubiquitous Inter-networking Operating System
CIOU	Communications Input/Output Unit
CIP	Cassette In Place
CIP	Central Input Program

CIP	Centralized Information Processing
CIP	Command Interface Port
CIP	Commerce Interchange Pipeline
CIP	Communications Interface Processor
CIP	Communications Interrupt Program
CIP	Complex Information Processing
CIP	Core Image Program
CIP	Country Implemented Program
CIP	Crystallized Information Processing
CIPAS	Communication Interference (jamming) Propagation Analysis System
CIPPC	Computer Integrated Production Planning and Control
CIPREC	Conversational and Interactive Project Evaluation and Control
CIPRESS	Cryptographic Intellectual Property Rights Enforcement Systems
CIPS	Canadian Information Processing Society
CIPS	Centralized Information Processing System
CIR	Central Input Recording
CIR	Circuit
CIR	Circular
CIR	Colour Infra-Red
CIR	Committed Information Rate
CIR	Communications Industry Researchers (US)
CIR	Computerized Information Retrieval
CIR	Current Instruction Register
CIRC	Centralized Information Reference and Control
CIRC	Circular Reference
CIRCA	Center for Instructional and Research Computing Activities
CIRCA	Computerized Information Retrieval and Current Awareness
CIRCAL	Circuit Analysis
CIRCUS	Circuit Simulator
CIRIL	Centre Interuniversitaire de Ressources Informatiques de Lorraine (FR)
CIRIS	Common Internal Reporting Information System
CIRM	Centre International de Rencontres Mathématiques
CIRO	Card Inductor Read-Only (memory)
CIRP	Committee International for the Research of Production engineering
CIRS	Communication Information Retrieval System
CIRS	Customer Information Record|Reference System
CIRT	Conference on Industrial Robot Technology
CIS	CASE Integration Services (CASE = Computer Aided Software Engineering)
CIS	Central Information System
CIS	Central Input|Interrupt System
CIS	Character Instruction Set
CIS	Client Information System
CIS	Commercial Information System

CIS	Commercial Instruction Set
CIS	Common Input System
CIS	Communications|Community Information System
CIS	Compatible Information System
CIS	Compuserve Information Service(s)
CIS	Computer (and) Information Services|Systems
CIS	Contact Image Sensor
CIS	Control Indicator Set
CIS	Correction Information System
CIS	Current|Customer Information System
CIS	Current Information Selection
CIS	Customized Intercept Service
CISA	Certified Information Systems Auditor
CISAM	Compressed Index Sequential Access Method
CISC	Complex Instruction Set Computer
CIS-COBOL	Compact Interactive Standard
CISD	Corporate Information Service Department
CISPR	Comité Internationale Special Perturbation Radioelectrique
CISS	Conference on Information Science and Systems
CISS	Consolidated Information Storage System
CIT	Cable Interface Tape
CIT	Call In Time
CIT	Command Interpretation Table
CIT	Computer Integrated Telephony
CITAP	Computerized Interactive Thermal Analysis Program
CITI	Center for Information Technology Integration
CITIS	Contractor Integrated Technical Information Services
CITS	Central Integrated Test Sets|System
CIU	Central|Channel Interface Unit
CIU	Communications Interface Unit
CIU	Computer Interface Unit
CIUP	Country Installed User Program
CIVNET	Civilization Net
CIX	Commercial Internet Exchange
CIX	Compulink Information Exchange
CJB	Cold Junction Box
CJL	Conversational Job Language
CJLI	Command Job Language Interpreter
CK	Check
CK	Clock
CK	Cook Islands
CK	Count Key
CKC	Check Character
CKD	Count-Key-Data (device)
CKD	Cryptographic Key Data (set)
CKMAS	Clock Master
CKO	Check Operator
CKO	Chief Knowledge Officer

CKPT	Check Point
CKT	Cryptographic Key Translation (center)
CL	Center|Central Line
CL	Chile
CL	Class
CL	Clear
CL	Command|Compiler Language
CL	Computer Linguistics|Link
CL	Configuration List
CL	Connectionless (mode)
CL	Control Language
CL	Control Leader|Line
CL	Correction List
CLA	Carry Look Ahead
CLA	Communications Line Adapter
CLA	Computer Law Association (US)
CLAC	Corporate Language And Communication (CH)
CLAD	Cover Layer Automated Design
CLAIMS/CLASS	Class codes, Assigned, Index, Method, Search/Classification
CLAMMS	Communication Level Analysis of Multi-Media Systems
CLAMP	Computer Listing and Analysis of Maintenance Programs
CLAR	Channel Local Address Register
CLASIC	Customized Logic Application Specific Integrated Circuit
CLASP	Circuit Layout, Automated Scheduling and Production
CLASP	Comprehensive Logistics Automated Support Program
CLASP	Connecting Link for Applications and Source Peripherals
CLASS	Capacity Loading And Scheduling System
CLASS	Centralized Local Area Selective Signalling
CLASS	Client Access to Systems and Services
CLASS	Closed Loop Accounting for Stores Sales
CLASS	Closed Loop Automatic Scheduling System
CLASS	Cooperative Library Agency for Systems and Services
CLASS	Custom Local Area Signalling Services
CLAT	Communications Line Adapter for Teletype
CLB	Central Logic Box
CLB	Clear Both
CLB	Common Logic Board
CLB	Communication services Local Block
CLB	Configurable Logic Block
CLB	Continuous Line Bucket
CLC	Clear Carry (flag)
CLC	Coil-Loaded Cable
CLC	Communications Line|Link Control(ler)
CLC	Compare Logical
CLC	Control Leader Cell
CLC	Current Leading Components
CLCB	Charged Liquid Cluster Beam
CLCC	Ceramic Leaded Chip Carrier

CLCM	Communication Line Concentrator Module
CLD	Called (line)
CLD	Clear Direction (flag)
CLD	Condensed Logic Diagram
CLD	Control Logic Diagram
CLD	Current Limiting Device
CLDAS	Clinical Laboratory Data Acquisition System
CLDATA	Cutter Location Data
CLDN	Calling Line Directory Number
CLE	Conservative Logic Element
CLEAR	Closed Loop Evaluation And Reporting
CLEAR	Compiler Executive program Assembler Routines
CLEAT	Computer Language for Engineers And Technologists
CLEC	Competitive Local Exchange Carrier
CLEF	Certified Licensed Evaluation Facility
CLEI	Common-Language Equipment Identification
CLEM	Code Learning Machine
CLEO	Clear Language for Expressing Orders
CLF	Command Language Facility
CLF	Current Level File
CLFC	Condensed Logic Flow Chart
CLFILE	Cutter Location File
CLG	Calling (line)
CLG	Compile, Load, and Go
CLI	Calling|Caller Line Identification
CLI	Call Level Interface
CLI	Clear Interrupt (flag)
CLI	Client Library Interface
CLI	Command Language Interpreter
CLI	Command Line Interface|Interpreter
CLI	Control Language Interpreter
CLIB	Command Library
CLIC	Command Language for Interrogating Computers
CLIC	Computer Layout of Integrated Circuits
CLIC	Conversational Language for Interactive Computing
CLICS	Categorical Logic In Computer Science (EU)
CLID	Caller Identification
CLID	Calling Line Identification
CLID	Clustering Identification
CLIM	Cellular Logic In Memory
CLIM	Common LISP Interface Management (LISP = List Processor)
CLIMATE	Computer and Language Independent Modules for Automatic Test Equipment
CLIN	Community Learning and Information Network
CLIO	Conversational Language for Input/Output
CLIP	Calling Line Identification Presentation
CLIP	Cellular Logic Image Processor
CLIP	Compiler Language for Information Processing

CLIP	Computer Layout Installation Planner
CLIP	Connectionless Integrated Production
CLIP	Coverage Line Inventory Profile
CLIPS	C Language Integrated Production System
CLIR	Calling Line Identification Restriction
CLIR	Council on Library and Information Resources
CLIRA	Closed Loop In Reactor Assembly
CLISP	Conversational List Processor
CLIST	Command List
CLK	Clock
CLKIN	Clock In
CLKOUT	Clock Out
CLLI	Common-Language Location Identification
CLLM	Consolidated Link-Layer Management
C/LM	Called/Left Message
CLM	Communications Line Multiplexer
CLNAP	Connectionless Network Access Protocol
CLNP	Connectionless Network Protocol
CLNS	Connectionless-mode Network Service
CLOB	Character Large Object
CLOB	Core Load Overlay Builder
CLODS	Computerized Logic Oriented Design System
CLOG	Computer Logic Graphics
CLOGS	Classification Of Goods and Services
CLOS	Common LISP Object System (LISP = List Processing language)
CLP	Cell Loss Priority
clp	clipboard (file extension indicator)
CLP	Communication Line Processor
CLP	Constraint Logic Programming
CLP	Current Line Point(er)
CLR	Cell Loss Ratio
CLR	Clear
CLR	Combined Line and Recording
CLR	Common Line Receiver
CLR	Computer Language Recorder|Research
CLRC	Circuit Layout Record Card
CLS	Clear Screen
CLS	Close
CLS	Closed Loop System
CLS	Communications Line Switch
CLSA	Collective Learning Stochastic Automation
CLSF	Connectionless Service Functions
CLSID	Class Identifier
CLSIM	Clocked Logic Simulation
CLT	Channel Load Table
CLT	Clock Track
CLT	Collect
CLT	Communication Line Terminal(s)

CLT	Component Library Tape
CLT	Computer Language Translator
CLT	Corporate Logic Tester
CLTP	Connectionless Transport Protocol
CLTS	Clear Task Switch Flag
CLTS	Connectionless Transport Service
CLU	Central Logic Unit
CLU	Circuit Line-Up
CLU	Close Up
CLU	Command Line Utility
CLUE	Compiler Language Utility Extension
CLUI	Command Line User Interface
CLUSAN	Cluster Analysis
CLUT	Colour Look-Up Table
CLX	Common LISP interface to X window (LISP = List Processor)
CM	Cameroon
cm	centimeter
CM	Central Memory
CM	Circuit Module
CM	Common Mode
CM	Communication Module|Multiplexer
CM	Communication(s) Management|Manager
CM	Communications Monitor
CM	Computer Module
CM	Configuration Management
CM	Connection Machines
CM	Continuous Media
CM	Control Mark|Memory
CM	Control Mode|Module
CM	Convergence Matrix
CM	Core Memory
CM	Corrective Maintenance
CM4D	Coordinate Measuring Machine Management Mechanism for Data
CMA	Cache Memory Array
CMA	Communication Managers Association (US)
CMA	Computerized Management Account
CMA	Computer Management Association (US)
CMA	Computer Monitor Adapter
CMA	Concert Multithread Architecture
CMA	Cross Mathematical Applications
CMAN	Coastal Marine Automated Network (US)
CMAP	Central Memory Access Priority
CMAP	Charge Materials Allocation Processor
CMAP	Colour Map
CMAR	Control Memory Address Register
CMARS	Cable Monitoring And Rating System
CMB	Code Matrix Block

CMB	Configuration Management Board
CMB	Core Matrix Block
CMC	Code Magnetic Character
CMC	Command Module Computer
CMC	Common Mail Call(s)
CMC	Common Messaging Calls
CMC	Communication Magnetic Card
CMC	Communications Management Configuration
CMC	Communications Mode Control
CMC	Comparison Measuring Circuit
CMC	Complement Carry (flag)
CMC	Computer-Mediated Communication
CMC	Customer Master Card
CMCA	Character Mode Communications Adapter
CMCC	Computer Monitor Control Console
CM/CCM	Counter Measures/Counter Counter-Measures
CMCOKP	Cybernetic Modelling and Control of One-of-a-Kind Production
CMCS	Computer Mediated Communication Systems
CMD	Centralized Message Distribution
CMD	Circuit Mode Data
cmd	command (file extension indicator)
CMDATA	Carrier Modulated Data
CMDF	Combined Main Distributing Frame
CMDIS	Computer Management Distributed Information Software
CMDR	Command Reject
CMDS	Centralized Message Data System
CMDS	Computer Management and Development Services
CME	Central Memory Extension
CME	Common Mode Error
CME	Computer Measurement (and) Evaluation
CMF	Capacity Management Facility
CMF	Common Mode Feedback circuit
CMF	Comprehensive Management Facility
CMF	Condensed Master File
CMF	Constant Magnetic Field
CMF	Creative Music Format
CMF	Cross-Modulation Factor
CMG	Computer Measurement Group
CMI	Coded Mark Inversion
CMI	Computer Managed Instruction
CMIB	Cellular Message Information Block
CMIP	Common Management Information Protocol
CMIPDU	Common Management Information Protocol Data Unit
CMIPM	Common Management Information Protocol Machine
CMIS	Common Management Information Services|System
CMIS	Common Manufacturing Information System
CMIS	Configuration Management Information System

CMIS	Controls|Corporate Management Information System
CMISE	Common Management Information Service Element
CMIST	Configuration Management Integrated Support Tool
CML	Common Machine Language
CML	Common Mode Logic
CML	Computer Managed Laboratory
CML	Computer Managed Learning
CML	Conceptual Modelling Language
CML	Current Mode Logic
CMM	Colour Matching Methods
CMM	Communications Multiplexer Module
CMM	Computer Main Memory
CMM	Coordinate Measuring Machine
CMMS	Computerized Maintenance Management Software
CMMU	Cache Memory Management Unit
CMND	Command
CMOI	Common Management information services and protocol Over IEEE 802
CMOS	Complementary Metal-Oxide Semiconductor
CMOT	Common Management information services and protocol Over TCP/IP
CMOV	Conditional Move
CMP	Chip Multi-Processor
CMP	Chromatographic Monitoring Program
CMP	Communications Management Processor
CMP	Compare
CMP	Complement
CMP	Computational
CMP	Console Message Processor
CMP	Control and Maintenance Processor
CMPLX	Complex
CMPP	Computer Managed Process Planning
CMPR	Compression
CMPS	Compare (word) String
CMPT	Computer
CMR	Code Matrix Reader
CMR	Colossal Magneto-Resistance
CMR	Common Mode Rejection
CMR	Compare Right (half word)
CMR	Customer Master Record
CMRR	Common Mode Rejection Ratio
CMS	Cable Management Software
CMS	Call Management System
CMS	Central Maintenance Support
CMS	Circuit Maintenance System
CMS	Code Management System
CMS	Communications Management Subsystem
CMS	Compiler Monitor System

CMS Computer Management System
CMS Continuous Media Server
CMS Conversation(al) Monitor(ing) System
CMSO CIM for Multi-Supplier Operations (CIM = Computer Integrated Manufacturing)
CMSR Core Memory Shift Register
CMT Cassette Magnetic Tape
CMT Cellular Mobile Telephone
CMT Change Management Tracking
CMT Circuit Master Tape
CMT Communications Maintenance Terminal
CMT Computer Managed Training
CMT Computer Mediated Teleconferencing
CMU COLT Measurement Unit (COLT = Computerized On-Line Testing)
CMU Control Maintenance Unit
CMV Circuit Mode Voice
CMV Common Mode Voltage
CMVC Configuration Management Version Control
CMW Compartmented Mode Workstation
CMX Character Multiplexer
CMX Customer Multiplexer
CMY Cyan Magenta Yellow
CMYK Cyan Magenta Yellow Black
C/N Carrier-to-Noise (ratio)
CN Change Notice
CN China
CN Combined Nomenclature
CN Common Name
CN Communications Network|News
CN Condensation Nucleus
CN Connect
CN Consignment Note
CN Customer Network
CNA Certified NetWare|Novell Administrator
CNA Combinations Not Allowed
CNA Communications Network Application|Architecture
CNAPS Co-processing Node Architecture for Parallel Systems
CNB Connected Network Backup
CNC Centre for Novel Computing
CNC Computer(ized) Numerical Control
CNC Concentrator
CNC Consecutive Number Control
CNCC Customer Network Control Center
CNCL Cancel
CND Condition(ed)
CNDL Conditional
CNDP Communication(s) Network Design Program

CNE	Certified NetWare Engineer
CNE	Communications Network Emulator
CNE	Compare Numerical Equal
CNEP	Cable Network Engineering Program
CNET	Centre National d'Etudes des Télécommunications (FR)
cnf	configuration (file extension indicator)
CNF	Conjunctive Normal Form
CNG	Calling (tone)
CNG	Collective Number Group
CNI	Certified NetWare|Novell Instructor
CNI	Changed Number Interception
CNI	Coalition for Networked Information
CNIDR	Clearinghouse for Network Information and Discovery and Retrieval
CNL	Cancel
CNL	Circuit Net Loss
CNL	Circuit Noise Level
CNLP	Connectionless Protocol
CNM	Communication Network Management|Manager
CNMA	Communications Network for Manufacturing Applications
CNMA	Computer Network for Manufacturing Applications
CNMC	Corporate Network Management Center
CNMI	Communications Network Management Interface
CNMS	Cylink Network Management System
CNN	Cable News Network
CNN	Composite Network Node
CNO	Computer Not Operational
CNOM	Committee on Network Operations and Management
CNOP	Conditional Non-Operation
CNOS	Change Number Of Sessions
CNOS	Computer Network Operating System
CNP	Communications Network Processor
CNP	Communications statistical Network analysis Procedure
CNR	Carrier-to-Noise Ratio
CNR	Common Network Representation
CNRCVUUCP	Compressed News Received Via Unix to Unix Copy Program
CNS	Central Network Server
CNS	Communications Network Simulator|System
CNS	Complementary Network Service(s)
CNS	Compuserve's Network Services
CNS	Computational Neuro-Science
CNS	Continuous Net Settlement
CNSD	Computer and Network Services Division
CNSR	Comet Nucleus Sample Return
CNSS	Core Nodal Switching Subsystem
CNT	Contents
CNT	Count(er)
CNTL	Control

CNTR	Counter
CNTRL	Control
CNU	Compare Numeric Unequal
CNV	Conventional
CNVT	Convert
CNX	Certified Network Expert
CO	Carry Out
CO	Certificate of Origin
CO	Change Order|Over
C/O	Close/Open
CO	Colombia
CO	Colorado (US)
CO	Command Output
CO	Computer Operator
CO	Connection Oriented
CO	Console Output
CO	Contact Operate
CO	Convert Out
CO	Coordinate
CO	Couple Output
CO	Cross Operation(s)
CO	Customer Owned
CO	Cut Off
COAD	Coordinate Adder Display
COADS	Comprehensive Ocean-Atmosphere Data Set
COADS	Conference On Application Development Systems
COAM	Customer Owned And Maintained
COAMP	Computer Analysis of Maintenance Policies
COAST	Computer Operations, Audit, and Security Technologies
COAT	Coherent Optical Adaptive Technique
COAT	Computer Operator Aptitude Test
COATS	Computer Aided Tool Selection System
COATS	Computer Operated Automatic Test System
COAX	Co-Axial (cable)
COB	Camp-On Busy
COB	Card-On-Board (logic)
COB	Chip-On-Board
cob	cobol (source code) (file extension indicator)
COB	Computation Over Bandwidth
COBCS	Computer Oriented Building Control System
COBI	Coded Biphase
COBLOS	Computer Based Loans System
COBOL	Common Business Oriented Language
COBRA	Consolidation Of Basic Records Audit
COBUS	Coaxial Bus
COC	Central Office Connections
CoC	Chain of Custody
COC	Change Of Control

COC	Character Oriented Communications controller
COC	Circuit Order Control
COC	Complete Operational Capability
COC	Cross-Over Connector
COCF	Connection Oriented Convergence Function
COCO	Collaborative Computing (group)
COCO	Communication Computer
COCO3	Color Computer 3
COCOL	COBOL Compiler Oriented Language
COCOT	Customer Owned Coin Operated Telephone
COCS	Container Operating Control System
cod	code (file extension indicator)
COD	Cross-Over Detector
CODA	Computer Oriented Data Acquisition
CODA	Configuration Design of Assemblies
CODAP	Comprehensive Occupational Data Analysis Program
CODAS	Customer Oriented Data System
CODASYL	Conference of Data Systems Language
CODCF	Central Office Data Connecting Facility
CODE	Client-server Open Development Environment
CODEC	Coder/Decoder
CODEC	Coding-Decoding device
CODEC	Compression/Decompression
CODEM	Coded Modulator-demodulator
CODES	Co-Design
CODEX	Coder-Decoder
CODIC	Computer Directed Communication
CODIL	Content Dependent Information Language
CODILS	Commodity Oriented Digital Input Label System
CODINE	Conversational Design Information Network
CODIS	Combined DNA Index System (DNA = Distributed Network Architecture)
CODLS	Connection-mode Data Link Service
CODP	Customer Order Decoupling Point
COE	CATIA's Operators Exchange
COE	Center Of Excellence
COE	Central Office Equipment
COE	Customer Order Engineering
COEA	Cost (and) Operational Effectiveness Analysis
COED	Computer Operated Electronic Display
COEES	Centre Of Excellence Engineering System
COEF	Coefficient
COEM	Commercial Original Equipment Manufacturer
COER	Central Office Equipment Report
COF	Cable Order Form
COF	Customer Order File
COFAD	Computerized Facilities Design
COFDM	Coded Orthogonal Frequency Division Multiplex

COFF	Common Object File Format
COFM	Cognitive Operator Function Model
COFO	Certificate Of Origin
COFOT	Coaxial to Fiber Optic Transceiver
COG	Computer Operations Group
COGAP	Computer Graphics Arrangement Program
COGECOM	Compagnie Générale des Communications
COGENT	Compiler and Generalized Translator
COGO	Coordinate Geometry Oriented (language/program)
COGS	Consumer Goods System
COHO	Coherent Oscillator
COIN	Community Of Interest Network
COINS	Computer and Information Sciences
COINS	Corporate Information Network System
COL	Collision
COL	Column
COL	Common Optical Language
COL	Communications Oriented Language
COL	Computer Oriented Language
COLA	Conversational Oriented Language
COLAMM	Computer Operated Laser Active Modelling Machine
CO-LAN	Central Office, Local Area Network
COLAS	Command Language System
COLD	Computer Output to Laser Disk
COLIDAR	Coherent Light Detection And Ranging
COLL	Collector
COLM	Collector Mesh
COLP	Collapse
COLP	Connected Line identification Presentation
COLR	Connected Line identification Restriction
COLT	Central Office Line Tester
COLT	Communication Line Terminator
COLT	Computerized On-Line Testing
COLT	Computer Oriented Logic Test(s)
COM	Cassette Operating Monitor
com	command (file extension indicator)
COM	Commercial
COM	Common Object Model
COM	Communication
COM	Compiler
COM	Component Object Model
COM	Computer Output Management
COM	Computer Output Micrographics
COM	Computer Output on Microfiche|Microfilm
COM	Continuation Of Message
COM	Control Oriented Microprocessor
COM	Conversational Operation Mode
COMA	Cache-Only Memory Architecture

COMAC	Continuous Multiple Access Collator|Comparator
COMACS	Common Manufacturing Accounting Control System
COMAL	Common Algorithmic Language
COMAR	Computer Aerial Reconnaissance
COMARS	Coordinate Measuring And Recording System
COMAS	Central Office Maintenance and Administration System
COMAS	Configuration Management System
COMAT	Computer Assisted Training
COMATS	Computer Oriented Manufacturing And Test System
COMB	Console Oriented Model Building
COMBS	Customer Oriented Message Buffer System
COMCODE	Commodity|Component Code
COMDEX	Computer Dealers Exposition
COMEF	Coherent Memory Filter
COMET	Chip On Module Evaluation Tester
COMET	Computer Message Transmission
COMET	Computer Operated Management Evaluation Technique
COMFACTS	Communications Facilities Testing System
COMFOR	Computer Forum and exhibition
COMFORT	Commercial Fortran
COMIC	Colorant Mixer Computer
COMICS	Computer Oriented Managed Inventory Control System
COMINT	Communication(s) Intelligence
COMIS	Communication Management Information System
COMIS	Compilation and Interpretation System
COMITT	Committee on Multimedia In Teacher Training
COMJAM	Communication Jamming
COMLINK	Communications Link
COMM	Communications
COMMANDS	Computer Operated Marketing, Mailing And News Distribution System
COMMCEN	Communications Center
COMMEND	Computer aided Mechanical Engineering Design
CommKit	Software to connect Unix SV hosts to Datakit VCS nodes (SV = Shared Virtual; VCS = Virtual Circuit Switched)
CommonLOOPS	Common LISP Object Oriented Programming System (LISP = List Processing language)
COMMSTOR	Communications Storage unit
COMMZ	Communications Zone
COMn	nth serial Port
COMOD	Communications Mode
COMP	Compare
COMP	Computer(s)
COMPAC	Computer Program for Automatic Control
COMPACS	Computer Oriented Manufacturing Production And Control System
COMPACT	Commercial Product Acquisition Team
COMPACT	Compatible Algebraic Compiler and Translator

COMPACT Computer Program for Automatically Controlled Tools
compander compressor/expander
COMPARE Compliance Progress And Readiness
COMPARE Computer Oriented Method of Program Analysis, Review and Evaluation
COMPAS Computer Acquisition System
COMPAS Computer Oriented Method for Payroll, Accounting and Statistics
COMPASS Compiler Assembler
COMPASS Computer Assisted classification and assignment System
COMPASS Computer Oriented Management Planning And Scheduling System
COMPAY Computer Payroll
COMPCON Computer Conference
COMPEC Computer Peripherals and small computer systems trade Exhibition and Conference
COMPENDEX Computerized Engineering Index
COMPILE Customs On-line Method of Preparing from Invoices Lodgeable Entries
COMPOOL Communications Pool
COMPROC Command Processor
COMPROSL Compound Procedural Scientific Language
COMPROT Communications Protocol
COMPSAC Computer Software and Applications Conference
COMPSCAN Computerized Scanner
COMPT Chip-On-Module Parameter Tester
COMPTEXT Computer Prepared Text
COMPTIA Computing Technology Industry Association
COMPULOG Computational Logic (EU)
COMPUSEC Computer Security
COMRADE Computer Aided Design Environment
COMREG Communication Region
COMS Computer-based Operations Management System
COMSAT Communications Satellite (corporation)
COMSEC Communications Security
COMSPEC Command Specifier
COMSYL Communications System Language
COMSYS Communications System
COMTEC Computer Micrographics Technology
COMTEX Communications Oriented Multiple Terminal Executive
COMTRAC Computer aided Traffic Control
COMTRAN Commercial Translator
COMX Communications Executive
CON Console
CON Constant
CONC Concentrate
CONCAM Committee On Computing in Applied Mechanics
CONCH Client Oriented Normative Control Hierarchy

CONCUR	Calculi and algebra's of Concurrency (EU)
COND	Condition
CONDL	Conditional
CoNDUIT	Cooperative Network for Dual-Use Information Technologies
CONET	Concentrator Network
CONF	Configure
CONFER	Concurrency and Functions: Evaluation and Reduction (EU)
CONFESS	Conduction Fingerprint Electro-optic Search System
CONFIG	Configuration
CONFIRM	Context Free Index Retrieval Method
CONIO	Console Input/Output
CONIT	Connector for Networked Information Transfer
CONMAN	Console Manager
CONN	Connection
CO/NO	Current Operator/Next Operator
CONOPS	Concept (of) Operations
CONS	Carrier Operated Noise Suppression
CONS	Connection mode Network Service
CONS	Connection Oriented Network Service
CONS	Console
CONS	Construct
CONSENS	Concurrent/Simultaneous Engineering System
CONSYS	Control System
CONT	Contact
CONT	Continue
CONTAC	Central Office Network Access
CONTONE	Continuous Tone
CONTRAST	Conditional Term Rewriting on Attribute Syntax Trees
CONV	Conversion
CONVR	Converter
COOL	Chorus Object Oriented Layer
COOL	Class Object-Oriented Library
COOL	C++ Object-Oriented Language
COOL	Combined Object-Oriented Language
COOTS	Conference on Object-Oriented Technologies and Systems
COP	Character Oriented Protocol
COP	Coaxial Output Printer
COP	Communication Output Printer
COP	Computer Optimization Package
COP	Continuous Optimization Program
COP	Control Optimization Program
COPD	Compound
COPE	Cassette Operating Executive
COPE	Centralized On-line Processing Environment
COPE	Communication Oriented Processing Equipment
COPE	Computerized Office Plans and Environment
COPE	Console Operator Proficiency Examination

COPICS	Communication Oriented Production and Inventory Control System
COPP	Connection Oriented Presentation Protocol
COPPA	Children's On-line Privacy Protection Racket
COPPL	Computer Process Planning Language
COPPS	Connection of Order and Planning Processing Systems
COPS	Calculator Oriented Processor System
CoPs	Cluster of Personal computers
COPS	Common Open Policy Service(s)
COPS	Computer Oracle and Password System
COPS	Connection Of Processing Systems
COPS	Connection Oriented Presentation Service
COPS	Control for Operations Programming and Systems
COPSYS	Copy System
COR	Command Register
COR	Correct
CORA	Canadian OSI Registration Authority (OSI = Open Systems Interconnect)
CORAL	Class Oriented Ring Associative Language
CORAL	Common Routine Autocoder Language
CORAL	Computer On-line Real-time Application Language
CORAL	Correlated Radio Lines
CORBA	Common Object (management) Request Broker Architecture
CORD	Computer with On-line Remote Devices
CORDIC	Coordinate Rotation Digital Computer
CORDIS	Community Research and Development Information Service
CORE	Comprehensive Reporting (system)
CORELAP	Computerized Relationship Layout Planning
CoREN	Corporation for Research and Enterprise Network
COREP	Code Repertory
CORMORANT	Cortical Maps Of Resistive Anisotropic Networks (EU)
CORN	Computer Resource Nucleus
CORNET	Corporate Network
COROS	Copper Oxide Read-Only Storage
CORR	Correspondent
CORRECT	Customized Optical Reader Random Error Correction Technique
CORREGATE	Correctable Gate
CORSAIR	Computer Oriented Reference System for Automatic Information Retrieval
COS	Call-Off Schedule
COS	Call Originate Status
COS	Central Office Switch
COS	Class Of Service
COS	Code Operated Switch
COS	Commercial Operating System
COS	Communications Operating System|Software
COS	Compatible Operating System

COS	Complementary Symmetry
COS	Concurrent Operating System
COS	Consumer Operating System
COS	Corporation for Open Standards|Systems
cos	cosine
COS	COS Object System (COS = Corporation for Open Systems)
COS	Current Output Station
COSAC	Canadian Open Systems Applications Criteria
COSACS	Computer Operations Scheduling Accounting and Control System
COSACS	Computer Oriented Scheduling And Control System
COSAM	COBOL Shared Access Method
COSCL	Common Operating System Control Language
COSE	Combined Office Standard Environment
COSE	Common Open Software|Systems Environment
COSE	Common Operating System Environment
COSEL	Coselector
cosh	cosinus hyperbolicus
COSINE	Cooperation for Open Systems Interconnection Networking in Europe
COSIP	Coherent Signal Processor
COSMIC	Common Systems Main Inter-Connection (frame system)
COSMIC	Computer Operated Sequential Memory and Integrated Circuits
COSMIC	Computer Software Management and Information Center
COSMOS	Complementary Symmetry Metal-Oxide Semiconductor
COSMOS	Computer Oriented System for Machine Order Synthesis
COSMOS	Computer Systems for Main frame Operations
COSMOS	Console Oriented Statistical Matrix Operator System
COSN	Consortium for School Networking
COSS	Common Object Services Specification
COSSO	Computer Service en Software Offices (vereniging) (NL)
COSSS	Committee on Open Systems Support Services
COST	Computer Operated Sequential Tester
COSY	Compiler System
COSY	Compressed Symbolic
COT	Central Office Terminal|Trunks
COT	Common Output Tape
COT	Customer Oriented Terminal
CO/TP	Connection Oriented Transaction Protocol
COTP	Connection Oriented Transport Protocol
COTS	Commercial Off-The-Shelf
COTS	Connection Oriented Transport Service
COTT	Computer Operated Thermal Tester
COU	Computer Operating Unit
COUNT	Computer Operated Universal Test (system)
COUPLE	Communications Oriented User Programming Language
CoVis	Collaborative Visualization
COW	Character Oriented Windows

COW	Cluster Of Workstations
CP	Call Processor
CP	Cancel Power
cp	candle power
CP	Central Processor
CP	Character Printer
CP	Check Point
CP	Circuitpack
CP	Circuit Packaging|Protector
CP	Clock Phase|Pulse
CP	Command Processor
CP	Communications Processor
CP	Comparison
CP	Component Panel
CP	Condition Precedent
CP	Connection Point
CP	Connect Presentation
CP	Continuous Progressive
CP	Control Part|Point
CP	Control Processor|Program
CP	Copy Protected
CP	Correspondence Printer
CP	Critical Path
CP	Current Process
CP	Customer Premises
CP	Cyclic Permuted
CPA	Channel Program Area
CPA	Colour Phase Alternation
CPA	Communications & Public Affairs
CPA	Computer Programming & Analysis
CPA	Connect Presentation Accept
CPA	Critical Path Analysis
CPAR	Customer Problem Analysis and Resolution
CPB	Channel Program Block
CPBX	Computerized Private Branch Exchange
CPC	Calling Party Control
CPC	Cellular Phone Company
CPC	Central Processing Complex
CPC	Circuit Provisioning Center
CPC	Clock Pulsed Control
CPC	Common Peripheral Channel
CPC	Computer(ized) Production|Process Control
CPC	Computer Power Center
CPC	Computer Program Component
CPC	Continuous Progression|Progressive Code
CPC	Cost Per Copy
CPC	Cycle Program Control|Counter
CPC	Cyclic Permuted Code

CPCEI	Computer Program Contract End Item
CPCI	Computer Program Configuration Item
CPCI	CPU Power Calibration Instrument (CPU = Central Processing Unit)
CPCS	Cheque Processing Control System
CPCS	Common Part Convergence Sublayer
CPCS	Computerized Parts Control System
CPD	Central Pulse Distributor
CPD	CICS Program Development
CPD	Compound
CPD	Control Panel Diagram(s)
CPDAMS	Computer Program Development And Management System
CPDD	Cellular Packet Digital Data
CPDD	Command Post Digital Display
CPDP	Computer Program Development Plan
CPDS	Commerce Procurement Data System
CPDS	Computer Program Design Specification
CPE	Central Processing Element
CPE	Central Program and Evaluator
CPE	Check Point Entry
CPE	Computer Performance Evaluation
CPE	Contractor Performance Evaluation
CPE	Control Processing Element
CPE	Convergence Protocol Entity
CPE	Counter Position Exit
CPE	Cover Page
CPE	Customer Premises|Provided Equipment
CPEUG	Computer Performance Evaluation User Group
CPF	Capacity Planning Finite (loading)
CPF	Central Processing Facility
CPF	Coded Print File
CPF	Control Program Facility
CPFM	Continuous Phase Frequency Modulation
CPFSK	Continuous Phase Frequency Shift Keying
CPG	Clock Pulse Generator
CPG	COBOL Program Generator
CPG	Communications Program Generator
CPGA	Control Program Generation and Analysis
cph	characters per hour
CPI	Capacity Planning Infinite (loading)
cpi	characters per inch
CPI	Clock Per Instruction
cpi	code page information (file extension indicator)
CPI	Common Programming Interface
CPI	Compiler Program Interrupt
CPI	Computer Private Interface
CPI	Computer to PABX Interface
CPI	Continuous Process Improvement

CPIC	Common Programming Interface for Communications
CPIM	Certified in Production and Inventory Management
CPIN	Computer Program Identification Number
CPIO	Copy In/Out
CPIP	Computer Pneumatic Input Panel
CPIS	Computerized Personnel Information System
CPIW	Customer Provided Inside Wiring
CPL	Central Program Library
cpl	characters per line
CPL	Check Point Label
CPL	Combined Programming Language
CPL	Common|Computer Program Library
CPL	Computer Programming Language
CPL	Control Panel
CPL	Conversational Programming|Planning Language
CPL	Current Privilege Level
CPL	Customized Plug List
CPLD	Completed
CPLD	Complex Programmable Logic Device
CPLD	Coupled
CPLG	Coupling
CPLMT	Complement
cpm	characters per minute
CPM	Chip Placement Machine
CPM	Computer Performance Management|Monitor
CPM	Computer Port Module
CPM	Connection Point Manager
CPM	Continuous Phase Modulation
CPM	Continuous Processing Machine
CPM	Control Program for Microcomputers
CPM	Control Program Monitor
cpm	cost per minute
CPM	Cost Per Thousand (Mille)
CPM	Critical Path Method
CPM	Current Processor Method
cpm	cycles per minute
CPMA	Central Processor Memory Address
CPMA	Computer Peripherals Manufacturers Association
CPMP	Carrier Performance Measurement Plan
CPMP	Control Program Management Package
CPMU	COSINE Project Management Unit
CPN	Card Part Number
CPN	Coloured Petri Net
CPN	Component Part Number
CPN	Computer Product News
CPN	Customer Premises Network
CPO	Code Practice Oscillator
CPO	Computer|Concurrent Peripheral Operations

CPOL	Communications Procedure Oriented Language
CPP	Cable Patch Panel
cpp	characters per pica
CPP	Command Processing Program
CPP	Concurrent Peripheral Processing
CPP	Control Panel Programming
CPP	C Pre-Processor
CPP	Critical Path Plan
CPP	Current Purchasing Power
CPPP	Computerized Production Process Planning
CPPS	Continuous Process Plant Scheduling
CPPS	Critical Path Planning and Scheduling
CPQN	Clocks Per Quarter Note
CPR	Certification Print Routine
CPR	Check Point Record
CPR	Connect Presentation Reject
CPR	Constant Percentage Resolution
CPR	Control Program Real-time
CPS	Cathode Potential Stabilization
CPS	Central(ized) Processing System
CPS	Central Processor Subsystem
CPS	Central Programming Service
cps	characters per second
CPS	COBOL Programming System
cps	columns per second
CPS	Computerized Publishing System
CPS	Console Programming System
CPS	Continuous Processing System
CPS	Controlled Path System
CPS	Control Program Services|Support
CPS	Conversational Programming System
CPS	Corporate Programming Standard
cps	counts per second
cps	cycles per second
CPSK	Coherent Phase Shift Keying
CPSR	Computer Professionals for Social Responsibility
CPSS	Computer Power Support System
CPST	Compensate
CPT	Chief Programmer Team
CPT	Colour Picture Tube
CPT	Command Pass Through
CPT	Computer Programming Training
CPT	Computer Technology
CPT	Continuous Performance Test
CPT	Current Potential Transformer
CPT	Customer Provided Terminal
CPT&E	Computer Program Test & Evaluation
CPTO	Chief Programmer Team Organization

CPTY	Capacity
CPU	Central Processing Unit
CPU	Communication Processor Unit
CPUE	Catch Per Unit Effort
CPUID	Central Processor Unit Identification
CPVA	Channel Program Variable Area
CPY	Copy
CQ	Seek You
CQA	Computer aided Question and Answering
CQMS	Circuit Quality Monitoring System
CQN	Corporate Quality Network
CR	Call Repetition|Request
CR	Carriage Return (character)
CR	Carrier's Risk
Cr	Chromium
CR	Classification Research
CR	Class Rate|Ratio
CR	Clock Register
CR	Command Register|Reject
CR	Communication|Compare Register
CR	Component Release
CR	Condition Register
CR	Connect Request
CR	Contact Resistance
CR	Continuously Running
CR	Control Register
CR	Costa Rica
CR	Critical Ratio
CR	Current Rate
CR	Customer Record
CRA	Catalog Recovery Area
CRA	Computer Retailers Association (US)
CRAC	Chatter Recognition And Control
CRAFT	Computerized Relative Allocation of Facilities Technique
CRAM	Computational Random Access Memory
CRAM	Computerized Reliability Allocation Method
CRAM	Condensed Random Access Memory
CRAM	Conditional Relaxation Analysis Method
CRAM	Cyberspatial Reality Advancement Movement
CRAMM	Coupon Reading And Marking Machine
CRAP	Completely Redundant Array of Pointers
CRAR	Control ROM Address Register (ROM = Read Only Memory)
CRAS	Cable Repair Administrative System
CRAYON	Create Your Own Newspaper
CRB	Complementary Return to Bias
CRB	Customer Reconfiguration Bandwidth
CRBE	Conversational Remote Batch Entry
CRC	CAD Resource Center (CAD = Computer Aided Design)

CRC	Camera-Ready Copy
CRC	Carriage Return Contact|Character
CRC	Central Read Control
CRC	Class Responsibility Collaborator
CRC	Column Reference Centerline
CRC	Constant Ratio Code
CRC	Cross Reference Code
CRC	Customer Record Center
CRC	Cyclic(al) Redundancy Character
CRC	Cyclic Redundancy Check(ing)
CRCB	Continuously Running Circuit Breaker
CRCC	Cyclic Redundancy Check Character
CRCG	Center for Research in Computer Graphics (DE)
CRCRD	Credit Card Reader
CRCT	Correct
CRD	Capacitor-Resistor-Diode (network)
crd	card file (file extension indicator)
CRD	Computer Read-out Device(s)
CRD	Critical Resource Diagram(ming)
CRD	Customer Required Date
CRDP	Certificate Revocation Distribution Point
CRDtools	Climate Research Data tools
CRE	Current Ring End
CREDFACS	Conduit, Raceway, Equipment Ducts & Facilities
CREDIT	Cost Reduction Early Detection Information Technique(s)
CREF	Connection Refused
CREF	Cross Reference
CREF	Cross Reference File
CREN	Computer Research Education Network
CREN	Corporation for Research and Educational Networking
CREOL	Center for Research in Electro-Optics and Lasers
CREST	Computer Routine for Evaluation of Simulation Tactics
CREWS	Cooperative Requirements Engineering With Scenarios (EU)
CRF	Cable Retransmission Facility
CRF	Checkpoint Restart Facility
CRF	Communication Related Function
CRF	Context Roll File
CRF	Control Relay Forward
CRF	Cross Reference File
CRFMP	Cable Repair Force Management Plan
CRF(VC)	Connection Related Functions (Virtual Channel)
CRG	Charge
CR-HI	Channel Request-High priority
CRI	Colour Rendering Index
CRI	Colour Reproduction Indices
CRI	Compare Register Immediate (instruction)
CRI	Computer Related Industries
CRIS	Command Retrieval Information System

CRIS Concentric Research Information Service
CRIS Current Research Information System
CRIS Customer Record Information System
CRISC Complex/Reduced Instruction Set Computer
CRISP Categories for Recursive Information Systems Postulation
CRISP Complex-Reduced Instruction Set Processor
CRISP Computer Resources Integrated Support Plan
CRISP-DM Cross Industry Standard Process for Data Mining
CRISTAL-ED Coalition on Reinventing Information Science, Technology And Library Education
CRJ Call Record Journaling
CRJ Command Reject
CRJE Conversational Remote Job Entry
CRK Cross Reference Key(s)
CRL Certificate Revocation List
CR-LDP Constraint-based Label Distribution Protocol
CR/LF Carriage Return/Line Feed
CR-LO Channel Request-Low priority
CRM Computer Resources Management
CRM Contra-lateral Remote Masking
CRM Control and Reproducibility Monitor
CRM Customer Relationship Management
CRMA Computer Resource Management Architecture
CR-MED Channel Request-Medium priority
CRMS Control and Reproducibility Monitor System
CRMT Circuit Rule Master Tape
CRN Correction
CRNO Command Register Not Operable
CRO Cathode Ray Oscilloscope
CROM Capacitor|Control Read-Only Memory
CROP Common Routing Output
CROPS Coherent Rules for On-line Production Scheduling
CROS Capacitor Read-Only Storage
CROTCH Computerized Routine for Observing and Testing the Channel Hardware
CRP Channel Request Priority
CRP Combined Refining Process
CRP Common Reflection Point
CRP Computer Reset Pulse
CRP Configuration Report Program
CRP Counter Rotation Platform
CRPS Common Release Processing System
CRQ Call Request
CRQ Console Reply Queuing
CRR Computer Restart and Recovery
CRR Condition Recall Register
CRR Constant Ratio Rule
CRR Credential Record Reference

CRRB–CS

CRRB	Change Request Review Board
CRRD	Customer Requested Removal Date
CRS	Command Retrieval System
CRS	Compound Regulating System
CRS	Computer Related Services
CRS	Computer Reservation System
CRS	Configuration Report Server
CRS	Customer Record Status|System
CRS	Customized Routing Selector|System
CRSA	Current Routine Starting Address
CRSO	Cellular Radio Switching Office
CRSS	Customer Record System Structure
CRT	Cathode Ray Tube
CRT	Channel Response Time
CRT	Choice Reaction Time
CRT	Computer Remote Terminal
CRTC	Cathode Ray Tube Controller
CRTS	Cathode Ray Tube Storage
CRTU	Combined Receiving and Transmitting Unit
CRU	Communications Register Unit
CRU	Control and Reporting Unit
CRU	Customer Replaceable Unit
CRU	Customer Responsibility Usage
CRUD	Create, Retrieve, Update, Delete
CRV	Contact Resistance Variation
CRV	Cryptography Request Verification
CrypTEC	Cryptographic Techniques & e-commerce (international workshop)
CRYPTO	Cryptographic subsystem
CRYPTONET	Crypto-communications Network
CS	Centralized Scheduling
C/S	Certificate of Service
CS	Channel Status
CS	Character Sense
CS	Check Sorter
CS	Chip Select
CS	Circuit Switching
CS	Clear to Send
C/S	Client/Server
CS	Clock Signal
CS	Code Segment
CS	Coding Specifications
CS	Cognition Science
CS	Column Split
CS	Command System
CS	Common Shelf
CS	Communication Scanner
CS	Communication Services|System

CS Communications Subsystem
CS Compatibility Support
CS Complementary Symmetry
CS Computer Science|Society
CS Computing Services
CS Conditioned Stimulus
CS Condition Subsequent
CS Context-Sensitive (grammar)
CS Context Switch
CS Continue Specific
CS Controlled Switch
CS Control Section|Set
CS Control Signal|Storage
CS Convergence Sublayer
CS Conversation System
CS Coordinated Single-layer
CS Core Shift|Storage
CS Counter Sign
CS Critical Section
CS Cross Simulator(s)
CS Current Schedule|State
CS Current Switch
CS Customer Service|System
CS Cycle Shift|Stealing
c/s cycles per second
CS Source Capacitance
CSA Calendaring and Scheduling API (= Application Program Interface)
CSA Canadian Standards Association
CSA Carrier Serving Area
CSA Carry Save Adder
CSA Client Server Agent
CSA Common Service|Storage Area
CSA Common Sub-Assembly
CSA Common System Area
CSA Compaq Solutions Alliance
CSA Computer Services Association (GB)
CSA Customer Service Administration
CSAB Computing Sciences Accreditation Board
CSACC Customer Service Administration Control Center
CSACS Centralized Status, Alarm and Control System
CSAJTL Current Switching Alloy Junction Transistor Logic
CSAM Circular Sequential Access Memory
CSAM Control Sequence Access Method
CSAP Control Systems Analysis Program
CSAR Centralized System for Analysis Reporting
CSAR Channel System Address Register
CSAR Communication Satellite Advanced Research

CSAR	Control Stor(ag)e Address Register
CSAR	Cycle Steal Address Register
CSB	Central System Bus
CSB	Channel Status Byte
CSB	Communication Scanner Base
CSC	Cell Site Controller
CSC	Circuit Switching Center
CSC	Command Scheduling Chain
CSC	Communication Systems Center
CSC	Computer Sciences Corporation
CSC	Computer Service Center
CSC	Computer Software Component
CSC	Computer Subsystem Controller
CSC	Consecutive Sequence Computer
CSC	Control Systems Character
CSC	Customer Support Center
C-SCAT	C-band Scatterometer
CSCB	Command Scheduling Control Block
C/SCC	Computer/Standards Coordinating Committee (IEEE Computer Society)
CSCC	Cumulative Sum Control Chart
CSCI	Computer Software Configuration Item
CSCL	Computer Supported Cooperative Learning
CSCS	Consolidated Scientific Computing System
CSCS	Cost, Schedule, and Control System
CSCW	Computer Supported Cooperative Work
CSD	Canonical Sign Digit
CSD	Circuit Switched Data (service)
CSD	Closed System Delivery
CSD	Computerized Standard Data
CSD	Constant Speed Drive(r)
CSD	Corrective Service Diskette
CSD	Current Schedule Date
CSD	Customer Service Division
CSDBTL	Current Switching Diffused Base Transistor Logic
CSDC	Circuit-Switched Data|Digital Capability
CSDD	Computer Subprogram Design Document
CSDF	Computer System Development Facility
CSDL	Conceptual Schema Definition Language
CSDM	Continuous Slope Delta Modulation
CSDN	Circuit-Switched Data|Digital Network
CSDR	Control Stor(ag)e Data Register
CSE	Certified System Engineer
CSE	Circuit Switch Exchange
CSE	Common Sub-expression Elimination
CSE	Computational|Computer Science and Engineering
CSE	Control Systems Engineering
CSE	Coordinated Single-layer Embedded

CSECT	Control Section
CSEE&T	Conference on Software Engineering, Education & Training
CSEF	Current Switch Emitter Follower
CSELT	Centro Studi e Laboratori Telecommunicazioni (IT)
CSEM	Centre Suisse d'Electronique et de Microtechnique
CSEP	Computational Science Education Project
CSERB	Computer, Systems, and Electronics Requirement Board (GB)
C-SET	Chip-Secure Electronic Transaction
CSF	Communications Serviceability Facilities
CSF	Created Shell Face
CSF	Critical Success Factor
CSF	Customer Structure File
CSFA	Cost Sensitivity Feature Analysis
CSFI	Communication Subsystem For Interconnection
CSFO	Chained Sequential File Organization
CSFS	Cable Signal Fault Signature
CSG	Clock Signal Generator
CSG	Computer Support Group
CSG	Constructive Solid Geometry
CSG	Consulting Services Group
CSG	Context Sensitive Grammar
CSG	Course Selection Guide
CSH	C-Shell
CSI	Character Set Identifier
CSI	Command Sequence Introducer
CSI	Command String Interpreter
CSI	Commercial Systems Integration
CSI	Compuserve Incorporated
CSI	Computer Security Institute (US)
CSI	Computer Service Industry
CSI	Conditional Stop Instruction
CSI	Conditional Symmetric Instability
CSIC	Customer Specific Integrated Circuit
CSIDC	Computer Society International Design Competition
CSIRO	Commonwealth Scientific and Industrial Research Organization (AU)
CSIS	Central Secondary Item Stratification
CSIT	Computer Science & Information Technologies (conference)
CSK	Countersink
CSL	Components Source List
CSL	Computer Sensitive Language
CSL	Computer Systems Laboratory
CSL	Context Sensitive Language
CSL	Control and Simulation Language
CSLI	Center (for the) Study (of) Language (and) Information
CSLIP	Compressed Serial Line Interface Protocol
CSM	Command and Service Module
CS&M	Common Systems & Methods

CSM	Communications and Systems Management
CSM	Communications Services Manager
CSM	Communication System Monitoring
CSM	Component and Supplier Management
CSM	Computer System Manual
CSMA	Carrier Sense, Multiple Access
CSMA	Communications Systems Management Association (US)
CSMA/CA	Carrier Sense Multiple Access/Collision Avoidance
CSMA/CD	Carrier-Sense Multiple Access/Collision Detect
CSMA/CP	Carrier Sense Multiple Access/Collision Prevention
CSMC	Communications Services Management Council
CSME	Core Shape Manipulation Engine
CSMP	Continuous Simulation|System Modelling Program
CSMS	Computerized Specification Management System
CSMS	Customer Support Management System
CS-MUX	Carrier-Switched Multiplexer
CSN	Carrier Service Node
CSN	Common Subset Node
CSN	Computer Systems News
CSNAP	Communications Statistical Network Analysis Procedure
CSNET	Computer (and) Science Network
CSNP	Communications Sub-Net Processors
CSNW	Client Services for NetWare
CSO	Central Services Organization
CSO	Central Switching Office
CSO	Chained Sequential Operation
CSO	Computing Services Office
CSO	Customer Set Organization
CSOP	Current Switch Optimization Program
CSOS	Complementary Silicon On Sapphire
CSP	Certified Systems Professional
CSP	Channel Scheduling Process
CSP	Chip Select Pin
CSP	Commerce Service Provider
CSP	Commercial Subroutine Package
CSP	Common Security Package
CSP	Common System Product
CSP	Communicating Sequential Processes
CSP	Communications System Programming
CSP	CompuCom Speed Protocol
CSP	Concurrent Structured Programming
CSP	Connection Setup Procedure
CSP	Constraint Satisfaction Problem
CSP	Control Switching Point(s)
CSP	Cross System Product
CSP	Current System Programs
C-space	Configuration space
CSP/AD	Cross System Product/Application Development

CSP/AE	Cross System Product/Application Execution
CSPAN	Cable-Satellite Public Affairs Network
CSPDN	Circuit-Switched Public Data Network
CSPE	Control System for Plan Execution
CSPM	Computer System Performance Measurement
CSPP	Computer Systems Policy Project
CSP/Q	Cross System Product/Query
CSQ	Control Section Qualification
CSQ	Cross System Query
CSR	Channel Service Register
CSR	Channel Status Routine
CSR	Closed Subroutine
CSR	Command Status Register
CSR	Control (and) Status Register
CSR	Core Storage Resident
CSR	Current Schedule for Removal
CSR	Customer Service Record
CSR	Customer Service Representation|Representative
CSR	Cycle Shift Register
CSRAM	Custom Static Random Access Memory
CSRC	Computer Security Resource Clearinghouse
CSRG	Computer Systems Research Group
CSRL	Character Spacing Reference Line
CSROEPM	Communication, System, Results, Objectives, Exception, Participation, Motivation
CSS	Cascading Style Sheet(s)
CSS	Central Support Services (US)
CSS	Character Start-Stop
CSS	Communications Service Satellite
CSS	Computer Scheduling System
CSS	Computer Sub-System
CSS	Computer System(s) Simulator
CSS	Conceptual Signalling and Status
CSS	Conceptual Sub-Schema
CSS	Contact Start-Stop
CSS	Continuous System Simulator (language)
CSS	Controlled Slip Second
CSS	Control System Service
CSS	Customer Switching System
CSSB	Compatible Single Side-Band
CSSF	Customer Software Support Facility
CSSFE	Controlled Slip Second, Far End
CSSL	Continuous System Simulation Language
CSSM	Client-Server Systems Management
CST	Channel Status Table
CST	Code Segment Table
CST	Computer System Training
CST	Connection Setup Time

CST	Console Terminal
CST	Consolidated Scheduling Technique
CST	Constant
CSTA	Circuit Switched Telephony Access
CSTA	Computer Supported Telecommunications Applications
CSTB	Computer Science and Telecommunications Board
CSTO	Computer Systems Technology Office
CSTR	Current Status Register
CSTS	Computer Sciences Teleprocessing System
CSTU	Currently Signed-on Terminal User
CSU	Central Switching Unit
CSU	Channel Service Unit
CSU	Circuit Switching Unit
CSU	Clear and Subtract
CSU	Communications System User
CSU	Communication Switching Unit
CSU	Computer Software Unit
CSU	Core Storage Unit
CSU	Customer Set Up
CSU	Customer Setup Unit
CSUDSU	Channel Service Unit Data Service Unit
CSV	Circuit-Switched Voice
CSV	Comma Separated Variables|Values
CSV	Common Services Verbs (interface)
CSW	Channel Status Word
CSW	Clear and Subtract Word
CT	Cable|Call Transfer
CT	Cassette Tape
CT	Change Tracker|Type
C&T	Chips & Technologies
CT	Circuit Theory
C&T	Classification & Testing
CT	Clock Track
CT	Cognitive Technology
CT	Collection Time
CT	Combination Tone
CT	Commercial Translator
CT	Communications and Tracking
CT	Communication(s) Terminal
CT	Compute(rize)d Tomography
CT	Computer Technology
CT	Connecticut (US)
CT	Console Typewriter
CT	Continuous Time|Tone
CT	Cordless Telephone
CT	Counter
CT	Critical Thinking
CT	Crosstalk

CT	Current Transformer|Type
CT1+	Enhanced analogue Cordless Telephony (1st generation)
CT2	Digital Cordless Telephony (2nd generation)
CT3	Digital Cordless Telephony (3rd generation)
CTAB	Commerce Technical Advisory Board
CTAK	Cipher Text Auto Key
CTAN	Comprehensive TeX Archive Network
CTAP	Circuit Transient Analysis Program
CTAPS	Contingency Theater Automated Planning System
CTAS	Central Tracon Automation System
CTB	Cipher Type Byte
CTB	Code Table Buffer
CTB	Concentrator Terminal Buffer
CTC	Central Test Center
CTC	Channel To Channel
CTC	Computer Test Console
CTC	Conditional Transfer of Control
CTC	Counter Timer Circuit
CTC	Customer Test Center
CTCA	Channel To Channel Adapter
CTCC	Central Terminal Computer Controller
CTCM	Computer Timing and Costing Model
CTCP	Client-To-Client Protocol
CTCPEC	Canadian Trusted Computer Product Evaluation Criteria
CTCS	Component Time Control System
CTD	Charge Transfer Device
CTD	Combined Transport Document
CTD	Completion Time Deviation
CTD	Computer to Terminal Demultiplexer
CTD	Computing and Telecommunications Division
CTD	Conductivity, Temperature, Depth
CTD	Cumulative Transit Delay
CTDL	Complementary Transistor Diode Logic
CTDR	Channel Tag Diagnostic Register
CTE	Computer Telex Exchange
CTE	Customer Terminal Equipment
CTERM	Command Terminal (protocol)
CTF	Contrast Threshold Function
CTFC	Central Time and Frequency Control
CTFT	Continuous-Time Fourier Transform
CTI	Charge Transfer Inefficiency
CTI	Colour Transient Improvement
CTI	Computers in Teaching Initiative (GB)
CTI	Computer Telephone Integration
CTIA	Cellular Telecommunications Industry Association (US)
CTIN	Continue
CTIP	Commission on Computing, Telecommunications, and Information Policies

CTIR-CEAL	Centre Technique Informatique du Réseau des Caisses d'Epargne d'Alsace et de Lorraine (FR)
CTISS	Computers in Teaching Initiative Support Service
CTL	Cassette Tape Loader
CTL	Checkout Test Language
CTL	Common Target Language
CTL	Compiler Target Language
CTL	Complementary Transistor Logic
CTL	Control
CTL	Core Transistor|Transmission Logic
CTLCD	Control Card
CTLP	Control Panel
CTLR	Control Register
CTM	Cable Television Modem
CTM	Communications Terminal Module
CTM	Composite Tape Memory
CTM	Contact Trunk Module
CTMC	Communications Terminal Multiplex Cabinet
CTMS	Carrier Transmission Measuring System
CTMS	Common Table Management System
CT/N	Counter, N stages
CTO	Call Transfer Outside
CTO	Combined Transport Operator
CTOC	Center To Center
CTOC	Compound file Table Of Contents
CTOS	Cassette Tape Operating System
CTOS	Computerized Tomography Operating System
CTOS	Convergent Technologies Operating System
CTP	Central Transfer Point
CTP	Computer To Plate
C/TP	Control/Test Panel
CTPA	Coax-to-Twisted-Pair Adapter
CTR	Center
CTR	Character Transfer Rate
CTR	Common Technical Regulation
CTR	Counter
CTRL	Complementary Transistor-Resistor Logic
CTRL	Control
CTRLR	Controller
CTRS	Computerized Test-result Reporting System
CTS	Carpal Tunnel Syndrome
CTS	Cartridge Tape Subsystem
CTS	Clear To Send
CTS	Coaxial Terminal Switch
CTS	Combination Tones
CTS	Communications and Tracking Subsystem
CTS	Communications Technology Satellite
CTS	Communications Terminal Synchronous|Station

CTS	Compatible Timesharing System
CTS	Computer Telegram System
CTS	Computer Test Stand
CTS	Conformance Testing Service
CTS	Consultative Tele-information Service
CTS	Consumer Transaction Systems
CTS	Contract Technical Services
CTS	Control Transfer Statement
CTS	Conversational Terminal|Test System
CTS	Conversational Time-Sharing
CTS-LAN	Conformance Testing Service Local Area Network
CTSS	Classroom Test Support System
CTSS	Compatible Time-Sharing System
CTSS	Computer Time-Sharing Service
CTS-WAN	Conformance Testing Service Wide Area Network
CTT	Cartridge Tape Transport
CTTC	Cartridge Tape Transport Controller
CTTD	Cute Things They Do
CTTN	Cable Trunk Ticket Number
CTU	Central Terminal Unit
CTV	Cable Television
CTV	Closed circuit TV
CTV	Colour TV
CTW	Console Typewriter
CU	Close Up
CU	Control Unit
Cu	Copper
CU	Correlation Unit
CU	Crosstalk Unit
CU	Cuba
CU	Customer Us(ag)e|Unit
CU	See You
CUA	Channel Unit Address
CUA	Circuit Unit Assembly
CUA	Common|Computer User Access
CUA	Computer Users Association (US)
CUAG	Computer Users Association Group (US)
CUB	Control Unit Busy
CUB	Cursor Backward
CUC	Computer Users Committee
CUCB	Control Unit Control Block
CUCUG	Champaign Urbana Commodore Users Group
CUD	Computer Underground Digest
CUD	Control Unit Description
CUD	Cursor Down
CUDN	Common User Data Network
CUE	Computer Updating Equipment
CUE	Computer Utilizing English

CUE	Configuration Utilizing Efficiency
CUE	Control Unit End|Error
CUE	Correlation Update Extension
CUE	Custom Updates and Extras (card)
CUEING	Control Unit Error Insertion Generator
CUEPEND	Control Unit End Pending
CUESTA	Communications User Emulated System for Traffic Analysis
CUF	Cursor Forward
CUG	Closed User Group
CUG	Critical Unit Gage
CUI	CADES User Interface (CADES = Computer Aided Design for Electronic Systems)
CUI	Character-oriented User Interface
CUI	Common User Interface
CUJT	Complementary Uni-Junction Transistor
CUL	See You Later
CULP	Computer Usage List Processor
CUM	Central Unit Memory
CUM	Cumulative
CUMULI	Computational Understanding of Multiple Images (EU)
CUNF	Customer Update Notification File
CUNT	Computer Users Network Team
CUNT	Coupled Unix Network Tree
CUP	Communication(s) User Program
CUP	Critical Unit Process
CUP	Cursor Position
CUPID	Completely Universal Processor Input/output Design
CUPID	Create, Update, Process, Interrogate, and Display
CUR	Complex Utility Routine
CUR	Current
cur	cursor (file extension indicator)
CURTS	Common User Radio Transmission Sounder
CUSA	Customer Address (line)
CUSCL	Customer Class
C-use	Computational use
CUSI	Customizable Unified Search Index
CUSNM	Customer Name
CUSNR	Customer Number
CUSP	Commonly Used System Program
CUT	Circuit|Computer Under Test
CUT	Control Unit Terminal
CUTE	Clarkston University Terminal Emulator
CUTS	Computer Users Tape System
CUU	Cursor Up
CUV	Current Use Value
CV	Cape Verde
CV	Code Violation
CV	Coefficient of Variation

CV	Constant Voltage
CV	Continuously Variable
CV	Control Vertice
CV	Conversion
CV	Converter
CVBS	Colour Video Blanking Synchronization
CVBS	Composite Video Blanking and Synchronization
CVC	Carrier Virtual Circuit
CVC	Current Voltage Characteristic
CVCC	Controlled Vortex Combustion Chamber
CVCP	Code Violation, CP-bit Parity (CP = Control Program)
CVCRC	Code Violation, Cyclical Redundancy Check
CVE	Circuit Verification Environment
CVE	Collaborative Virtual Environment
CVF	Compressed Volume File
CVFE	Code Violation, Far End
CVGA	Colour Video Graphics Array
CVIA	Computer Virus Industry Association
CVIS	Computerized Vocational Information System
CVL	Continuous Velocity Log
CVM	Cisco Voice Manager
CVM	COBOL Virtual Machine
CVMP	Committed Volume/Mix Performance
CVP	Code Violation, P bit
CVP	Committed|Confirmed Volume Performance
CVPR	Computer Vision & Pattern Recognition (conference)
CVR	Collaborative Virtual Environment
CVR	Compass|Computer Voice Response
CVR	Continuous Video Recorder
CVS	Computer Vision Syndrome
CVS	Concurrent Version(s) System
CVS	Constant Volume Sampling
CVS	Conversational System
CVSD	Continuous(ly) Variable Slope Delta modulation
CVSN	Conversion
CVT	Communications Vector Table
CVT	Constant Voltage Transformer
CVT	Continuous Variable Transmission
CVT	Convert
CVTC	Conversational Voice Technologies Corporation
CVU	Constant Voltage Unit
CVW	Code View for Windows
CW	Call Waiting
CW	Clear and Write
CW	Clockwise
CW	Command Word
CW	Continuous Wave
CW	Control Word

CWA	Communication Workers of America
CWA	Control Word Address
CWAN	Corporate Wide Area Network
CWAR	Control Word Address Register
CWARC	Canadian Workplace Automation Research Center
CWC	Cable & Wireless Communications (GB)
CWC	Carrier Wave Connection
CWD	Change Working Directory
CWD	Clerical Work Data
CWD	Convert Word to Double word
CWE	Cleared Without Examination
CWF	Customer Work File
CWI	Centrum voor Wiskunde en Informatica (NL)
CWI	Change Window Indicator
CWIS	Campus-Wide Information System
CWIS	Community Wide Information System
CWM	Carrier Wave Modulation
CWMI	Common Warehouse Meta-data Interchange
CWO	Continuous Wave Oscillation
CWP	Communicating Word Processor
CWP	Current Word Pointer
CWPS	Communicating Word Processor System
CWQ	Channel Waiting Queue
CWS	Carrier Wave Shift
CWS	Compiler Writing System
CWT	Command Word Trap
CWU	Character Word Unit
CWV	Continuous Wave Video
CWW	Cisco Works for Windows
CWYL	Chat With You Later
CX	Central Exchange
CX	Christmas Island
CX	Communications Exchange
CX	Composite Signalling
CX	Cycle Crossover
CXA	Central Exchange Area
CxAM	Context-Addressable Memory
CXI	Common X-window Interface
CXML	Commerce Extended Markup Language
CXS	Coaxial Shield
CY	Calendar Year
CY	Capacity
CY	Cycle
CY	Cyprus
Cyborg	Cybernetic Organism
CYC	Cycle
CYC	Cyclic (selection)
CYCT	Cycle Count

CYCTL	Cycle Control
CYDEL	Cycle Delay
CYL	Cylinder
CYMAJ	Cycle Major
CYMI	Cycle Minor
CYMK	Cyan-Yellow-Magenta-Black
CYSH	Cycle Shift
CZ	Control Zone
CZ	Czech Republic
CZE	Compare Zone Equal
CZRT	Count Zero Refill Trigger
CZU	Compare Zone Unequal

D

D	Data(line)
d	day
d	deci (prefix)
d	degree
D	Deleted
D	Dielectric shift
d	difference
d	differential
D	Diffusion (constant)
D	Digit(al)
D	Diode
D	Direct
D	Displacement
D	Display
D	Dissipation (factor)
d	distance
d	drain
d	drive
D	Symbol for electrostatic flux Density
D2B	Domestic Digital Bus
D2C	Decimal To Character
D2T2	Dye Diffusion Thermal Transfer (printing)
D2X	Decimal To Hexadecimal
D3D	Direct 3-Dimension(al)
DA	Data Acquisition|Address
DA	Data Administrator
DA	Data Automation|Available
da	deca
DA	Decimal Add
DA	Define Area
DA	Demand Assignment
DA	Design Automation
DA	Destination Address

DA	Device Address
DA	Differential Analyzer
DA	Digital Audio
D/A	Digital to Analog
DA	Direct Access|Address
DA	Discrete Address
DA	Disk Action
DA	Display Adapter
DA	Distribution Amplifier
DA	Domain Analysis
DA	Don't Answer
DAA	Data Access Arrangement
DAA	Deadlock Avoidance Algorithm
DAA	Decimal Adjust for Addition
DAA	Distributed Application Architecture
DAA	Dual Address Adapter
DAAC	Data Acquisition and Analysis Complex
DAAC	Distributed Active Archive Center
DAAM	Dynamic Associative random Access Memory
DAB	Demand Adjustment Bucket
DAB	Digital Audio Broadcasting
DAB	Display Assignment|Attention Bit
DABAS	Data Base Access Service
DABAS	Data Base System
DABS	Discrete Address Beacon System
DAC	Data Accepted
DAC	Data Acquisition|Analysis and Control
DAC	Data Authentication Code
DAC	Demand Assignment Controller
DAC	Design Augmented by Computer(s)
DAC	Design Automation Conference
DAC	Digital/Analog Converter
DAC	Digital Arithmetic Center
DAC	Digital to Analog Converter
DA/C	Directory Assistance/Computerized
DAC	Disk Access Control (program)
DAC	Dual-Attachment Concentrator
DACAP	Discrete And Continuous Analysis Program
DACBU	Data Acquisition Control and Buffer Unit
DACC	Data Acquisition and Checkout Computer
DACCS	Digital Access Cross Connect System
DACD	Directory Access Control Domain
DACE	Data Acquisition and Control Executive
DACI	Direct Adjacent Channel Interference
DACL	Diablo Application Compiler Language
DACOM	Datascope Computer Output Microfilmer
DACOR	Data Correction|Correlation
DACOS	Data Communication Operating System

DACPO	Data Count Print Out
DACR	Delete And Complete Reinstall
DACS	Data Access Control System
DACS	Data Acquisition and Control|Conversion System
DACS	Design Aided by Computer System
DACS	Digital Access and Cross-connect System
DACS	Digital Access & Control System
DACU	Data Acquisition Control Unit
DACU	Device Attachment Control Unit
DACU/AT	Device Attachment Control Unit for Asynchronous Terminals
DAD	Database Action Diagram
DAD	Desktop Application Director
DAD	Digital Animation Dream (machine)
DAD	Document Access Definition
DAD	Drum And Display
DADB	Data Analysis Data Base
DADEC	Design And Demonstration Electronic Computer
DADIOS	Direct Analog to Digital Input/Output System
DADS	Data Access and Dissemination System
DADS	Data Acquisition and Display System
DADSM	Direct Access Device Space Management
DAE	Data Acquisition Equipment
DAE	Distributed Automation Edition
DAEDR	Delimitation, Alignment and Error Detection in Receive direction
DAF	Data Acquisition Facility
DAF	Destination Address Field
DAF	Direct Access File
DAF	Directory Authentication Framework
DAF	Dispersion Attenuation Factor
DAF	Distributed Application Framework
DAFA	Data Accounting Flow Assessment
DAFC	Digital Automatic Frequency Control
DAFGD	Drag And File Gold Desktop
DAFM	Direct Access File Manager
DAFT	Digital-to-Analog Function Table
DAG	Directed Acyclic Graph
DAGC	Delayed Automation Gain Control
DAI	Direct Access Information
DAI	Distributed Artificial Intelligence
DAIR	Dynamic Allocation Interface Routine
DAIS	Digital Avionics Information System
DAIS	Direct Access Intelligence System
DAIS	Distributed Applications and Interoperable Systems (conference)
DAIS	Distributed Automatic Intercept System
DAISY	Data Analysis of the Interpreter System
DAISY	Decision Aiding Information System

DAISY	Double-precision Automatic Interpretative System
DAK	Data Acknowledge
DAL	Data Access Language|Line
DAL	Data Acquisition Language
DAL	Data Address Line
DAL	Dedicated Access Line
DAL	Direct Attached Drop
DAL	Disk Access Lockout
DAL	Document Address Lister
DALC	Divided Access Line Circuit
DALC	Dynamic Asynchronous Logic Circuit
DALIB	Direct Access Library
DAM	Data Access Manager
DAM	Data Acquisition and Monitoring
DAM	Data Addressed Memory
DAM	Data Association Manager|Message
DAM	Descriptor Attribute Matrix
DAM	Diagnostic Acceptability Measure
DAM	Direct Access(ible) Memory
DAM	Direct Access Method
DAM	Dummy Average Machine
DAMA	Data Administration Management Association
DAMA	Data Assigned Multiple Access
DAMA	Demand Assigned Multiple Access
DAMC	Digital Automation Map Compilation
DAME	Digital Automatic Measuring Equipment
DAMOD	Direct Access Module
DAMOS	Data Moving System
DAMPS	Data Acquisition Multi-Programming System
DAMPS	Digital Advanced Mobile Phone Service
DAMQAM	Dynamically Adaptive Multicarrier Quadrature Amplitude Modulation
DAMS	Direct Access Management System
DAN	Departmental Area Network
DANDELION	Discourse functions And representation: an Empirically and Linguistically motivated Interdisciplinary approach On Natural language texts (EU)
DANDR	Diagnostic And Repair
DANTE	Decision Aiding with New Technologies
DANTE	Delivery of Advanced Network Technology to Europe
DANTE	Deutsche Anwendervereinigung TeX (DE)
DAO	Data Access Object(s)
DAOB	Digital and Analog Output Basic
DAOM	Direct Access Organization Methods
DAOR	Discriminator Average Output Race
DAP	Data Access Protocol
DAP	Data Acquisition and Processing
DAP	Developer Assistance Program

DAP	Diagnostic Assistance Program
DAP	Diffused Alloy Power
DAP	Direct Access Processing|Photo-memory
DAP	Directory|Distributed Access Protocol
DAP	Distributed Array Processor
DAP	Document Application Profile
DAPL	Direct Access Programming Language
DAPPO	Device And Program Performance Optimization
DAPS	Data processing Automatic Publications Service
DAPS	Direct Access Programming System
DAPS	Distributed Application Processing System
DAPU	Data Acquisition and Processing Unit
DAR	Data Access Register
DAR	Delay Address Register
DAR	Dynamic Address Register|Relocation
DARB	Data Access Request Block
DARC	Device for Automatic Remote data Collection
DARDO	Direct Access to Remote Databases Overseas
DARE	Data Analysis and Reduction
DARE	Differential Analyzer Replacement
DARI	Database Application Remote Interface
DARIC	Data Reduction In Columns
DARPA	Defense Advanced Research Projects Agency
DARS	Data Acquisition and Reduction System
DARS	Design of Advanced Robotics Systems (international conference)
DART	Daily Automatic Rescheduling Technique(s)
DART	Data Analysis Recording|Reduction Tape
DART	Data Automatic Rescheduling Technique
DART	Data Reduction Translator
DART	Design Automation Routing Tool
DART	Detection, Action, and Response Technique
DART	Digital Audio Reconstruction Technology
DART	Diode Automatic Reliability Tester
DARTNET	Defense Advanced Research Testbed Network
DARTS	Distributors Automated Real-Time System
DARU	Distributed Audio Response Unit
DAS	Data Access Security
DAS	Data Acquisition|Administration System
DAS	Data Automation|Analysis System
DAS	Decimal Adjust for Subtraction
DAS	Design Analysis System
DAS	Digital-Analog Simulator
DAS	Digital Attitude Simulation
DAS	Direct Access Storage|System
DAS	Directory Assistance System
DAS	Disk Array Subsystem
DAS	Distributor And Scanner
DAS	Documentation Aid System

DAS	Dual-Attached Station
DAS	Dynamically Assigned Sockets
DASC	Design Automation Standards Subcommittee
DASD	Direct Access Storage Drive
DASDDR	Direct Access Storage Device Dump/Restore
DASDI	Direct Access Storage Device Initialization
DASDL	Data And Structure Definition Language
DASDM	Direct Access Storage Data Management
DASDR	Direct Access Storage Dump Restore
DASE	Distributed Application Support Environment
DASEL	Data Analysis and Statistical Experimental Language
DASF	Direct Access Storage Facility
DASFAA	Database Systems For Advanced Applications (international conference)
DASH	Data Acquisition Sequential Histogram
DASH	Direct Access Storage Handling
DASL	Direct Access System Language
DASM	Direct Access Storage Media|Model
DASP	Datapoint Attached Support Processor
DASP	DOS Automatic Spooling Program (DOS = Disk Operating System)
DASS	Design Automation Standards Subcommittee
DASS	Digital Access Signalling System
DASSY	Data transfer and interfaces for open and very large integrated Semiconductor Systems
DASTAR	Data Storage And Retrieval
DASY	Design Automation System
DAT	Data Abstract Type
dat	data (file extension indicator)
DAT	Desktop Analysis Tool
DAT	Diagnostic Address Translation
DAT	Diffused-Alloy Transistor
DAT	Digital Audio Tape
DAT	Disk Allocation Table
DAT	Disk Array Technology
DAT	Duplicate Address Test
DAT	Dynamic Address Translation
DATA	Decision Analysis by TreeAge
DATA	Dictionary of Acronyms and Technical Abbreviations
DATA	Direct Access Terminal Application
DATACOM	Data Communication
DATACOMM	Data Communications
DATACON	Data Conference
DATACOR	Data Correction
DatAn	Data Analysis
DataNet	Data Network
Datapak	National Packet-switched network of Denmark, Norway, Sweden and Finland

DATAR	Digital Automatic Tracking And Ranging
DATAS	Data in Associative Storage
DATC	Design Automation Technical Committee
DATE	Design Automation and Test in Europe (conference)
DATEL	Data Telecommunications|Telephone
DATEX	Data And Telex network
Datex-P	German and Austrian Packet-switched service
DATICO	Digital Automatic Tape Intelligence Check-Out
DATIME	Date And Time
DATOR	Digital Auxiliary Track Output Recording
DaTran	Data Transmission
DATS	Dynamic Accuracy Test Set
DATV	Digital Assisted Television
DAU	Data Access|Adapter Unit
DAU	Display Adapter Unit
DAV	Data Above Voice
DAV	Data Available
DAV	Data Valid
DAV	Digital Audio-Video
DAVB	Digital Audio and Video Broadcasting
DAVIC	Digital Audio Visual Council
DAVID	Digital Audio/Video Interactive Decoder
DAVIE	Digital Alphanumeric Video Insertion Equipment
DB	Data Bank|Base
DB	Data Bit
DB	Data Bus|Buffer
dB	decibel
DB	Device Busy
DB	Dial Box
DB	Diffused Base
DB	Double Biased (relay)
DB	Double Break
DB2	Data Base 2
DBA	Data Base Administration
dBa	decibels adjusted
DBAAM	Disk Buffer Area Access Method
DBAC	Data Base Administration Center
DBAM	Data Base Access Method
DBAS	Data Base Administration System
dBASE	data Base relational management system (Borland)
DBASM	Data Base Access Service Module
DBB	Detector Balanced Bias
DBC	Data Base Computer
dBc	decibels (referred) to carrier
DBC	Decimal-to-Binary Conversion
DBCB	Data Base Control Block
DBCCP	Data Base Command and Control Processor
DBCL	Data Base Command Language

DBCLOB	Double-Byte Character Large Object
DBCS	Data Base Control System
DBCS	Double-Byte Character Set
DBCTG	Data Base Concepts Task Group
DBD	Data Base Definition|Design
DBD	Data Base Directory
DBD	Distribution Board
DBD	Double Base Diode
DBDA	Data Base Design Aid
DB/DC	Data Base/Data Communication
DBDD	Data Base Design Document
DBDE	Data Base Design Evaluator
DBDGEN	Data Base Description Generator
DBDL	Data Base Definition Language
DB/DM	Data Base/Data Management
DBDM	Data Base Development Methodology
DBDNAME	Data Base Description Name
DBE	Data Bus Enable
DBF	Data Base Facility
dbf	database file (file extension indicator)
DBF	Data Base Format
DBG	Data Base Generator|Group
DBI	Data Base Interactive (system)
DBI	Data Bus In(terleave)
DBI	Double Byte Interleaved
DBIN	Data Bus In
DBIO	Data Base Input/Output
DBIOC	Data Base Input/Output Control
dBj	radio frequency signal level relative to 1 millivolt
dBk	radio frequency signal level relative to 1 kilowatt
DBL	Data Base Language
DBL	Data Block
DBL	Double
DBLIB	Data Base Library
DBLK	Data Block
DBM	Data Base Machine|Management
dbm	database manager (file extension indicator)
DBM	Data Base Module
dBm	decibel/milliwatt
dBm	decibels referenced to one milliwatt
DBM	Direct Branch Mode
dBm0	dBm relative to zero transmission level
dBm0p	dBm0 with psophometric weighting
DBMC	Data Base Management Computer
DBME	Data Base Management Environment
DBML	Data Base Management Language
dBmp	dBm with psophometric weighting
DBMS	Data Base Management Subsystem|Software

dBmV decibels relative to 1 millivolt
DBO Data Bus Out
DBOE Data Base Operating Environment
DBOM Driver Block Output Mode
DBOMP Data Base Organization and Maintenance Processor
DBOS Disk Based Operating System
DBP Data Base Processor
dBp decibel/picowatt
DB-PCB Data Base Program Communication Block
DBQ Data Base for Quality
DBR Data Base Recovery|Representation
DBR Data Base Retrieval
dBr dBm relative to zero transmission level
DBR Descriptor Base Register
DBR Dynamic Base Relocation
DBRAD Data Base Relational Application Directory
dBrap decibels above reference acoustic power
DBRC Data Base Recovery and Control
DBRG Data Base Requirements Group
DBRN Decibels above Reference Noise
dBrn0 noise in DBRN relative to zero level
DBRNC Decibels away from the Reference Noise when measured with a C-message filter
DBS Data Base Server|Service
DBS Data Base System
DBS Direct Broadcast Satellite|System
DBS Distributor Bus
DBS Duplex Bus Selector
DBS Duplicate Buffer Storage
DBT Data Base Tools
DBTG Data Base Task Group
DBU Data Base User
DBU Dial Back-up Unit
dBV decibels relative to 1 Volt
dBW decibels relative to 1 Watt
dBw decibel/watt
dBx decibels above reference coupling
DBX Digital Branch Exchange
dBXL dBASE III plus compatible Extra Large data base management system
DC Data Cartridge|Cassette
DC Data Center|Channel
DC Data Check|Classifier
DC Data Code|Collection
DC Data Communication|Control
DC Data Counter|Cycle
DC Define Constant
DC Desktop Calculator

DC Device Control (character)
DC Digital Comparator|Computer
DC Digital Control(ler)
DC Direct Coupled|Cycle
dc direct current
DC Disconnect Confirm
DC Disk Controller
DC Display Console|Control
DC Distribution Center
DC District of Columbia (US)
DC Double Column
DC Downward Compatible
DCA Data Communications Administrator|Architecture
DCA Data Concentration Adapter
DCA Digital Communications Associates
DCA Digital Computer Association
DCA Direct Coupled|Current Amplifier
DCA Distributed Communication(s) Architecture
DCA Document Content(s) Architecture
DCA Driver Control Area
DCAA Dual Call Auto Answer
DC-AC Direct Current to Alternating Current
DCAF Distributed Console Access Facility
DCAM Data Collection Access Method
DCAM Digital Camera
DCAM Direct Chip Attach Module
DCAP Dynamic Checkout Assistance Program
DCARE Driver Control Area Region Extension
D/CAS Data/Cassette
DC&AS Digital Control & Automation System
DCB Data and Control Bus
DCB Data Communication Bank
DCB Data Control Block
DCB Define|Device Control Block
DCB Directory Cache Buffer
DCB Disk Control Block
DCB Disk Coprocessor Board
DCBD Define Control Block Dummy
DCC Data Circuit Concentrator
DCC Data Collection Center
DCC Data Communications Channel|Controller
DCC Data Converter Check
DCC Data Country Code
DCC Decimal Coded Character
DCC Development Computer Center
DCC Device Cluster Controller
DCC Device Control Character
DCC Digital Command Control

DCC	Digital Compact Cassette
DCC	Digital Cross Connect
DCC	Direct Calculating Capability
DCC	Direct Client Connection
DCC	Direct Control Channel|Connection
DCC	Direct Coupled System
DCC	Direct Current Clamp
DCC	Disk Controller Channel
DCC	Display Combination Code
DCC	Display Control Computer|Console
DCC	Document Control Center
DCCA	Dependable Computing for Critical Applications (conference)
DCCB	Digital Connectivity Control Bus
DCCH	Disk Cartridge Channel
DCCS	Distributed Capability Computing System
DCCS	Distributed Computer Control System
DCCU	Data Communications Control Unit
DCCU	Displays and Controls Control Unit
DCD	Data Carrier Detect
DCD	Data Cell Drive
DCD	Data Correlation and Documentation (system)
DCD	Decode
DCD	Double Channel Duplex
DC/DC	Direct Current to Direct Current
DC/DKI	Disk Controller/Data-Kit Interface
DCDL	Data Communications for Distance Learning
DCDR	Decoder
DCDS	Digital Control Design System
DCDU	Data Collection and Distribution Unit
DCE	Data Circuit(-terminating) Equipment
DCE	Data Communication(s) Equipment
DCE	Digital Computer Equipment
DCE	Discrete|Display Control Equipment
DCE	Displays Common Equipment
DCE	Distributed Computing Environment|Equipment
DCEC	Defense Communications Engineering Center
DCE-RPC	Distributed Computing Environment – Remote Procedure Call
DCF	Data Communication(s) Facility|Function
DCF	Data Compression Facility
DCF	Data Count Field
DCF	Direct Control Feature
DCF	Disk Control Field
DCF	Disk Controller/Formatter
DCF	Distributive Computing Facility
DCF	Dot Clock Frequency
DCF	Driver Configuration File
DCG	Definite Clause Grammar
DCH	Data Channel

DCH	Data Communication Handbook
DCI	Data Collection Interface
DCI	Data Communication Interface|Interrogate
DCI	Desktop Color Imaging
DCI	Direct Channel Interface
DCI	Direct Coupled Inverter
DCI	Display Color|Control Interface
DCIA	Digital Card Inverting Amplifier
DCIO	Direct Channel Interface Option
DCIP	Disk Cartridge Initialization Program
DCIU	Data Communications Interface Unit
DCK	Data Check
DCL	Data Command Language
DCL	Declare
DCL	Device Clear
DCL	Dialog Control|Command Language
DCL	Digital Control Logic
DCL	Diode-less Core Logic
DCL/DCS	Delay Calculation Language/Delay Calculation System
DCLU	Digital Carrier Line Unit
DCM	Data Channel Multiplexer
DCM	Data Communications Module|Multiplexer
DCM	Diagnostic Controlled Modem
DCM	Diagnostic Control Manager
DCM	Diffused Current Mode
DCM	Digital Capacitance Meter
DCM	Digital Carrier Module
DCM	Digital Circuit Multiplication
DCM	Direct Channel with Memory
DCM	Direct Current Modulation
DCM	Display Control Module
DCME	Digital Circuit Multiplication Equipment
dcml	decimal
D/CMOS	DMOS combined with Complementary Metal-Oxide Semiconductor (DMOS = Double-diffused Metal-Oxide Semiconductor)
DCMS	Data Capture and Management System
DCMS	Dedicated Computer Message Switching
DCMS	Distributed Call Measurement System
DCMT	Decrement
DCMT	Document
DCMU	Digital Concentrator Measurement Unit
DCN	Data Communication Network
DCn	Device Control n
DCN	Distributed Computer Network
DCNA	Data Communications Network Architecture
DCO	Digitally Controlled Oscillator
DCOM	Distributed Common|Component Object Model

DCOS Data Collection Operating System
DCOS Direct Coupled Operating System
DCP Data Ciphering Processor
DCP Data Collection Program
DCP Data Communication Processor|Protocol
DCP Data Control|Communications Program
DCP Digital Communications Protocol|Processor
DCP Digital Contour Processor
DCP Distributed Communications Processor
DCP Duplex Central Processor
DCPBX Digitally Connected Private Branch Exchange
DCPC Dual Channel Port Controller
DC-PCB Data Communication Program Communication Block
DCPCM Differentially Coherent Pulse Code Modulation
DCPP Data Communications Pre-Processor
DCPSK Differentially Coherent Phase Shift Keying
DCR Data Collection Routine
DCR Data Conversion Receiver
DCR Digital Cassette Recorder
DCR Digital Conversion Receiver
DCR Direct Current Resistance|Restorer
DCR Disk Capture/Restore
DCRABS Disk Copy Restore And Backup System
DCRS Decrease
DCS Data Carrier System
DCS Data Center Scheduler|Services
DCS Data Circuit Switches
DCS Data Clarification System
DCS Data Collection System
DCS Data Communication Subsystem|Software
DCS Data Consolidation Simulation
DCS Data Control System
DCS Defined Context Set
DCS Desktop Colour Separation
DCS Diagnostic Compiler System
DCS Diagnostic Control Store
DCS Digital Command|Control System
DCS Digital Correcting Subranging
DCS Digital Cross-connect System
DCS Direct Coupled System
DCS Distributed Communications System
DCS Distributed Computer|Control System
DCS Document Control System
DCS1800 Digital Cellular System working at 1800 MHz
DCSL Deterministic Context Sensitive Language
DCS/MIP Dynamic Computer System/Multi-purpose Information Processor
DCSS Discontinuous Shared Segments

DCSSC	Data Center Services and Support Center
DCT	Data Collection Terminal
DCT	Data Communication(s) Terminal
DCT	Destination Control Table
DCT	Device Characteristics Table
dct	dictionary (file extension indicator)
DCT	Digital Carrier Trunk
DCT	Digital Conversion Terminal
DCT	Direct Coupled Transistor
DCT	Dispatcher Control Table
DCTA	Documentation Control Testing Application
DCTC	Documentation Control Testing Center
DCTE	Data Circuit Terminating Equipment
DCTL	Data Control
DCTL	Direct Coupled Transistor Logic
DCTN	Defense Commercial Telecommunications Network
DCTS	Dimension Custom Telephone Service
DCTS	Document Control and Testing System
DCTU	Dominant Certified Telecommunications Utility
DCU	Data-Cache Unit
DCU	Data Communications|Control Unit
DCU	Data Conversion Utility
DCU	Decade Counting Unit
DCU	Device Control Unit
DCU	Diagnostic Control Unit
DCU	Disk|Display Control Unit
DCU	Drum Control Unit
DCUTL	Direct Coupled Unipolar Transistor Logic
DCV	Data Converter
DCV	Digital(ly) Coded Voice
dcv	direct current voltage
DCVG	Display Control Vector Generator
DCW	Data Communication Write
DCW	Data Control Word
DCW	Distributed Collaborative Workbench
DCWV	Direct Current Working Volts
DCY	Delay Cycle
DD	Data Definition|Description
DD	Data Demand|Division
DD	Data Dictionary
DD	Dataset Definition
dd	day
DD	Days after Date
DD	Decimal Decode|Display
DD	Definition Direction
DD	Deformation Dipole
DD	Delay Driver
DD	Destination Determination

DD	Device Driver
DD	Digital Data|Display
DD	Disk Drive
DD	Disk to Disk (copy)
DD	Display Description
DD	Double Density|Deck
D&D	Drag & Drop
DDA	Design Decision Analysis
DDA	Digital Data Acquisition
DDA	Digital Differential Analyzer
DDA	Digital Display Alarm
DDA	Direct Disk Attachment
DDA	Distributed Data Access
DDA	Domain Defined Attribute
DDAM	Dynamic Design Analysis Method
DDAS	Design of Data Acquisition Subsystem
DDAS	Digital Data Acquisition System
DDB	Device Dependent Bitmap
DDB	Device Descriptor Block
DDB	Distributed Data Base
DDBMS	Distributed Data Base Management System
DDBS	Development Data Bases Service
DDBS	Distributed Data Base System
DDC	Data Distribution Center
DDC	Digital Data Channel|Converter
DDC	Direct Data Channel
DDC	Direct Digital Control(ler)
DDC	Display Data Channel
DDC	Divisional Distribution Center
DDC	Dual Directional Coupler
DDC1	Display Data Channel One
DDCE	Dynamic Dispatching Control Element
DDCMP	Digital Data Communications Message Protocol
DDCP	Digital Data Communication(s) Protocol
DDCP	Digital Dynamic Convergence and Picture
DDCS	Distributed Database Connection Services
DDCU	Dc-to-Dc Converter Unit (Dc = Direct current)
DD/D	Data Dictionary/Directory
DDD	Design Description Document
DDD	Direct Distance Dialling
DDDL	Double Diffused Diode Logic
DD/DS	Data Dictionary/Directory System
DDDU	Display Data Distribution Unit
DDE	Digital Definition Exchange
DDE	Direct Data Entry
DDE	Dynamic Data Exchange
DDECS	Design and Diagnostics of Electronic Circuits and Systems
DDEML	Dynamic Data Exchange Manager Library

DDES	Digital Data Exchange Specifications
DDES	Digital Data Exchange Standard
DDF	Data Description File
DDF	Data Dictionary Facility
DDF	Display Data File
DDF	Dynamic Data Formatting
DDFF	Distributed Disk File Facility
DDFF	Distributed Document Formatting Facility
DDFII	Data Description File for Information Interchange
DDG	Data Dependence Graph
DDG	Digital Display Generator
DDGE	Digital Display Generator Element
DDGL	Device Dependent Graphics Layer
DDI	Device Driver Interface
DDI	Direct Dial(ling) In
DDIC	Digital Data Input Converter
DDIE	Digital Display Indicator Element
DDIF	Digital Document Interchange Format
DDIS	Digital Display Indicator Section
DDIT	Diagnostic Data Interface Tapes
DDK	Device Driver Kit
DDK	Driver Development Kit
DDL	Data Definition|Description Language
DDL	Descriptive Design Language
DDL	Digital Data Link
DDL	Digital Design Language
DDL	Document Description Language
DDLC	Database Development Life Cycle
DDLC	Data Description Language Committee
DDLC	Digital Data Link Connection
DDLCN	Distributed Double-Looped Computer Network
DDM	Data Description Modification|Module
DDM	Development Data Management
DDM	Device Descriptor Module
DDM	Difference in Depth of Modulation
DDM	Distributed Data Management
DDN	Data Definition Name
DDN	Defense|Digital Data Network
DDN	Dotted Decimal Notation
DDNAME	Data Definition Name
DDN-NIC	Defense Data Network – Network Information Center
DDO	Direct Dialing Out
DDO	Dynamic Drive Overlay
DDOC	Digital Data Output Converter
DDP	Data Description Point
DDP	Datagram Delivery Protocol
DDP	Decentralized Data Processing
DDP	Diagnostic Disk Pack

DDP	Digital Data Processor
DDP	Distributed Data Processing
DDPEX	Device Dependent Packet Exchange (protocol)
DDPP	Direct Digital Printing Plate
DDPREP	Device-Dependent Parameter conversion and Replacement
DDR	Data Definition|Descriptive Record
DDR	Data Dependent Routing
DDR	Data Direction Register
DDR	Digital Data Receiver
DDR	Digital Disk Recorder
DDR	Double Data Rate
DDR	Dynamic Device Reallocation|Reconfiguration
DDRA	Decimal Divide Restore Answer
DDS	DASD Dump Store
DDS	Data Dependent System
DDS	Data Description Specification(s)
DDS	Dataphone Digital Service
DDS	Decision Support System(s)
DDS	Design Data Sheet
DDS	Digital Dataphone Service
DDS	Digital Data Service|System
DDS	Digital Data Storage
DDS	Digital Display Scope
DDS	Direct Data Set
DDS	Direct Digital Service
DDS	Display Data Subsystem
DDS	Distributed Database Services
DDS	Double Density Storage
DDSA	Digital Data Service Adapter
DDSA	Direct-Digitally-Synthesized Audio
DDSD	Dynamic Data Set Definition
DDS-SC	Dataphone Digital Service with Secondary Channel
DDT	Data Description Table
DDT	Design and Debug Tool
DDT	Digital Data Transceiver
DDT	Dynamic Debugging Technique|Tool
DDT	Dynamic Decision Table
DDT&E	Design, Development, Test & Evaluation
DDTL	Diffused Diode Transistor Logic
DDTL	Double Diffused Transistor Logic
DDTL	Double Diode Transistor Logic
DDU	Data Dictionary Utilities
DDUMP	Disk Dump
DDX	Device Dependent X-window
DDX	Digital Data Exchange
DDX	Distributed Data Exchange
DDX-PS	Digital Data Exchange – Packet Switch
DE	Data Element|Entry

DE	Decision Element
DE	Delaware (US)
DE	Device Emulation|End
DE	Differential Equation
DE	Digital Element
DE	Disk Enclosure
DE	Display Element|Equipment
DE	Division Entry
DE	Germany
DEA	Data Encryption Algorithm
DEA	Double-Ended Amplifier
deac	deaccentuator
DEACON	Direct English Access & Control
DEACTLH	Deactivate Line Halt
DEACTLO	Deactivate Line Orderly
DEAL	Design And Logistics
DEB	Data Event|Extent Block
DEB	Directory Entry Block
DEBASS	Debug Assembler
DEBUT	Delay (line) Buffered Terminal
DEC	Data Evaluation Center
DEC	Decimal
DEC	Decimal Error Correcting code
DEC	Decision
DEC	Declination
DEC	Decrement
DEC	Device Clear
DEC	Digital Equipment Corporation
DEC	Document Effected Code
DEC	Document Evaluation Center
DECACC	Decimal Accumulator
DECB	Data Event Control Block
DECCA	Defense Commercial Communication Activity
DECIS	Data Entry Communication Interface Set
DECMOV	Decimal Move
DECNET	Digital Equipment Corporation Networking (protocol)
DECNOSYS	Decimal Number System
DECOM	Digital Equipment Corporation broadband Ethernet transceiver
DECOMP	Decomposition
DECOR	Digital Electronic Continuous Ranging
DECPT	Decimal Point
DECR	Decrement
DECT	Digital Enhanced|European Cordless Telecommunications
DECTAT	Decision Table Translator
DED	Data Element Descriptor|Dictionary
DEDAS	Direct Entry Dispatching Audio System
DEDB	Data Entry Data Base
DEDUCOM	Deductive Communicator

DEE	Data Encryption Equipment
DEE	Digital Electronic Exchange
DEE	Digital Evaluation Equipment
DEE	Direct Entry Equipment
DEE	Discrete Events Evaluator
def	defaults (file extension indicator)
def	definitions (file extension indicator)
DEF	Desktop Functional equivalent
DEF	Direct Equipment Failure
DEFL	Deflection
DEFL	Diode Emitter Follower Logic
DEFMEF	Definition language for MEBAS and EFFORD
DEFRAG	Defragment(ation)
DEFT	Dynamic Error-Free Transmission
deg	degree(s)
DEGC	Degrees in Celsius
DEGF	Degrees in Fahrenheit
DEGK	Degrees in Kelvin
DEI	Display Evaluative Index
DEK	Data Encryption Key
DEL	Delete (character)
DEL	Direct Exchange Line
DELS	Diagnostics through Error & Logic Simulation
DELSTR	Delete String
DELTRAC	Delay Line Transmission Converter
DELUA	Digital's Ethernet Low-power Unibus network Adapter
DEM	Demodulator
dem	demonstration (file extension indicator)
DEM	Digital Echo Modulation
DEM	Digital Elevation Map|Model
DEM	Dynamic Enterprise Modelling
DEMA	Data Entry Management Association (US)
DEMF	Display Exception Monitoring Facility
DEMOD	Demodulator
DEMON	Decision Mapping via Optimum go-nogo Networks
DEMS	Digital Electronic Message Service
DEMSE	Dynamic Enterprise Modelling Strategy-Execution
DEMUX	Demultiplex(er)
DEN	Density
DEN	Directory Enabled Network
DEN	Document Enabled Networking
DENIM	Directory-Enabled Network Infrastructure Model
dens	density
DEP	Data Entry Processor
DEP	Display Executive Program
DEP	Document Evaluation Program
DEPEX	Delft Product data Exchange center (NL)
DEPR	Depress

DEPS	Departmental Entry Processing System
DEPSK	Differentially Encoded Phase-Shift Keying
DEPT	Department
DEQ	Double-Ended Queue
DEQNA	Digital's Ethernet Q-bus Network Adapter
DER	Digital Event Recorder
DER	Disk Entry Record
DER	Distinguished Encoding Rules
DES	Data Encryption Standard
DES	Data Entry Sheet
DES	Data Entry Station|System
DES	Data Exchange System
DES	Descending
des	description (file extension indicator)
DES	Design and Evaluation System
DES	Differential Equation Solver
DES	Digital Editing|Expansion System
DES	Digital Encryption Standard
DES	Display Editing System
DES	Distributed End System
DES	Domino Extended Search
DESC	Dependency Selection Criteria
DESIRE	Design and Specification of Interacting Reasoning
DESIRE	Directory of European Information Security standard Requirements
DEST	Destination
DESTA	Digital's Ethernet thin-wire Station Adapter
DET	Detection
DET	Determinator
DET	Device Execute Trigger
DET	Direct Entry Terminal
DET	Directory Entry Table
DETAB	Decision Table(s)
DETD	Detected
DETOC	Decision Table To Cobol (translator)
DETR	Detector
DETRAN	Decision table Translator
DETS	Default Entry Type Standard
DEU	Data Encryption Unit
DEU	Data Entry|Exchange Unit
DEU	Direct Entry Unit
DEUNA	Digital's Ethernet to Unibus Network Adapter
DEV	Device
DEVAR	Device Address Register
DEVD	Device Description
DEVNO	Device Number
DEVS	Discrete-Event System Specification
DEXEC	Diagnostic Executive (software)

DF	Data Fetch|Field
DF	Data Flag
DF	Destination Field
D&F	Determination & Findings
DF	Device Flag
DF	Disk File|Free
DF	Dominance Factor
DF	Don't Fragment
DF	Double Frequency|Flag
DF	Drag Factor
DF	Drop Forge
DFA	Design For Assembly
DFA	Deterministic Finite Automaton
DFAD	Digital Feature Analysis Data
DFB	Distributed Feedback
D-FBM	Declarative Feature Based Modeller
DFBS	Daily Finish Build Schedule
DFC	Data Flow Chart|Control
DFC	Defect
DFC	Disk File Check
DFC	Dual Feed Carriage
DFCHIP	Data Flow Chip
DFCNV	Data File Conversion (program)
DFCU	Disk File Control Unit
DFD	Data Flow Diagram
DFD	Deadly Frequency Distortion
DFDS	Data Facility Device Support
DFDSS	Data Facility Data Set Services
DFE	Data Flow Editor
DFEF	Data Facility Extended Function
DFEP	Diagnostic Front End Processor
DFF	Display Format Facility
DFG	Diode Function Generator
DFHSM	Data Facility Hierarchical Storage Manager
DFI	Data File Interrogate
DFI	Digital Facility Interface
DFICT	Design For In-Circuit Testability
DFID	Data Format Identifier
DFL	Distributed Feedback Laser
DFL	Dynamic Function Language
DFLD	Device Field
DFLL	Data Flow Like Language
DFMS	Database/File Management System
DFMS	Digital Facility Management System
DFN	Deutsches ForschungsNetz (DE)
DFO	Direct File Operation
DFP	Data Facility Product
DFP	Digital Flat Panel (interface)

DFR	Decreasing Failure Rate
DFR	Disk File Read
DFR	Document Filing and Retrieval
DFR	Double Frequency Recording
DFS	Depth-First Search
DFS	Distributed File System
DFSG	Data Flow Sub-Graph
DFSK	Differential Frequency Shift Keying
DFSMS	Data Facility Storage Management Subsystem
DFSORT	Data Facility Sort
DFSU	Disk File Storage Unit
DFT	Defect and Fault Tolerance in very large scale integrated systems (international conference)
DFT	Destination Fetch Trigger
DFT	Diagnostic Fault|Function Test
DFT	Digital Facility Terminal
DFT	Discrete Fourier Transform(ation)
DFT	Distributed Function Terminal
DFU	Data File Utility
DFW	Disk File Write
DFWMAC	Distributed Foundation Wireless Medium Access Control
DG	Data General
DG	Display Gate
DGA	Direct Graphics Access
DGBC	Digital Geo-Ballistic Computer
DGBIT	Disagreement Bit
DGC	Diagnostic
DGCT	Direción General de Correos y Telegrafos (ES)
DG/DP	Differential Gain/Differential Phase
DGIS	Direct Graphics Interface Standard
DGL	Data Generation Language
DGL	Distributed Graphical Language
DGNS	Diagnosis
DGP	Dissimilar Gateway Protocol
DGPS	Differential Global Positioning System
DGSS	Distributed Group Support Systems
DGT	Degate
DGT	Digit
DH	Design Handbook
DH	Distribution Heterogeneity
DH	Document Handling
DHA	Destination Hardware Address
DHCF	Distributed Host Command Facility
DHCP	Dynamic Host Configuration Protocol
DHG	Digital Harmonic Generation
DHI	Disk Head Interference
DHL	Dynamic Head Loading
dhp	dr. halo pic(ture) (file extension indicator)

DHP	Drum Head Plug
DHS	Data Handling System
DHTML	Dynamic Hyper Text Markup Language
DI	Data Identifier
DI	Data In
DI	Data Interface|Item
DI	Delete Inhibit
DI	Destination Index
DI	Device Independent
DI	Digital Input|Impulse
DI	Direct Issue
DI	Discrete Input
DI	Distributed Intelligence
DI	Double Indexing
DIA	Document Interchange Architecture
DIAC	Diode A-C (switch)
DIAD	Drum Information Assembler/Dispatcher
DIA/DCA	Document Interchange Architecture/Document Content Architecture
DIAG	Diagnostic
DIAG	Diagram
DIAL	Direct Information Access Line
DIALIST	Diagnostic Listing
DIAM	Data Independent Accessing Model
DIAN	Digital Analog
DIANA	Descriptive Intermediate Attributed Notation for ADA
DIANA	Diagnostic Analyzer
DIANA	Digital and Analog Analysis program
DIANE	Direct Information Access Network for Europe
DIB	Data Input Bus
DIB	Data Inspection Board
DIB	Data Integrity Block
DIB	Device Independent Bitmap
DIB	Digital Input Basic
DIB	Directory Information Base
DIB	Disk Information Block
DIB	Dual Independent Bus
DIBI	Device Independent Backup Interface
DIBOL	Digital Business Oriented Language
DIC	Data Input Control
DIC	Data Interchange Code
DIC	Development Information Center
dic	dictionary (file extension indicator)
DIC	Digital Incremental Computer
DIC	Digital Input Channel
DIC	Digital Integrating Computer
DIC	Digital Interface Controller
DIC	Distributed Communications

DIC	Double Index Control
DICAM	Datasystem Interactive Communications Access Method
DICOM	Digital Imaging and Communications in Medicine
DICORTS	Digital Compare Recirculating Test System
DICOST	Diagnostic Control System
DICT	Dictionary
DID	Data Item Description
DID	Destination Identification|Identifier
DID	Digital Information Display
DID	Direct Inward Dial(ling)
DIDAC	Digital Data Communications
DIDAS	Digital Data Acquisition System
DIDC	Depository Institutions Deregulatory Committee
DI/DO	Data Input/Data Output
DIDOCS	Device-Independent Display Operator Console Support
DIDS	Decision Information Distribution System
DIDS	Digital Information Display System
DIDU	Disaster Dump
DI/E	Direct Import/Export
DIF	Data Interchange Facility|Format
DIF	Device Input Format
DIF	Difference
DIF	Differential
DIF	Digital Interface Frame
DIF	Directory Interchange Format
DIF	Display Information Facility
DIF	Distributed Interchange Facility
DIF	Document(ation) Interchange Format
DIF	Document Interchange Facility
DIF	Drawing Interchange Format
DIFF	Differential
DIFFSENS	Differential Sense
DiffServ	Differentiated Services
DIFMOS	Double Injection Floating gate Metal Oxide Silicon
DIFRACC	Digital Fractional Count Computer
DIG	Digit(al)
DIG	Digit Input Gate
DIGEM	Dig-It Emitter
DIGI	Deutsche Interessengemeinschaft Internet
DIGICOM	Digital Communication
DIGL	Device-Independent Graphics Layer
DIGS	Device-Independent Graphics Services
DIIC	Dielectrically Isolated Integrated Circuit
DIIG	Digital Information Infrastructure Guide
DIIP	Direct Interrupt Identification Port
DIL	Dual In-Line
DILOG	Distributed Logic
DIM	Data In the Middle

DIM	Dimension
DIMASZ	Dimension of Arrowhead Size
DIMDI	Deutsches Institut für Medizinische Dokumentation und Information
DIME	Dual Independent Map Encoding
DIMM	Dual In-line Memory Module(s)
DIMS	Data Information and Manufacturing System
DIMS	Document Image Management System
DIMSCALE	Dimension of overall Scale
DIMTXT	Dimension of Text height
DIMUS	Digital Multibeam Steering
DIN	Data Identification Number
DIN	Data In
DIN	Deutsches Institut für Normung
DINA	Direct Noise Amplification
DIN-NAM-IA	Deutsches Institut für Normung - Normenausschuss Maschinenbau - fachbereich Industrielle Automatisierung (DE)
DIO	Data Input-Output
DIO	Diode
DIO	Do It Ourselves
DIOCB	Device Input/Output Control Block
DIOF	Display Input/Output Facility
DIOS	Distributed Input/Output System
DIP	Diagnostic Interrogation Program
DIP	Dial-up Internet Protocol
DIP	Digital Imaging Processing
DIP	Digitizer Input
DIP	Display Information Processor
DIP	Distributed Processing
DIP	Document Image Processing
DIP	Dual In-line Package|Pin
DIPE	Distributed Interactive Processing Environment
DIPS	Development Information Processing System
DIPSTC	Datakit Internet Protocol/Streams Converter
DIR	Direct(ory)
DIRAC	Direct Access
DIRAM	Digital Range Machine
DIREC	Digital Rate Error Computer
DIRENT	Direct Entry
DIRS	Database Information Retrieval System
DIS	Design Information System
DIS	Digital Instrumentation Subsystem
DIS	Direct Information System
DIS	Distributed Intelligence System
DIS	Distributed Interactive Simulation
DIS	Divisional Information System
DIS	Document(al) Information System
DIS	Double Index Selection

DIS	Draft International Standard
DIS	Dynamic Impedance Stabilization
DISA	Data Interchange Standards Association
DISA	Direct Inward Switch |Systems Access
DISAC	Digital Simulator And Computer
DISAM	Direct Indexed-Sequential Access method
DISASM	Disassemble
DISBL	Disable
DISC	Disconnect
DISC	Disconnect Command
DISC	Dissemination Center
DISCA	Departament de Informatica de Sistemes, Computadors i Automatica (ES)
DISCUS	Data Interchange and Synergistic Collateral Usage System
DISEGS	Diagnostic Segments
DISERF	Data Interchange Standards Education and Research Foundation
DISH	Data Interchange for Shipping
DISH	Discrete Identifiable Silicone Handler
DISKUS	Dynamic Interactive Scheduling and Knowledge System
DISLAN	Display Language
DISM	Display Monitor
dism	dissimilar
DISN	Defense Information Systems Network
DISOCS	Distributed Office Communication System
DISOPE	Dynamic Integrated System Optimization and Parameter Estimation
DISOSS	Distributed Office Support(ed) System
DISP	Directory Information Shadow Protocol
DISP	Dispatcher
DISP	Displacement
DISP	Display
DISPN	Disposition
DISRP	Double Index Selection Register Party
DIST	Distributor
DIT	Data Input Technician
DIT	Directory Information Tree
DITRAN	Diagnostic Fortran
DITTO	Data Interfile Transfer, Testing and Operations
DIU	Digital Interface Unit
DIV	Data In Voice
DIV	Divide
DIVCHK	Divide Check
DIVE	Direct Interface Video Extension
DIVE	Distributed Interactive Virtual Environment
DIVOT	Digital-to-Voice Translator
DIW	D Inside Wire
DIY	Do It Yourself

DIZ	Description In Zip
dj	diffused junction
DJ	Djibouti
DK	Denmark
DK	Disk
DK	Don't Know
DKAP	Datakit Application Processor
DKB	Diagnostic Knowledge-Based system
DKDK	Don't Know that you Don't Know
DKK	Don't Know that you Know
DKS	Data Key System
DL	Data Language
DL	Data Length|Lock
DL	Data Link|List
DL	Delay Line
DL	Distribution List
DL	Disturbance Lines
DL	Download
DL	Dummy Load
DL/1	Data (manipulation) Language 1
DLA	Data Link Adapter
DLA	Direct Line Attachment
DLAB	Disk Label
DLAT	Directory Look-Aside Table
DLBB	Data Layer Building Block
DLC	Data Link Control(ler)|Connection
DLC	Digital Leased Circuit
DLC	Digital Loop Carrier
DLC	Duplex Line Control
DLCI	Data Link Connection|Control Identifier
DLCN	Distributed Loop Computer Network
DLC-RT	Digital Loop Carrier Remote Terminal
DLCS	Data Link Control System
DLCU	Digital Line Carrier Unit
DLD	Delay Line Driver
DLD	Display List Driver
DLE	Data Link Escape (character)
DLE	Digital Local Exchange
DLE	Distributed LAN Emulation
DLFM	Data Link File Manager
DLG	Digital Line Graph
DLI	Data Link Interface
DLIB	Distribution Library
DLL	Data Link Layer
DLL	Dial Long Lines
dll	dynamic link(ed) library (file extension indicator)
DLM	Data Line Monitor
DLM	Data Link Management

DLM	Distributed Lock Manager
DLM	Distributed Logic Memory
DLM	Dynamic Link Module
DLO	Data Line Occupied
DLOS	Dynamic Logic Simulation
DLP	Digital Light Processing
DLP	Distributed Logic Programming
DLP	Dynamic Limit Programming
DLPDU	Data Link Protocol Data Unit
DLPI	Data-Link Provider Interface
DLR	Delay Line Register
DLR	DOS LAN Requester (DOS = Disk Operating System)
DLS	Data Link Services
DLS	Data Logging System
DLS	Delay Line Storage
DLS	Device Level Selection
DLS	Digital Link Service
DLS	Digital Local Switch
DLSAP	Datalink Layer Service Access Point
DLSDU	Datalink Layer Service Data Unit
DLSE	Device Level Selection Enhanced
DLSW	Data Link Switch(ing)
DLT	Data Link Transceiver
DLT	Decision Logic Table|Translator
DLT	Digital Linear Tape
DLT	Disk Latency Time
DLT	Distributed Language Translation
DLT	Drum Latency Time
DLTU	Digital Line & Trunk Unit
DLUPG	Digital Line Unit-Pair Gain
DLY	Delay
DLYD	Delayed
DM	Data Management|Manager
DM	Data Mining
dm	decimeter
DM	Decision Maker
DM	Deferred Maintenance
DM	Delay|Delta Modulation
DM	Depeche Mode
DM	Design Manual
DM	Desktop Manufacturing
DM	Diagnostic Monitor
DM	Dialog Manager
DM	Differential Modulation
DM	Digital Modulation
DM	Disconnected Mode
DM	Distributed Memory
DM	Dominica

DMA	Digital Model Assembler
DMA	Direct Memory Access|Address(ing)
DMA	Document Management Alliance
DMA	Drum Memory Adapter|Assembly
DMA	Dynamic Memory Access
DMAC	Direct Memory Access Channel|Controller
DMACS	Distributed Manufacturing Automation and Control Software
DMAPS	Digital Manufacturing Process System
DMARA	Dynamic Multicast Address Relay Agent
DMAS	Distributors Management Accounting System
DMB	Data Management Block
DMC	Data Management Component|Control
DMC	Discrete Memory-less Channel
DMCA	Digital Millennium Copyright Act
DMCC	Dual Multiple Column Control (language)
DMCL	Device Media Control Language
DMD	Deformable Mirror Device
DMD	Digital Micro-mirror Device
DMD	Directory Management Domain
DMDC	Dual Module Display and Control (unit)
DMDD	Distributed Multiplexing Distributed Demultiplexing
DME	Distributed Management Environment
DMED	Digital Message Entry Device
DMEDS	Distributed Multimedia Electronic Document System
DMF	Development Master File
DMF	Distribution Media Format
DMH	Device Message Handler
DMI	Definition of Management Information
DMI	Desktop Management Interface
DMI	Digital Multiplexed Interface
DMI	Dual Mode Ignition
DMIG	Data Migration Interface Group
DMIS	Dimensional Measurement Interface Standard
DMIS	Dimensional Measuring Interface Specification
DML	Data Management Language
DML	Data Manipulation Language|Logic
DML	Digital Monitor Logic
DM-M	Delay Modulation Mark
DMM	Digital Multi-Meter
DMM	Domain Modelling Module
DMMP	Distributed Memory, Message Passing (system)
DMMS	Dynamic Memory Management System
DMMU	Data Memory Management Unit
DMN	Digital Multimedia Network
DMO	Data Management Officer
DMO	Domain Management Organization
DMOS	Diffusion Metal Oxide Semiconductor
DMOS	Double diffused Metal-Oxide Semiconductor

DMOSFET	Depletion Metal Oxide Silicon Field Effect Transistor
DMOX	Dynamic Modelling Open Extensions
DMP	Data Mapping Program
DMP	Diagnostic and Monitoring Protocol
DMP	Disk Management Program
DMP	Dot Matrix Printer
DMPC	Distributed Memory Parallel Computer
DMPDU	Derived Medium access control Protocol Data Unit
DMPL	Digital Microprocessor Plotter Language
DMPS	Data Management Programming System
DMQS	Display Mode Query and Set
DMR	Data Management Routine(s)
DMR	Deuteron Magnetic Resonance
DMRS	Database Management Retrieval System
DMS	Data Management Software|System
DM-S	Delay Modulation Space
DMS	Desktop Management Services
DMS	Development Management System
DMS	Dialog Management Services
DMS	Digital Multiplexed System
DMS	Diskette Mass Storage
DMS	Diskless Management Services
DMS	Disk Management|Monitor System
DMS	Display Management System
DMS	Distribution Management System
DMSD	Digital Multi-Standard Decoding
DMSP	Distributed Mail Service Protocol
DMSU	Digital Main Switching Unit (GB)
DMT	Device Mask Table
DMT	Digital|Discrete Multi-Tone
DMT	Discrete Multitone Technology
DMTF	Desktop Management Task Force
DMTS	Dynamic Multi-Tasking System
DMU	Data Management|Manipulation Unit
DMU	Digital Mock-Up
DMU	Display Monitor Unit
DMUX	Demultiplexer
DMV	Digital Multiplex Vocoder
DMX	Demultiplexing
DMX	Direct Music Express
DMY	Day, Month, Year
DN	Data Name|Number
DN	Data Net(work)
DN	Diagnostic (programs)
DN	Digital Number
DN	Directory Number
DN	Down
DN-1	Datanet 1

DNA	Digital Network Architecture
DNA	Distributed (inter)Net Applications
DNA	Distributed Network Architecture
DNA-M	Distributed (inter)Net Applications for Manufacturing
DNC	Direct Numerical Control
DNC	Distributed Network Control
DNC	Dynamic Network Controller
DNDG	Dynamic Network Data Generator
DNDS	Distributed Network Design System
DNF	Disjunctive Normal Form
DNF	Do Not Fill
DNHR	Dynamic Non-Hierarchical Routing
DNIC	Data Network Identification Code
DNIC	Destination Network Identification Code
DNIS	Dialled Number Identification Service
DNL	Do Not Load
DNL	Dynamic Noise Limiter
DNP	Distributed Network Processing
DNR	Data Network Routing
DNR	Device Not Ready
DNR	Dialled Number Recorder
DNRC	Domain Name Right Coalition
DNS	Data Network Supervisor
DNS	Direct Numerical Simulation
DNS	Distributed Name Service
DNS	Domain Name Server
DNS	Domain Name Service(s)|System
DNS	Domain Naming Server|System
DNS	Dynamic Noise Suppression
DNT	Device Name Table
DNX	Data Net Exchange
DNX	Dynamic Network X-connect
DO	Data Out
D/O	Delivery Order
DO	Digital Output
DO	Discrete Output
DO	Distributed Objects
DO	Dominican Republic
DO	Drop Out
DOAM	Distributed Office Applications Model
DOAS	Differential Optical Absorption Spectroscopy
DOB	Disbursed Operating Base
DOC	Display Operator Console
DOC	Distributed Object Computing
doc	document(ation) (file extension indicator)
DOC	Dynamic Overload Control
DOCS	Display Operator Console System
DOCS	Dynamic Operations Control System

DOCSIS	Data Over Cable System Interface Specification
DOCUS	Display Oriented Computer Usage System
DOD	Department Of Defense (US)
DOD	Digital Optical Device
DOD	Direct Outward Dialling
DOD	Drop On Demand
DOE	Data Origination Event
DOE	Distributed Object Environment|Everywhere
DOES	Direct Order Entry System
DOF	Depth Of Field
DOF	Device Output Format
DOFICS	Domain-Originated Functional Integrated Circuits
DOFLT	Date Of Last (issue)
DOGS	Drawing Office Graphics System
DOHS	Disconnected Operation Handling System
DOIMP	Drop-Out Impulse
DO/IT	Digital Output/Input Translator
DO-IT	Disabilities, Opportunities, Internetworking and Technology
DOL	Design Oriented Language
DOLAP	Desktop On Line Analytical Processing
DOLARS	Disk On-Line Accounts Receivable System
DOLPRO	Designer Oriented Language Program
DOLS	Domino Off-Line Services
DOLT	Delay Oriented Logic Tester
DOM	Data On Master group
DOM	Data Output Multiplexer
DOM	Distributed Operations Manager
DOM	Document Object Model
DOM	Domestic
DOMAIN	Distributed Operating Multi-Access Interactive Network
DOMF	Distributed Object Management Facility
DOMINA	Distribution Oriented Management Information Analyzer
DOMS	Distributed Object Management System
DOMSAT	Domestic Satellite (service)
DONACS	Department Of the Navy Automation and Communication System
DONUT	Digitally Operated Network Using Threshold
DOOD	Deductive Object-Oriented Database
DOORS	Dynamic Object Oriented Requirements System
DOP	Directory Operational Protocol
DOPA	Dynamic Output Printer Analyzer
DOPIC	Documentation Of Programs In Core
DOPLOC	Doppler Phase Lock
DOPS	Distributed Office Processing System
DOQ	Digital Orthophoto Quadrangle
DOR	Digital Optical Recorder|Recording
DOR	Digital Output Relay
DORACE	Design Organization, Record, Analyze, Charge, Estimate

DORE	Dynamic Object Rendering Environment
DORI	Displace on Order – Replace Installed
DORIS	Direct Order Recording and Invoicing System
DORO	Displace on Order – Replace on Order
DoS	Denial of Service
DOS	Disk Operating|Oriented System
DOS	Distributed Office System
DOSEM	Disk Operating System Emulation
DOSES	Development Of Statistical Expert Systems
DOSF	Distributed Office Support Facility
DOSP	Disk Operating System Prime
DOSPT	Disk Operating System Performance Tool
DOSS	Disk Operating System – Standard
DOSS	Distributed Office Support System
DOS/VS	Disk Operating System with Virtual Storage
DOS/VSE	Disk Operating System with Virtual Storage Extended
DOT	Data Organizing Translator
DOT	Domain Tips
DOTS	Digital Office Timing Supply
DOUT	Data Out
DOV	Data Over Voice
DOVE	Data Over Voice Equipment
DOW	Day Of Week
DOY	Day Of Year
DOZ	Dozen
DP	Data Print
DP	Data Processing|Processor
DP	Decimal Packed
DP	Diagnostic Power
DP	Dial Pulse
DP	Diametral Pitch
DP	Display Postscript
DP	Distribution Point
DP	Documentation Program
DP	Domain Process
DP	D-Pulse
DP	Draft Proposal
DP	Dynamic Program(ming)
DPA	Data Processing Application
DPA	Demand Protocol Architecture
DPA	Digital Pulse Analyzer
DPA	Display/Printer Adapter
DPA	Document Printing Application
DPAGE	Device Page
DPAIS	Data Processing Advanced Information System
DPAM	Demand Priority Access Method
dpANS	draft proposed American National Standard
DPAO	Data Processing Administrative Operation

DPAP	Data Processing Administrative Procedure
DPAP	Data Processing Asset Protection
DPAREN	Data Parity Enable
DPB	Drive Parameter Block
DPB	Dynamic Pool Block
DPC	Data Processing Center
DPC	Deferred Procedure Call
DPC	Destination Point Code
DPC	Direct Program Control
DPC	Distributed Processing Computer
DPCC	Data Processing Control Center
DPCE	Data Processing Customer Engineering
DPCM	Differential Pulse Code Modulation
DPCS	Data Processing and Communications System
DPCX	Distributed Processing Control Executive
DPD	Data Processing Division
DPD	Decimal Point Digit
DPD	Digit Plane Driver
DPDT	Double Pole Double Throw
DPE	Data Path Extender
DPE	Data Processing Equipment
DPED	Data Processing Education Department
DPEM	Data Processing Equipment Manufacturers
DPF	Detailed Program Flowchart
DPF	Dual Programming Feature
DPG	Dedicated Packet Group
DPG	Diagnostic Programming Group
DPGA	Dynamically reconfigurable field-Programmable Gate Array
dpi	dots per inch
DPIC	Data Processing Installation|Inventory Control
DPIN	Drop Pin
DPIO	Data Processing Input/Output
DPL	Dataless Programming Language
DPL	Data Processing Language
DPL	Decision Programming Language
DPL	Dedicated Private Line
DPL	Descriptor Privilege Level
dpl	dots per line
DPL	Dynamic Program Loading
DPLIS	Development Pilot Line Information System
DPLL	Digital Phase-Locked Loop
DPLY	Display
DPM	Data Processing Machine|Manager
dpm	defects per million
DPM	Digital Panel|Power Meter
DPM	Directional Policy Matrix
dpm	disintegrations per minute
DPM	Distributed Presentation Management

dpm	documents per minute
DPMA	Data Processing Management Association (US)
DPMI	DOS Protected Mode Interface (DOS = Disk Operating System)
DPMIS	Data Processing Management Information System
DPMO	Data Processing Machine Order
DPMS	Display Power Management Support
DPMS	Distributed Plant Management System
DPMS	DOS Protected Mode Services (DOS = Disk Operating System)
DPN	Dual Processing Node
DPNPH	Data Packet Network-Packet Handler
DPNSS	Digital Private Network Signalling System
DPO	Data Phase Optimization
DPO	Data Processing Operation(s)
DPO	Dial Pulse Originating
DPO	Digital Processing Oscilloscope
DPOM	Data Processing Orders and Movements
DPOW	Data Processing Order Worksheet
DPP	Data Processing Policy
DPP	Divisional Programming Practice
DPP	Dynamic Production Planning
DPPO	Data Processing Productivity Office
DPPX	Distributed Processing Programming Executive
DPPX/SP	Distributed Processing Programming Executive/Systems Product
DPR	Digit Present
DPR	Double Pulse Recording
DPR	Dual Port RAM (= Random Access Memory)
DPR	Dynamic Path Reconnect
DPR	Dynamic Program Relocation
DPS	Data Presentation|Processing System
DPS	Decentralized|Departmental Processing System
DPS	Digital Processing System
dps	disintegrations per second
DPS	Disk Programming System
DPS	Display PostScript
DPS	Distributed Parameter|Processing System(s)
DPS	Distributed Presentation Services
DPS	Divisional Programming Specification|Standard
DPS	Document Processing System
DPS	Dynamic Path Selection
DPSC	Data Processing Service Center
DPSK	Differential Phase Shift Keying
DPSPAR	Digital Processing Systems Personal Animation Recorder
DPSS	Data Processing Sub-System
DPSS	Data Processing Support Services
DPSS	Data Processing Systems Support
DPSS	Diagnostic Program Sub-System
DPST	Double Pole Single Throw

DPT	Data Processing Technique|Terminal
DPT	Dial Pulse Terminating
DPT	Dioptre
DPtoTP	Display Points to Tablet Points (converting)
dpu	defects per unit
DPU	Digit Pick-Up
DPU	Display Processing Unit
DQ	Dormant Queue
DQA	Data Quality Assessment
DQCB	Disk Queue Control Block
DQDB	Distributed Queue Dual Bus
DQE	Descriptor Queue Element
DQL	DataEase Query Language
DQM	Dormant Queue Manager
DQO	Data Quality Objectives
DR	Data Rate
DR	Data Ready|Receive(d)
DR	Data Record(ing)
DR	Data Register|Report
DR	Dead Reckoning
DR	Deficiency Report
DR	Delivery Report
DR	Density Ratio
DR	Device Ready
DR	Diagnostic Resolution
DR	Diode Rectifier
D/R	Direct or Reverse
DR	Direct Resist
DR	Disconnect Request
dr	dram
DR	Drive
DRA	Digital Read-in Assembly
DRA	Drum Read Amplifier
DRAC	Dell Remote Assistant Card
DRAM	Dynamic Random Access Memory
DRAMI	Digital Range Measuring Instrument
DRAPE	Data Reduction And Processing Equipment
DRAPE	Digital Recording And Playback Equipment
DRAW	Direct Read After Write
DRBOND	Dial-up Router Bandwidth On Demand
DRC	Data Recording Control
DRC	Data Recovery Center
DRC	Directory Reporting Console
DRCS	Dynamically Redefinable Character Set
DRD	Data Reading|Recording Device
DRD	Demand Return Disposal
DRDA	Distributed Relational Data Architecture
DRDBMS	Distributed Relational Data Base Management System

DRDOS	Digital Research Disk Operating System
DRDS	Dynamic Reconfiguration Data Set
DRDW	Direct Read During Write
DREAC	Drum Experimental Automatic Computer
DREAM	Digital Recording And Measurement
DRI	Data Reduction Interpreter
DRILL	Direct Routing Investigation of Line Layouts
DRL	Divisional Records List
DRL	Double Rail Logic
DRM	Diagnostic Record Matching
DRM	Dynamic Resources Management
DRMS	Digital Radiation Monitoring System
DRMU	Digital Remote Measurement Unit
DRMV	Digital Rights Management for Video
DRN	Data Reference Number
DRN	Data Routing Network
DRO	Data Request Output
DRO	Destructive Read-Out
DRO	Digital Read-Out
DRON	Data Reduction
DROS	Disk Remote Operating System
DRP	Directory Replication Protocol
DRP	Distribution Resource Planning
DRPF	Decimal Reference Publication Format
DRQ	Data Ready Queue
DRR	Data Recall Register
DRR	Display Refreshing Rate
DRRS	Direct Reading Ratio Set
DRS	Data Rate Select(or)
DRS	Data Receiving|Retrieval System
DRS	Deficiency Reporting System
DRS	Direct Release System
DRS	Distributed Resource System
drs	driver resource (file extension indicator)
DRTS	Distributed Real-Time System
DRUMS	Defeasible Reasoning and Uncertainty Management Systems (EU)
drv	device driver (file extension indicator)
DRV	Drive
DRW	Drawing (exchange format)
drw	drawing (file extension indicator)
DS	Dansk Standardiseringsrad
DS	Data Segment|Send
DS	Data Server|Storage
DS	Data Set|Stream
DS	Data Specification|Structure
DS	Data Synchronization|Strobe
DS	Data System(s)

DS	Days after Sight
DS	Demand Scanner
DS	Design Specification
DS	Detailed Routings
d/s	dhrystone per second
DS	Digital carrier Span
DS	Digital Select(or)|Section
DS	Digital Service|System
DS	Digital Signal|Signature
DS	Diode Switch
DS	Directory Service
DS	Direct Signal
DS	Disk Storage
DS	Display Station
DS	Distributed Single layer
DS	Distributed System
DS	Document Storage
DS	Double Sided
DS	Draft Standard
DS	Dynamic Slack
DSA	Data Set Adapter
DSA	Data Stack Address
DSA	Data Systems Automation
DSA	Dedicated Switched Access
DSA	Demand Statement Analysis
DSA	Destination Software Address
DSA	Dial Service|System Assistance
DSA	Digital Signature Algorithm
DSA	Digital Storage Architecture
DSA	Directory Service Agent
DSA	Distributed Systems Architecture
DSA	Dynamic Storage Allocation|Area
DSAB	Distributed Systems Architecture Board
DSAC	Data Set Authority Credential
DSAD	Direct Storage Access Device
DSAF	Destination Sub-Area Field
DSAM	Direct Sequential Access Method
DSAP	Datalink Service Access Point
DSAP	Data Structures and Application Protocols
DSAP	Destination Service Access Point
DSB	Data Set Block
DSB	Double Side Band
DSBAM	Double Side Band Amplitude Modulation
DSBFC	Double Side Band Full Carrier
DSBS	Dedicated Sensor-Based System
DSBSC	Double Side Band Suppressed Carrier
DSC	Data Set Control
DSC	Data Stream Compatibility

DSC	Data Streaming Channel
DSC	Data Synchronizing Channel
DSC	Design Stability Code
DSC	Digital Selective Calling
DSC	Digit Select Character
DSC	Digit Selector Common
DSC	Direct Satellite Communications
DSC	Distributed Service Coordinator
dsc	double silk covered
DSC	Dynamic Standby Computer
DSCA	Default System Control Area
DSCB	Data Set Control Block
DSCN	Discontinued
DSCP	Data Services Command Processor
DSCP	Disk System Control Processor
DSD	Data Set|Structure Definition
DSD	Data Structure Diagram
DSD	Direct Store Delivery
DSDC	Direct Service Dialling Capability
DS/DD	Double Sided/Double Density
DSDL	Data Storage Description Language
DSDM	Dynamic Systems Development Method
DSDP	Deformographic Storage Display (tube)
DSDS	Dataphone Switched Digital Service
DSDT	Data Set Definition Table
DSE	Data Set Extension
DSE	Data Storage Equipment
DSE	Data Structure Editor
DSE	Data Switching Equipment|Exchange
DSE	Digit Selector Emitter
DSE	Distributed Single layer Embedded
DSE	Distributed System Environment
DSE	DSA Specific Entry (DSA = Dedicated Switched Access)
DSEA	Display Station Emulation Adapter
DSECT	Dummy control Section
DS/ED	Double Sided/Enhanced Density
DSEE	Domain Software Engineering Environment
DSEL	Deselect
DSELCY	Deselect Cycle
DSENQ	Data Set Enquire (table)
DSERV	Directory Service
DSGN	Design
DSH	Digital Synchronous Hierarchy
DS/HD	Double Sided/High Density
DSI	Data Stream Interface
DSI	Digital Speech Interpolation
DSIA	Digital Systems Interconnect Architecture
DSID	Data Set Identification

DSIM	Diagnostic fault Simulation (procedure)
DSIS	Distributed Support Information Standard
DSK	Down Stream Keyer
DSL	Data Set Label
DSL	Data Specification Library
DSL	Data Structure|Simulation Language
DSL	Delivered Source Lines
DSL	Design Language
DSL	Digital Simulation Language
DSL	Digital Subscriber Line
DSL	Dynamic Simulation Language
DSLAM	Digital Subscriber Line Access Multiplexer
DSLO	Distributed System License Option
DSM	Data Services Manager
DSM	Data Specification Methodology
DSM	Deep Sub-Micron
DSM	Delta Sigma Modulation
DSM	Design Specification Model
DSM	Design Station Manager
DSM	Digital Scope Multimeter
DSM	Direct Swap Memory
DSM	Discrete Source with Memory
DSM	Disk Sort/Merge
DSM	Disk System Management
DSM	Distributed Switching Matrix
DSM	Dynamic Scattering Mode
DSMA	Digital Sense Multiple Access
DSMA	Distributed Scheduling Multiple Access
DSMCC	Digital Storage Management Command and Control
DSMCC	Digital Storage Media Command (and) Control
DSN	Data Set Name
DSN	Deep Space Network
DSN	Digital Signal (level) N
DSN	Distributed Systems Network
DS/ND	Double Sided/Normal Density
DSNT	Data Set Name Table
DSNX	Distributed System Node Executive
DSO	Data Set Optimizer
DSO	Digital Storage Oscilloscope
DSO	Direct System Output
DSOM	Distributed System Object Model
DSOM	Distributed Systems, Operations and Management (workshop)
DSP	Digital Signal Processor
DSP	Directory System Protocol
DSP	Display
DSP	Display System Protocol
DSP	Distributed System Program
DSP	Domain Specific Part

DSP	Double Silver Plated
DSP	Dynamic Support Program
DSPAR	Distributed System Partition
DSPChip	Digital Signal Processing Chip
DSPICE	D-variant of SPICE
DSPMT	Displacement
DSPN	Deterministic Stochastic Petri Net
DSPT	Decision Support Problem Technique
DSPU	Downstream Physical Unit
DSQD	Double Sided, Quad Density
DSR	Data Set Ready
DSR	Data Signalling|Set Rate
DSR	Device Service Routine
DSR	Device Status Register|Report
DSR	Digital Standard Runoff
DSR	Dynamic Segment Relocation
DSR	Dynamic Service Register
DSR	Dynamic Status Recording
DSR	Dynamic Storage Relocation
DSRB	Data Services Request Block
DSRI	Data Set to Record Interface
DSRI	Digital Standard Relational Interface
DSRS	Data Signalling Rate Selector
DSS	Data Specification System
DSS	Data Station Selector
DSS	Data Storage Subsystem
DSS	Decision Support System
DSS	Deliverable Specification Statement
DSS	Device Support Station
DSS	Digital Satellite System
DSS	Digital Signal|Signature Standard
DSS	Digital Spread Spectrum
DSS	Digital Subscriber Service
DSS	Direct Station Selection
DSS	Direct Synchronous Sensor
DSS	Disk Support System
DSS	Display Sub-System
DSS	Distribution Scheduling System
DSS	Distribution System Simulator
DSS	Dynamic Support System
DSSA	Distributed Systems Security Architecture
DSS/BLF	Direct Station Selection/Busy Lamp Field
DSSC	Double Sideband, Suppressed Carrier
DS/SD	Double Sided/Single Density
DSSI	Digital Storage System Interconnect
DSSS	Direct-Sequence Spread Spectrum
DSSSL	Document Style, Semantics, and Specification Language
DST	Data Services Task

DST	Dedicated Service Tools
DST	Destination
DST	Device Service Task
DSTE	Data Subscriber Terminal Equipment
DSTN	Destination
DSTN	Double Super Twisted Nematic
DSTR	Distort
DSU	Data Sequentializer Unit
DSU	Data Service(s) Unit
DSU	Digital Service Unit
DSU	Disk Subsystem Unit
DSU	Drum Storage Unit
DSU/CSU	Data Service Unit/Channel Service Unit
DSVD	Digital Simultaneous Voice and Data
DSW	Data Status Word
DSW	Device Status Word
DSx	Digital Signal, level x (x = 0, 1, 1C, 2, 3, or 4)
DSX	Digital Signal|System Cross-connect
DSX	Distributed Systems Executive
DSX1/3	Digital Signal Cross-connect between Levels 1 and 3
DT	Data
DT	Data Table|Terminal
DT	Data Text|Transfer
DT	Data Transmission|Transmit
DT	Detection Threshold
DT	Dial Tone
dt	differential of time
DT	Diffusion Transfer
DT	Di-group Terminal
DT	Direct Transfer
DT	Disk to Tape
DT	Documentation Terminology
DT	Double Throw
dta	data (file extension indicator)
DTA	Differential Thermal Analysis
DTA	Disk Transfer Area
DTAM	Document Transfer, Access, and Manipulation
DTAS	Data Transmission And Switching system
DTAS	Digital Test Access System
DTB	Data Transmission Block
DTB	Digit Transfer Bus
DTC	Data Terminal Configuration
DTC	Desk Top Computer
DTC	Digital Trunk Controller
DTC	Di-group Terminal Controller
DTC	Direct Traffic Control
DTC	Distributed Terminal Controller
DTC	Distributed Transaction Controller|Coordinator

DTC	Distribution Traffic Control
DTC	Dynamic tape Tension Control
DTCU	Data Transmission Control Unit
DTD	Disk To Disk
DTD	Document Type Definition
DTE	Data Terminal Equipment
DTE	Data Transmission Equipment|Exchange
DTE	Deep Texture Editor
DTE	Display Tester Element
DTE	Dumb Terminal Emulator
DTF	Define The File|Format
DTF	Definite Type File
DTF	Distributed Test Facility
DTF	Document Transmission Facility
DTFT	Discrete-Time Fourier Transform
DTG	Date-Time Group
DTG	Direct Trunk Group
DTH	Direct To Home
DTI	Desktop Imaging
DTI	Direct Trader Input
DTI	Display Terminal Interchange
DTI	Distributed Target Intelligence
DTIF	Digital Transmission Interface Frame
DTL	Detail
DTL	Dialog Tag Language
DTL	Diode-Transistor Logic
DTL	Disburse To Location
DTLS	Descriptive Top-Level Specification language
DTM	Digital Terrain Model
DTM	Dynamic synchronous Transfer Mode
DTM	Dynamic Transient Master
DTMCB	Dynamic Transient Master Control Block
DTMF	Data Tone Multiple Frequency
DTMF	Dial Tone Multi-Frequency (signalling)
DTMS	Database and Transaction Management System
DTn	Double Tinned
DTP	Data(gram) Transfer Protocol(s)
DTP	Desk-Top Publishing
DTP	Discrete Transient Protection
DTP	Disk To Printer
DTP	Distributed Transaction Process(ing)
DTP	DocuTech Printer
DTP	Drum Timing Pulse
DTPA	Dynamic Transient Pool Area
DTPD	Divider Time Pulse Distributor
DTPL	Domain Tip Propagation Logic
DTPM	Distributed Transaction-Processing Middleware
DTPM	Dynamic Transient Pool Management

DTQP Desk-Top Quality Publishing

DTR Data Terminal Ready

DTR Data Transfer|Transmission Rate

DTR Disposable|Distribution Tape Reel

DTRT Do The Right Thing

DTS Data Terminal Set|System

DTS Data Transformation Services

DTS Data Transmission|Transfer System

DTS Defect Tracking System

DTS Digital Telemetry|Transmission System

DTS Digital Termination Service|Systems

DTS Dynamic Transient Segment

DTSC Digital Transit Switching Centre

DTSC Drum Test Self Check (routine)

DTSR Datakit Terminal Send/Receive

DTSRS Dynamic Transient Segment Register Save

DTSS Dartmouth Time-Sharing System

DTSX Data Transport Station for X.25

DTT Digital Terrestrial Television

DTT Disk To Tape

DTT Domestic Transit Time

DTTDD Disk-to-Tape and Tape-to-Disk Dump

DTTU Data Transmission Terminal Unit

DTU Data Transfer|Transformation Unit

DTU Data Transmission Unit

DTU Di-group Terminal Unit

DTU Download The Universe

DTUL Deflection Temperature Under Load

DTUPC Design-To-Unit Production Cost

DTV Desktop Video

DTV Digital Television

DTVC Desktop Video Conferencing

DTW Data Word

DTX Data Throughput

DU Data Unit

DU Decision Unit

DU Destination Unknown

DU Disk Usage

DU Dobson Unit

DU Duty (cycle)

DUA Directory User Agent

DUAT(S) Direct User Access Terminal (System)

DUC Data Unit Control

DUCS Display Unit Control System

DUE DOD User Environment (DOD = Department Of Defense)

DUF Diffusion Under epitaxial Film

DUI Duration of Unscheduled Interrupt(ion)

DUMPGEN Dump Generation (program)

DUNS	Data Universal Numbering System
DUP	Data User Part(s)
DUP	Device|Disk Utility Program
DUP	Duplication
DUS	Diagnostic and Utility System
DUS	Disk Utility System
DUSD	Data Users Services Division
DUV	Data Under Voice
DV	Dependent Variable
dv	desk view (script) (file extension indicator)
DVACS	Data Verification, Access and Control System
DVB	Device Vector Base
DVB	Digital Video Broadcast
DVB/DAVIC	Digital Video Broadcasting/Digital Audio-Visual Council
DVC	Digital|Desktop Video Conferencing
DVC	Digital Video Cartridge|Cassette
DVC	Digital Voice Card
DVCDN	Device Down (console command)
DVCID	Device Identifier
DVCP	Digital Videotex Communication Processing unit
DVCUP	Device Up (console command)
DVD	Digital Versatile|Video Disc
DVD	Direct View Display
DVD	Dividend
DVD	Dope Vector Description
DVD-R	Digital Versatile Disc – Recordable
DVD-RAM	Digital Versatile Disc – Random Access Memory
DVD-ROM	Digital Versatile Disc – Read Only Memory
DVD+RW	Digital Versatile Disc + Rewritable
DVE	Device End
DVE	Digital Video Effect(s)
dvi	device-independent (file extension indicator)
DVI	Digital Video(tex) Interactive
DVIPS	Device-Independent (format to) PostScript
DVMA	Direct Virtual Memory Access
DVMRP	Distance Vector Multi-cast Routing Protocol
DVOM	Digital Volt-Ohm-Milliammeter
dvr	device driver (file extension indicator)
DVR	Digital Video Recording
DVR	Divisor
DVS	Digital Voltage Source
DVS	Dynamic Volatile Storage
DVST	Direct View Storage Tube
DVT	Device Vector Table
DVTR	Digital Video Tape Recorder
DVX	Digital Voice Exchange
DW	Daisey Wheel
DW	Data Warehousing|Word

DW	Display Write(r)
DW	Double Word
DW	Drum Writer
DWB	Documenter's Work-Bench
DWC	Data Word Cycle
DWDM	Dense Wave(length) Division Multiplexing
DWEL	Discrete Wire Equivalence List
DWG	Drawing
DWIM	Do What I Mean
DWMT	Discrete Wavelet Multi-Tone
DWP	Daisy Wheel Printer
DWR	Display Writer
DWS	Display Writer System
DWT	Discrete Wavelet Transformation
DX	Data Exchange|Explorer
DX	Direct Current (signalling)
DX	Distance
DX	Duplex
DXAM	Distributed Indexed Access Method
dxb	drawing interchange binary (file extension indicator)
DXC	Data Exchange Control
DXC	Digital|Direct Cross-Connect
DXF	Data|Design Exchange Format
DXF	Digital|Drawing Exchange Format
DXP	Data Exchange Program
DXS	Data Exchange System
DXT	Data Extract (facility)
DXU	Data Exchange Unit
DY	Day
DYANA	Dynamic interpretation of Natural language (EU)
DYCMOS	Dynamic Complementary Metal-Oxide Semiconductor
DYN	Did You Notice
dyn	dyne
DYNDADIS	Dynamic Data Display System
DYSAC	Digitally Simulated Analog Computer
DYSTAL	Dynamic Storage Allocation
DZ	Algeria
DZ	Decimal Zoned
DZ	Dozen
DZ	Drop Zone
DZT	Digit(al) Zero Trigger

E

E	Electrical charge
e	emitter
E	Enable
E	Energy

e	exponent
E1	European standard transmission speed of 2048 Mb/s
E2PROM	Electrically Erasable Programable Read Only Memory
EA	Effective Address
EA	Element|Enterprise Activity
EA	Energy Analysis
EA	Event Action
EA	Extended Attribute
EAASY	Educator's Automated Authoring System
EAB	Effective Address Buffer
EABM	Electronically Addressable Bulk Memory
EAC	End-Around Carry
EACL	European chapter of the Association for Computational Linguistics
EAD	Estimated Availability Date
EADAS	Engineering and Administration Data Acquisition System
EADASNM	Engineering and Administrative Data Acquisition System/Network Management
EAE	Extended Arithmetic Element
EAEO	Equal Access End Office
EAGLES	Expert Advisory Group on Language Engineering Standards
EAI	Enterprise Application Integration
EAL	Electrical Analysis Laboratory
EAL	Electromagnetic Amplifying Lens
EALI	Electrical Attraction of Liquid Inks
EAM	Electronic Accounting Machine
EAM	Enterprise Asset Management
EAN	European Academic Network
EAN	European Article Number(ing)
EAO	Extended Architecture Offering
EAP	Emulator Application Program
EAP	Extended Arithmetic Processor
EAPROM	Electrically Alterable Programmable Read Only Memory
EAR	Effective|Extended Address Register
EAR	External Access Register
EARL	Easy Access Report Language
EARN	European Academic and Research Network
EAROM	Electrically Alterable Read-Only Memory
EARS	Electronic Access to Reference Services
EAS	Electronic Accounting System
EAS	Engineering Administration System
EAS	Extended|Extensive Area Service
EASAL	Easy Application Language
EASD	Equal Access Service Date
EASE	Energy Analysis Scattered Electrons
EASE	Engineering Automatic systems for Solving Equations
EASE-Grid	Equal Area SSM/I Earth Grid
EASI	Electronic Accounting for the Securities Industries

EASI	Enhanced Asynchronous SCSI Interface
EASI	Epioptics Applied to Semiconductor Interfaces (EU)
EASI	European Academic Supercomputer Initiative
EASINet	European Academic Supercomputer Initiative Network
EASY	Efficient Assembly System
EASY	Exception Analysis System
EAT	Entity Alignment Time
EATA	Enhanced AT-bus Attachment (AT = Automatic Test)
EAU	Erase All Unprotected
EAU	Extended Arithmetic Unit
EAV	Extended Application Verification
EAX	Electronic Automatic Exchange
EB	Electron Beam
EB	Elementary Block
EB	End Bracket
EB	End of Block
EB	Entry Block
EB	Erlang B
EB	Extended Bintree
EBAM	Electron Beam Addressable Memory
EBAM	Electron Beam Addressable Metal-oxide semiconductor
EBAN	Electronic Business Assurance Network
EBC	EISA Bus Controller (EISA = Extended Industry Standard Architecture)
EBC	Electronic Business Communications
EBCD	Extended Binary-Coded Decimal
EBCDIC	Extended Binary-Coded Decimal Interchange Code
EBCS	Electronic Business Communications System
EBDI	Electronic Business Data Interchange
EBEAM	Electron Beam
EBES	Electron Beam Exposure System
EBHC	Equated Busy Hour Calls
EBI	Equivalent Background Input
EBIC	Electron Beam Internal Compressed
EBIOS	Extended Basic Input/Output System
EBIP	Enterprise Business Intelligence Portals
EBIT	Electron Beam Injection Transistor
EBIT	European Broadband Intercommunication Trial
EBL	Explanation Based Learning
EBM	Electron Beam Machine|Multiplier
EBM	Extended Branch Mode
EBNF	Extended Backus Naur Form
EBONE	European Backbone
EBP	Error Back Propagating
EBPA	Electron Beam Parametric Amplifier
EBPG	Electron Beam Pattern Generator
E-BPR	Enhanced Bottom Pressure Recorder
EBR	Electron Beam Recorder|Recording

EBRSC	Electronic Bulletin of the Rough Set Community
EBS	Electron Beam Scanlaser
EBS	Electron Bombarded Semiconductor
EBS	Emergency Broadcast System
EBT	Electronic Benefits Transfer
EBU	European Broadcasting Union
EBV	Extended Binary Vectors
EBX	Electronic Branch Exchange
EC	Echo Check
EC	Ecuador
EC	Edge Connector
EC	Electronically Coupled
EC	Electronic Commerce|Computer
EC	Electronic Conductivity
ec	enamel covered
EC	Enterprise Characterization
EC	Equipment Check
EC	Erase Character
EC	Error Control
EC	Error Correcting|Correction
EC	European Commission
EC	Evaluation Center
EC	Event Condition
EC	Exchange Carrier
EC	Extended Control
EC	External Computer
ECA	Electrical Components Application
ECA	Electronics Control Assembly
ECA	Event-Condition-Action
ECAC	Electromagnetic Compatibility Analysis Center
ECAD	Electronic Computer Aided Design
ECAD	Error Check Analysis Diagram
ECAI	European Conference on Artificial Intelligence
ECAL	Enjoy Computing And Learn
ECAM	Extended Content-Addressable Memory
ECAP	Electronic Circuit Analysis Program
ECAP	Electronic Customer Access Program
ECAPS	Error Check Analysis Programs
ECARS	Electronic Coordinatograph And Read-out System
ECAS	External Compliance Assessment Survey
ECASS	Electronically Controlled Automatic Switching System
ECASS	Export Control Automated Support System
ECAT	Electronic Card Assembly and Test
ECB	Electronic Code|Cook Book
ECB	Event Control Block
ECBS	Engineering of Computer Based Systems
ECC	Eccentric
ECC	Edit Control Code

ECC	Elliptic Curve Cryptography
ECC	Enter Cable Change
ECC	Error Check Code
ECC	Error Check(ing) and Correcting|Correction
ECC	Error Correcting Code(s)
ECC	Error Correcting Control|Circuitry
ECC	Error Correction Code|Control
ECC	Exchange Control Copy
ECC	Execute Control Cycle
ECC	Executive Computer Course
ECCM	Electronic Counter-Counter Measure
ECCNP	European Conference on Computer Network Protocols
ECCO	Error Checking and correcting Coder
ECCS	Economic C (hundred) Call Seconds
ECCS	Electronically Changeable Control Stores
ECD	Electro-Chromatic Display
ECD	Electron Capture Detector
ECD	Enhanced Colour Display
ECD	Error Correction Decoder
ECDB	Engineering Change Data Base
ECDC	Electro-Chemical Diffused-Collector
ECDD	Exceeded
ECDIN	Environmental Chemical Data and Information Network
ECDIS	Electronic Chart Display Information Systems
ECDL	European Computer Driving Licence
ECDO	Electronic Community Dial Office
ECDR	Encoder
ECDT	Electro-Chemical Diffused-Transistor
ECE	Economic Commission for Europe
ECE	Executive Communications Exchange
ECF	Enhanced Connectivity Facilities
ECF	Error Correction Feature
ECF	European CATIA Forum (CATIA = Computer graphics Aided Three-dimensional Interactive Application)
ECF	Expanded Code File
ECG	Electro-Chemical Grinding
EC-GC	Electron Capture – Gas Chromatograph
ECHO	Electronically Controlled High Output
ECHO	Electronic Computing Hospital Oriented
ECHO	European Commission Host Organization
ECHT	European Conference on HyperText
ECI	European Cooperation in Informatics
ECI	External Call Interface
ECIF	Electronic Components Industry Federation
ECIM	Electronic Computer Integrated Manufacturing
ECIP	European CAD Integration Project (CAD = Computer Aided Design)
ECIS	European Committee for Interoperable Systems

ECITC	European Committee for Information-technology Testing and Certification
ECITC	European Committee for IT Testing and Certification (IT = Information Technology)
ECK	Equipment Check
ECL	Emitter Coupled Logic
ECL	End Communication Layer
ECL	Equipment Component List
ECL	Error Correction Logic
ECL	Establishment Communications Link
ECL	Executive|Extended Control Language
ECLO	Emitter Coupled Logic Operator
ECM	Electric Coding Machine
ECM	Electronic Control Module
ECM	Electronic Counter Measure(s)
ECM	Enterprise Commerce Management
ECM	Enterprise Configuration Manager
ECM	Error Correcting Memory
ECM	Extended Control Mode
ECM	Extended Core Memory
ECM	Externally Controlled Machine
ECMA	European Computer Manufacturers Association
ECME	Electronic Counter Measures Equipment
ECmode	Extended Control mode
ECMS	Enterprise Communication Messaging Solutions
ECN	End Connector
ECN	Engineering Computer Network
ECN	Explicit Congestion Notification
ECNE	Enterprise Certified NetWare Engineer
ECO	Electronic Central Office
ECO	Electronic Contact Operate (signal)
ECO	Electronic Coupled Oscillator
ECOD	Encoder
ECOL	Education Computer Language
ECOM	Electronic Commerce Operations Management
ECOM	Electronic Computer-Originated Mail
ECOMA	European Computer Measurement Association
ECON	Equipment Close-Out Notice
ECON	Extended Console (system)
ECOOP	European Conference on Object Oriented Programming
ECOS	Extended Communications Operating System
ECP	Electronic Coding Path
ECP	Emulation|Emulator Control Program
ECP	Enhanced|Extended Capabilities Port
ECPA	Electronic Communication Privacy Act
ECPC	Economic Classification Policy Committee
ECPM	Environmental Control and Processing Modules
ECP.NL	Electronic Commerce Platform Nederland

ECPP	Enterprise Collaborative Processing Portals
ECPS	Extended Control Program Support
ECPT	Electronic Coin Public Telephone
ECR	Electronic Cash Register
ECR	Electronic Character Recognition
ECR	Embossed Character Reader
ECR	Enterprise Customer Resource
ECR	Error Cause Removal
ECRC	Electronic Commerce Resource Center
ECRC	Electronic Component Reliability Center
ECRC	European Computer-industry Research Centre
ECREEA	European Conference of Radio and Electronic Equipment Association
ECRM	Electronic Character Recognition Machine
ECRV	Extended Curve
ECS	Electronic Communication Society
ECS	Electronic Cross-connect System
ECS	Embedded Control Software
ECS	Environmental Control System
ECS	European Common Standard
ECS	European Communications Satellite
ECS	Extended Common Services
ECS	Extended Control|Core Storage
ECSA	European Computing System Analyzer
ECSA	Exchange Carriers Standard Association (US)
ECSA	Extended Character Set Adapter
ECSD	Electronic Corrective Service Diskette
ECSL	Extended Computer Simulation Language
ECSPC	Electronic Communications Service Provider Committee
ECSS	Extendable Computer System Simulator
ECSW	Extended Channel Status Word
ECT	European Combined Terminals
ECT	Explicit Call Transfer
ECTA	Error Correcting Tree Automata
ECTEI	European Conference of Telecommunications and Electronics Industries
ECTEL	European Communications, Telecommunications and professional Electronics industry
ECTF	Engineering Change Tracking File
ECTF	Enterprise Computer Telephony Forum
ECTL	Electronic Communal Temporal Lobe
ECTL	Emitter Coupled Transistor Logic
ECTP	Ethernet Configuration Test Protocol
ECTS	Engineering Changes Transmission System
ECTUA	European Council of Telecommunication Users Associations
ECU	EISA Configuration Utility
ECU	Electronic Control Unit
ECU	External Cache Unit

ECU	Extreme Close-Up
ECV	Extended Content Verification
ECVT	Electronic Continuously Variable Transmission
ECWD	Error Channel Word
ECX	Electronically Controlled (telephone) Exchange
ED	Edge
ED	Edit(or)
ED	Electrical Differential
ED	Electro-Dynamic
ED	Electronic Device
E/D	Encode/Decode
ED	Encryption Device
ED	End Delimiter
ED	End of Data
ED	Erase Display
ED	Error Detection|Diagnostics
ED	EWOS Document (EWOS = European Workshop for Open Systems)
ED	Exception Data
ED	Expanded Display
ED	Expedited Data
ED	Exponent Difference
ED	External Delay|Device
EDA	Electrical|Electronic Design Automation
EDA	Embedded Document Architecture
EDA	Error Data Analysis
EDA	External Device Address
EDAC	Electromechanical Digital Adapter Circuit
EDAC	Error Detection And Correction
EDB	Engineering Data Base
EDB	Epic Data Blank
EDBMS	Engineering Data Base Management System
EDBS	Educational Data Base (management) System
EDC	Early Display Configuration
EDC	Education Center
EDC	Electronic|Emergency Digital Computer
EDC	Engineering Data Control
EDC	Enhanced Data Correction
EDC	Error Detection and Correction
EDC	Error Detection Code
EDC	Extended Device Control
EDC	External Disk|Drum Channel
EDCW	External Drive Control Word
EDD	Electronic Document Distribution
EDD	Envelope Delay Distortion
EDD	Executive Development Department
EDD	Expert Database Designer
EDD	Expert Dungeons & Dragons

EDDA	Electronic Demand Deposit Accounting
EDDC	Extended Distance Data Cable
EDE	Emitter Dip Effect
EDEC	Electronic Design Education Consortium (GB)
EDF	Engineering Data Form
EDF	Execution Diagnostic Facility
EDFM	Extended Disk File Management system
EDGAR	Electronic Data Gathering And Retrieval
EDGE	Electronic Data Gathering Equipment
EDGE	Experimental Display Generation
EDGE	Extendible Graph Editor
EDHE	Experimental Data Handling Equipment
EDI	Electronic Data|Document Interchange
EDI	Enhanced Definition Interlaced
EDI	External Device Interrupt
EDIA	Electronic Data Interchange Association
EDIAC	Engineering Decision Integrator And Communicator
EDICON	Electronic Data Interchange for the Construction industry
EDICT	Engineering Document Information Collection Technique
EDICUSA	Electronic Data Interchange Council of the United States of America
EDIF	Electronic Design Interchange Format
EDIFACT	Electronic Data Interchange for Finance, Administration, Commerce and Transport
EDIFICE	Electronic Data Interchange For the Industrial Community of Electronic firms
EDILIBE	Electronic Data Interchange between Libraries and Booksellers in Europe
EDIM	Electronic Data Interchange user agent Message
EDIME	Electronic Data Interchange Messaging Environment
EDIMG	Electronic Data Interchange Messaging
EDIMS	Electronic Data Interchange Messaging System
EDIMS	Environmental Data and Information Management Systems
EDIN	Electronic Data Interchange Notification
EDINET	Educational Instruction Network
EDIPAS	Engineering Data Interactive Presentation and Analysis System
EDIS	Engineering Data Information System
EDIS	Environmental Data and Information Service
EDIT	Edit(or)
EDIUA	Electronic Data Interchange User Agent
EDL	Edit Decision List
EDL	Electromagnetic Delay Line
EDL	Emulation Design Language
EDL	Event Driven Language
EDLC	Ethernet Data Link Control
EDLIN	Editor for Lines of text
EDM	Electronic Document Management
EDM	Engineering Data|Document Management

EDM	Event Driven Monitor
EDM	Evolutionary Design Methodology
EDMA	Extended Direct Memory Access
EDMD	Electronic Document Message Directory
EDMK	Edit and Mark
EDMS	Electronic Document Management System
EDMS	Engineering Database Modelling System
EDMS	Engineering Data Management System
EDMS	Engineering Document|Drawing Management System
EDMS	Enterprise Document Management Systems
EDMS	Extended Database Management System
EDN	Expedited Data Negotiation
EDO	Enhanced|Extended Data Out
EDOC	Enterprise Distributed Object Computing (international conference)
EDORAM	Enhanced Data Output Random Access Memory
EDOS	Enhanced|Extended Disk Operating System
EDP	Electronic Data Processing
EDP	Enhanced Definition Progressive
EDPC	Electronic Data Processing Center
EDPE	Electronic Data Processing Equipment
EDPEP	Electronic Data Processing Education Program
EDPM	Electronic Data Processing Machine|Manager
EDPS	Electronic Data Processing System
EDQNM	Eddie-Dampened Quasi-Normal Markovian (equation)
EDR	Early Developer|Device Release(s)
EDR	Effective Data (transfer) Rate
EDR	Error Detection and Recovery
EDR	Exception Disposition Report|Routine
EDR	Executive Diagnostic Routine
EDRAM	Erasable|Extended Dynamic Random Access Memory
EDRC	Engineering Design Research Center
EDRS	Endorse
EDS	Electronic Data Switching|System
EDS	Electronic Document Storage
EDS	Emergency Detection System
EDS	Engineering Data Sheet|System
EDS	Engineering Design System
EDS	Environmental Data Service (US)
EDS	Exchangeable Disk Store
EDSAC	Electronic Delayed Storage Automatic Calculator
EDSD	Electronic Document Segment Directory
EDSI	Enhanced Small Device Interface
EDSW	External Device Status Word
EDSX	Electronic Digital Signal Cross-connect
EDT	End Data Transmission
EDT	Engineering Description Tape
EDT	Estimated Down Time

EDTC	Electronic Desktop Computer
EDTCC	Electronic Data Transmission Communication Center
EDU	Edit Display Unit
EDU	Education
EDU	Electronic Display Unit
EDU	Error Detection Unit
EDV	Elektronische Daten Verarbeitung (DE)
EDVAC	Electronic Discrete Variable Automatic Calculator
EDW	Enterprise Data Warehouse
EDWA	Enterprise Data Warehouse Architecture
EDX	Event Driven Executive
E/E	Electrical/Electronic
EE	Electrical|Electronic Engineer(ing)
E-E	End-to-End
EE	Equation Error
EE	Errors Excepted
EE	Estonia
EE	Execution Element|Environment
EE	Extended Edition
EE	External Environment
EEA	Electronic Engineering Association
EEAPROM	Electrically Erasable And Programmable Read-Only Memory
EEB	Extended Erlang B
EEBIC	Emergent and Evolutionary Behaviour, Intelligence, (and) Computation
EEC	Electronic Engine Control
EEC	European Economic Commission|Community
EEC	Extended Error Correction
EECA	European Electronic Component manufacturers Association
EECL	Emitter to Emitter Coupled Logic
EECM	End Effector Change-out Mechanism
EECS	Electrical Engineering and Computer Science
EECT	End-to-End Call Trace
EED	Electro-Explosive Device(s)
EED	Extended Entity Data
EEDP	Expanded Electronic (tandem switching) Dialing Plan
EEE	Exchange End Equipment
EEI	Equipment-to-Equipment Interface
EEI	Essential Elements of Information
EEI	External Environment(al) Interface
EEIC	Elevated Electrode Integrated Circuit
EEL	Epsilon Extension Language
EEM	Empirical Evaluation Methods (workshop)
EEM	External Expansion Module
EEMA	Electrical and Electronic Manufacturing Association (US)
EEMA	European Electronic Mail Association
EEMAC	Electrical and Electronic Manufacturing Association of Canada
EEMS	Enhanced Extended Memory Specification(s)

EEP	Electromagnetic Emission Policy
EEP	Electronic Evaluation and Procurement
EEPROM	Electrically Erasable Programmable Read-Only Memory
EER	Etch Epitaxial Refill
EER	Extended Entity-Relationship model
EEROM	Electrically Erasable Read-Only Memory
EES	Electromagnetic Environment Simulator
EES	Escrowed Encryption Standard
EESP	Enterprise Extended Services Portals
EET	Electrical Engineering Technology
EETDN	End-to-End Transit Delay Negotiation
EF	Emitter Follower
EF	Execution Function
EF	Extended Facility
EF	External Flag
EFA	Extended File Attribute
EFA	Extended Finite Automation
EFC	Enterprise Fabric Connectivity
EFCI	Explicit Forward Congestion Indicator
EFD	Event Forwarding Discriminator
EFF	Effective
EFF	Efficiency
EFF	Electronic Freedom Foundation
EFF	Expandable File Family
EFFORD	Environment For Fortran Development
EFI	Electromechanical Frequency Interference
EFI	Electronics For Imaging
EFIRS	Electronic Filing of Information Returns System
EFL	Emitter Follower Logic
EFL	Equivalent Focal Length
EFL	Error Frequency Limit(s)
EFM	Eight to Fourteen Modulation
EFOP	Expanded Function Operator Panel
EFP	Expanded Function (operator) Panel
EFRAP	Electronic Feeder Route Analysis Program
EFRU	Electronic Field-Replaceable Unit
EFS	Electronic Foodservice Network
EFS	Electronic Form Systems
EFS	End Frame Sequence
EFS	Error Free Second(s)
EFS	Extended Facility Set
EFS	Extended Function Store
EFS	External File System
EFS	External Function Store
EFT	Effect
EFT	Electronic Financial Transaction
EFT	Electronic Funds Transfer
EFTA	Electronic Funds Transfer Association

EFTA	European Free Trade Association
EFTAM	Electronic File Transfer Access Method
EFTPOS	Electronic Funds Transfer at Point Of Sale
EFTPS	Electronic Federal Tax Payment System (US)
EFTS	Electronic Funds Transfer Service|System
EG	Egypt
EG	End Group
EG	Experts Group
EGA	Enhanced Graphics Adapter
EG-CAE	Experts Group for Command Application Environment
EGCR	Extended Group Coded Recording
EGCS	Extended Graphic Character Set
EG-CT	Experts Group for Conformance Testing
EG-DIR	Experts Group on Directory
EG-FT	Experts Group on File Transfer
EGIU	Error Generator Injection Unit
EG-LL	Experts Group on Lower Layers
EGM	Enhanced Graphics Module
EGM	Ethernet Gateway Module
EG-MHS	Experts Group for Message Handling Service
EG-MMS	Experts Group Manufacturing Message Specification
EG-NM	Experts Group for Network Management
EG-ODA	Experts Group for Office Document Architecture
EGP	Exterior|External Gateway Protocol
EGS	Extended Graphics Subsystem
EG-TP	Experts Group on Transaction Processing
EG-VT	Experts Group on Virtual Terminals
EH	Engineering Hardware (test)
EH	Error Handler
EH	Western Sahara
EHA	European Harmonization Activity
EHCI	Engineering for Human-Computer Interaction (conference)
EHCN	Experimental Hybrid Computer Network
EHF	Extremely High Frequency (30-300 GHz)
EHLLAPI	Emulator High Level Language Application Programming Interface
EHP	Error Handling Package
EHPM	Electro-Hydraulic Pulse Motor
EHT	Extremely High Tension
EHTS	Emacs HyperText System
EHV	Extra High Voltage
EI	Electrical Iron
EI	Enable Interrupt
EI	Engineering Index
EI	Enterprise Integration
EI	Error Indicator
EI	External Interrupt
EIA	Electronic Industries Association

EIA	Energy Information Administration
EIA	Enterprise Integration of Applications
EIA	Extended Interaction Amplifier
EIA-J	Electronic Industries Association of Japan
EIB	Enterprise Information Base
EIB	Error Information Block
EIB	Error Interrupt Buffer
EIB	Execution Interface Block
EIC	Electronic Instrument Cluster
EIC	Equipment Identification Code
EICAR	European Institute for Computer Anti-virus Research
EIDAP	Emitter Isolated Difference Amplifier Paralleling
EIDE	Enhanced Integrated Drive Electronics
EIEMA	Electrical Installation Equipment Manufacturers Association
EIES	Electronic Information Exchange System
EIF	Enterprise Integration Framework
EIGRP	Enhanced Interior Gateway Routing Protocol
EIK	Extended Interaction Klystron
EIM	Enterprise Integration Modelling
EIM	Express Instant Manager
EIN	European Informatics Network
EINET	Enterprise Integration Network
EINOS	Enhanced Interactive Network Optimization System
EIO	Execute Input/Output
EIO	Extended Interaction Oscillator
EIOS	Extended Input/Output System
EIP	Enterprise Information Portal
EIR	Error Interrupt Request
EIRENE	European Information Researchers Network
Eirepac	Irish packet-switched network
EIRP	Effective Isotropic Radiated Power
EIRV	Error Interrupt Request Vector
EIS	Early Installation Support
EIS	Electrolytic Insulator Semiconductor
EIS	Electronic Information Security|Service
EIS	End Interruption Sequence
EIS	Engineering|Equipment Information System
EIS	Executive Information System
EIS	Expanded In-band Signalling
EIS	Extended Instruction Set
EISA	Extended Industry Standard Architecture
EISB	Electronic Imaging Standards Board
EISS	European Informatics Skills Structure
EIT	Encoded Information Type
EITA	Enterprise Information Technology Architecture
EITIRT	European Information Technology Industry Round Table
EITO	European Information Technology Observatory
EITS	Encoded Information Types

EITS	Express International Telex Service
EITT	Export/Import Transit Time
EIUF	European ISDN Users' Forum (ISDN = Integrated Services Digital Network)
EIX	Enterprise Information Exchange
EJ	End of Job
Ejava	Embedded java
EJB	Enterprise Java Beans
EJN	Ejection
EJOR	European Journal of Operational Research
EJT	Eject
EJX	Eject X
EKP	Enterprise Knowledge Portal
EKTS	Electronic Key Telephone System
EKW	Electrical Kilowatts
EL	Electro-Luminescent
EL	End of the Line
EL	Erase Line
EL	Error Loop
EL	External Label|Link
EL	Extra Line
ELA	Extended Line Adapter
ELAN	Educative Language
ELAN	Error Logging and Analysis
ELAN	European Local Area Network
ELAP	EtherTalk Link Access Protocol
ELARD	Electronic Analogous Computer (Rechner) Darmstadt (DE)
ELC	Embedded Linking and Control
ELCSS	Electronic Collateral Support System
ELD	Edge-Lighted Display
ELD	Electro-Luminescent Display
ELEC	Electric
ELECD	Electrode
ELECN	Electronic
ELECP	Electro-Plating
ELEM	Element
ELEPL	Equal Level Echo Path Loss
ELF	Executable and Linking Format
ELF	Extensible|Extension Language Facility
ELF	Extremely Low Frequency
ELFEXT	Equal-Level Far-End Cross-Talk
ELG	Even Linear Grammars
ELIAS	Entry Level Interactive Application System
ELIB	Environmental Legal Information Base
ELIM	Eliminate
ELINT	Electronic Intelligence
ELM	Electronic Mailer
ELM	Element Load Module

ELM	Error Log Manager
ELONG	Elongate
ELOT	Hellenic Organization for Standardization
ELP	Electronic Line Printer
ELPC	Electro-Luminescent Photo-Conductor
ELPC	Even Longitudinal Parity Check
ELR	Error Logging Register
ELR	Execution Local/Remote (routine)
ELS	Entry Level System
ELSEC	Electronic Security
ELSI	Extremely Large Scale Integration
ELSIE	Electronic Letter Sorting Indicator Equipment
ELSNET	European Language and Speech Network
ELT	Electrometer
ELT	Electronic Locator Transmitter
ELT	Emergency Locator Transmitter
ELVIS	Expanding Linear Visualization Information Structure
EM	Efficiency Modulation
EM	Electro-Magnetic(s)
EM	Electro-Mechanical
EM	Electro-Microscope
EM	Electromotive
EM	Electronic Mail
EM	Emphasized
EM	Emulated Machine
EM	End of Medium|Message
EM	Enterprise Model(ling)
EM	Error Message
EM	Expanded Memory
EM	Exposure Meter
EM	Extended Memory
EMA	Editor Macros
EMA	Electronic Mail Association
EMA	Enterprise Management Architecture
EMA	Enterprise Marketing Automation
EMA	Extended Memory Access|Area
EMACS	Editing Macros
EMACS	Emacs Makes A Computer Slow
EMACS	Escape Meta Alt Control Shift
e-mail	electronic mail
EMAR	Experimental Memory Address Register
EMAT	Electro-Magnetic Acoustic Transducer
EMATS	Experimental Message Automatic Transmission System
EMB	Embedded Memory Block
EMB	Emulator Board
EMB	Extended Memory Block
EMBARC	Electronic Mail Broadcast to A Roaming Computer
EMC	Electro-Magnetic Compatibility

EMC	Electronic Media Center
EMC	E-Mail Connection
EMC	Emitter Coupled Logic
EMC	End of Medium Character
EMC	Extended Mathematical Coprocessor
EMC	External Multiplexer Channel
EMC	Eye-Movement Camera
EMD	Element Merge and Distribution
EMD	Eye Movement Desensitization
EMDB	Engineering Master Data Base
EMDIR	Electronic Mail Directory
EMDR	Emulated Machine Description Record
EMEIS	Enterprise Modelling Execution and Integration Services
EMER	Emergency
EMER	Express Master (file)
emf	electromotive force
EMF	Engineering Master File
EMF	Enterprise Modelling Framework
EMFT	Extended Multiprogramming with a Fixed number of Tasks
EMH	Expedited Message Handling
EMI	Electro-Magnetic Interference
EMIC	Engineering Management Inquiry Console
EMIM	Executive Master in Information Management
EMIND	European Modular Interactive Network Designer
EMIT	Embedded Micro Interface Technology
EMIT	Emitter
eml	electronic mail (file extension indicator)
EML	Elementary Mathematical Library
EML	Emulator Machine Language
EMLTN	Emulation
EMLTR	Emulator
EMM	Expanded|Extended Memory Manager
EMMP	Enterprise Mission Management Portals
EMMS	Electronic Mail and Message Service
EMOD	Erasable Memory Octal Dump
EMOSFET	Enhanced Metal Oxide Semiconductor Field Effect Transistor
EMP	Electro-Magnetic Propagation|Pulse
EMP	Electro-Mechanical Power
EMPL	Extensible Micro-Programming Language
EMPM	Electronic Manuscript Preparation and Markup
e-MPS	e-Mail Preference Service
EMR	Electro-Magnetic Radiation|Response
EMR	Enhanced Metafile Record
EMR	Equipment Maintenance Report
EMR	Error Monitor Register
EMS	Education Management System
EMS	Electronic Mail|Management System
EMS	Electronic Measurement|Medical System

EMS	Electronic Message Service|System
EMS	Element Management System
EMS	Emulator Machine Support
EMS	Energy Management System
EMS	Engineering|Enterprise Modelling System
EMS	Environmental Management System(s)
EM&S	Equipment Maintenance & Support
EMS	Expanded Memory Specification
EMS	Extended Main Stor(ag)e
EMS	Extended Memory Store
EMS	Extended Monitor System
EMSC	Engine Monitoring System Computer
EMSS	Electronic Message Service System
EMT	Emulator Trap
EMT	Express Master Tape
EMU	Electro-Magnetic Unit
EMU	Extended Memory Unit
EMUL	Emulate
EMW	Electro-Magnetic Wave
EMWAC	European Microsoft Windows Academic Consortium
EMX	Electronic Mobile Exchange
EN	European Norm
ENA	Extended Network Addressing
ENABL	Enable
ENALIM	Evolving Natural Language Information Model
ENAP	Economic Network Analysis Program
ENBL	Enable
enc	encoded (file extension indicator)
ENCDR	Encoder
ENDEC	Encoder/Decoder
ENDEXE	End Execute
ENDOC	European directory of environmental information and Documentation Centers
ENDOP	End Of Operations
ENDOR	Electron Nuclear Double Resonance
ENDS	End Segment
ENDSR	End Subroutine
ENE	Enterprise Networking Event
ENFIA	Exchange Network Facilities for Interstate Access
ENG	Electronic News Gathering
ENG	Engineer(ing)
ENG	Engrave
ENGY	Energy
ENIAC	Electronic Numerical Integrator And Calculator
ENIC	Voltage-Negative Impedance Converter
ENLG	Enable Level Group
ENN	Expand Nonstop Network
ENQ	Enquire (character)

ENR	External Number Repetition
E-NRZ	Enhanced Non-Return-to-Zero
ENS	Ensign
ENS	Enterprise Naming|Network Service
ENS	Extended Network Service
ENSA	Enterprise Network Storage Architecture
ENSDU	Expedited Network Service Data Unit
ENSI	Equivalent Noise Sideband Input
ENSS	Exterior Nodal Switching Subsystem
ENST	Ecole Normale Supérieure des Télécommunications (FR)
ENT	Enter
ENT	Equivalent Noise Temperature
ENTAB	Entry Table
ENTELEC	Energy Telecommunications and Electrical association
ENTR	Entrance
ENV	draft European Norm
ENV	Envelope
EO	Emergency Order
EO	Enable Output
EO	End Office
EO	End of Operation
EO	Erasable Optic(al)
EO	Extend and Offset
EOA	Effective On or About
EOA	End Of Address (character)
EOAP	Event Oriented Application Program
EOB	End Of Block (character)
EOB	End Of Buffer|Burst
EOC	Embedded Operations Channel
EOC	End Of Card|Character
EOC	End Of Content
EOC	End Of Conversation|Conversion
EOD	End Of Data|Discussion
EOD	Entry On Duty
EOD	Erasable Optical Disk
EODI	End-Of-Data Indicator
EOE	Electronic Order Exchange
EOE	End Of Extent (mark)
EOE	Errors and Omissions Excepted
EOF	Empirical Orthogonal Function
EOF	End Of Field|File
EOF	Enterprise Objects Framework
EOI	End Of Identity|Idles
EOI	End Of Inquiry|Injury
EOJ	End Of Job
EOL	End Of Line|List
EOL	Expression Oriented Language
EOLIS	Emission Of Light In Silicon (EU)

EOLM	Electro-Optical Light Modulator
EOLM	End Of Line Marker
EOM	End Of Medium|Month
EOM	End Of Message (character/code)
EOMI	End Of Message Incomplete|Indicator
EON	Edge Of Network
EON	End Of Number
EOP	Electrical-Optical (converter)
EOP	End (of) Operation (trigger)
EOP	End Of Page|Process
EOP	Engineering Operating Procedure
EOQ	End Of Query
EOR	Electro-Optical Reconnaissance
EOR	End Of Record (mark)
EOR	End Of Reel|Run
EOR	Exclusive OR
EOS	Electronic Office System
EOS	Electro-Optical System(s)
EOS	Element Of Service
EOS	End Of Screen|Segment
EOS	End Of Sequence
EOS	End Of Step|String
EOS	End Operation Suppress
EOS	Enterprise Optimization System
EOS	Equation-Of-State
EOS	Extended Operating System
EOSDIS	Earth Observing System Data and Information System
EOSEQ	End Order Sequence
EOSF	Electro-Optical Simulation Facility
EOT	End Of Tape (marker)
EOT	End Of Task|Test
EOT	End Of Text (character)
EOT	End Of Track (mark)
EOT	End Of Transmission (character)
EOT	Error On Tape
EOTC	European Organization for Testing and Certification
EOTM	End-Of-Tape Marker
EOTR	End Of Transmission (character)
EOTS	Electro-Optical Tracking System
EOTT	End Office Toll Trunking
EOUG	European Oracle Users Group
EOV	End Of Volume
EOW	End Of Word
EOWG	Engineering and Operations Working Group
EOY	End Of Year
EP	Electrically Polarized
EP	Electronic Purchasing
EP	Electro-Pneumatic

EP	Emulation Program|Package
EP	Emulator Program
EP	End of Program
EP	Entry Point
EP	Error Print
EP	Experience Points
EP	Extend(ed) Play
EP	Externally Pressured
EP	Extreme Power
EPA	Enhanced Performance Architecture
EPA	Estimated Profile Analysis
EPA	Extended Performance Analysis
EPA	External Page Address
EPABX	Electronic Private Automatic Branch Exchange
EPAD	Error protecting Packet Assembler/Disassembler
EPAI	Exchange of Publicity Available Information
EPAM	Elementary Perceiver And Memorizer
EPBX	Electronic Private Branch Exchange
EPC	Easy Processing Channel
EPC	Editorial Processing Center
EPC	Electronic Page Composition
EPC	Electronic Program Control
EPC	Even Parity Check
EPC	Extended Plotter Code|Command
EPCI	Entry Point Control Item
EPCS	Experimental Physics Control Systems
EPD	Electronic Product Definition
EPD	Erasable Programmable Device
EPEX	European Product (definition) Exchange
EPG	Edit Program Generator
EPG	Electronic Programme Guide
EPHOS	European Procurement Handbook for Open Systems
EPIC	Electronic Patient Information Control (system)
EPIC	Electronic Properties|Privacy Information Center
EPIC	Engineering Planning Information Coordination (system)
EPIC	Epitaxial Passivated and Isolated (integrated) Circuit
EPIC	European PDES Incorporated Center (PDES = Product Data Exchange Standard)
EPIC	Explicit(ly) Parallel Instruction Computing
EPICS	Extended Purpose Inline Console System
EPIL	Entirely Pushed-In Length
EPIRB	Electronic Personal Identification Radio Beacon
EPIX	Emergency Preparedness Information Exchange
EPL	Effective Privilege Level
EPL	Electronic switching systems Programming Language
EPL	Encoder Programming Language
EPL	European Program Library
EPLAN	Econometric Planning Language

EPLANS	Engineering, Planning, and Analysis System
EPLD	Electrically Programmable Logic Device
EPLD	Erasable Programmable Logic Device
EPM	Eight Phases Modulation
EPM	Electronic Photocomposing|Printing Machine
EPM	Enhanced editor for Presentation Manager
EPM	Enterprise Performance Management
EPM	Enterprise Process|Product Management
EPM	Evolutionary Project Management
EPN	European Pulsar Network
EPN	External Priority Number
EPNL	Effective Perceived Noise Level
EPO	Emergency Power Off (switch)
EPOL	Elementary Procedure Oriented Language
EPOL	Entirely Pulled-Out Length
EPOS	Electronic Point Of Sale
EPOS	Engineering and Project-management Oriented Support (system)
EPOS	European PPT Open learning Service (PPT = Programmer Productivity Techniques)
EPP	Electron Pair Production
EPP	Electrostatic Printer Plotter
EPP	Enhanced Parallel Port
EPPL	Environmental Planning and Programming Language
EPPS	Electronic Pre-Press Systems
EPPS	Enterprise Production Planning System
EPR	Einstein, Podolsky, Rosen (pairs of objects)
EPR	Electron Paramagnetic Resonances
EPR	Error Pattern Register
EPROM	Electrically|Erasable Programmable Read Only Memory
EPS	Electronic Payments System
EPS	Electronic Prepress Solutions
EPS	Electronic Publishing Service
eps	encapsulated postscript (file extension indicator)
EPS	Engineering Process Specification
EPS	Entry Power Subsystem
EPS	Equipment Process Specification
EPS	Even Parity Select
EPS	Exceptional Packaging System
EPS	Extended Program Stor(ag)e
EPS	External Page Storage
EPSCS	Enhanced-Private Switched Communications Service
EPSE	Electronic Power Supplies Equipment
EPSF	Encapsulated PostScript Files|Format
EPSI	Encapsulated PostScript Interchange
EPSM	External Page Storage Management
EPSS	Electronic Performance Support System
EPST	Extended Partition Specification Table

EPT	Engineering Programming and Technology
EPT	Executive Process Table
EPT	External Page Table
EPTS	Electronic Problem Tracking System
EPU	Electrical|Emergency Power Unit
EPU	Execution|Extended Processing Unit
EPU	External Programmer Unit
EPUB	Electronic Publication
EQ	Equalizer
EQ	Equal to
EQ	Equation
EQ	Equivalent
EQEEB	Equivalent Queue Extended Erlang B
EQG	Equalizing
EQP	Equipment
EQU	Equate
EQUATE	Electronic Quality Assurance Test Equipment
EQUIP	Equation Input Processor
EQUIV	Equivalent
E-R	Entity-Relationship
ER	Eritrea
ER	Error
ER	Error Recovery|Register
ER	Established Reliability
ER	Evaluation Report
ER	Exchange Register
ER	Expected Result
E/R	Expense to Revenue
ER	Explicit Route
ER	Exponent Register
ER	External Reference
ERA	Earned Run Average
ERA	Electronically Reconfigurable Array
ERA	Electronic Reading Automation
ERA	Engineering Records Automation
ERA	Entity Relationship Attribute (approach)
ERAR	Error Return Address Register
ERAS	Electronic Routing and Approval System
ERAS	Erase
ERB	Execution Request Block
ERC	Equipment Request Code
ERCC	Error Checking and Correcting
ERCIM	European Research Consortium on Informatics and Mathematics
ERCR	Electronic Retina Computing Reader
ERD	Effective Recording Density
ERD	Entity Relation(ship) Diagram
ERD	Event Report Discriminator

ERDA	Electronic Research and Development Activity
ERE	Echo Return Error
ERE	Entity Relationship Editor
ERE	Event Recorder Evaluator
EREP	Environmental Recording, Editing, and Printing
EREP	Error Recording, Edit, and Print
EREP	Error Recovery Executive Program
ERF	Error Function
ERF	Event Report Function
ERFPI	Extended Range Floating Point Instruction
ERG	Erase Gap
ERIC	Educational Resources Information Center
ERICA	Eyegaze Response Interface Computer Aid
ERIN	Environmental Resources Information Network
ERIS	Equal Rank-Intervals Set
ERJE	Extended Remote Job Entry
ERL	Echo Return Loss
ERL	Electronic Reference Library
ERLL	Enhanced Run Length Limited
ERM	Electronics Right Management (group)
ERM	Entity Relation Model
ERM	Equipment Reference Manual
ERM	Error-Recovery Manager|Module
ERMA	Electronic Recording Machine Accounting
ERMF	Event Report Management Function
ERN	Explicit Route Number
EROCP	Extended Remote Operator Control Panel
EROM	Erasable Read Only Memory
ERP	Effective Radiated Power
ERP	Enterprise Resource Planning
ERP	Equipment Requirements Planning
ERP	Error-Recovery Procedure(s)
ERR	Error
ERRAN	Error Analysis
ERRC	Error Character|Correction
ER/RC	Extended Result/Response Code
ERS	Experimental Retrieval System
ERT	Estimated Removal Time
ERT	Expected Run Time
ERU	Error Return address Update
ERU	External Run Unit
ES	Earliest Start (time)
ES	Electromagnetic Storage
ES	Electronic Switch
ES	Electro-Static
ES	Emulator System
ES	End System
ES	Engineering Specification

ES	Errored Second
ES	Expert System
ES	Extended Support
ES	External Store
ES	Extra Segment
ES	Spain
ESA	End-of-Storage Area
ESA	Engineering Service Agreements
ESA	Enterprise Systems Architecture
ESA	European Space Agency
ESA	Expanded Save Area
ESA	Externally Specified Address
ESAC	Electronic Surveillance Assistance Center
ESAC	Electronic Systems Assistance Center
ESAFE	Errored Second, type A, Far End
ESB	Errored Second, type B
ESBAR	Epitaxial Schottky Barrier
ESBFE	Errored Second, type B, Far End
ESBT	Expert System Building Tool
ESC	Educational Scientific Computer
ESC	Effective Store Capacity
ESC	EISA System Component (EISA = Extended Industry Standard Architecture)
ESC	Electronic Spark Control
ESC	Electronics System Center
ESC	Emulator Sub-Channel
ESC	Escape (key) (character)
ESCAPE	European Symposium on Computer Aided Process Engineering
ESCAPE	Expansion Symbolic Compiling Assembly Program for Engineering
ESCD	Extended System Configuration Data
ESCD	Extended System Contents Directory
ESCE	Expert System Consultation Environment
ESCM	Extended Services Communications Manager
ESCON	Enterprise System Connection (architecture)
ESC/P	Epson Standard Code for Printers
ESCP	Errored Second, CP-bit Parity (CP = Control Program)
ESCRC	Errored Second, Cyclic Redundancy Check
ESCS	Emergency Satellite Communications System
ESD	Electronic Software Distribution
ESD	Electro-Static Discharge
ESD	Electrostatic Storage Deflection
ESD	Element Status Display
ESD	Elongated Single Domain
ESD	Energy Storage Device
ESD	External Symbol Dictionary
ESDA	Electronic System Design Automation
ESDD	Electro-Static Discharge Damage

ESDE	Expert System Development Environment
ESDI	Enhanced Small|Standard Device Interface
ESDIM	Earth System Data and Information Management (US)
ESDM	Expert System Development Methodology
ESDP	Evolutionary System for Data Processing
ESDR	External System Design Report
ESDS	Entry Sequenced Data Set
ESDT	Electrostatic Storage Display Tube
ESDTR	Electronic Selector Dropout Tape Read
ESDX	Environment & Safety Data Exchange
ESE	Electrical Support Equipment
ES-ES	End System to End System
ESE/VM	Expert System Environment/Virtual Memory
ESF	Extended Spooling Facility
ESF	Extended Super Frame
ESF	Extended Superframe Format
ESFE	Errored Second, Far End
ESFK	Electrostatically Focused Klystrons
ESG	Electronic Sweep Generator
ESH	End-System Hello
ESI	Electro-Static Interference
ESI	End System Identifier
ESI	Enhanced Serial Interface
ESI	European Software Institute
ESI	Experiment Information System
ESI	Externally Specified Index(ing)
ESI	External Systems Interface
ESID	External Symbol Identification
ESIM	Externally Specified Index Mode
ESIOA	Extended Serial Input/Output Adapter
ES-IS	End System to Intermediate System
ESL	Electronic Software Licensing
ESL	Electronic Systems Laboratory
ESL	Emergency Stand-Alone
ESL	Enhanced Signalling Link
ESL	European Systems Language
ESM	Electronic Support Measures
ESM	Electro-Static Memory
ESM	Enterprise Solution Module
ESM	European Satellite Multimedia services
ESM	External Storage Module
ESMD	Enhanced Storage Module Device
ESML	Extended Systems Modelling Language
ESMS	European Satellite Multiservice System
ESMST	Extended Switching Module System Test
ESMTP	Extended Simple Mail Transfer Protocol
ESN	Electronic Security|Serial Number
ESN	Electronic Solutions Now

ESN	Electronic Switched Network
ESN	External Segment Name
ESnet	Energy Sciences network
ESOFTA	Embedded Software Association (US)
ESP	Early Support Program
ESP	E-commerce Service Provider
ESP	Electro-Sensitive Paper|Programming
ESP	Emulation Sensing Processor
ESP	Encapsulating Security Protocol
ESP	Enhanced Serial Port
ESP	Enhanced Service Provider
ESP	Entire Shape Plan
ESP	Errored Second, P-bit
ESP	Externally Stored Program
ESPEC	Ethernet Specification
ESPER	Easy Simple Programming by Expert system
ESPICE	Extended Simulation Program with an Integrated Circuit Emphasis
ESPL	Electronic Switching Programming Language
ESPOL	Executive Systems Problem Oriented Language
ESPRIT	European Strategic Project for Research on Information Technology
ESPS	Education Systems Partition Supervisor
ESPS	Entropic Signal Processing System
ESR	Effective Signal Radiated
ESR	Electronic Send/Receive
ESR	Electron Spin Resonance
ESR	Electron Storage Ring
ESR	Equivalent Series Resistance
ESR	Event Service Routine
ESRI	European Space Research Institute
ESRO	European Space Research Organization
ESS	Electronic Sales System
ESS	Electronic Spark Selection
ESS	Electronic Still Store
ESS	Electronic Switching System
ESS	Establishment Sub-System
ESS	Event Scheduling System
ESSEX	Experimental Solid State Exchange
ESSID	Extended Service Set Identification|Identifier
ESSL	Engineering and Scientific Subroutine Library
ESSM	Electronically Saturated Symmetric Molecules
ESSS	Engineering Scientific Support Switch
ESSU	Electronic Selective Switching Unit
ESSX	Electronic Switching System Exchange
EST	Element Simulation Technique
ES+T	Entry Sequence and Timer
EST	Estimated

ESTAE Extended Specify Task Abnormal Exit
ESTEC European Space research and Technology Centre
ESTI Estimation
ESTV Error Statistics by Tape Volume
esu electrostatic unit(s)
ESV Error Statistics by Volume
ESW Error Status Word
E&T Education & Training
ET Electr(on)ic Typewriter
ET Emerging Technology
ET Ethiopia
ET Exchange Termination
ETACS Extended Total Access Communications System
ETAM Entry Telecommunication Access Method
ETANN Electrically Trainable Analog Neural Network
ETAS Emergency Technical Assistance
ETB End of Transmission Block (character)
ETC Electronic Temperature Control
ETC Electronic Time Clock
ETC Electronic Toll Collection
ETC Enhanced Throughput Cellular
ETC Estimated Time of Compilation
ETC Excess-Three Code
ETCO European Telecommunications Consultancy Organization
ETCOM European Testing for Certification of Office and Manufacturing equipment
ETDE Energy Technology Data Exchange
ETDI Electronic Trade Data Interchange
ETDL Electronics Technology and Devices Laboratory
ETE Estimated Time En-route
ETF Electronic Toll Fraud
ETF Error Transfer Function
ETFA Emerging Technologies and Factory Automation (conference)
ETFD Electronic Toll Fraud Device
ETG EWOS Technical Guide
ET/GTS Electronic Text and Graphics Transfer System
ETI Elapsed Time Indicator
ETIM Elapsed Time
ETIM Electro-Technical Information Model
E-TIME Execution Time
ETIP Electrical Tool Interface Processor
ETIS Electronic Telephone Inquiry System
ETK Embedded Tool Kit
ETL Electronic Testing Laboratory
ETL Emitter Transistor Logic
ETL Epitaxial Transistor Logic
ETL Etching by Transmitted Light
ETL Extract(ion)/Transform(ation)/Load(ing)

ETLG	Enable This Level Group
ETLT	Equal To or Less Than
ETM	Elapsed Time Meter
ETMF	Elapsed Terminal Measurement Facility
ETMF	Extended Telecommunications Modules Feature
ETN	Electronic Tandem Network
ETN	Equivalent To New
ETNO	European public Telecommunications Network Operators association
ETO	Earth-To-Orbit
E-to-B	Emulsion to Base
E-to-E	Electronics to Electronics
E-to-E	Emulsion to Emulsion
ETOM	Electron-Trapping Optical Memory
ETOS	Extended Tape Operating System
ETP	Electrical Tough Pitch
ETR	Early Token Release
ETR	Expected Time of Response
ETRI-PEC	Electronics and Telecommunications Research Institute - Protocol Engineering Center
ETRO	Estimated Time of Return to Operation
ETS	Econometric Time Series
ETS	Electronic Tandem Switch(ing)
ETS	Electronic Translation|Translator System
ETS	Electronic Typing System
ETS	Elementary Transition System
ETS	Entry Terminal System
ETS	European Telecommunication Standard|System
ETS	European Teleprocessing System
ETS	Executable Test Suite
ETSACI	Electronic Tandem Switching Administration Channel Interface
ETSDU	Expedited Transport Service Data Unit
ETSI	European Telecommunications Standards Institute
ETSO	Extended Time-Sharing Option
ETSPL	Extended Telephone Systems Programming Language
ETSS	Electronic Transmission and Switching System
ETSS	Entry Time-Sharing System
ETSS	Experimental Time-Sharing System
ETSSP	Entry Terminal System Status Panel
ETT	Expected Test Time
ETT	Export Transit Time
ETTM	Electronic Toll and Traffic Management
ETX	End of Text|Transmission
ETXR	End-of-Task Exit Routine
EU	End User
EU	European Union
EU	Execution|Extension Unit
EU	Exposure Unit

EUA	Electrical Utilities Application
EUA	Erase Unprotected to Address
EUADC	End User Application Development Center
EUC	End User Computing
EUC	Extended Unix Code
EUCLID	Easily Used Computer Language for Illustration and Drawing
EUDG	European Data manager user Group
EUF	End User Facility
EUL	End User Layer gateway
EULA	End User Licensing Agreement
EUNET	European Unix Network
EURAM	European Research in Advanced Materials
EURATOM	European Atomic energy agency
EUREKA	European Research Cooperation Agency
EURESCOM	European Institute for Research and Strategic studies in (tele)Communications
EUROCOMP	European Computing congress
EUROCON	European Conference on electronics
EURODICAUTOM	European on-line Automated Dictionary and terminology databank
EUROMICRO	European association for Microprocessing and Microprogramming
EURONET	Packet-switched data network of the European Community
EuroPACE	European Professional and Academic Channel for Europe 2000
EUROSINET	European Open System(s) Interconnect Network
EUROSTAT	European Statistical (office)
EUSIDIC	European association of Information services
EUT	Express Update Tape
EUTELSAT	European Telecommunications Satellite
EUUG	European Unix Users Group
EUV	Extreme Ultra-Violet
EUVE	Extreme Ultra-Violet Explorer
eV	Electron-Volt(s)
EV	End Vector
EV	Extreme Value
EVA	Electron Vote Analysis
EVA	Error Volume Analysis
EVAL	Evaluate
EVAS	Electroformed Vibration Absorbing Structure
EVBDD	Edge Valued Binary Decision Diagram
EVDS	Electronic Visual Display Subsystem
EVE	European Video-conferencing Equipment|Experiment
EVFU	Electronic Vertical Format Unit
EVGA	Extended Video Graphics Adapter|Array
EVIL	Extensible Video Interactive Language
EVITA	Extended Videotext Intelligent Terminal Adapter
EVL	Electronic Visualization Laboratory
EVM	Extended Virtual Machine

EVMA	Expanded Virtual Machine Assist
EVN	European VLSI Network
EVPC	Even Vertical Parity Check
E-VPN	Enterprise Virtual Private Network
EVPREP	Event Preparation
EVR	Electronic Video Recording
EVS	Electro-optical Viewing System
EVT	Embedded Visual Tool(s)
EVX	Electronic Voice Exchange
EW	End of Work
EWA	Engineering Work Authorization
EWA	Erase/Write Alternate
EWAN	Emulator Without A (good) Name
EWICS	European Workshop on Industrial Computer Systems
EWM	End-of-Warning Marker
EWOS	European Workshop for Open Systems
EWS	Employee Written Software
EWS	Engineering Work Station
EWSD	Electronic Worldwide Switch Digital
EX	Exchange key
EX	Execut(iv)e
EX	Exit
EX	Expedited
EXACT	Exchange of Authenticated electronic Component technology and Test data
EXAM	Experimental Aerospace Multiprocessor
EXC	Excess
exc	excitation
exc	exciter
ExCA	Exchangeable Card Architecture
EXCD	Exceed
EXCH	Exchange
EXCLU	Exclusive
EXCP	Exception
EXCP	Execute Channel Program
EXCPAM	Execute Channel Program Access Method
EXCPN	Exception
EXCV	Excessive
EXD	External Device
EXDAMS	Extendable Debugging And Maintenance System
EXDT	Expedite
exe	executable (file extension indicator)
EXEC	Execute
EXECN	Execution
EXERPS	Extended Error Recovery Procedures
EXET	Execution Time
EXF	External Function
EXH	Extension Hunting

EXLIST	Exit List
EXOR	Exclusive OR
EXP	Expand
exp	experimental
EXP	Exponent
exp	exponential
EXPAND	Extensive Processing of Alpha-Numeric Data
EXPED	Expedite
EXPLOR	Explicit Patterns, Local Operations, and Randomness
EXPO	Exponent(ial)
EXPO	Exposition
EXPRESS	Expanded Parts usage Records and Structure System
EXPRESS	Expression of STEP (= Standard for The Exchange of Product model data)
EXPRESS-G	EXPRESS-Graphical version
EXPRESSNET	Express Network
EXPT	Export
EXR	Exception Request
EXSR	Exit Subroutine
EXT	Extension
EXT	External
EXTM	Extended Telecommunications Module
EXTRAN	Expression Translator
EXTREM	Extended Relational Model
EXTRN	External Reference
EXTSN	Extension
EXUG	European X User Group
E/Z	Equal Zero
E-ZINE	Electronic Magazine

F

f	factor
F	Fahrenheit
F	Farad
F	Feedback
f	femto
F	Field
F	Filter
F	Finish
F	Flag
F	Frequency
F	Function
F1	Frequency shift keying
FA	Factor Automation
FA	Field|Final Address
FA	Floating Add
FA	Full Adder

FAA	Flexible Automatic Assembly
FAAB	Floating Add Absolute
FAATE	Fault Analyzing Automatic Test Equipment
FAB	Fabrication plant of computer chips
FAB	Feature-Advantage-Benefit
FAC	Factor
FAC	Features for Attaching Communications
FAC	File Access Channel|Code
FAC	Floating Accumulator
FACCS	Flexible Assembly Cell Control System
FACD	Foreign Area Customer Dialing
FACE	Federation (international) of Associations of Computer users in Engineering
FACE	Field Alterable Control Element
FACE	Framed Access Command Environment
FACE	Functional Automatic Circuit Evaluator
FACOM	Fujitsu Automatic Computer
FACS	Facilities Assignment and Control System
FACS	Factory|Final Assembly Control System
FACS	Failure Analysis Control System
FACS	Financial Accounting and Control System
FACS	Floating-decimal Abstract Coding System
FACT	Facility for Automation, Control, and Test
FACT	Factory Automation, Control, and Test facility
FACT	Factual (compiler)
FACT	Federation of Automated Coding Technologies
FACT	Fully Automatic Compiling Technique
FACTS	Fully Automated Computer Testing System
FAD	Failure Activity Determination
FAD	Floating Add Double
FADEC	Full Authority Digital Engine Control
FADP	Federal Automatic Data Processing
FADPUG	Federation of Automatic Data Processing Users Group(s)
FADS	Force Administration Data System
FADS	Fortran Automatic Debugging System
FADSN	Floating Add Double Suppress Normal
FADU	File Access Data Unit
FAG	Face Adjacency Graph
FAH	Face Adjacency Hypergraph
FAHQ	Fully Automatic High Quality
FAIL	Fast Artificial-Intelligence Language
FAIM	Fairchild Artificial Intelligence Machine
FAIM	Flexible Automation & Intelligent Manufacturing (international conference)
FAIR	Failure Analysis Incident Report
FAIR	Failure Analysis Information Retrieval (program)
FAIR	Fully Automatic Information Retrieval
FAIRS	Failure Analysis Information Retrieval System

FAIRS	Federal Aviation Information Retrieval System (US)
FAIS	Factory Automation Interconnection System
FAITH	Forming And Intelligently Testing Hypotheses
FAL	File Access Listener
FALC	Final Assembly Logistics Control
FAM	Factory Automation Model
FAM	Fast Auxiliary Memory
FAM	File Access Manager
FAM	Floating Add Magnitude
FAM	Fuzzy Associative Memory
FAME	Flexible Automated Manufacturing Environment
FAMIS	Financial And Management Information System
FAMOS	Floating-gate Avalanche-injection Metal-Oxide Semiconductor
FAMP	Ferrite Aperture Memory Plate
FAMPR	Feature Assembly Manufacturing Process Record
FAMS	Flexible Automated Modelling and Scheduling
FAMS	Forecasting And Modelling System
FAMT	Fully Automatic Machine Translation
FAN	Facility Area Network
FAP	Failure Analysis Program
FAP	File Access Procedure|Protocol
FAP	Financial Analysis Program
FAP	Floating point Arithmetic Package
FAP	Format And Protocol
FAP	Fortran Assembly Program
FAP	Free Access Provider
FAP	Functional Assignment Panel
FAPI	Family Application Program Interface
FAPL	Format And Protocols Language
FAPM	Functional Activity Program Manager
FAPS	Financial Application Preprocessor System
FAQ	Frequently Asked Questions
FAQL	Frequently Asked Question List
FAQS	Fast Queuing System
FAR	Failure Analysis Report
FAR	Field Altering and Reconditioning
FAR	File Activity Ratio
FAR	File Address Register
FARADA	Failure Rate Data
FARNET	Federation of American Research Networks
FARS	Financial Accounting and Reporting System
FAS	Fast Access Store
FASCIA	Fixed Asset System Control Information and Accounting
FASNET	Fast Network
FASR	Forward Acting Shift Register
FASST	Flexible Architecture Standard System Technology
FAST	Factory Automation Support Technology
FAST	Fast Access Storage Technology

FAST	Field Activities Simulation Tool
FAST	File Analysis and Selection Technique
FAST	Final Automated System Test(s)
FAST	Fine-grained Architecture and Software Technologies
FAST	First Application System Test
FAST	Flexible Algebraic Scientific Translator
FAST	Flexible Automation in Shipbuilding Technology
FAST	Forecasting and Assessment in Science and Technology
FAST	Formula Automatic Scalar Translator
FAST	Functional Analysis Specification Tree
FASTER	Filing And Source data entry Techniques for Easier Retrieval
FASTOR	Fast Access Storage
FAT	File Access Table
FAT	File Allocation Table
FATAR	Fast Analysis of Tape And Recovery
FATS	Fast Analysis of Tape Surfaces
FAUST	Fault Analysis Using Simulation and Testing
FAVER	Fast Virtual Export/Restore
FAW	Final Associated With
fax	facsimile
fax	fax (file extension indicator)
FB	File|Fixed Block
FB	Finish Build
F/B	Foreground/Background
FB	Fuse Box
FBA	Fixed Block Architecture
FBB	Fiber Base Board
FBB	Full-Baud Bipolar
FBC	Feedback Balanced Code
FBDS	Feature Based Design Model
FBE	Free Buffer Enquiry
FBF	Fixed Block Format
FBFA	Feature-Based Fixturing Analysis
FBLK	Functional Block
FBM	Feature Based Modeller|Modelling
FBM	File Base Maintenance
FBM	Fixed Block Mode
FBM	Foreground and Background Monitor
FBO	Fixed-Base Operator
FBP	Fall Back Procedure
FBQE	Free Block Queue Element
FBS	Finish Build Schedule
FBSB	Forward Biased Second Breakdown
FBW	Fly By Wire
FC	Feed Check
FC	Fill Code
FC	Flow Controller
FC	Font|Flux Change

fc	foot-candle
FC	Frame Control
FC	Front-end Computer
FC	Full Colour
FC	Functional Character
FC	Function Code
FCA	Fiber Channel Association
FCA	Free Carrier
FCA	Functional Configuration Audit
FCAL	Fiber Channel Arbitrated Loop
FCAP	Floating point Commercial Arithmetic Processor
FCAR	Field Change Activity Report
FCAR	Forms Control Address Register
FCB	Feed Circuit Breaker
FCB	File Cache Buffer
FCB	File Control Block(s)
FCB	Forms Control Buffer
FCB	Function Control Block
FCC	Federal Communications Commission (US)
FCC	Federal Computer Conference
FCC	File Carbon Copy
FCC	Flight Control Computer
FCC	Font Change Character
FCC	Forbidden Combination Check
FCC	Formulae Categorizing Concept
FCC	Forward Command Channel
FCC	Full Custom Chip
FCC	Function Class Code
FCCFF	First Check Character Flip-Flop
FCCH	Frequency Correction Channel
FCCM	FPGA-based Custom Computing Machines
FCD	Failure Correction Decoding
FCDR	Failure Cause Data Report
FCFC	First Character Forms Control
FCFO	Full Cycling File Organization
FCFS	First Come, First Served
FCG	Facing
FCG	False Cross or Ground
FCG	Format Computer Graphics
FCH	Fetch
fci	flux changes per inch
FCI	Frame Copied Indicator (bit)
FCIA	Fiber Channel Industry Association
FCIC	Field Change Identification Code
FCIM	Flexible Computer Integrated Manufacturing
FCL	FAITH Control Language
fcl	flux changes per length
FCL	Format Control Language

FCL	Functional Capabilities List
FCM	Feature Code Master
FCM	Firmware Control Memory
FCM	Fuzzy Cognitive Map
FCO	Field Change Order
FCO	Field Check-Out (time)
FCO	Fixed Cycle Operation
FCP	Feed Control Panel
FCP	Fibre Channel Protocol
FCP	File Control Processor|Program
FCP	Floating-point Co-Processor
FCP	Function Control Program
fcpi	flux changes per inch
FC/PM	Facility Control/Power Management
FCPU	Flexible Control Processing Unit
FCR	File Component Rules
FCR	France Câbles et Radio
FCRAM	File Create And Maintenance
FCS	Federation of Communication Services (GB)
FCS	Fiber Channel Standard
FCS	Field Computing Services
FCS	File Control System
FCS	Finance Communication System
FCS	First Customer Ship(ment)
FCS	Fixed Control Storage
FCS	Flight Control System(s)
FCS	Frame Check(ing)|Control Sequence
FCS	Frame Check Sum
FCS	Function Control Sequence
FCS	Future Communications Systems
FCSI	Fiber Channel Systems Initiative
FCSI	Field Change Shipping Instruction
FCSS	Force Cooled Superconducting System
FCST	Forecast
FCT	File Control Table
FCT	Four Corner Test
FCT	Function
FCTS	Federal Compiler Testing Service
FCU	File Control Unit
FCUR	Field Change Uninstalled Report
FCVS	Fortran Compiler Validation System
FCW	Flag and Control Word
FC-x	Fiber Channel, level x (x = 0, 1, 2, 3, or 4)
FD	Feed
FD	Field Definition
FD	File Definition|Descriptor
FD	Flexible Drive|Diskette
FD	Floating Divide

FD	Floppy Drive|Diskette
FD	Frequency Divider
FD	Full Duplex
FD	Functional Decomposition
FD0	Floppy Drive 0
FDAD	Full Disk Address
FDAd	Functional Data Administrator
FDB	Field Descriptor Block
FDB	File Data Block
FD/BE	Finite Difference/Boundary Element
FDC	Field Distribution Center
FDC	Floppy Disk Control(ler)
FDC	Frame Dependent Control (mode)
FDCS	Functionally Distributed Computing System
FDCT	Factory Data Collection Terminal
FDD	Field Data Department
FDD	Flexible Disk Drive
FDD	Floating Divide Double
FDD	Floppy Disk Drive
FDD	Four Double|Dual Diodes
FDD	Frequency Divider and Distribution
FDDI	Fiber Digital Device Interface
FDDI	Fiber Distributed Data Interface
FDDL	Fiberoptics Distributed Data Link
FDDL	File Data Description Language
FDDN	Fiber Digital Data Network
FDEP	Formatted Data Entry Program
FDF	Fiber Distribution Frame
FDGS	Factory Data Gathering System
FDH	Floppy Disk Handler
FDHD	Floppy Drive High Density
FDHM	Full Duration Half Maximum
FDI	Format Directory
FDIS	Final Draft International Standard
FDISK	Fixed Disk
FDISS	Full Dielectric Silicon Substructure
FDJ	Fortuitous Distortion Jitter
FDL	Facility Data Link
FDL	Flexible Distance Learning
FDL	Forms Description Language
FDM	Factory Data Model
FDM	Finite Difference Method
FDM	Frequency Division Multiplex(ing)
FDMA	Frame-Division Multiple-Access
FDMA	Frequency Division, Multiple Access
FDMI	Function Data Management Interpreter
FD:OCA	Formatted Data: Object Content Architecture
FDOS	Floppy Disk Operating System

FDOS	Functional Disk Operating System
FDP	Fast Digital Processor
FDP	Field Developed|Development Program
FDP	Flat Domains Propagation
FDP	Flow Diagram Processor
FDP	Forms Description Program
FDP	Full Drive Pulse
FDR	Fast Dump Restore
FDR	Feeder
FDR	File Data Register
FDR	Floating Divide Remainder
FDR	Frequency Doubling Recording
FDS	Field Separator
FDS	Fire Detection System
FDS	Flexible Disk System
FDS	Floppy Disk System
FDSR	Floppy Disk Send/Receive
FDT	Formal Description Technique
FDT	Functional Description Table
FDT	Function Data Table
FDTD	Finite Difference Time Domain
FDU	Fixed Disk Unit
FDU	Form Description Utility
FDX	Full Duplex
F&E	Facilities & Equipment
FE	Feature Extraction
FE	Ferro-Electric
FE	Field Editor
FE	Field Erase|Error
FE	Format Effector (character)
FE	Framing Error
FE	Front End
FEA	Finite Element Analysis
FEA	Functional Economic Analysis
FEA	Functional Entity Actions
FEALD	Field Engineering Automated Logic Diagram
FEANLPI	Functional Entity Actions Network Layer Protocol Identifier
FEAT	Feature
FEAT	Frequency of Every Allowable Term
FEB	Functional Electronic Block
FEBE	Far End Block Error
FEBP	Fuzzy Error Back Propagation
FEC	Forward Error Correction
FEC	Forwarding Equivalent Class
FEC	Front-End Computer
FECN	Forward Explicit Congestion Notification
FECP	Front-End Communications Processor
FED	Far End Data

FED	Field-Emitter Display
FED	Format Element Descriptor
FEDLINK	Federal Library and Information Network (US)
FEDNET	Federal Information Network
FEDS	Fixed/Exchangeable Disk Store
FEDSIM	Federal computer performance evaluation and Simulation center
FEDSIM	Federal Systems Integration and Management center
FEE	Find End Entry
FE-EL	Ferro-Electric Electro-Luminescence
FEF	Factory Express File
FEFET	Ferro-Electric Field-Effect Transistor
FEFO	First-Ended, First Out
FEL	Free Electron Laser
FELS	Field Engineering Logistics System
FEM	Finite Element Method|Model(ling)
FEMA	Failure Mode and Effects Analysis
FEMAP	Finite Element Modelling And Pictorialization
FEMF	Foreign Electro-Motive Force
FEMOD	Feature based Modeller
FEN	Front-End Network
FENP	Front-End Network Processor
FEP	Front-End Processor
FEPI	Front End Programming Interface
FEPS	Facility and Equipment Planning System
FER	Front-End Rejection
FERF	Far End Receive Failure
FERMI	Formalization and Experimentation on the Retrieval of Multimedia Information (EU)
FERPM	FTAM Error Recovery Protocol Machine (FTAM = File Transfer and Access Method)
FERR	Ferrite
ferri	denotes magnetic properties
ferro	denotes magnetic properties
FERS	Financial Engineering Reporting System
FES	Forms Entry System
FESDK	Far East Software Development Kit
FESR	Finite Energy Sum Rule
FET	Field-Effect Transistor
FETT	Field-Effect Tetrode Transistor
FETVOM	Field-Effect Transistor Volt-Ohm-Milliammeter
FEU	Forty foot Equivalent Unit
FEUS	File Enquiry and Update System
FEV	Far End Voice
FEXT	Far End Cross-Talk
FF	Fan Fold
FF	Fixed Format
FF	Flat Film

FF	Flip-Flop
FF	Form Feed (character)
FF	Free Float (time)
FFA	Factory Flow Analysis
FFA	Field Failure Analysis
FFC	Flat Field Conjugate
FFC	Flat Flexible Cable
FFDC	First Failure Data Capture
FFE	Furniture, Fittings and Equipment
FFF	Ferro-resonant Flip-Flop
FFI	Feature File Index
FFL	Flip-Flop Latch
FFL	Front Focal Length
FFM	Fixed Format Message(s)
FFM	Flat Film Memory
FFM	Fundamental Frequency Modulator
FFMED	Fixed Format Message Entry Device
FFMM	Fundamental Frequency Magnetic Modulator
FFN	Full Function Node
FFP	Fast Fortran Processor
FFP	Formal Functioning Program
FFPI	Flip-Flop Position Indicator
FFRNS	Flat-Field Response, Noise Spectrum
FFS	Fast|Flash File System
FFS	Floating Foil Security
FFS	Formatted File System
FFST	First Failure Support Technology
FFT	Fast Fourier Transform(ation)
FFT	Final Form Test
FFTF	Fast Flux Test Facility
FFVUR	Front Feed Visual Unit Record
FG	Fiber Glass
FG	Foreground
FG	Forward Gate
FG	Frame Ground
FG	Functional Group
FGC	Fifth Generation Computer
FGCS	Fifth Generation Computer System
FGD	Fine Grain Data
FGDC	Federal Geographic Data Committee (US)
FGM	Functionally Graded Materials
FGN	Foreign
FGND	Frame Ground
FGPC	Fifth Generation Personal Computer
FGREP	Fixed Global Regular Expression Print
FGS	Fine Guidance Sensors
FGS	Fortran Graphics Support
FGS	Fourth|Fifth Generation Software

FGT	Foreground Table
FGU	Floating point/Graphics Unit
FH	Field|File Handler
FH	Flat|Fixed Head
FHD	Fixed Head Disk
FHDS	Fixed Head Disk|Drum Store
FHF	Fixed Head File
FHP	Fixed Header Prefix
FHSF	Fixed Head Storage Facility
FHSRB	File History Selection Request Block
FHSS	Frequency Hopping Spread Spectrum
FI	Field Intensity
FI	File Interchange|Initialization
FI	Finland
FI	Fixed Interval
FI	Front-end processor Interface
FI	Function Interpreter
FIA	Formal Interaction Analysis
FIAP	Fellow (of the) Institution (of) Analysts (and) Programmers
FIAR	Fault Isolation Analysis Routine
FIB	Fiber
FIB	File Information Block
FIB	Forwarding Information Base
fiberoptronics	fiberoptics and optoelectronics
FIBS	Field by Information Blending and Smoothing
FIC	Field Information Center
FIC	Field Installation Charge
FIC	First In Chain
FIC	Fixed Installation Charge
FICC	Freon Isopropyl Circuit Cleaner
FICON	Fibre Connection (channel)
FICON	File Conversion
FICS	Factory Information Control System
FICS	Financial Information and Control System
FICS	Forecasting and Inventory Control System
FID	Fédération Internationale de la Documentation
FID	Format Identification
FIDAC	Film Input to Digital Automatic Computer
FIDAS	Forms-oriented Interactive Database System
FIDE	Formally Integrated Data Environment (EU)
FIDES	Forecaster's Intelligent Discussion Experiment System
FIDIS	Financial Data Information System
FIDO	Finite Domain(s)
FIDO	Functions Input Diagnostic Output
FIDS	Flight Information Display System
FIDT	Forced Incident Density Testing
FIF	Family Information Facility
FIF	Financial Information File

FIF	Fractal Image Format
FI-FMD	Function Interpreter for Function Management Data
FIFO	First In, First Out
FIFO	Floating Input, Floating Output
fig	figure
FIGS	Figures Shift
FII	Field Installation Instruction
FIIT	Fault Isolation Interface Test
fil	filament
FIL	Fillister
FILEX	File Exchange
FILH	Fillister Head
FILL	Fillet
FILO	First In, Last Out
FILSYS	File System
FILU	Four-bit Interface Logic Unit
FIM	Fiber Interface Module
FIM	Flexible Intelligent Manufacturing
FIMA	Financial Management (system)
FIMAS	Financial Institution Message Authentication System
FIMBE	Focused Ion Molecular Beam Epitaxy
FIML	Full Information Maximum Likelihood
FIMM	Flexible Intelligent Microelectronics Manufacturing
FIMS	Federated|Financial Information Management System
FIMS	Form(s) Interface Management System
FIMS	Functionally Identification Maintenance System
FIN	Field Information Notice
FIN	Final
FIN	Finance
FINAC	Fast Interline Non-active Automatic Control
FINAR	Financial Analysis and Reporting
FINDES	Feature Integrated Design System
FINIS	Financial Industry Information Service
FINMAN	Financial Management
FINOS	Financial Operating System
FINSTAT	Financial times database of key Statistical information
FINTR	Financial Transaction
FINUFO	First In, Not Used, First Out
FIO	For Information Only
FIO	Frequency In and Out
FIOA	File Input/Output Area
FIOC	Frame Input/Output Controller
FIP	Facility Interface Processor
FIP	Fault Isolation Programs
FIP	File Processor (buffering)
FIP	Finance Image Processor
FIP	Fixed Interconnection Pattern
FIPS	Federal Information Processing Standard(s) (US)

FIPS-O	General description of the Federal Information Processing Standards register
FIQ	Frequently Invented Questions
FIR	Far Infra-Red
FIR	Field Information Report
FIR	File Indirect Register
FIR	Finite Impulse Response
FIRE	Factor Information Retrieval system
FIRL	Fiber-optic Inter-Repeater Link
FIRM	Financial Information for Resource Management
FIRM	Flowcharting Is Realistic Management
FIRP	Federal Internetworking Requirements Panel (US)
FIRST	Financial Information Reporting System
FIRST	For Inspiration and Recognition of Science and Technology
FIS	Factory Information System
FIS	Feasible Ideal System
FIS	Federated Information System
FIS	Field Instruction System
FIS	Financial Information System
FIS	Fixed Instruction System
FIS	Floating (point) Instruction Set
FISH	First In, Still Here
FISH	Forensic Information System for Handwriting
FISS	Finance Inter-System Support
FISSL	Finite State Specification Language
FIST	Fault Isolation by Semiautomatic Techniques
FIST	First In, Still There
FISUS	Fill In Signal Units
FIT	Field Installation Time
FIT	File Inquiry Technique
FIT	Fixed Installation Time
FIT	Functional Industrial Training
FITAL	Financial In Terminal Application Language
FITB	Fill In The Blank
FITCE	Fédération des Ingenieurs des Télécommunications de la Communauté Européenne (EU)
FITL	Fiber In The Loop
FITNR	Fixed In The Next Release
FITS	Flexible Image Transport System
FITS	Flexible Integrated Tool(ing) System
FITS	Functional Interpolating Transformational System
FIU	Federation of Information Users
FIX	Federal Information|Internet Exchange
FIX	Fixture
FIXBLK	Fixed Blocked
FIXML	Financial Information Exchange Markup Language
FIXUNB	Fixed Unblocked
FJ	Fiji

FJ	Fixed Jack
FJA	Functional Job Analysis
FK	Falkland Islands (Malvinas)
FKA	Formerly Known As
FKN	Fraunhofer Knowledge Network (DE)
FKT	Fyns Kommunale Telefonselskab (DK)
F/L	Fetch/Load
FL	Field Length
FL	File Label
FL	Filler
FL	First Level
FL	Fixed Length
FL	Flip Latch
FL	Florida (US)
FL	Fluid
fl	foot-lambert
FL	Functional Learning
FL	Liechtenstein
FLAG	Fiber-optic Link Around the Globe
FLAG	Fortran Load And Go
FLAME	Fault Location Automated by Monitored Emulation
FLAP	FDDI Link Access Protocol (FDDI = Fiber Distributed Data Interface)
FLAP	Flow Analysis Program
FLBIN	Floating-point Binary
FLC	Ferro-electric Liquid Crystal
FLC	Fixed Length Code
FLCN	Field Length Condition register
FLD	Field
FLD	Flux Line Dislocation
FLDEC	Floating-point Decimal
FLDL	Field Length
FLDS	Fault Location Diagnostics
FLE	Free List Exhausted
FLEA	Flux Logic Element Array
FLEA	Flux Logic Evaluation Assembly
FLEE	Fast Linkage Editor
FLEXER	Fundamental Loop Exerciser
FLEXIMIS	Flexible Management Information System
Flexo	Flexography
FLEXQUAR	Flexible high-Quality And Reliable production
FLFL	Flip-Flop
FLG	Flag
FLIC	Fortran Language Industrial Control
FLIH	First-Level Interrupt Handler
FLIM	Fast Library Maintenance
FLINT	Facilities Loading Investigation New Technique
FLINT	Floating Interpretive (language)

FLIP	Floating Indexed Point (arithmetic)
FLIP	Floating Instrument Platform
flips	fuzzy logic inferences per second
FLIR	Forward Looking Infrared Radar
FLIRT	Free Language Information Retrieval Tool
FLIS	Federal Logistics Information System (US)
FLIS	Flexible Interruption System
FLIT	Fault Location by Interpretive Testing
FLITE	Flight Information Test Element
FLL	Fixed Loss Loop
FLL	FoxPro Link Library
FLN	Functional Link Net
FLoC	Federated Logic Conference
FLOP	Floating Octal Point
FLOP	Floating point Operation
flops	floating-point operations per second
FLOTOX	Floating-gate Tunnel Oxide
FLOWGEN	Flowchart Generator
FLP	Floating-Point
FLPAU	Floating-Point Arithmetic Unit
FLPC	Floating-Point Computer
FLPROM	Fusible Link Programmable Read Only Memory
FLR	Filter
FLR	Flag Register
FLR	Floating-point Register
FLS	Field Logistic System
FLS	Floating License Server
FLSF	Font Library Service Facility
FLT	Fault Location Test
FLT	Flat(ness)
FLT	Floating
FLT	Front Load Tape
FLT	Functional Logic Trend
FLT	Function Linkage Table
FLTG	Floating
FLTS	FASTER Language Translation System (FASTER = Filing And Source data entry Techniques for Easier Retrieval)
FLTSATCOM	Fleet Satellite Communications system
FLU	Fluid
FLUID	Facility for Listing, Updating & Interpreting Decks
FLV	Finite Logical View
FLX	Field-effect Liquid X-tal
FM	Facility|Fault Management
FM	Feasibility Model
FM	Feedback Mechanism
FM	Fiduccia-Mattheyses
FM	Field Manual|Mark
FM	Field Merge

FM	File Manager
FM	Floating Multiply
FM	Form
FM	Format Manager|Mechanism
FM	Frequency Modulation
FM	Functional Module
FM	Function Management
FM	Micronesia (Federal States of)
FMA	Fault Modus Analysis
FMA	Federated Management Architecture
FMA	Frequency Measurement Adapter
FMAC	Facility Maintenance And Control
FMBS	Frame-Mode Bearer Service
FMC	Flexible Machining Center
FMC	Flexible Manufacturing Cell
FMCA	Failure Mode Criticality Analysis
FMCU	File Memory Control Unit
FMCW	Frequency-Modulated Continuous Wave
FMD	Function Management Data
FME	Frequency Measuring Equipment
FMEA	Failure Mode and Effect Analysis
FMECA	Fault Mode Effect and Criticality Analysis
FMEVA	Floating point Means and Variance
FMFB	Frequency Modulation Feedback
FMH	Function Management Header
FMI	Field Maintenance Instructions
FMIR	Frustrated Multiple Internal Reflectance
FMIS	Financial Management Information System
FMISC	Field Measurement Information System Center
FML	File Management Language
FML	File Manipulation Language
FMLF	File Management Loading Facility
FMLI	Form (and) Menu Language Interpreter
FMM	Frequency Multiplex Modulation
FMP	Full Multi-Programming
FMPP	Flexible Multi-Pipeline Processor
FMPS	Functional Mathematical Programming System
FMR	Facility Management Reporting
FMR	Feature Model Reconstruction
FMR	Ferro-Magnetic Resonance
FMR	Former
FMR	Frequency Modulation Recording
FMS	File Maintenance|Management System
FMS	Financial Management Service|System
FMS	Flexible Machining|Manufacturing System
FMS	Forms Management System
FMSK	Form Skip
fmt	format (file extension indicator)

FMV	Full Motion Video
FMVFT	Frequency Modulation Voice Frequency Telegraph
FMVT	Failure Mode Verification Testing
FN	Form Number
FN	Functional Network
FNA	Free Network Architecture
FNB	File Name Block
FNC	Federal Networking Council (US)
FND	Found
FNF	First Normal Form
FNP	Front-end Network Processor
FNPA	Foreign Numbering Plan Area
FNR	File Next Register
FNRC	Financial Network Readiness Consortium
FNS	Federated Naming Service
FNS	Feedback Node Set
FNT	File Name Table
fnt	font (file extension indicator)
FO	Faroer Islands
FO	Fiber Optics
FO	File Organization
FO	Follow On
FO	Forward Observer
FO	Frame Outstanding
FOAC	Federal Office Automation Center (US)
FOAF	Friend Of A Friend
FOC	Faint Object Camera
FOC	Fiber Optic(s) Communications|Cable
FOC	First Of Chain
FOC	Flow Of Control
FOC	Focal (point)
FOC	Free Of Charge
FOCA	Font Object Content Architecture
FOCAL	Formula Calculator
FOCIS	Fiber Optic Communications and Information Society
FOCIS	Financial On-line Central Information System
FOCS	Foundations Of Computer Science (conference)
FOCUS	Federation On Computing in the United States
FOCUS	Forum Of Control data Users
FOD	Fax On Demand
FOD	Format Office Document
FOD	Functional Operational Design
FODA	Feature-Oriented Domain Analysis
FODATEC	Feasibility demonstration of Office Document Architecture for Technical documents
FOF	Factory Of the Future
FOF	Friend Of (a) Friend
FOIL	File Oriented Interpretive Language

FOIL	First Outside, Inside Last
FOIL	Foil On Incandescent Light
FOIMS	Field Office Information Management System
FOIRL	Fiber Optic Inter-Repeater Link
FOK	Fill Or Kill
FOL	Field Operations Leader
FOL	Function Of Lines
FOM	Fiber Optic Modem
FOM	Functional Operating Module
FOMOT	Four Mode Ternary code
FOMS	Fiber Optic Myocardium Stimulator
FON	Fiber Optic Network
fon	font (file extension indicator)
fon	phone directory (file extension indicator)
FONPA	Finding Of No Practical Alternative
FONTGEN	Font Generation (program)
FOOBAR	FTP Operation Over Big Address Records (FTP = File Transfer Process)
FOOF	Fanout Observed Output Function
FOOT	Forum for Object Oriented Technology
FOP	Fade-Out Point
FOP	Flight Operations Planning
FOPC	First Order Predicate Calculus
FOPG	Federal networking council OSI Planning Group (OSI = Open Systems Interconnect)
FOPIC	Fiber Optic Input to Computers
FOPL	First Order Predicate Logic
FOPS	Forecast Operating System
FOPT	Fiber Optic Photo Transfer
FOR	Fewest Operations Remaining
for	fortran source code (file extension indicator)
FORAST	Formula Assembler Translator
FORC	Formula Coder
FORCK	Format Checking
FORD	Fix Or Repair Daily
FORD	Flip Over and Read Directions
FORDAP	Fortran Debugging Aid Program
FORDAP	Fortran Dynamic Analyzer Program
FOREM	File Organization Evaluation Model
FORGE	File Organization Generator
FORGO	Fortran load and GO
FORMAC	Formula Manipulation Compiler
FORMAN	Formula Manipulation
FORMDEF	Format Definition
FORML	Formal Object Role Modelling Language
FORMS	File Organization Management System
FORMS	Forms Management System
FORMTL	Form Tool

FORT	Formula Translation
FORT	Fortran
FORTE	File Organization Techniques
FORTE	Formal description Techniques
FORTE	Formal (description) Techniques on distributed systems and communications protocols (conference)
FORTE/PSTV	Formal description Techniques and Protocol Specification, Testing, and Verification
FORTH	Fourth generation language
Fortran	Formula Translation
FOS	Filed Oriented Support
FOS	File Operating System
FOS	Fortran Operating System
FOS	Function Operational Specification
FOSDIC	Film Optical Scanning Device for Input to Computers
FOSI	Format(ting) Output Specification Instance
FOSSIL	Fido/Opus/Seadog Standard Interface Layer
FOT	Fiber Optic Transceiver
FOT	Frequency of Optimum Traffic
FOTFL	Falls On The Floor Laughing
FOTS	Fiber Optic Transmission System
FOU	Field Operating Unit
FOV	Field Of View
FOX	Field Operational X.500
FP	Faithful Performance
FP	Fast Path
FP	File Processor|Protect
FP	Floating-Point
FP	Frame Pointer
FP	Functional Profile|Properties
FP	Function Processor
FPA	Flexible Production Automation
FPA	Floating Point Accelerator
FPA	Floating-Point Adder|Arithmetic
FPAL	Floating-Point Arithmetic Library
FPASD	Facsimile Packet Assembler/Disassembler
FPB	Floating-Point Board
FPB	Full Power Bandwidth
FPC	Fixed Program Computer
FPC	Flexible Program Computer
FPC	Floating Point Calculation|Coprocessor
FPC	Functional Processor Cluster
FPCE	Floating-Point C Extension (specification)
FPDU	FTAM Protocol Data Unit
FPDU	Function Protocol Data Unit
FPE	Floating Point Engine
FPE	Formal Program Error
FPF	Facility Parameter Field

FPF	Fast Path Feature
FPGA	Field Programmable Gate Array
FPI	Family Programming Interface
FPI	Fast Probability Integration
fpi	frames per inch
FPLA	Field Programmable Logic Array
FPLF	Field Programmable Logic Family
FPLMTS	Future Public Land Mobile Telecommunication Systems
FPLS	Field Programmable Logic Sequencer
FPM	Fast Page Mode
fpm	feet per minute
FPM	File Protect Memory
FPM	Floating-Point Multiplexer
FPM	Four Phase Modulation
FPNW	File and Print services for NetWare
FPO	Fixed Point Operation
FPO	For Position Only
FPODA	Fixed(-contention) Priority-Oriented Demand Assignment
FPP	Fixed Path Protocol
FPP	Floating-Point Package|Processor
FPPU	Floating-Point Processor Unit
FPR	File Protection Ring
FPR	Fixed Program Receive
FPR	Floating-Point Register|Representation
FPR	Forms Printing Requisition
FPROM	Field Programmable Read Only Memory
FPRP	Field Programmable Read-only memory Patch
FPS	Fast Packet Switching
fps	feet per second
FPS	Fiber Placement System
FPS	Field Programming System
FPS	Fixed Program Send
FPS	Flexible Production System
FPS	Floating-Point System
FPS	Focus Projecting and Scanning
fps	foot-pound-second
fps	frames per second
FPS	Future Programming System
FPSK	Fixed Phase-Shift Keying
FPSNW	File and Print Service for NetWare
FPSR	Floating-Point Status Register
FPT	First Pass Trigger
FPT	Forced Perfect Termination
FPU	File Processing Unit
FPU	Floating-Point Unit
FPV	Fixed Point Verifier
FQDN	Fully Qualified Domain Name
FQE	Free Queue Element

FQK	Fully Qualified Key
FQL	Formal Query Language
FQR	Formal Qualification Review
FQT	Formal Qualification Test
FQT	Functional Qualification Test
F/R	Failure and Recovery
FR	Field Requisition
FR	File Recovery|Register
FR	Final Report|Request
FR	Floating-point Register
FR	Frame Relay|Reprint
FR	France
fr	frankline
FR	Front
FR	Functions of Reads
FRA	Floating Reset Add
FRAC	Fraction(al)
FRACA	Failure Reporting, Analysis, and Corrective Action
FRACAS	Failure Reporting And Corrective Action System
FRACTAL	Fractional (dimensional)
FRAD	Frame Relay Access Device
FRAG	Fragment(ation)
FRAMEWORK	Formal Risk Assessment, Millennium Engineers, Workaround Options, Replacement policy, Keep going
FRAN	Framed Structure Analysis
FRAS	Feature Ratio Analysis System
FRB	Formal Reference Base
FRC	Front-Range Consortium
FRC	Functional Redundancy Checking
FRCS	Feature Ratio Control System
FRD	Floating Round
FRD	Functional Requirements Document
FRDY	Front Panel Ready
FRE	Full Radiance Equation
FRED	Frame Editor
FRED	Friendly Robotic Educational Device
FRED	Front-End for Databases
FRED	Front-End to Dish
FREDMAIL	Free Educational electronic Mail
FRENA	Frequency and Amplitude (system)
FREND	Fast Running program Enabling a Natural Diagnosis for the programming language ADA
FREQ	Frequency
FRET	Functional Reliability End Test
FRET	Functional Reliability Evaluation Technique
FRF	Frame Relay Forum
fri	flux reversals per inch
FRICC	Federal Research Internet Coordination Committee

FRIEND	Fast Running Interpreter Enabling Natural Diagnosis
FRIMP	Flexible Reconfigurable Interconnected Multi-Processor system
FRINGE	File and Report Information processing Generator
FRISC	Formally Reduced Instruction Set Computer
FRISCO	Framework for Integrated Symbolic/numeric Computation (EU)
FRL	Frame Representation Language
FRM	Fixed Range Marks
FRM	Frame
FRM	Functional Requirements Model
FRMG	Framing
FRMR	Frame Reject
FRN	Fixed Radix Notation
FROG	Free Ranging On Grid
FROLIC	Formal Retrieval Oriented Language for Indexing Context
FROM	Fusible Read Only Memory
FROOM	Features and Relations used in Object Oriented Modelling
FROSS	Fast Read-Only Storage Simulator
FRP	Fast Response Processing
FRP	File Reconstruction Procedure(s)
FRP	File Rules Pointer
frpi	flux reversals per inch
FRR	Functional Recovery Routine(s)
FRS	Fast Reporting System
FRS	Feature Recognition System
FRS	Flexible Route Selection
frs	graphics driver for wordperfect (file extension indicator)
FRSS	Financial Results Simulator System
FRTISO	Floating point Root Isolation
FRU	Failing Replaceable Unit
FRU	Field|Fixed Replaceable Unit
FRUGAL	Fortran Rules Used as a General Applications Language
FRUMP	Fast Reading and Understanding Memory Program
FRV	Format Restoring Vocoder
FRZ	Freeze
FS	Factor Storage
FS	Fail Safe
FS	Far Side
FS	Fetch and Store (instruction)
FS	Field Separator|Service
FS	File Separator (character)
FS	File Server|System
FS	Finite State
FS	Floating Subtract
FS	Floppy System
FS	Frame Status
FS	Frequency Shift
FS	Full Scale

FS	Functional Specification|Standard
FS	Functional Symbol
FS	Function Select
FS	Future System(s)
FSA	Field Search Argument
FSA	Finite State Automaton
FSA	Fixed Slot Acknowledgment
FSA	Framed Structure Analysis
FSAB	Floating Subtract Absolute
FSB	Forward Space Block
FSB	Fractional Sampling Bit
FSC	Field System Center
FSC	Fixed Self Contacting
FSC	Full Scale
FSCB	File System Control Block
FSCK	File System Check
FSCR	Field Select Command Register
FSCS	Functional Standard Conformance Statement
FSD	File System Driver
FSD	Full Scale Deflection
FSD	Functional Sequence Diagram
FSE	Field Service Engineer
FSE	Full Screen Editor
FSEC	Federal Software Exchange Center
FSF	Fixed Sequential Format
FSF	Forward Space File
FSF	Free Software Foundation
FSG	Format Standard Generalized markup language
FSG	Free Standards Group
FSIM	Functional Simulation
FSIP	Fast Serial Interface Processor
FSK	Frequency Shift Keying
FSL	Forecast Systems Laboratory
FSL	Formal Semantic Language(s)
FSLT	Full System Life Testing
FSM	Field-Strength Meter
FSM	Finite State Machine
FSN	File Server Network
FSN	Frame Sequence Number
FSN	Full Service Network
FSOS	Free Standing Operating System
FSP	File Security Period
FSP	File Service Process|Protocol
FSP	Forward Signalling Path
FSP	Full Screen Processing
FSR	Feedback Shift Register
FSR	File Space Rules
FSR	File Storage Region

FSR	Forward Space Record
FSR	Free System Resource(s)
FSR	Full Scale Range
FSS	Fast System Switch
FSS	Field System Support
FSS	Fixed Satellite Service(s)
FSS	Flying Spot Scanner
FSSA	Flight Service Station Automation
FSSM	Flying-Spot Scanner Memory
FST	Fast
FST	File Status Table
FST	Flat Square Tube (monitor)
FSU	Facsimile Switching Unit
FSU	Field Select Unit
FSU	File Support Utility
FSURAM	Functional Storage Unit Random Access Method
FSUROS	Functional Storage Unit Read-Only Storage
FSV	Floating-point Status Vector
FSW	Final Status Word
FT	Feature Translation
ft	feet
FT	Field|File Transfer
FT	Flow|Float Time
ft	foot
FT	Format Type
FT	Fourier Transformation
FT	France Télécom
FT	Frequency and Time
F/T	Full Time
FT2	Fixed Thermal Transfer
FTA	Fault Tree Analysis
FTAB	Field Tab
FTAM	File Transfer and Access Method
FTASB	Faster Than A Speeding Bullet
FTB	Filter Transmission Band
FTC	Fast Time Constant
FTC	Fault Tolerant Computer
FTC	Form-Throw Character
FTC	Fractal Transform Compression
FTC	Frequency Time Control
FTCS	Fault Tolerant Computing Symposium
FTCS	Future Technology Computer Systems
FTD	Fluorescent Tube Display
FTE	Frame Table Entry
FTE	Full-Time Equivalent
FTF	Face To Face
FTF	Factory Terminal Facility
FTF	File To File

FTF	File Transfer Facility
FTF	Flux Transition Frequency
FTG	Final Trunk Group
FTH	Fetch
FTI	Fixed Time Interval
FTIR	Fourier Transform Infrared Radiometer
FTL	Faster Than Light
FTL	Fast Transient Loader
FTL	Flash Translation Layer
FTL	Flexible Transfer Line
ft-L	foot-Lambert
ft-lb	foot-pound
FTM	File Transfer Method
FTM	Flat Tension Mask
FTM	Frequency Time Modulation
ftn	fortran source code (file extension indicator)
FTP	Fiber Termination Point
FTP	File Transfer Program|Protocol
FTP	Fixed Term Plan
FTP	Folded, Trimmed, and Packed
FTP	Fourier Transform Processor
FTPD	File Transfer Protocol Daemon
ftpi	flux transitions per inch
FTR	Factor
FTS	Fast Time Scale
FTS	Fault Test
FTS	Favorite Track Selection
FTS	Federal Telecommunications Standard|System
FTS	File Transfer Service
FTS	Financial Terminal|Transaction System
FTS	Free Time System
FTS	Frequency and Timing Subsystem
FTSA	Fault Tolerant Systems Architecture
FTSC	Federal Telecommunications Standards Committee
FTSPS	Field Technical Support Programming System(s)
FTT	Fault Test
FTT	Financial Transaction Terminal
FTTC	Fiber To The Curb
FTTH	Fiber To The Home
FTTO	Fiber To The Office
FTTP	Fault Tolerant Test Plan
FTTT	From Time To Time
FTTZ	Fiber To The Zone
FTW	File Tree Walk
FTZ	Fernmeldtechnische Zentralamt (DE)
FTZ	Flush-To-Zero
FU	Field Unit
FU	File Updating

FU	Fouled Up
FU	Functional Unit
FUBAR	Failed Unibus Address Register
FUBAR	Fouled Up Beyond All Recognition
FUBB	Fouled Up Beyond Belief
FUCT	Failed Under Continuous Testing
FUD	Fear, Uncertainty and Doubt
FUD	Field Use Date
FUD	Fire-Up Decoder
FUDD	Frequently Updated Distributed Data
FUI	File Update Information
FUIF	Fire Unit Integration Facility
FUNC	Function
FUNET	Finnish University (and research) Network
FUNLOC	Function Location
FUO	Final Used On
FURPSI	Functionality, Usability, Reliability, Performance, Supportability, Integratability
FUS	Fortran Utility System
F/V	Frequency to Voltage (converter)
FVN	File Version Number
FVO	For Valuation Only
FVT	Full Video Translation
FVU	File Verification Utility
FVWM	Feeble Virtual Window Manager
FW	Failure Warning
FW	First Word
FW	Full Wave
FWA	File Work Area
FWA	First Word Address
FWC	Four-Wire Circuit
FWCM	Fast Wavelet Collocation Method
FWD	Forward
FWHM	Full Width at Half Maximum
FWIW	For What It's Worth
FWL	Fixed Word Length
FWP	First Word Pointer
FWP	Full Write Pulse
FWP	Functional Wiring Principle(s)
FWT	Flexible Working Time
FX	Effects
FXBIN	Fixed Binary
FXCH	Floating-point (register) Exchange
FXD	Fixed
FXP	Fixed Point
FXPALU	Fixed Point address Arithmetic Logic Unit
FY	Fiscal Year
FYA	For Your Action

FYA	For Your Amusement|Attention
FYC	For Your Consideration
FYE	For Your Entertainment
FYI	For Your Information
FYIG	For Your Information and Guidance
FYS	For Your Signature
FZ	Frozen Zone

G

g	conductance
G	Deflection factor
G	Gate
G	Gauss
G	Generate
G	Generator
G	Giga (prefix)
g	gram
G	Graphics
g	gravity
g	grid
G	Ground
G2B	Government-To-Business
G2C	Government-To-Citizen
G2G	Government-To-Government
GA	General Availability
GA	Genetic Algorithm
GA	Go Ahead
GA	Graphic Adapter
GaAs	Gallium Arsenide
GAB	Graphic Adapter Board
GAB	Group Audio Bridging
GAC	General Access Copy
GAC	Geometric Adaptive Control
GAC	Global Area Coverage
GACB	Graphic Attention Control Block
GAD	Gate Anomaly Detector
GAD	Graphic Active Device
GADDR	Group Address
GADS	Geographical Analysis and Display System
GAG	Ground-Air-Ground
GAGS	General Application Guidance System
GAIA	GUI Application Interoperability Architecture (GUI = Graphical User Interface)
GAIC	Gallium Arsenide Integrated Circuit
GAL	Generalized Assembly Language
GAL	Generic Array Logic
GAL	Get A Life

GAL	Global Address List
GALESIA	Genetic Algorithms in Engineering Systems: Innovations and Applications (international conference)
GALPAT	Galloping Pattern
GAM	Generic Action|Activity Model
GAM	Graphic Access Method
GAN	Generalized Activity Network
GAN	Generating and Analyzing Networks
GAN	Global Area Network
GAP	General Assembly Program
GAP	Graphics Application Program
GAP	Groupe d'Analyse et de Prévision (FR)
GAP	Guide for Application Programming
GAPE	Graphics Aids to Packaging Engineering
GAPI	Gateway Application Programming Interface
GAPM	Generalized Access Path Method
GAPSS	Graphical Analysis Procedure for System Simulation
GARLIC	General Analysis Regardless of Logical Interconnections of Circuits
GARM	General AEC Reference Model (AEC = Architecture, Engineering, and Construction)
GARMAC	Garment Mechanization And Control
GAS	Get-Away Special
GAS	Global Address Space
GAS	Graphics Application System
GASP	General Activity Simulation Program
GASP	General Analysis of System Performance
GASP	Generalized Audit Software Package
GASP	Graph Algorithm Software Package
GASP	Graphic Application Subroutine Package
GAT	Gallium Arsenide Technology
GAT	Generalized Algebraic Translator
GAT	Georgetown Automatic Translator
GATD	Graphic Analysis of Three-dimensional Data
GATED	Gateway Daemon
GATS	Generalized Acceptance Test Software
GAUGE	General Automation Users Group Exchange
GB	Gain Bandwidth
GB	General Business
Gb	Gigabit(s)
GB	Gigabyte(s)
GB	Great Britain
GB	Grid Bias
GB	Group Band
GBCS	Global Business Communications Systems
GBF	Geographic Base File
GBGB	Graded Band Gap Base
GBH	Group Busy Hour

GBIS	Geographic Base Information System
GBIT	GigaBit
GBM	General Bookkeeping Machine
GBM	Generalized Bridge Method
GBMP	General Benchmark Program
GBP	Gain Bandwidth Product
Gbps	Gigabits per second
GBps	Gigabytes per second
GBS	General Business System
Gb/s	Gigabits per second
GBT	Great Big Table
GBTS	General Banking Terminal System
GBU	Graphics Business Unit
GBYTE	Gigabyte
GC	Garbage Collection
GC	Gate Connector
GC	Graphic(s) Context|Console
GCA	Game Control Adapter
GCA	General Communications Architecture
GCA	Global Chipcard Alliance
GCAP	Graphic Curve Analysis Program
GCB	Gate Control Block
GCB	General Circuit Breaker
GCB	Graphic Control Byte
GCC	General Communications Controller
GCC	Geometric Control Constraint
GCC	GNU C Compiler
GCC	GNU Compiler Connection
GCC	Group Control Change
GCCA	Graphic Communications Computer Association (US)
GCCS	General Consolidating Computing System
GCD	Greatest Common Divisor
GCDIS	Global Change Data and Information System
GCE	Gate Count Error
GCE	Ground Communication Equipment
GCF	Greatest Common Factor
GCF	Ground Communication Facility
GCHQ	General Communications Headquarters
GCI	General Circuit Interface
GCI	Generalized Communication Interface
GCL	Global Commerce Link
GCL	Graphic Control|Card Language
GCL	Graphics Command Language
GCLA	Group Carry Look-Ahead
GCM	Gateway Channel Module
GCM	General Circulation Model
GCM	Global Circulation Model
GCMD	Global Change Master Directory

GCOS	General Comprehensive Operating Supervisor
GCOS	Global Change Observation System
GCP	Guidance Computer
GCR	Graphic-Coded Recording
GCR	Gray Component Removal|Replacement
GCR	Group(-)Code(d) Recording
GCRA	Generic Cell Rate Algorithm
GCRP	Global Change Research Plan
Gc/s	Gigacycle per second
GCS	Graphics Compatibility System
GCS	Group Control System
GCSC	Guidance Control and Sequencing Computer
GCSMP	Graphic Continuous System Modelling Program
GCT	Graphics Communication Terminal
GD	Gate Driver
GD	Global Data
GD	Graphics Display
GD	Grenada
GD	Guide
GDA	Global Data Administrator|Area
GDB	Global Data Base
GDB	Gnome|Geometric Data Base
GDB	GNU Debugger
GDBMS	Generalized Data Base Management System
GDBS	Geofacilities Data Base Support
GDBS	Global Data Base System
GDC	Guidance Display Computer
GDDL	Graphical Data Definition Language
GDDM	Graphical|Graphics Data Display Manager
GD/DS	Generalized Dictionary/Directory System
GDE	Generalized Data Entry
GDF	Graphics Data File
GDF	Group Distribution Frame
GDG	Generation Data Gap|Group
GDH	Global Digital Highway
GDHS	Ground Data Handling System
GDI	Generalized Database Interface
GDI	Graphical|Graphics Device Interface
GDL	Graphic Display Library
GDLC	Generic Data Link Control
GDM	Global Data Manager
GDMI	Generic Definition of Management Information
GDMO	Guidelines for the Definition of Managed Objects
GDMS	Generalized Data Management System
GDN	Global Data Network
GDO	Gate-Dip Oscillator
GDO	Grid-Dip Oscillator
GDOA	Graphic Data Output Area

GDOS	Graphic Device Operating System
GDP	Gas Discharge Panel
GDP	Generalized Database|Data Processor
GDP	Geometric Data Processing
GDP	Geometric Design Processor
GDP	Goal Directed Programming
GDP	Graphic Data Processing
GDP	Graphic Display Program
GDP	Graphic Draw Primitive
GDPT	Graphical Database Presentation Tool
GDQF	Graphical Display and Query Facility
GDR	Geophysical Data Record
GDR	Group Decision Room
GDS	Generalized Data Stream
GDS	Graphic Data Syntax|System
GDS	Graphic Design System
GDS	Graphics Data Syntax
GDSDF	Generalized Data Structure Definition Facility
GDSIB	Global Digital Sea Ice data Bank
GDSS	Group Decision Support Systems
GD&T	Geometrical Design & Tolerancing
GDT	Geometric Dimensioning and Tolerancing
GDT	Global Descriptor Table
GDU	Graphic(al) Display Unit
GDV	Graphic Deflection Vector
GE	General Engineering
GE	Georgia
Ge	Germanium
GE	Greater than or Equal to
GEADAC	General Automatic Data Circulation
GECOM	General Compiler
GECOS	General Electric Comprehensive Operating System
GEDAN	Gerät zur Dezentrentralen Anrufweiterschaltung im Netzknoten (DE)
GEDI	Group (on) Electronic Document Interchange
GEDIT	General purpose text Editor
GEDW	Global Enterprise Data Warehouse
GEEM	Generic Enterprise Engineering Methodology
GEF	Global Environmental Facility
GEFF	Global Engineering Fact File
GEIMS	Generalized Inventory Management System
GEIS	General Electric Information Service
GEL	General Emulation Language
GEM	General Evaluation Model
GEM	General Event Monitor
GEM	Generic Enterprise Models
GEM	Global Enterprise Management

GEM	GRAI Evaluation Method (GRAI = Groupe de Recherche en Automatisation Intégrée) (FR)
GEM	Graphic(al) Environment Manager
GEM	Graphic(al) Extension Method
GEM	Graphic Expression Machine
GEM	Graphics Environment Manager
GEMBITS	Generic Electro-Mechanical Burn In Test System
GEMCOS	Generalized|Generative Message Control System
GEMDES	Government Electronic Messaging and Document Exchange Service
GEMINI	Government Expert systems Method Initiative
GEMS	Global Environment Monitoring System
GEMT&L	Generic Enterprise Modeling Tools & Languages
GEN	General
GEN	Generator
GENCMM	Generic Enterprise Change Management Methodology
GENDRA	Generalized Data Reduction and Analysis
GENIE	General Electric Network for Information Exchange
GENIE	General Information Extractor
GENISYS	Generalized Information System
genlock	generator locking
GENU	Generated Non-elementary Unit
GEO	Geostationary|Geosynchronous Earth Orbit
geocode	geographical code
GEOPS	Geodesy Program System
GEOS	Graphic Environment Operating System
GEP	General Extraction Program
GEPAC	General Purpose Automatic Checkout (system)
GEPPCOM	General-Purpose Parallel Computing (EU)
GERA	Generic Enterprise Reference Architecture
GERAM	Generic Enterprise Reference Architecture Methodologies
GERP	Global Environment Research Project
GERT	Graphic(al) Evaluation and Review Technique(s)
GERTS	General Remote Terminal System
GEST	Graphic Evaluation Systems Technique
GET	Get Execute Trigger
GET	Gross Error Test
GET	Ground Elapsed Time
GeV	Giga-electron-Volt
GF	French Guiana
GF	Gap Filler (radar)
GFA	Gust Front Algorithm
GFC	Generic Flow Control
GFCI	Ground Fault Circuit Interrupter
GFDA	Gust Front Detection Algorithm
GFDL	Geophysical Fluid Dynamics Laboratory
GFG	General Function Generator
GFI	Gap Filler Input

GFI	General Format Identifier
GFI	Ground Fault Interrupter|Interruption
GFI	Guided Fault Isolation
GFID	General Format Identifier
GFIML	Gap Filler Input Message Label
GFLP	General Facility Layout Problem
GFM	Gradient decent based Fiduccia-Mattheyses
GFMS	Generalized File Management System
GFP	Generalized File Processor
GFP	Green Fluorescent Protein
GFR	Guaranteed Frame Rate
GFT	Grant Functional Transmission
G/G	Ground-to-Ground
GGI	Generalized Graphical Input
GGN	Gotta Go Now
GGP	Gateway-to-Gateway Protocol
GH	Ghana
GHC	Gating Half Cycle
GHCN	Global Historical Climate Network
GHN	Get Hold Next
GHNP	Get Hold Next within Parent
GHPIN	Global Health Public Information Network
GHU	Get Hold Unique
GHz	Gigahertz
GI	Galvanized Iron
GI	General Information|Issue
GI	Gibraltar
GI	Graded Index
GI	Group Indicate
GIAM	Graphic Interactive Application Monitor
GIC	General Input/output Channel
GICS	Graphic Input/output Command System
GID	Group Identity
GIDEP	Government Industry Data Exchange Program
GIE	Group Indication Elimination
GIF	General Image Format
GIF	General Information File
gif	graphic interchange format (file extension indicator)
GIFS	Generalized Interrelated Flow Simulation
GIFT	General Information File Tester
GIGO	Garbage In, Garbage Out
GIHS	Generalized Information Handling System
GII	Global Information Infrastructure
GIL	General Instruction Logic
GIL	General-purpose Interactive programming Language
GIM	General Information Manual
GIM	Generalized Information Management
GIM	Global Integration Methodology

GIM	Graphic Integrated Manual
GIM	Group Information Mark
GIMIC	Guard-ring Isolated Monolithic Integrated Circuit
GIMP	GNU Image Manipulation Program
GIN	Graphics Input
GIOP	General-purpose Input/Output Processor
GIOP	Generic Inter-ORB Protocol (ORB = Object Request Broker)
GIOS	Graphic Input/Output System
GIP	Graphic Input Program
GIP	Graphic Interactive Processing
gips	giga instructions per second
GIPS	Ground Information Processing System
GIPSY	General Information Processing System
GIR	Generalized Information Retrieval
GIRL	General Information Retrieval Language
GIRLS	Generalized Information Retrieval and Listing System
GIRLS	Graphical data Interpretation and Reconstruction in Local Satellite
GIRS	Generalized Information Retrieval System
GIS	General Inquiry System
GIS	Generalized Information System
GIS	Generalized Initialization Sequencer
GIS	Geographic Information System
GIS	Global Information Solutions|System
GIS	Government|Graphical Information System
GIS	Guidance Information System
GISA	Geographic Information Systems Association
GIS/VS	Generalized Information System/Virtual Storage
GIT	Group Information Table
GIU	Guidance Interface Unit
GIWIST	Gee, I Wish I'd Said That
GIX	Global Internet Exchange
GJ	GigaJoule
GJ	Graphic Job (processor)
GJD	Germanium Junction Diode
GJP	Graphic Job Processor
GKB	Graphical Knowledge Base
GL	General Ledger
GL	Graphics Language
GL	Greenland
GL/1	Graphics Language/1
GLAP	Generalized Life Analysis Program
GLAS	General Ledger Accounting System
GLAS	General Logic Analysis Simulator|System
GLC	Gas Liquid Chromatography
GLEAM	Graphic Layout and Engineering Aid Method
GL/FICS	General Ledger/Financial Information and Control System
GLIN	Great Lakes Information Network

GLIS	Global Land Information System
GLIT	Ground Loop Impedance Tester
GLM	General Linear Models
GLOBE	Global Learning by Observations to Benefit the Environment
GLOMR	Global Low-Orbiting Message Relay
GLOS	General Ledger Operating System
GLP	Graphic Language Processor
GLPL	GNU Library general Public Licence
GLPPS	Graphical Lathe Part Programming System
GLS	Generalized Logic Simulator
GLU	Gluing
GLU	Green Lay-Up
GLUC	Green Lay-Up Composites
GLUT	General Ledger Utility Toolkit
gly	glossary (file extension indicator)
GM	Gambia
GM	Gated Memory
GM	General Macro-assembly
GM	General Manager
GM	General MIDI (= Musical Instrument Digital Interface)
GM	Generic Module
gm	gram
GM	Graphic Machine
GM	Group Mark
GMAP	Generalized Macro-Processor
GMAP	General Macro-Assembly Program
GMD	Gesellschaft für Mathematik und Datenverarbeitung (DE)
GMDH	Group Method of Data Handling
GME	Group Modulation Equipment
GMIDI	General Musical Instrument Digital Interface
GMIS	Generalized Management Information System
GMIS	Government Management Information Sciences
GMIS	Grants Management Information System
GML	Generalized Markup Language
GML	General Modelling Language
GML	Graphic Machine Language
GMM	Glass-Metal Module
GMMRIT	Glass-Metal Module Release Interface Tape
GMN	Global Manufacturing Network
GMR	General Modular Redundancy
GMR	Giant Magneto-Resistance
GMRAE	Geometric Mean Relative Absolute Error
GMRS	General Mobile Radio Service
GMRT	Giant Meter-wave Radio Telescope
GMS	Generalized Main Scheduling
GMS	General Maintenance System
GMS	Global Management System
GMS	Global Messaging Service

GMSK	Gaussian (prefiltered) Minimum Shift Keying
GMSS	Graphical Modelling and Simulation System
GMSTC	General Motor's standard for STEP Translation Center (STEP = Standard for The Exchange of Product data)
GMT	Generic Mapping Tools
GMTA	Great Minds Think Alike
GMV	Guaranteed Minimum Value
GMWM	Group Mark with Word Mark
GMX	Generalized Monitor Experimental
GN	General Network
GN	Get Next
GN	Given Name
GN	Green
GN	Guinea
GNA	Graphic Network Architecture
GNC	Generative|Geometric Numerical Control
GND	Ground(ing)
GND	Ground (signal/system)
GNIN	General Inquiry
GNL	General
GNM	Generic Network Model
GNMP	Government Network Management Profile (US)
GNN	Global Network Navigator
GNOME	GNU Network Object Model Environment
GNP	Get Next within Parent
GNS	Global Network Services
GNSS	Global Navigation Satellite System
GNU	GNU's Not Unix operating system (recursive acronym)
GO	General Operations|Order
GO	General Order|Output
GO	Generated Output
GOAL	Generator for Optimized Application Languages
GOAL	GIS-Oriented Analysis Language (GIS = Graphical Information System)
GOCA	Graphics Object Content Architecture
GOES	Geostationary Operational and Environmental Satellite
GOL	Goal Oriented Language
GOLD	Global On-Line Database
GOMAC	Government Microcircuit Applications Conference
GOMS	Goals Objectives Methods Selections
GONG	Global Oscillation Network Group
GOP	Group Of Pictures
GOR	General Operational|Operations Requirement
GOS	Grade Of Service
GOS	Graphics Operating System
GOSET	Groupe Opérationelle pour le Standard d'Echange et de Transfert (FR)
GOSIP	Government Open Systems Interconnect Profile

GOSN	Goal Objective Strategy Need
GOSS	Ground Operational Support System
GOTH	Gothic
gov	government
GOWON	Gulf Offshore Weather Observing Network
GP	Gang Punch
GP	Generalized Programming
GP	General Processor|Products
GP	General Purpose
GP	Graphic Package
GP	Graphic(s) Processor
GP	Group Processor
GP	Guadeloupe
GPA	General Procurement Agreements
GPA	General Purpose Array
GPA	Grade Point Average
GPAC	General Purpose Analog Computer
GPAC	Graphics Package
GPACK	General utility Package
GPAR	Generalized Performance Analysis Reporting
GPATS	General Purpose Automatic Test System
GPBEST	General Purpose Boundary Element Software Technology
GPC	General Peripheral Controller
GPC	General Purpose Computer
GPC	Germanium Point-Contact
GPCA	General Purpose Communications Adapter
GPCB	General Purpose Communications Base
GPCF	General Purpose Computing Facility
GPCI	Graphics Processor Command Interface
GPCP	Generalized Process Control Programming
GPC/P	General Purpose Controller/Processor
GPD	General Protocol Driver
GPD	General Purpose Data|Discipline
GPDC	General Purpose Device Controller|Computer
GPDS	General Purpose Discrete Simulator
GPDS	General Purpose Display System
GPF	General Protection Fault
GPG	Graphics Program Generator
GPGS	General Purpose Graphic System
GPI	Generalized Packaging Interface
GPI	General Package|Process Interface
GPI	General Purpose Interface
GPI	Global Project for Internet
GPI	Graphical|Graphics Programming Interface
GPI	Gross Points Installed
GPI	Ground Position Indicator
GPIA	General Purpose Interface Adapter
GPIB	General Purpose Interface Bus

GPIBA	General Purpose Interface Bus Adapter
GPIC	General Purpose Intelligent Cable
GPIF	General Purpose Interface
GPIO	General Purpose Input/Output
GPL	Gap Length
GPL	Generalized Parameter List
GPL	General Purpose Language
GPL	GNU Public License
GPL	Graphics Programming Language
GPLAN	Generalized Plan
GPM	General Purpose Module|Macro(generator)
GPM	Geometric Product Modelling
GPM	Gross Processing Margin
GPN	General Performance Number
GPN	Government Packet Network
GPOS	General Purpose Operating System
GPP	General Purpose Processor|Program
GPP	Generative Production Process(es)
GPR	General Purpose Register
gPROMS	generalized Process Modelling System
GPRS	General Packet Radio Service
GPS	General Problem Solver|System
GPS	General Programming Subsystem
GPS	Global Positioning Satellite|System
GPS	Global Professional Services
GPS	Graphic Programming Services
GPS	Great Plains Software
GPSCS	General Purpose Satellite Communications System
GPSI	Graphics Processor Software Interface
GPSN	General purpose Packet Satellite Network
GPSS	General Purpose Simulation System
GPSS	General Purpose Systems Simulator
GPT	General-Purpose Terminal
GPU	General|Graphical Processing Unit
GPU	Ground Power Unit
GPUCP	General Purpose User Control Program
GPX	Generalized Programming Extended
GQ	Equatorial Guinea
GQ	Generation Qualifier
GQL	Graphical Query Language
GQM	Goal/Questions/Metrics
GQU	Generalized Queue Entry
GR	General Records|Register
GR	Grade
GR	Greece
GR	Group
GRAD	Generalized Remote Access Database
GRADS	Generalized Remote Access Database System

GRAF	Graphic Additions to Fortran
GRAFCET	Graphe de Commande Etape-Transition (FR)
graftal	graphical fractal
GRAI	Graph with Results and Activities Interrelated
GRAI	Groupe de Recherche en Automatisation Intégrée (FR)
GRAIL	Graphic Input Language
GRAIN	Graphics-oriented Relational Algebraic Interpreter
GRAM	Globus Resource Allocation Manager
GRAMPA	General Analytical Model for Process Analysis
GRANADA	Grammatical Non-Algorithmic Data description
GRANIS	Graphical Natural Inference System
GRAPH	Graphic(s)
GRAPHMOD	Graphic Modification (module)
GRAS	Graphical Representation of Algorithms and Structures
GRASP	Generalized Read And Simulate Program
GRASP	Generalized Remote Acquisition and Sensor Processing
GRASP	Generalized Restarting And Scheduling Procedure
GRASP	General Resource Allocation and Scheduling Program
GRASP	Graphical Representation of Algorithms, Structures and Processes
GRASP	Graphics Subroutine Package
GRASP	Graphic System for (on-line) Plotting
GRASS	Geographic(al) Resources Analysis Support System
GRCAL	Group Calculate
GRCTRL	Group Control
GRD	GOSIP Register Database
GRD	Grind(ing)
GR&D	Grinning, Running & Ducking
GRE	Generic Routing Encapsulation
GRE	Graphics Engine
GREG	Groupement pour la Recherche et l'Etude des Genomes (FR)
GREP	Global Regular Expression Print
GRF	Geographic Reference File
grf	graph (file extension indicator)
GRG	Generalized Reduced Gradients
GRG	Graphical Rewriting Grammar
GRIB	Gridded Binary (data format)
GRIB	Grid In Binary
GRID	Global Resource Information Database
GRIN	Graded Indices
GRIN	Gradient Index
GRIND	Graphical Interpretive Display
GRINDER	Graphical Interactive Network Designer
GRINS	General Retrieval Inquiry Negotiation Structure
GRIP	Graphics Interactive Programming
GRIPS	Government Raster Image Processing Software
GRIS	Graphics System group
GRN	Grain

GRN	Green
gro	gross
GROPE	General Reconnaissance Of Peripheral Equipment
GROPE	Graphical Representation Of Protocols in Estelle
GROS	Goods Receiving On-line System
GRP	Gaussian Random Process
grp	group (file extension indicator)
GRPMOD	Group Modulator
GR/PS	Graphical Representation/Phrase Representation
GRR	Guidance Reference Release
GRRA	Guidance Reference Release Alert
GRS	General Records Schedule
GRS	General Register Stack
GRS	General Reporting|Retrieval System
GRS	Goal Representation Set
GRSUP	Group Suppression
GRT	General Recomplement Trigger
GRT	Greater Than
GRTS	General Remote Terminal Supervisor
GS	Gateway Switch
GS	General|Graphics System
GS	General Schedule|Storage
GS	General Synthesizer
GS	Gray-Scale
GS	Group Separator (character)
GSAM	Generalized Sequential Access Method
GSAT	Global Satellite data Acquisition Team
GSC	Ground Support Configuration
GSC	Group Switching Center
GSD	General Situation Display
GSDB	Genome Sequence Data Base
GSDC	Geodetic Satellites Data Service
GSE	Gate Substitution Error
GSF	General Source File
GSI	General Server Interface
GSI	Grand Scale Integration
GSIU	Ground Standard Interface Unit
GSKT	Gasket
GSLIB	Geo-Statistical Software Library
GSM	Generalized Sequential Machine
GSM	Generative Shape Modelling
GSM	Global Shared Memory
GSM	Global System for Mobile communications (network)
GSM	Graphics System Module
GSN	Gigabyte System Network
GSN	Government Satellite Network
GSOH	Good Sense Of Humour
GSOS	Graphics Operating System

GSP	General Systems of Preference
GSP	Global Strategy Planner
GSP	Graphics Subroutine Package
GSP	Graphics System Program
GSP	Group Step Pulse
GSPAN	Graphic S Plane Analysis
GSPN	Generalized Stochastic Petri Net
GSS	Graphics Support System
GSS	Graphics Symbol Set
GSSAPI	Globus Security System Application Programming Interface
GSSC	Ground Support Simulation Computer
GST	Global Storage Table
GSTDN	Ground Spacecraft Tracking and Data Network
GSTN	General Switch(ed) Telephone Network
GSTS	Ground-based Surveillance and Tracking System
GSVC	Generalized Supervisor Call(s)
GSW	Graphics System Workstation
G/T	Gain-to-noise Temperature ratio
GT	Game Theory
GT	Gate
GT	Generic Theory
GT	Give Token
GT	Graphics Terminal
GT	Greater Than
GT	Group Technology
GT	Guatemala
GTA	Give Token Acknowledgment
GTA	Government Telecommunications Agency
GTC	Give Token Confirm
GTC	Good Till Canceled
GTD	Gated
GTD	Geometrical Theory of Diffraction
GTD	Graphic Table Display
GTDPL	Generalized Top-Down Parsing Language
GT+E	General Telephone + Electronics
GTF	General(ized) Trace Facility
GTF	Greater Than Flag
GTG	Gating
GTL	Gunning Transceiver Logic
GTMOSI	General Teleprocessing Monitor for Open Systems Interconnect
GTN	Global Transportation|Trends Network
GTN	Government Telecommunications Network
GTNI	Global Transportation Network Interface
GTO	Gate Turn-Off (thyristor)
GTO	Guide To Operations
GTOSCR	Gate Turn-Off Silicon Controlled Rectifier
GTP	Gap Time Pulse
GTP	General Tracking Program

GTP	Graphic Transform Package
GTS	Global Telecommunications System
GTS	Graphic Terminal Services
GTS	Ground Test Subsystem
GTT	Gate Terminal
GTT	Global Title Transmission
GTTP	Get To The Point
GTXT	Generate Text (character set)
GU	Get Unique
GU	Guam
GUB	Generalized Upper Bounding
GUDT	Ground type Uni-Directional Transducer
GUF	General Utility Functions
GUI	Generalized User Interface
GUI	Graphical User Interface
GUID	Globally Unique Identification
GUIDE	Graphical User Interface Development Environment
GUIDE	Guidance to Users of Integrated Data processing Equipment
GUINS	Global Unit Identification Numbering System
GURU	General User Repair Utility
GUS	Guide for/to the Use of Standards
GUSIT	General Usage Shorts and Impedance Tests
GUT	Grand Unified Theory
GUUG	German Unix User Group
GVPN	Global Virtual Private Network
GVT	Global Virtual Time
GVU	Graphics, Visualization & Usability
GW	Gamma World
GW	Graphics Workstation
GW	Guinea-Bissau
GW-BASIC	Gee Whiz BASIC
GWEN	Ground Wave Emergency Network
GWM	Generic Window Manager
GWPAS	General Workforce Performance Appraisal System
GWS	Graphics Work Station
GY	Gray|Grey
GY	Guyana
GZ	Galvanized
GZIP	GNU Zip

H

H	Half adder
H	Harmonic
h	header (file extension indicator)
h	hecto (prefix)
H	Henry
H	Hexadecimal

H	High
H	Horizontal
h	hour
H	Hue
H	Hundred(s)
H	Hydrogen
H	Magnetic field strength
HA	Half Adder|Adjust
HA	Header Authentication
HA	Home Address|Automation
HAA	Human Action Analysis
HAB	Home Address Back
HAC	Hierarchical Abstract Computer
HAC	Hydraulic Analog Computer
HACD	Home Area Customer Dialing
HACE	High-order Automatic Cross-connect Equipment
HACMP	High Availability Cluster Multi-Processing
HACU	Handling And Conditioning Unit
HACWS	Highly Available Control Workstation
HAD	Herein After Described
HADA	High Availability Disk Array
HAG	Home Address Gap
HAK	Hugs And Kisses
hal	halogen
HAL	Hard Array Logic
HAL	Hardware Abstraction|Adaptation Layer
HAL	Heuristically programmed Algorithmic (computer)
HAL	High Automated Logic
HAL	Highly Automatic Logic
HAL	House-programmed Array Logic
HALT	Highly Accelerated Life Testing
HAM	Hierarchical Address(ing) Method
HAM	Housing Assembly Machine
HAM	Hybrid Access Method
HAMA	High Accuracy Mathematics Algorithm
HAMOTS	High Accuracy Multiple Object Tracking System
HAMT	Human Aided Machine Translation
HAP	Host Access Protocol
HAPUB	High-speed Arithmetic Processing Unit Board
HAR	Home Address Register
HARDMON	Hardware Monitor
HARM	High Availability, Reliability, and Maintainability
HART	Highway Addressable Remote Transducer
HASP	High-level Automatic Scheduling Program
HASP	Houston Attached Support Processor
HASP	Houston Automatic Spooling Program|Process
HASQ	Hardware Assisted Software Queue
HASS	High Availability Sub-System

HASS	Host Access Sub-System
HATRS	High Altitude Transmit/Receive Satellite
HAU	Horizontal Arithmetic Unit
HAU	Hybrid Arithmetic Unit
HAVEN	Hyper-programmed Agents for Virtual Environment
HAVI	Home Audio/Video Interoperability
HBA	Host Bus Adapter
HBAR	Head Bar Address Register
HBCI	Home Banking Communication|Computer Interface
HBD	Half-Baud Bipolar
HBD	Half Byte Decimal
HBEN	High Byte Enable
HBR	High Bit Rate
HBS	High Byte Strobe
HC	Hamming Code
HC	Handling Capacity
HC	Held Code|Covered
HC	Hierarchical Classification
HC	Hold(ing) Coil
HC	Host Command|Computer
HC	Hybrid Computer
HC	Hyper Channel
HCA	Head of Contracting Activity
HCD	Hard-Copy Device
HCD	Hot Carrier Diode
HCF	Hash Coding Function
HCF	Host Command Facility
HCI	Host|Human Computer Interface
HCI	Hybrid Computer Interface
HCL	Hardware Compatibility List
HCL	Hop Count Limit
HCM	Half-Cycle Magnetizer
HCM	Human Capital and Mobility
HCMTS	High Capacity Mobile Telecommunications System
HCO	Head of Contracting Office
HCP	Hard Copy Printer
HCP	High-speed Channel Processor
HCP	Host Command|Communications Processor
HCP	Host Computer Processor
HCR	Hardware Check Routine
HCR	Hybrid Communication Routine
HCRST	Hardware Clipping, Rotation, Scaling, and Translation
HCS	Hard Copy System
HCS	Header Check Sequence
HCS	Heterogeneous Computer System
hcs	hundred calls per second
HCSDS	High-Capacity Satellite Digital Service
HCSS	High Capacity Storage System

HCT	Hard Copy Task
HCT	Hook Control Table
HCU	Home Computer User
HCV	High Capacity Voice|Voltage
HD	Half Duplex
HD	Hard
HD	Hard Disk
H/D	Head/Disk
HD	Head (driver)
HD	Heading to Detail
HD	Heavy Density|Duty
HD	Hierarchic(al) Direct
HD	High Density
HD	Human Dialogue service
HDA	Hail Detection Algorithm
HDA	Head (and) Disk Assembly
HDAM	Hierarchical Direct Access Method
HDAS	Hybrid Data Acquisition System
HDB	High Density Bipolar|Buffer
HDB3	High Density Bipolar-3
HDBC	High Density Bipolar Coding
HDBV	Host Data Base View
HDC	High-speed Data Channel
HDCD	High Definition Compatible Digital
HDD	Hard Disk Drive
HDD	Hybrid Decision Diagram
HDDI	Host Displaywriter Document Interchange
HDDR	High Density Digital Recording
HDDS	High Density Data Stream
hdf	hierarchical data format (file extension indicator)
HDF	High Density Flexible
HDG	Heading
HDH	HDLC Distance Host protocol (HDLC = High-level Data Link Control)
HDI	Head (to) Disk Interference
HDI	Heidi Device Interface
HDI	Help-Desk Institute
HDI	High Definition Interlaced
HDL	Handle
HDL	Hardware Description|Design Language
HDL	High Density Lipoprotein
HDL	High-level Description Language
HDL	Host Document Library
HDLA	High-level Data Link control Adapter
HDLC	High-level|High-speed Data Link Control
HDLCM	High Density Line Conditioning Module
HDLM	High-level Data Linkage Module
HDLR	Hexadecimal symbolic Loader

HDM	Hardware Device Module
HDM	Hierarchical Development Methodology
HDM	High Density Modem
HD-MAC	High Definition – Multiplexed Analogue Components
HDML	Handheld Device Markup Language
HDMR	High Density Multitrack Recording
HDN	Harden
HDNG	Heading
HDOS	Hard Disk Operating System
HDP	High Definition Progressive
HDPA	Help-Desk Professional Association
HDPE	High Density Polythene
HDQ	High Definition Quincunx
HDR	Header
HDR	High Density Recording
HD-ROM	High Density – Read Only Memory
HDS	Hierarchical Distributed System
HDS	Hybrid Development System
HDSC	High Density Signal Carrier
HDSL	High-bit-rate Digital Subscriber Line
HDSL	High (data) rate Digital Subscriber Line
HDT	High Definition Television
HDT	High Density Tape
HDT	Host Digital Terminal
HDTC	Hold, Drop, Transfer, Conference
HDTL	Hybrid Diode Transistor Logic
HDTR	High Density Tape Recorder
HDTV	High Definition Television
HDVD	High Definition Volumetric Display
HDW	Hardware
HDWDM	High Density Wavelength-Division Multiplexing
HD-WDM	High Duplex Wave Division Multiplexing
HDX	Half Duplex
He	Helium
HEAD	Heterogeneous Extensible And Distributed database management system
HEANET	Higher Education Authority Network
HEAT	Higher Education and Advanced Technology
HEC	Header Error Check
HEDM	High Energy Density Matter
HEEB	High Energy Electron Beams
HEED	High Energy Electron Diffraction
HEL	Hardware Emulation Layer
HEL	Helical
Hellaspac	The Greek packet-switched network
HELP	Header Listing Program|Printing
HEM	High-level Entity Management
HEM	Hybrid Electro-Magnetic (wave)

HEMS	High-level Entity Management|Monitoring System
HEMT	High Electron Mobility Transistor
HENNA	Home Executives National Networking Association
HENSA	Higher Education National Software Archive (US)
HEO	High Earth|Elliptical Orbit
HEP	Heterogeneous Element Processor
HEP	High Energy Physics
HEPA	High Efficiency Particulate Air
HEPDB	High Energy Physics Data Base
HEPiX	High Energy Physics – Unix oriented
HEPNET	High Energy Physics Network
HEPVM	High Energy Physics – Virtual Machine implementations
HER	Human Error Rate
HERF	Hazards of Electromagnetic Radiation to Fuel
HERF	High Energy Radio Frequency
HERF	High Energy Rate Forming
HERMES	High-performance Multimedia information management Systems (EU)
HEST	High-Explosive Simulation Technique
HETP	High Energy Theoretical Physics
HETS	High Environmental Test System
HEX	Hexadecimal
HEX	Hexagon
HEXCALC	Hexadecimal Calculator
HF	Half
HF	Height Finder
H/F	Held For
HF	High Frequency
HFC	Hybrid Fiber Coax(ial cable)
HFDF	High Frequency Distribution Frame
HFE	Human Functional Entity
HFN	Home Financial Network
HFO	Human Functional Operation
HFP	Host-to-Front-end Protocol
HFR	High Flux Reactor
HFR	Hold For Release
HFS	Height Finder Supervisor
HFS	Hierarchical File System
HFT	High Function Terminal
HFT	Host File Transfer
HFWD	Halfword
hg	hectogram
Hg	Mercury
HGA	Hercules Graphics Adapter
HGA	High-Gain Amplifier|Antenna
HGC	Hercules Graphics Card
HGCEP	Hierarchical Gradual Constraint-Enforced Partitioning
HGCP	Hercules Graphics Card Plus

HGML	Hypertext General Markup Language
HGT	Height
HH	Heading to Heading
H/H	Host to Host
hh	hour
HI	Hawaii (US)
Hi8	High-band 8 mm
HIAVA	High Availability
HIBC	Health Industry Bar Code
HIBS	Health Information Base System
HIC	Human Interaction Component
HIC	Hybrid Integrated Circuit
HICAP	Hierarchical Interactive Computer-Aided Placement
HICAPOM	High-Capacity Communication(s)
HICOM	High technology Communication
HICS	Hierarchical Information Control System
HIDAM	Hierarchical Indexed Direct Access Method
HIDES	Highway Design System
HID/LOD	High Density/Low Density
HIDM	High Information Delta Modulation
HIF	Hyper-g Interface Format
HIFD	High-density Floppy Disk
HiFi	High Fidelity
HIFT	Hardware Implemented Fault Tolerance
HIGFET	Heterostructure Insulated Gate Field Effect Transistor
HIGZ	High-level Interface tot Graphics and Zebra
HIL	Human Interface Link
HILAPI	High Level Application Programming Interface
HILI	Higher Level Interface
HIM	Hardware Interface Module
HIM	Hierarchy of Interpretive Modules
HIMAC	Hierarchical Management And Control in manufacturing systems (EU)
HIMEM	High Memory
HIMSS	Healthcare Information Management Systems Society
HINIL	High Noise Immunity Logic
HIO	Halt|High Input/Output
HIP	Host Interface Processor
HIPO	Hierarchy, Input, Process, Output
HIPOT	High Potential
HIPPARCH	High Performance Protocol Architecture (EU)
HIPPI	High Performance Parallel|Peripheral Interface
HIRES	High Resolution
HIRLAM	High-Resolution Limited Area Model
HIROS	Hitachi Industrial Real-time Operating System
HIS	Hardware Interrupt System
HIS	Headquarters Information System
HIS	Homogeneous Information Sets

HIS	Hospital Information System
HISAM	Hierarchical Indexed Sequential Access Method
HISDAM	Hierarchical Indexed Sequential Direct Access Method
HISS	Hospital Information Support System
HISTORIC	Hurlsey Information System Terminal Oriented Retrieval Information Center
HIT	High Isolation Transformer
HIT	Human Interface Technology
HI-TECH	High Technology
HITS	Hyperlink Induced Topic Research
HIVOS	High Vacuum Orbital Simulator
H&J	Hyphenation & Justification
HK	Hook
hl	hectoliter
HL	High Level
HL	Host Language
HLA	High Level Assembler
HLA	High-speed Line Adapter
HLAF	Higher Level Arithmetic Function
HLAIS	High Level Analog Input System
HLAN	Huge Local Area Network
HLAPI	High Level Application Programming Interface
HLCO	High Low Close Open
HLD	Hardware Logic Diagram
HLD	Hold
HLDA	Hold Acknowledge
HLDTL	High Level Diode Transistor Logic
HLHSR	Hidden Line & Hidden Surface Removal
HLI	Hard Limited Integrator
HLI	High Level Interface
HLI	Host Language Interface
HLL	High Level Language|Logic
HLLAPI	High Level Language Application Program Interface
HLML	High Level Microprogramming Language
HLMPL	High Level Micro-Programming Language
hlp	help (file extension indicator)
HLP	High Level Protocol
HLPI	High Level Programming Interface
HLPIU	High Level Process Interface Unit
HLQ	High Level Qualifier
HLQL	High Level Query language
HLR	High Level Representation
HLR	Home Location Register
HLS	High Level Synthesis
HLS	Host Language System
HLS	Hue, Luminance, Saturation
HLSE	High Level, Single Ended
HLT	Halt

HLT	Highly Leveraged Transaction
HLTA	Halt Acknowledge
HLTU	Hierarchical Threshold Logic Unit
HLU	High Level User
HM	Heard and McDonald Islands
hm	hectometer
HM	High-resolution Monochrome
HM	Hypothetical Machine
HMA	High Memory Area
HMD	Head Mounted Display
HME	Hierarchical Modelling Environment
HMI	Hardware Monitor Interface
HMI	Hub Management Interface
HMI	Human-Machine Interface
HMIS	Headquarters Manufacturing Information System
HML	Human-Machine Language
HMLC	High-speed Multi-Line Controller
HMLK	Hammer Lock
HMM	Hardware Multiply Module
HMM	Hidden Markov Model(s)
HMMWV	High-Mobility Multipurpose Wheeled Vehicle
HMO	Hardware Microcode Optimizer
HMOS	High density|speed Metal Oxide Semiconductor
HMOS	High performance Metal Oxide Semiconductor
HMP	Host Monitoring Protocol
HMPL	High-level Micro-Program(ming) Language
HMPY	Hardware Multiplier
HMR	Hardware Malfunction Report
HMR	Hybrid Modular Redundancy
HMS	Homogeneous Multiprocessor System
hms	hours, minutes, seconds
HMUX	Hybrid Multiplexer
HN	Honduras
HN	Host to Network
HNA	Hierarchical Network Architecture
HNA	Hydraulic Network Analysis
HNDS	Hand Set
HNDS	Hybrid Network Design System
HNDT	Holographic Non-Destructive Testing
HNIC	Head Newbie In Charge
HNIL	High Noise Immunity Logic
HNN	Headline News Network
HNPA	Home Numbering Plan Area
HNPL	High-level Network Processing Language
HNS	Hospitality Network Service
HO	Hand Operated
HO	High Order
HOB	Hierarchical Operational Binding

HOBIC	Hotel Billing Information Center
HOBIS	Hotel Billing Information System
HOCS	Home Office Communication System
HO-DSP	Higher Order Domain Specific Part
HOE	Holographic Optical Element
HOL	High(er)-Order Logic|Language
HOL	Human Oriented Language
HOLAP	Hybridization of On-Line Analytical Processing
HOLDA	Hold Acknowledge
HOLDET	High Order Language Development and Evaluation Tool
HOLIDAY	Halliburton On-Line Information Discovery and Access System
HOLWG	High Order Language Working Group
HomePNA	Home Phoneline Networking Association
HOMPR	Hang On, Mobile Phone's Ringing
HONE	Hands-On Network Environment
HOOD	Hierarchical Object Oriented Design
HOP	Hopper
HOPSTOP	Hopper Stop
HOQ	High Order Quotient
HOR	Horizontal
HOS	High(er) Order Statistics
HOS	High Order Software
HOSED	Hardware Or Software Error Detected
HOST	Houston Operating Simulation Technique(s)
HOT	Horizontal Output Transformer|Transistor
HOT	Horizontal Output Tube
HOTT	Hot Off The Tree
HP	Hewlett-Packard
HP	High-Pass (filter)
HP	High Positive|Power
h-p	high-pressure
HP	Hit Points
hp	horsepower
HP	Host Processor
HP	Hundreds Position
HPA	Heuristic Path Algorithm
HPAA	High Performance Antenna Assembly
HPAD	Host Packet Assembler/Disassembler
HPC	Handheld Personal Computer
HPC	High Performance Computing
HPCA	High Parallel Computer Architecture
HPCA	High Performance Communications Adapter
HPCA	High Performance Computer Architecture
HPCA	High Performance Computing Act
HPCB	Hundreds Position C Bit
HPCC	High Performance Computing and Communications
HPCN	High Performance Computing and Networking
HPCS	High Performance Computing Systems

HPD	High Power Driver
HPDC	High Performance Distributed Computing (conference)
HPDJ	Hewlett-Packard Desk Jet
HPDL	High-Power Diode Laser
HPDR	High Performance Doppler Radar
HPE	High Performance Equipment
HPEX	High Priority Exit
HPF	Heterogeneous Packet Flow
HPF	Highest Priority First
HPF	High Pass Filter
HPF	High Performance File
HPF+	High Performance Fortran (EU)
HPF	High Possible Frequency
HPF	Host Preparation Facility
HPF	Hot Pressed Ferrite
HPFS	High Performance File System
HPGE	High Purity Germanium
HPGL	Hewlett-Packard Graphics Language
HPGS	High Performance Graphics System
HPIB	Hewlett-Packard Interface Bus
HPL	Hardware|High-level Programming Language
HPL	High Performance Language
HPL	High Production Loader
HPLJ	Hewlett-Packard Laser Jet
HPLT	High-Productivity Languages/Tools
HPM	High Power Microwave
HPMS	High Performance Main Storage
HPN	High Performance Network
HPO	High Performance Option
HPP	High Performance Processor
HPPA	Hewlett-Packard Precision Architecture
HPPCL	Hewlett-Packard Printer Control Language
HPPF	Host Program Preparation Facility
HPPI	High Performance Parallel Interface
HPR	High Performance Routing
HPR	High Power Rocketry
HPS	High Primary Sequence
HPSS	High Performance Switching System
HPT	Head Per Track
HPTLC	High Performance Thin-Layer Chromatography
HPUX	Hewlett-Packard Unix
HPW	High Performance Workstation
HQ	High Quality
hqx	binhex (file extension indicator)
HR	Croatia
HR	High Resolution
HR	Hit Ratio
HR	Holding Register

hr	hour
HR	Human Resources
HRC	Horizontal Redundancy Checking
HRC	Hybrid Ring Control
HRCB	Hammer Circuit Breaker
HRD	Hard
HREQ	Hold Request
HRF	Human Resources Function
HRG	High Resolution Graphics
HRH	High-Resistance Hold (relay)
HRI	Hardware RAID controller Interface (RAID = Redundant Array of Independent Disks)
HRI	High Resolution Imager
HRIM	High Resolution Infrared Measurement(s)
HRIS	Highway Research Information Service|System
HRIS	Human Resource Information System
HRM	Hardware Read-in Mode
HRM	Holistic Resource Management
HRMR	Human Read/Machine Read
HRMS	Human Resource Management System
HRN	Highest Response-ratio Next
HRNES	Host Remote Node Entry System
HRPF	Hexadecimal Reference Publication Format
HRPM	High Resolution Permanent Magnet
HRPT	High-Resolution Picture Transmission
HRRTC	High Resolution Real-Time Clock
HRSS	Host Resident Software System
HRT	High Rate Telemetry
HRT	High Resolution Timer
HRTF	Head Related Transfer Function
HRV	High-Resistance Value
HS	Half Subtracter
HS	Harmonized System
HS	Heat Seeking
HS	Hierarchic(al) Sequential
HS	High Speed
HS	Home System
HSA	High-Speed Adapter
HSAC	High-Speed Analog Computer
HSADC	High Speed Analog-to-Digital Converter
HSAI	High-Speed Analog Input
HSAM	Hierarchical Sequential Access Method
HSB	High-Speed Buffer
HSB	Hue, Saturation, Brightness
HSBA	High-Speed Buffer Adapter
HSC	Hierarchical Storage Controller
HSC	High-Speed Carry|Channel
HSC	High-Speed Computer|Counter

HSCS	High-Speed Contact Sense
HSCS	High-Speed Core Storage
HSCSD	High-Speed Circuit-Switched Data
HSDA	High-Speed Data Acquisition
HSDAS	High-Speed Data Acquisition System
HSDB	High-Speed Data Buffer
HSDC	High-Speed Data Card
HSDCA	High-Speed Data Channel Adapter
HSDL	High-Speed Data Line
HSDMS	Highly Secure Database Management System
HSE	High-Speed Exchange
HSE	High-speed Signal control Equipment
HSEL	High-speed Selector Channel
HSF	Horizontal Scan Frequency
HSFG	High Strength Friction Grip
HSFS	High Sierra File System
HSGT	High-Speed Ground Transport
HSI	Harsh Squirrel Interaction
HSI	High-Speed Interface
HSI	Horizontal Situation Indicator
HSI	Hue, Saturation, Intensity
HSIS	Hierarchical Sequential Interactive System
HSIT	Hypersonic Strong Interaction Theory
HSL	Hue, Saturation, Lightness
HSLAN	High-Speed Local Area Network
HSLC	High-speed Single Line Controller
HSLM	High-Speed Line Manager
HSLN	High-Speed Local Network
HSM	Hierarchical Storage Management
HSM	High-Speed Memory|Modem
HSML	High-Speed Modular Logic
HSP	Hard Storage Protection
HSP	Highest Significant Position
HSP	High-Speed Photometer|Printer
HSPICE	H-variant of SPICE
HSPN	Hierarchically Significant Part Numbering
HSR	High-Speed Reader
HSRJE	High-Speed Remote Job Entry
HSRP	High-Speed Research Program (US)
HSRP	Hot Standby Router Protocol
HSS	Hierarchical Sequential Structure
HSS	Hierarchical Service System
HSS	High-Speed Skip|Storage
HSS	Host Support Services
HSSCT	High-Speed Shorts and Continuity Tester
HSSDS	High-Speed Switched Digital Service
HSSI	High-Speed Serial Interface
HSSR	Hybrid Solid-State Relay

HSSTR	High-Speed Synchronous Transmitter/Receiver
HST	High-Speed Technology|Terminal
hst	history (file extension indicator)
hst	host (file extension indicator)
HSV	Hue, Saturation, Value
HSVMA	High-Speed Video Motion Analyzer
HSYNC	Horizontal Synchronization
HT	Haiti
HT	Half Tone|Total
HT	Handheld Terminal
HT	Hardware Test
H/T	Head per Track
HT	Home Terminal
HT	Horizontal Tabulation (character)
HTB	Hexadecimal To Binary
HTC	Height-to-Time Converter
HTC	High Technology Council
HTC	Horizontal Tabulation Character
HTC	Hybrid Technology Computer
HTF	Host Transaction Facility
HTH	Hope That Helps
HTL	High-Threshold Logic
htm	hypertext markup language (file extension indicator)
HTML	HyperText Markup Language
HTPLT	High-Temperature Power Life Tester
HTRB	High-Temperature Reverse Bias
HTS	Head, Track, and Sector
HTS	Host To Satellite
HTSC-GBJ	High-Temperature Superconductors in Grain Boundary Josephson junctions and circuits (EU)
HTTL	High(speed) Transistor-Transistor Logic
HTTP	HyperText Transport Protocol
HTTPD	HyperText Transfer Protocol Daemon
HTTP-NG	HyperText Transport Protocol – Next Generation
HTTPS	HyperText Transport Protocol Secure
HTTY	Happy Trails To You
HU	High Usage
HU	Horizontal arithmetic Unit
HU	Hungary
HUD	Head-Up Display
HUGE	Hewlett-Packard Unsupported GNU EMACS (EMACS = Editing Macros)
HULA	High-density Unit Logic Array
HULA	Highly integrated Unit Logic Assembly
HUMINT	Human Intelligence
HURD	Head Up Range Device
HUSH	Hyper Utility Shell
HUT	Helsinki University of Technology

HUT	High Usage (intertoll) Trunk
HUTG	High Usage Trunk Group
HV	High Voltage
H/V	Horizontal/Vertical
HVDI	High Voltage Dielectric Isolation
HVLP	High Velocity Low Pressure
HVP	Horizontal & Vertical Position
HVPS	High-Voltage Power Supply
HVR	Hardware Vector to Raster
HVSP	High Volume Stock Product
HVTS	High Volume Time-Sharing
HVY	Heavy
HW	Hardware
hW	hectowatt
HWCI	Hardware Configuration Item
HWCP	Hardware Code Page
HWD	Height-Width-Depth
HWI	Hardware Interpreter
HWL	Hard-Wired Logic
HWP	Hard-Wired Program
HWP	Height (and) Weight Proportionate
HWY	Highway
HYCOTRAN	Hybrid Computer Translator
HYD	Hydraulic
HYDAC	Hybrid Digital Analog Computer
HYDRA	Hybrid Document Reproduction Apparatus
hyp	hyphenation (file extension indicator)
HYP	Hyphen (character)
HYPER	Hypertape
HY-SPLIT	Hybrid Single Particle Lagrangian Integrated Trajectories (model)
HYTELNET	Hypertext browser for Telecommunications Network
HyTime	Hypermedia/Time-based (structuring language)
HyVIS	Hypermedia and Visual Information Systems
Hz	Hertz

I

I	Immediate
I	Incidence angle
I	Index
I	Indicator
I	Information
I	Input
I	Integrated
I	Interrupt
I	Inverter
I	symbol for Current

i	vector parallel to x-axis
I2-DSI	Internet2 – Distributed Storage Infrastructure
I2L	Integrated Injection Logic
I2O	Intelligent Input/Output
IA	Immediately Available
IA	Implementation Agreement
IA	Indirect Address(ing)
IA	Inspection Authorization
IA	Instruction Address
IA	Integrated Adapter|Attachment
IA	Intelligent Agent
IA	International Alphabet
IA	Interval Arithmetic
IA	Iowa (US)
IAB	Initiation Area Discriminator
IAB	Interactive Application Builder
IAB	Internet Activities|Architecture Board
IAB	Internet Advertising Bureau
IAB	Interrupt Address to Bus
IABI	Intel Application Binary Interface
IAC	Immediate Access Computer
IAC	In Any Case
IAC	Installation, Alteration, Cancellation
IAC	Integration, Assembly, Check
IAC	Interactive Array Computer
IAC	Inter-Application Communication
IAC	International Advisory Committee
IAC	International Association for Cybernetics
IAC	Interpret As Command
IACP	International Association of Computer Programmers
IACR	International Association for Cryptologic Research
IACS	Integrated Access and Cross-connect System
IAD	Initial Address Designator
IAD	Integrated Access Device
IAD	Integrated Automatic Documentation
IADR	Instruction Address
IAE	In Any Event
IAF	Initial Approach Fix
IAF	Interactive Facility
IAFA	Internet Anonymous FTP Archives (FTP = File Transfer Protocol)
IAFIS	Integrated Automated Fingerprint Identification System
IAG	Instruction Address Generation
IAGC	Instant(aneous) Automatic Gain Control
IAI	International Alliance for Interoperability
IAK	Internet Access Kit
IAL	International Algebraic Language
IALC	Instantaneous Automatic Level Control

IALE	Instrumented Architectural Level Emulation
IAM	Immediate Access Memory
IAM	Indexed Access Method
IAM	Innovation Access Method
IAM	Intelligent Actuation and Measurement
IAM	Interactive Algebraic Manipulation
IAMACS	International Association for Mathematics And Computers in Simulation
IANA	Internet Assigned Numbers Authority
IANET	Integrated Access Network
IAP	Image Array Processor
IAP	Industry Applications Program
IAP	Instrument Approach Procedures (automation)
IAP	Internet Access Provider
IAPC	Instantly Available Personal Computer
IAPD	International Association (of) Product Developers
IAPP	Industrial Automation Planning Panel
IAPR	International Association for Pattern Recognition
IAPS	Interim Antenna Pointing Subsystem
IAPS	International ASCII Publication Standard
IAR	Industry|Instruction Address Register
IAR	Interrupt Address Register
IAS	Immediate Access Storage
IAS	Interactive Application Supervisor|System
IAS	Intermediate Access Storage
IAS	International Accounting System
IASC	International Association for Statistical Computing
IASCA	International Auto Sound Challenge Association
IASSIST	International Association for Social Science Information Service and Technology
IAT	Import Address Table
IAT	Indexable Address Tag
IAT	Instance-As-Type
IAT	Intelligent Agent Technology (conference)
IAT	International Atomic Time
IATP	Installation and Acceptance Test Plan
IATUL	International Association of Technological University Libraries
IAU	Information Access Unit
IAU	Interface Adapter Unit
IAUP	Internet Account for Users Provider
IAVC	Instantaneous Automatic Volume Control
IAVQ	Image Adaptive Vector Quantization
IAW	In Accordance With
IAW	In Another Window
IAYF	Information At Your Fingertips
ib	ibidem (in the same place)
IB	Identifier Block
IB	In Between

IB	Incentive Base
IB	Information Bus
IB	Input Buffer|Bus
IB	Instruction Buffer|Bus
IB	Interface|Internal Bus
IBC	Ignore Block Character
IBC	Inside Back Cover
IBC	Instrument Bus Computer
IBC	Integrated Block Channel
IBC	Integrated Broadband Communications
IBCN	Integrated Broadband Communication Network
iBCS	Intel Binary Compatibility Specification
Iberpac	The Spanish packet-switched network
IBF	Incomplete Beta Forecast|Function
IBF	Input Buffer Full
IBG	Inter-Block Gap
IBGS	Interactive Business Graphics System
IBI	I Believe It
IBM	Individual-Based Model
IBM	International Business Machine (corporation)
IBM-GL	IBM Graphics Language
IBMNM	IBM Network Management
IBN	Integrated Business Network
IBOL	Interactive Business Oriented Language
IBOLS	Interactive Business Oriented Language Support
IBP	Incremental Bar Printer
IBR	It's Been Real
IBS	Intelsat Business Services
IBS	Internet Book Shop
IBSYS	Initial Basic System
IBT	Integrated Business Terminal
IBT	Interrupt Bit Table
IBT	Ion-implant Base Transistor
IBTD	I Beg To Differ
IBTS	International Bulk data Transmission System
IBU	Independent Business Unit
IBU	Internet Business Unit
IBX	Integrated Business Exchange
IC	Identification Code
IC	Ignore Character
I/C	Incoming
IC	Independent Carrier
IC	Indication Cycle
IC	Inductive-Capacitive (circuit)
IC	Information Center|Circular
IC	Input Channel|Circuit
IC	Input Controller
IC	Insert Character|Cursor

IC	Installation Center|Confirmation
I&C	Installation & Checkout
IC	Instruction Counter
IC	Instrumentation Control
IC	Integrated Circuit
IC	Interexchange Carrier
IC	Internal Connection
IC	Interrupt Controller
ICA	Independent Computing Architecture
ICA	Information Connection Architecture
ICA	Input Communications Adapter
ICA	Integrated Communications Adapter|Attachment
ICA	Intelligent Communications Adapter
ICA	Intelligent Console Architecture
ICA	Inter-Computer Adapter
ICA	International Callers Association (US)
ICA	International Communication(s) Association
ICA	Interrupt Communications Area
ICA	Intra-application Communications Area
ICAD	Integrated Control And Display
ICAD	Intelligent Computer Aided Design
ICADD	International Committee for Accessible Document Design
ICAE	Integrated Communications Adapter Extension
ICAI	Intelligent Computer Aided Instruction
ICAIL	International Conference on Artificial Intelligence & Law
iCAIR	International Center for Advanced Internet Research
ICALP	International Colloquium on Automata, Languages, and Programming
ICAM	Integrated Computer Aided Manufacturing
ICAN	Individual Circuit Analysis
ICANN	Internet Corporation for Assigned Names and Numbers
ICAP	Intermediate Course Applications Programming
ICAP	Internet Content Adaption Protocol
ICAPA	Integrated Circuit Active Phased array Antennas
I-CAPP	Ingersoll's Computer Aided Process Planning system
ICARO	Italian Computer Antivirus Research Organization
ICAS	Intel Communicating Applications Specifications
ICASAV	Intelligent Components for Autonomous and Semi-Autonomous Vehicles (international workshop)
ICB	Information Collection Budget
ICB	Internal Common Bus
ICB	Internet Citizen's Band
ICB	Interrupt Control Block
ICB	Interstate Computer Bank
ICBS	Interconnected Business System
ICBW	I Could Be Wrong
ICC	Integrated Circuit Computer
ICC	Interface Control Check

ICC	International Communication Conference
ICC	International Computer|Control Center
ICCA	Independent Computer Consultants Association
ICCAD	International Conference on Computer Aided Design
ICCB	Internet Configuration Control Board
ICCC	Inter-Client Communication Convention
ICCC	International Council for Computer Communication(s)
ICCCM	Inter-Client Communication Conventions Manual
ICCCM	International Conference on Computer Capacity Management
ICCDP	Integrated Circuit Communications|Communicators Data Processor
ICCF	Interactive Computing and Control Facility
ICCHP	International Conference on Computers Helping People
ICCIM	International Conference on Computer Integrated Manufacturing
ICCP	Information, Computer and Communications Policy
ICCP	Institute for the Certification of Computer Professionals
ICCS	Inter-Computer Communications System
ICCU	Inter-Computer Communications Unit
ICCU	Inter-Computer Control Unit
ICD	Index to Class Directory
ICD	Installation Confirmation Date
ICD	Interactive Call Distribution
ICD	Interface Control Document|Drawing
ICD	Internal Checking Document
ICD	International Code Designator
ICD	International Congress for Data processing
ICD	Internet Database Connector
ICDB	Integrated Corporate Data Base
ICDCS	International Conference on Distributed Computer Systems
ICDDB	Internal Control Description Data Base
ICDE	International Conference on Data Engineering
ICDES	Item Class Description
ICDIA	International Compact Disk Interactive Association
ICDL	Integrated Circuit Description Language
ICDL	Internal Control Description Language
ICDLA	Internal Control Description Language Analyzer
ICDM	Interdisciplinary Concurrent Design Methodology
ICDP	International Conference on Distributed Platforms
ICDR	Inward Call Detail Recording
ICDS	Input Command Data Set
ICDS	Integrated Circuit Design System
ICDS	International Conference on Digital Satellite communications
ICE	Illinois Computing Educators
ICE	In-Circuit Emulator
ICE	Information and Content Exchange
ICE	Information Collecting Equipment
ICE	Input Checking Equipment

ICE	Input Count Error
ICE	Interactive Collaborative Environment
ICE	Interconnection Equipment
ICE	Intrusion Countermeasure Electronics
ICEA	International Consumer Electronics Association
ICECCS	International Conference on Engineering of Complex Computer Systems
ICED	International Conference on Engineering Design
ICEF	Interactive Composition and Editing Facilities
ICES	Integrated Civil Engineering System
ICEX	Integrated Civil Engineering Executive
ICF	Interactive Command Facility
ICF	Interactive Communications Feature
ICF	Inter-Communication Flip-flop
ICF	Item Control File
ICG	Interactive Computer Graphics
ICGA	International Conference on Genetic Algorithms
ICHC	International Conference on the History of Computing
ICI	Image Component Information
ICI	Incoming Call Identification
ICI	Intelligent Communications Interface
ICI	Interexchange Carrier Interface
ICI	Interface Control Information
ICIIS	International Conference on Information, Intelligence & Systems
I-CIMPRO	International Conference on Integrated Manufacturing in the Process industries
ICIMS	Intelligent Control and Integrated Manufacturing System(s)
ICIP	Improved Control Interval Processing
ICIP	International Conference on Information Processing
ICIS	International Conference on Intelligent Systems
ICL	Incoming Line
ICL	Inrush Current Limiter
ICL	Instrument Controlled Landing
ICL	Inter-Communication Logic
ICL	Intercomputer Communication Link
ICL	Interface Clear
ICL	Interrupt Class List
ICLID	Incoming-Call Line Identification
ICLID	Individual Calling Line Identifier
ICM	Image Colour Matching
ICM	In-Cycle Measurement
ICM	Instruction Control Memory
ICM	Integrated Call Management
ICM	Internally Controlled Machine
ICMCS	International Conference on Multimedia Computing & Systems
ICME	International Conference on Multimedia & Exposition
ICMP	Internet Communication|Control Message Protocol

ICMS	Integrated Circuit and Message Switch
ICMUP	Instruction Control Memory Update Processor
ICMV	Input Common Mode Voltage
ICN	Indicator Coupling Network
ICN	Integrated Computer Network
ICN	International Cooperating Network
ICN	Inter-university Computer Network
I-CNOS	International Customer Network Organization System
ICNP	International Conference on Network Protocols
ico	icon (file extension indicator)
ICO	Internet Connectivity Option
ICOM	Input, Control, Output, Mechanism
iCOMP	Intel Comparative Microprocessor
ICOMS	Inputs, Controls, Outputs and Mechanisms
ICON	Image Converter
ICONF	Incremental Constraint Facility
ICONIP	International (annual) Conference On Neural Information Processing
ICOS	Interactive Cobol Operating System
ICOS	International Club for Open Systems
ICOT	Institute for new generation Computer Technology (US)
ICOT	ISDN Conformance Testing (ISDN = Integrated Services Digital Network)
ICP	Image Co-Processor
ICP	Independent Communication Provider
ICP	Independent Control Point
ICP	Initial Connection Protocol
ICP	Installation Confirmation Period
ICP	Instituto das Comunicacóes de Portugal
ICP	Integrated Channel Processor
IC/P	Intelligent Copier/Printer
ICP	Internet Control Protocol
ICP	Interprocess Communications Protocol
ICP	Inventory Control Program
ICPA	Interactive Customer Problem Analysis
ICPL	Initial Control Program Load
ICPS	Interim Critical Parts planning System
ICPT	Intercept Tone
ICQ	I Seek You
ICR	Impulse Check Routine
ICR	Independent Component Release
ICR	Indirect Control Register
ICR	Initial Cell Rate
ICR	Input Control Register
ICR	Installation Confirmation Review
ICR	Intelligent Character Recognition
ICR	Intermediate Code for Robots
ICR	Interrupt Code|Control Register

ICR	Interrupt Control Routine
ICR	Item Control Record
ICRA	International Conference on Robotics and Automation
ICRB	Inscribe
ICRP	International Commission (on) Radiological Protection
ICRT	Incorrect
ICS	Include Segment
ICS	Industrial Control System
ICS	Information Control|Collection System
ICS	Institute of Computer Science (US)
ICS	Integrated Communication|Computer System
ICS	Interactive Communications Software
ICS	Interactive|Internal Control System
ICS	International Computer Symposium
ICS	Internet Connection Sharing
ICS	Interpretive Computer Simulation
ICS	Intuitive Command Structure
ICS	Inventory Control System
ICSA	International Computer Security Association
ICSC	Inter-LATA Customer Service Center
ICSE	International Conference on Software Engineering
ICSI	International Computer Science Institute (at Berkeley)
ICSL	Internal Customer Service Level
ICSM	International Conference on Software Maintenance
ICST	Institute for Computer Science and Technology (US)
ICSU	Internal Channel Service Unit
ICT	In-Circuit Tester
ICT	In-Coming Trunk
ICT	Incremental Change Type
ICT	Information and Computer Technology
ICT	Inhibit Channel Trap
ICT	Initiation Control and Termination
ICT	Installation Confirmation Time
ICT	Integrated Circuit Technology
ICT	Integrated Computer Telemetry
ICT	Interactive Consumer Terminal
ICT	Inter-Company Transfer
ICT	Internal Cycle Time
ICTAI	International Conference on Tools with Artificial Intelligence
ICTL	Input Control
ICTP	Inter-Company Transfer Price
ICU	Image Converter Unit
ICU	Industrial Control Unit
ICU	Instruction Cache|Control Unit
ICU	Integrated Control Unit
ICU	Interactive Chart Utility
ICU	Interface|Interrupt Control Unit
ICV	Initial Chaining Value

ICW	Initial Condition|Control Word
ICW	Initialization Command Word
ICW	Interface Control Word
ICW	Interrupt(ed) Continuous Wave
ICWF	Interactive Computer Worded Forecast
ICX	International Customer Executive
ICXC	International Customer Executive Course
ICXP	International Customer Executive Program
ICY	Instruction Cycle
ID	Idaho (US)
ID	Identifier
ID	Illegal Direct
ID	Indicator Driver
ID	Indonesia
ID	Information Distributor
ID	Inside Diameter
ID	Installation Date
I/D	Instruction/Data
ID	Instruction Decoder
ID	Integration Domain
ID	Intelligent Digitizer
ID	Internet Draft
ID	Item Description|Descriptor
IDA	Indirect Data Addressing
IDA	Industrial Development Authority
IDA	Information Distribution Application
IDA	Integrated Data Analysis
IDA	Integrated Digital Access
IDA	Intelligent Data Access
IDA	Intelligent Disk|Drive Array
IDA	Interactive Data Analysis
IDA	Interactive Debugging Aid
IDA	Intercommunication Data Areas
IDA	Interconnect Device Arrangement
IDA	Inter-Data Access
IDA	International Data Administration
IDA	International Development Association
IDA*	Iterative Deepening A* (algorithm)
IDAC	Instant Data Access and Control
IDAC	Interactive Design for Analog integrated Circuits
IDAL	Indirect Data Address List
IDAM	Indexed Direct Access Method
IDAP	Intelligence Data Acquisition and Processing (system)
IDAPI	Integrated Database Application Programming Interface
IDAS	Industrial Data Acquisition System
IDAS	Interchange Data Structure
IDAS	International Database Access Service
IDAT	International Data and Application Technology

IDB	Information Descriptor Board
IDB	Input Data Buffer
IDB	Integrated Data Base
IDB	Interoperable Data Base
IDB	Inverted Data Base
IDBMS	Integrated Data Base Management System
IDBMS/R	Integrated Data Base Management System/Relational
IDBR	Input Data Buffer Register
IDC	Indirect Digital Control
IDC	Information and Documentation Center
IDC	Installation Date Confirmation
IDC	Integrated Desktop Connector
IDC	Internal Data Channel
IDC	International Data Corporation
IDC	Internet Database Connector
IDCB	Immediate Device Control Block
IDCC	Integrated Data Communications Controller
IDCMA	Independent Data Communication Manufacturers' Association
IDCSS	Integrated Design Collaboration Support System
IDCTR	Inductor
IDD	Installation Due Date
IDD	Integrated Data Dictionary
IDD	Interface Definition Dialog
IDD	International Direct Dialling
IDDD	International Direct Distance Dialling
IDDE	Integrated Development Debugging Environment
IDDL	Integrated Data Description Language
IDDL	Interactive Database Design Laboratory
IDDS	International Data Distribution System
IDDS	International Digital Data Service
IDDU	Interactive Data Definition Utility
IDE	Imbedded Drive Electronics
IDE	Integrated Development Environment
IDE	Integrated Drive Electronics
IDE	Intelligent Drive Electronics
IDE	Interactive Data Entry
IDE	Interactive Design and Engineering
IDE	Interface Design Enhancement
IDE	Interface for Data Exchange
IDEA	Industrial Design Exploiting Automation
IDEA	Interactive Data Entry/Access
IDEA	Interactive Data Exploration and Analysis
IDEA	International Data Encryption Algorithm
IDEA	Internet Design, Engineering, and Analysis notes
IDEAS	Integrated Design and Analysis System
IDEAS	Integrated Design and Engineering Automated System
IDEAS	Integrated Development (and) Engineering Application System
IDEAS	Interactive Database Easy Access System

IDEF	Integration Definition for Function modeling
IDEF0	ICAM Definition language 0
IDEFX	ICAM Definition language Extended
IDEM	Infrastructure Development Methodology
IDEM	Inter-Departmental Electronic Mail service
IDEN	Interactive Data Entry Network
IDENT	Identify
IDEP	Inter-service Data Exchange Programming
IDES	Interactive Data Entry System
IDF	Image Description File
IDF	Ink Donor Film
IDF	Inquiry and Development Facility
IDF	Integrated Data File
IDF	Interactive Data Facility
IDF	Intermediate Distribution Frame
IDG	Integrated Drive Generator
IDG	Inter-Dialog Gap
IDHS	Intelligent Data Handling System
IDI	Immediate Data Input
IDI	Improved Data Interchange
IDI	Initial Domain Identifier
IDI	Intelligent Dual Interface
IDIC	Infinite Diversity in Infinite Combinations
IDIDAS	Interactive Digital Image Display and Analysis System
IDIQ	Indefinite Delivery – Indefinite Quantity
IDIR	Indirect
IDITEM	Identifier Item
IDIV	Integer Divide
IDKH	I Don't Know How
IDL	Idle
IDL	Indicator Driver Lamp
IDL	Information Description Language
IDL	Instruction Definition Language
IDL	Interactive Data Language
IDL	Interface Definition|Description Language
IDL	Interface Definition Level
IDL	Intermediate Data description Language
IDL	Intruder Detection/Lockout
IDLIB	Item Description Library
IDLJava	Interface Definition Language of Java
IDM	Image Data Manager
IDM	Implementation Description Model
IDM	Instruction Diagram Manual
IDM	Integrated Diagnostic Modem
IDM	Intelligent Database Machine
IDM	Interactive Decision Making
IDMAS	Interactive Database Manipulator And Summarizer
IDMH	Input Destination Message Handler

IDMME	Integrated Design and Manufacturing in Mechanical Engineering (international conference)
IDMR	Inter-Domain Multicast Routing
IDMS	Integrated Data(base) Management System
IDN	Integrated Data|Digital Network
IDN	Intelligent Data Network
IDN	International Directory Network
IDN1	Integrated Digital Network 1*64 kbits/sec connection
IDN30	Integrated Digital Network 30*64 kbits/sec connection
IDNO	Identification Number
IDNX	Integrated Digital Network Exchange
IDO	Interpretatively Driven Operation(s)
IDO	Isolated Digital Output
IDOL	Improved Disk time Overlap
IDOS	Image Display Operating System
IDOS	Interrupt Disk Operating System
IDOT	Instrumentation Data Online Transcriber
IDP	Implied Decimal Point
IDP	Industrial Data Processing
IDP	Initial Domain Part
IDP	Integrated Data|Decision Processing
IDP	Integrated Detector Preamplifier
IDP	Interactive Database Processing
IDP	Intermodulation-Distortion Percentage
IDP	International Data Processing
IDP	Internet Datagram Protocol
IDPM	Institute of Data Processing Management (GB)
IDPR	Inter-Domain Policy Routing
IDPS	Interactive Direct Processing System
IDR	Identification Record
IDR	Information Descriptor (board)
IDR	Input Data Register
IDR	Intelligent Document Recognition
IDRP	Inter-Domain Routing Protocol
IDRS	Integrated Data Retrieval System
IDS	Identification Section
IDS	Image Distribution System
IDS	Information Display System
IDS	Integrated Database System
IDS	Integrated Data Store
IDS	Intelligent|Interactive Display System
IDS	Interactively Displayed Structures
IDS	Intermediate|Internal Data Structure
IDS	Internal Directory System
IDSN	Integrated Dyslexic Service Network
IDSS	Intelligent Decision Support System
IDT	Identification Table
IDT	Integrated Device Technology

IDT	Intelligent|Interactive Data Terminal
IDT	Inter-Digital Transducer
IDT	International Discount Telecommunications
IDT	Interrupt Descriptor|Dispatch Table
IDTF	Interactive Display Text Facility
IDTS	IBM Data Transfer Service
IDTV	Improved Definition Television
IDU	Interactive Database Utility
IDU	Interface Data Unit
IDVC	Integrated Data/Voice Channel
IDVM	Integrated Digital Voltammeter
idx	index (file extension indicator)
IE	Industrial Electronics|Engineering
IE	Information Economics|Exchange
IE	Information Engineering
IE	Instruction Element
IE	Inter Ethernet
IE	Internet Explorer
IE	Interrupt Enable
IE	Ireland
IEA/AIE	Industrial and Engineering Applications of Artificial Intelligence and Expert systems (conference)
IEAK	Internet Explorer Administration Kit
IEC	Incremental Engineering Change
IEC	Integrated Electronic Component
IEC	Inter-Exchange Carrier
IEC	International Education Center
IEC	International Electrotechnical Committee
IECI	Industrial Electronics and Control Instrumentation
IEEE	Institute of Electrical and Electronic Engineers
IEEE	Integrated Enterprise Engineering Environment
IEEE-488	IEEE committee for a digital interface for programmable instrumentation standard
IEEE-583	IEEE committee for modular instrumentation and design interface system standard
IEEE-802	IEEE committee on wireless technologies
IEEE-CS	Institute of Electrical and Electronics Engineers – Computer Society
IEEE-USA/CCIP	IEEE-USA Committee on Communications and Information Policy
IEF	Information Engineering Facility
IEF	Internet Equity Fund
IEICE	Institute of Electronics, Information and Communication Engineers (JP)
IEIF	Information Exchange Interface
IEM	Illuminated Entry Module
IEM	Integrated Enterprise Modelling
IEMSI	Interactive Electronic Mail Standard Identification

IEN	Individualized Electronic Newspaper
IEN	Internet Engineering Note(s)
IEN	Internet Experiment Notebook
IEOE	Integrated Enterprise Operation Environment
IEP	Inter-divisional Engineering Practices
IERE	Institution of Electronic and Radio Engineers
IES	Implemented External Schemata
IES	Information Exchange System
IESE	Institute for Experimental Software Engineering
IESG	Internet Engineering Steering Group
IETF	Internet Engineering Task Force
IETM	Interactive Electronic Technical Manual
I-ETS	Interim European Communication Standard
IEU	Integer Execution Unit
IEW	Information Engineering Workbench
IF	Indexed File
IF	Instruction Field
I/F	Interface
IF	Intermediate Frequency
IFA	Illegal File Access
IFA	Inductive Fault Analysis
IFA	Information Flow Analysis
IFA	Integrated File Adapter
IFA	Intensive Flux Array
IFAC	International Federation for Automatic Control
IFAM	Inverted File Access Method
IFB	Interruptible Fold-Back
IFC	Interface Clear
IFC	Internet Foundation Class(es)
IFCB	Interrupt Fan Control Block
IFCR	Interface Control Register
IFD	Image File Directory
IFD	Information Flow(s) Diagram
IFD	Interactive Feature Definition
IFDL	Independent Form Description Language
IFDR	Interface Data Register
IFEN	Intercompany File Exchange Network
IFF	Identification Friend or Foe
IFF	Image File Format
IFF	Interchange(able) File Format
IFF	Interrupt-enable Flip-Flop
IFF	Iterative Function Fractal
IFG	Incoming Fax Gateway
IFG	Inter-Frame Gap
IFI	Inter-Fault Interval
IFID	International Federation for Information and Documentation
IFIP	International Federation for Information Processing
IFIS	International Forwarding Information Structure

IFLA	International Federation of Library Associations (and institutions)
IFM	Interactive File Manager
IFMA	International Facility Management Association
IFMS	Integrated Financial Management System
IFOR	Implementation Force
IFORS	International Federation of Operations Research Societies
IFP	IMS/VS Fast Path
IFP	Instruction Fetch Pipeline
IFP	Integrated File Processor
IFPEC	Improved Floating-Point Engineering Change
IFPG	Intermediate Frequency Pulse Generator
IFPS	Interactive Financial Planning System
IFPUG	International Function Point User Group
IFR	Impulse Frequency Rate
IFR	Interface Register
IFR	Internal Function Register
IFR	Intrinsic Failure Rate
IFRA	Increasing Failure Rate Average
IFRB	International Frequency Registration Board
IFRU	Interference Rejection Unit
IFS	Installable File System
IFS	Intelligent Fixturing System
IFS	Interactive File Sharing
IFS	Interactive Financial System
IFS	Internal File System
IFSC	Information Field Separator Character
IFSM	Information Systems Management
IFSMGR	Installable File System Manager
IFT	Integrating Fabricating Technologies
IFT	Internal Function Test
IFT	Inverse Fourier Transform
IFU	Instruction Fetch Unit
IFU	Interworking Functional Unit
IG	Instructor Guide
IGA	Integrated Graphics Array
IGBT	Insulated Gate Bipolar Transistor
IGC	Idiographic
IGC	Institute for Global Communications
IGC	Integrated Graphics Controller
IGCT	Insulated Gate Commutable Thyristor
IGCT	Integrated Genomic Database
IGDS	Interactive Graphic(s) Design System|Software
IGES	Initial Graphics Exchange Specification
IGFET	Insulated Gate Field Effect Transistor
IGG	Inhibit Gate Generator
IGGI	Interdepartmental Group on Geographic Information
IGL	Index-Guided Laser

IGL	Interactive Graphics Language|Library
IGM	Interactive Guidance Mode
IGMP	Internet Group Management Protocol
IGN	Ignore
IGNN	I Got No Name
IGOSS	Industry and Government Open Systems Specification
IGP	Interior Gateway Protocol
IGRP	Internet Gateway Routing Protocol
IGS	Information Group Separator (character)
IGS	Integrated Graphics System
IGS	Interactive Graphics Software|System
IGS	Interchange Group Separator
IGS	Internet Go Server
IGT	Interactive Graphics Terminal
IGT	Internally Generated Transaction
IGU	Inertial Guidance Unit
IH	Interrupt Handler
IHAC	I Haven't A Clue
IHADSS	Integrated Helmet And Display Sight System
IHD	Integrated Help Desk
IHF	Image Handling Facility
IHF	Inhibit Halt Flip-flop
IHF	Institute of High-Fidelity
IHFM	Institute of High-Fidelity Manufacturers (US)
IHL	Interactive Hidden Line
IHL	Internet Header Length
IHP	Interrupt Handling Process
IHS	Information Handling Service
IHS	Integrated Hospital Support
IHV	Independent Hardware Vendor
IHY	I Heard You
II	Interrupt Inhibit
II	Isolating Inverter
IIA	Information Industries Association
IIA	Information Interchange Architecture
IIAS	Interactive Instructional Authoring System
IIASA	International Institute for Applied Systems Analysis
IICS	International Interactive Communications Society
IIE	Institute of Industrial Engineers
IIE	Institute of International Education
IIEP	International Institute for Educational Planning
IIF	Immediate If (instruction)
IIF	Input Interface Facility
IIF	Intermedia Interchange Format
IIIL	Isoplanar Integrated Injection Logic
IIJ	Internet Initiative Japan
IIL	Integrated Injection Logic
IIM	Inventory Information Management

IIMA	Irish Interactive Multimedia Association
IIN	Integrated Information Network
IIOP	Integrated Input/Output Processor
IIOP	Internet Inter-Operability Protocol
IIOP	Internet Inter ORB Protocol (ORB = Object Request Broker)
IIP	Installation Integration Period
IIPACS	Integrated Information Presentation And Control System
IIPRC	International Intellectual Property Rights Committee
IIPS	Interactive Instructional Presentation System
IIR	Immediate Impulse Response
IIR	Infinite Impulse Response
IIR	Interactive Image Regeneration
IIRC	If I Remember Correctly
IIS	Individual Information System
IIS	Installation Information System
IIS	Integrating Infrastructure
IIS	Interactive Instruction(al) System
IIS	Internet Information Server
IIS	Internet Interface Systems
IISPB	Image and Information Standards Policy Board
IITA	Information Infrastructure Technology Applications
IITF	Information Infrastructure Task Force
IITI	International Information Technology Institute
IIU	Instruction Input Unit
IIUG	International Informix Users Group
IIUP	International Installed User Program
IIW	ISDN Implementors Workshop (ISDN = Integrated Services Digital Network)
IIWM	If It Were Me|Mine
IJC	Inter-Job Communications
IJC	International Joint Commission
IJCAI	International Joint Conference on Artificial Intelligence
IJP	Ink-Jet Printer
IJP	Internal Job Processing
IJQ	Input Job Queue
IJS	Interactive Job Submission
IKA	Informatics Knowledge Areas
IKB	Intelligent Keyboard
IKBS	Intelligent Knowledge Based Systems
IKE	Internet Key Exchange
IKIWISI	I'll Know It When I See It
IKM	Institute for Knowledge Management
iKP	Internet Keyed payment Protocol
IKTT	Informations- und Kommunikations Technologie Transfer (DE)
IL	Idle character
IL	Illinois (US)
IL	Imaging Language
IL	Incorrect Length

IL	Information Language
IL	Initial Library
IL	Instruction List
IL	Intermediate Language
IL	Internal Label
IL	Interrupt List
IL	Israel
ILA	Insurance Logistics Automated
ILA	Intelligent Line Adapter
ILA	Intermediate Level Amplifier
ILAN	Input Language
ILAN	Israeli Academic Network
ILAR	Interrupt List Address Register
ILB	Initial Load Block
ILBM	Interleaved Bitmap
ILBT	Interrupt Level Branch Table
ILC	Initial License Charge
ILC	Input Language Converter
ILC	Instruction Length Code|Counter
ILC	Instruction Location Counter
ILCD	Interest/Late Charge Code
ILCSO	Illinois Library Computer Systems Office
ILD	Injection Laser Diode
ILD	Instructional|Intermediate Logic Diagram
ILD	Intersection Loop Detection
ILE	Interface Latching Element
ILEC	Incumbent Local Exchange Carrier
ILF	Infra-Low Frequency
ILGL	Illegal
ILLIAC	Illinois Automatic Computer
ILM	Inbound Logistics Management
ILM	Intermediate Language Machine
ILMI	Interim Local Management Interface
ILO	Individual Load Operation
ILO	Injection Locked Oscillator
ILO	International Labour Organization
ILOC	Internal Location
ILP	Individual Learning Package
ILP	Inductive Logic Programming (EU)
ILP	Instrument Landing Program
ILP	Interactive Learning Package|Program
ILPS	Interactive Linear Programming System
ILS	Integrated Logistics Support
ILS	International Language Support
ILS	International Line Selector
ILS	Interrupt Level Subroutine
ILSP	Integrated Logistic Support Plan
ILSW	Interrupt Level Status Word

ILT	In-Line Test
ILTMS	International Leased Telegraph Message Switching
ILY	I Love You
IM	Imaging Model
IM	Index Marker
IM	Information Management
IM	Inside Macintosh
IM	Installation & Maintenance
IM	Installation Manual
IM	Instant Messaging
IM	Instruction Memory
IM	Instrumentation and Measurement
IM	Integrated Modem
IM	Intelligent Messaging|Miner
IM	Intensity Modulation
IM	Interface Module
IM	Intermediate Modeling
IM	Intermodulation
IM	Interrupt Mask
IM	Inventory Master
IMA	Input Message Acknowledgment
IMA	Institute for Manufacturing Automation (US)
IMA	Intelligent Media Adapter
IMA	Interactive Multimedia Association
IMA	International MIDI Association (MIDI = Musical Instrument Digital Interface)
IMA	International Multimedia Association
IMA	Invalid Memory Address
IMA	Inverse Multiplexing over ATM (= Asynchronous Transfer Mode)
IMAC	Isochronous Media Access Control
IMACS	Image Management And Communication System
IMACS	International association for Mathematics And Computers in Simulation
IMADE	Integrated Manufacturing And Design Environment
IMAG	Image
IMAGE	International Multichannel Action Group for Education
IMAGO	Integral Management of Goods flow
IMAN	Information Manager
IMAP	Internet Message Access Protocol
IMAPx	Interactive Mail Access Protocol (version) x
IMARS	Information Management And Retrieval System
IMAS	Integrated Mass Announcement System
IMAX	Image which is Maximal
IMB	Intermode Bus
IMB	Inter-Module Bus
IMBE	Improved Multi-Band Excitation
IMBI	Inter-Module Bus Interface

IMC	Image Motion Compensation
IMC	Information Management Concepts
IMC	Integrated Multiplexer Channel
IMC	Interactive Module Controller
IMC	Internal Model Control
IMC	International Micrographics Congress
IMC	Internet Marketing Council
IMCS	Interactive Manufacturing Control System
IMD	Immediate
IMD	Information Management Department
IMD	Interactive Map Definition
IMD	Inter-Modulation Distortion
IMDB	Installed Machine Data Base
IMDR	Intelligent Mark Document Reader
IMDS	Image Data Stream (format)
IME	In My Experience
IME	Input Method Editor
IME	Interior Merging and Embedding
IME	International Microcomputer Exposition
IMEKO	International Measurement Confederation
IMF	IMS Management Facilities (IMS = Information Management System)
IMF	Item Master File
IMFM	Inverted Modified Frequency Modulation
IMG	Image
IMG	International Mailgram
IMH	Individual Machine History
IMH	Internodal Message Handler
IMHO	In My Humble Opinion
IMI	Intermediate Machine Instruction
IMIA	International Medical Informatics Association
IMIL	International Management Information Library
IMIS	Integrated Management Information System
IMKA	Initiative for Managing Knowledge Assets
IML	Inductive Machine Learning
IML	Information Manipulation Language
IML	Initial Machine Load
IML	Initial Microcode|Microprogram Load
IML	Interactive Mainframe Link
IML	Intermediate Machine Language
IMM	Immediate
IMM	Input Message Manual
IMM	Integrated Memory Module
IMM	Intelligent Memory Manager
IMMEX	Interactive Multi-Media Exercises
immitance	impedance/admittance
IMMM	International conference on Microcomputers, Minicomputers, Microprocessors

IMMU	Instruction Memory Management Unit
IMO	In My Opinion
IMOC	Inventory Management Order Control
IMOS	Interactive Multiprogramming Operating System
IMOS	Ion implantation Metal Oxide Semiconductor
IMP	Improved Multi-Processor
IMP	Impulse
IMP	Indicator Maintenance Panel
IMP	Information Management Package|Plan
IMP	Information Message Processor
IMP	Integrated Maintenance Package
IMP	Integrated Manufacturing Planning
IMP	Interface Message Processor
IMP	Interplanetary Monitoring Platform
IMP	Interpretive Master Program
IMP	Inventory Management Package
IMP	Irreversible Magnetic Process
IMPA	Intelligent Multi-Port Adapter
IMPAC	Industrial Multilevel Process Analysis and Control
IMPAC	Information for Management Planning Analysis and Coordination
IMPACS	Integrated Manufacturing And Control System (EU)
IMPACT	Image Processing for Automated Cartographic Tools (EU)
IMPACT	Improved Manufacturing Planning and Assembly Control Technique
IMPACT	Information Market Policy Actions (EU)
IMPACT	Integrated|Inventory Management Planning And Control Technique(s)
IMPATT	Impact (ionization) Avalanche (and) Transit Time
IMPC	Implicit
IMPL	Initial Micro-Program Load
IMPPACT	Integrated Modelling of Products and Processes using Advanced Computer Technologies
IMPR	Impression
IMPS	Integrated Modular Push-button Switch
IMP&S	Intelligent Manufacturing Processes & Systems (world congress)
IMPS	Interface Message Processors
IMR	Internet Monthly Report
IMR	Interrupt(ion) Mask Register
IMR	Inventory Management Report
IMRADS	Information Management, Retrieval And Dissemination System
IMRL	Intermediate Maintenance Requirements List
IMS	Information Maintenance System
IMS	Information Management Services|System
IMS	Installation Measurement System
IMS	Instructional Management System
IMS	Integrated Management|Manufacturing System

IMS	Integrated Message Services
IMS	Intelligent Manufacturing System
IMS	Interactive Multimedia Services
IMS	Intermediate Maintenance Standards
IMS	Intermediate Multiprocessing System
IMS	International Measurement System
IMS	Internetwork Management System
IMS	Inventory Management Simulator|System
IMS/DB	Information Maintenance System/Data Base
IMSE	Integrated Modelling Support Environment
IMSI	Information Management System Interface
IMSI	International Mobile Subscriber Identity
IMSL	International Mathematical Subroutine Library
IMSL	International Mathematics and Statistics Library
IMSO	Integrated Micro Systems Operation
IMSP	Integrated Mass Storage Processor
IMSTW	International Mixed Signal Testing Workshop
IMS/VS	Information Management System/Virtual Storage
IMT	Image Management Terminal
IMT	Intelligent Micro-image retrieval Terminal
IMT	Inter-Machine Trunk
IMTC	International Multimedia Teleconferencing Consortium
IMTI	Integrated Manufacturing Technologies Institute (CA)
IMTR	Integrated Manufacturing Technology Roadmapping
IMTS	Improved Mobile Telephone Service
IMTV	Interactive Multimedia Television
IMU	Increment Memory Unit
IMU	Inertial Measurement Unit
IMU	Instruction Memory Unit
IM-UAPDU	Interpersonal Messaging User Agent Protocol Data Unit
IMUL	Integer Multiply
IMUX	Inverse Multiplexer
IMX	In-line Multiplexer
IMX	Inquiry Message Exchange
in	inch(es)
IN	Index Number
IN	India
IN	Indiana (US)
IN	Information Network
IN	Input
IN	Intelligent|Interconnecting Network
IN	Internal Node
INA	Integrated Network Architecture
INACV	Inactive
InARP	Inverse Address-Resolution Protocol
INC	Improved Navigation Computer
INC	Inclinable
INC	Incoming

INC	Increase
INC	Increment
INC	Inertial Navigation Computer
INC	Integrated Network Connection
INC	International Carrier
INCA	Integrated Network Communication Architecture
INCH	Integrated Chopper
INCL	Included
INCOS	Integrated Control System
INCOSE	International Council Of System Engineers
INCOTERMS	International Commercial Terms
INCPLT	Incomplete
INCR	Increment
INCRB	Inscribe
INCWAR	Inbound Control Word Address Register
IND	Index
IND	Indicator
IND	Indirect
ind	inductance
INDAC	Industrial Data Acquisition and Control
INDC	Information Network and Data Communication (international conference)
INDCD	Industry Code
INDIS	International Distribution System
INDN	Indication
INDU	Induction
INEI	Instituto Nacional (de) Estadistica (e) Informatica (PE)
INEX	Information Exchange
inf	information (file extension indicator)
INF	Irredundant Normal Form
INFANT	Interactive Network Functioning on Adaptive Neural Topologies
INFN	Infinite
INFO	Information
INFODEV	Information for Development
INFOL	Information Oriented Language
INFONET	Information Network
INFOR	Information Network and File Organization
INFORM	Information Network For information Retrieval Management
INFORM	Information Network For On-line Retrieval and Maintenance
INFORMS	Information Organization Reporting and Management System
INFORMS	Institute For Operations Research and Management Science
INFOSEC	Information systems Security
INFOTEX	Information via Telex
INFRAL	Information Retrieval Automatic Language
INFUT	Information Utility
INGA	Interactive Graphic Analysis
INGRES	Interactive Graphic and Retrieval System

INH	Inhibit
INHYB	Inhibit Y Bit
ini	initialize (file extension indicator)
I-NIC	Industrial Network and Information support Center
INI-GraphicsNet	International Network of Institutions for computer Graphics
INIT	Initialize
INIT	Initiate|Initiator
INJ	Injection
INL	Inter-Node Link
INLETS	In-Line Execution Tests
INLK	Interlock
INLV	Interleave
INM	Integrated Network Management
INMARSAT	International Maritime Satellite (service)
INN	Intermediate Network
INN	Inter-Node Network
INND	Internet News Daemon
INOC	Internet Network Operations Center
INOPBL	Inoperable
INP	Input
INP	Installation Planning
INP	Integrated|Intelligent Network Processor
INPO	In No Particular Order
INQ	Inquiry
INR	Inner
INRECA	Induction and Reasoning from Cases
INREQ	Information Request
INRIA	Institut National de Recherche en Informatique et Automatique (FR)
INS	Information Network Services|System
INS	Input String
INS	Insert
INS	Integrated Network Server
INS	International Navigation System
INSAR	Instruction Address Register
INSCHAR	Insert Character
INSCRB	Inscribe
INSEA	Institut National (de) Statistique (et) d'Economie Appliquée (MA)
INSEE	Institut National de la Statistique et des Etudes Economiques (FR)
INSEM	Inter-institutional Electronic Mail system (EU)
INSF	Indexed Non-Sequential File
IN-SITU	In Place
INSN	Instruction
INSNT	Instruction Time
INSP	Inspect(ion)
INSPEC	Information Service in Physics, Electrotechnology and Control

INST	Instruction
INSTL	Installation
INSUL	Insulator
INSYD	Instantaneous Systems Display
INT	Integer
INT	Intermediate
INT	Internal
INT	International
INT	Interpret(er)
INT	Interrogate
INT	Interrupt(ion)
INT	Intersection
INTA	Interrupt Acknowledge
INTAP	Interoperability Technology Association for information Processing
INTCH	Interchange
INTCP	Intercept
INTCR	Input Tape Cartridge Reader
INTE	Interrupt Enable
INTEL	Integrated Electronics
INTELCOM	International Telecommunications exposition
INTELEC	International Telecommunications Energy Conference
INTELSAT	International Telecommunications Satellite organization
INTENT	Information Technology for National Transformation
INTERCOM	Interlanguage Communications
INTEREST	Integrated Retrieval and Statistic (program)
INTERMARC	International Machine Readable Cataloguing
INTERNEPCON	International Electronics Production Conference
InterNIC	Internet Network Information Center
InterNOC	Inter-Network Operating Company
INTERPERS	Interactive Personal System
INTERTEST	Interactive Test controller
INTF	Interface
INTFER	Interference
INTFU	Interface Unit
INTG	Integrated
INTGEN	Interpreter Generator
INTGTR	Integrator
INTIP	Integrated Information Processing
INTIPS	Intelligence Information Processing
INTIS	International Transport and Information System
INTL	International
INTLK	Interlock
INTNL	Internal
INTO	Interrupt if Overflow occurs
INTP	Interpret(er)
INTR	Interrupt Register|Request
INTRAN	Input Translator

INTRDR	Internal Reader
INTREX	Information Transfer Experiments
INTRG	Interrogate
INTRM	Intermittent
INTRO	Introduction
INTRPT	Interrupt
INTRQ	Interrupt Request
INTRVN	Intervention
INTUG	International Telecommunications Users' Group
INTV	Interval
INTVN	Intervention
INV	Inventory
inv	inverse
INV	Inverter
INVAL	Invalid
INVL	Involute
INVN	Intervention
INVT	Invert(er)
INWATS	Inward Wide Area Telephone Service
INWG	International Network Working Group
INX	Index (character)
IO	British Indian Ocean Territory
IO	Immediate Order
I/O	Input/Output
IO	Input/Output
IO	Interpretive Operation
IO	Inward Operator
IOA	Input/Output Adapter|Address
IOA	Input/Output Attachment
IOAU	Input/Output Access Unit
IOB	Input/Output Block|Buffer
IOBS	Input/Output Buffering System
IOC	Immediate-Or-Cancel order
IOC	Initial Operational Capacity
IOC	Input/Output Channel|Computer
IOC	Input/Output Control(ler) (module)
IOC	Input/Output Converter
IOC	Instruction Operation Code
IOC	Integrated Operating Capability
IOC	Integrated Optical Circuit
IOC	International Operating Center
IOC	Internet On Cable
IOC	Internet Operations Center
IOC	Inter-Office Channel
IOCB	Input/Output Control Block
IOCC	Input/Output Channel Converter
IOCC	Input/Output Control Command
IOCC	Input/Output Controller Chip

IOCCC	International Obfuscated C Code Contest
IOCON	Input/Output Converter
IOCP	Input/Output Configuration|Control Program
IOCR	Input/Output Control Routine
IOCS	Input/Output Control Subroutine|System
IOCTL	Input/Output Control
IOCU	Input/Output Control Unit
IOD	Identified Outward Dialling
IOD	Input/Output Device
IOD	Input/Output Dump (program)
IOD	Integration Object Definition
IODB	Input/Output Distribution Board
IODC	Input/Output Data Control
IOF	Input/Output Front end
IOFU	Instruction and Operand Fetch Unit
IOGEN	Input/Output Generation
IOH	Integrated Open Hypermedia
IOHO	In Our Humble Opinion
IOHS	Insufficient Operator Head Space
IOI	Input/Output Interruption
IOIH	Input/Output Interrupt Handler
IOIM	Input/Output Interrupt Message
IOIRV	Input/Output Interrupt Request Vector
IOLA	Input/Output Link Adapter
IOLC	Input/Output Link Control(ler)
IOLS	Input/Output Label(ing) System
IOLS	Integrated Online Library System(s)
IOM	Input/Output Microprocessor
IOM	Input/Output Module
IOM	Input/Output Multiplexer
IOMS	Input/Output Management System
ION	Integrated On-demand Network
IONL	Internal Organization of the Network Layer
IOO	Input/Output Operation
IOP	Input/Output Package|Processor
IOPC	Input/Output Processor Chip
IOPCB	Input/Output Program Communication Block
IOPE	Input/Output Parity Error
IOPKG	Input/Output Package
IOPL	Input/Output Privilege Level
iops	input/output (operations) per second
IOPS	Input/Output Processing|Processor System
IOPS	Input/Output Program(ming) System
IOQ	Input/Output Queue
IOQE	Input/Output Queue Element
IOR	Input/Output Read|Register
IOR	Institute for Operational Research
IOR	Interoperable Object Reference

IORC	Input/Output Read Control
IORCB	Input/Output Record Block
IOREQ	Input/Output Request
IORS	Input/Output Request Subroutine
IORT	Input/Output Remote Terminal
IOS	Input/Output Sense (line)
IOS	Input/Output Strobe
IOS	Input/Output Supervisor|System
IOS	Interactive Operating System
IOS	Intermediate Open System
IOS	Internetwork(ing) Operating System
IOSA	Input/Output Systems Association
IOSAP	Input/Output Subordinate Application Program
IOSGA	Input/Output Support Gate Array
IOSIM	Input/Output Simulator
IOSR	Input/Output Service Routine
IOSS	Intelligent Operation Support System(s)
IOSYS	Input/Output System
IOT	Input/Output Transfer|Trap
IOT	Inter-Office Trunk
IOTA	Incremental Operation Tape Adapter
IOTA	Input/Output Transaction Area
IOTB	Input/Output Transfer Block
IOU	Immediate Operation Use
IOU	Input/Output Unit
IOU	I Owe You
IOUGA	International Oracle Users Group Americas
IOW	In Other Words
IOW	Input/Output Wire|Write
IOWQ	Input/Output Wait Queue
IOX	Input/Output Executive
IP	Image Processing|Processor
IP	Immediate Processing
IP	Impact Printer
IP	Inbound Processor
IP	Index of Performance
IP	Information Planning|Processor
IP	Information Provider
IP	Initial Permutation
IP	Initial Phase|Point
IP	Installation Planning
IP	Instruction Processor|Pointer
IP	Instruction Pulse
IP	Instructor Pilot
IP	Integrated Processor
IP	Intellectual Property
IP	Intelligent Peripheral
IP	Interchangeable Parts

IP	Interface Processor
IP	Intermediate Point
IP	Internet Protocol|Provider
IP6	Internet Protocol (version) 6
IPA	Immediate Power Amplifier
IPA	Information Processing Architecture
IPA	Integrated Peripheral|Printer Adapter
IPA	Integrated Printer Attachment
IPA	Intelligent Protocol Analyzer
IPA	Interactive Pattern Analysis
IPA	Interactive Product Animator
IPA	Intermediate Power Amplifier
IPA	International Phonetic Alphabet
IPAD	Internet Protocol Adapter
IPAM	Integrated Program Activity Monitor
IPAM	Intellectual Property Asset Management
IPARS	International Passenger programmed Airlines Reservation System
IPB	Integrated Processor Board
IPB	Inter-Processor Buffer
IPBX	Intranet Private Branch Exchange
IPC	Industrial Process Control
IPC	Industrial Product(ion) Center
IPC	Information Processing Center|Code
IPC	Institute for Interconnection and Packaging of electrical/electronic Components (US)
ipc	instructions per clock
IPC	Integrated Peripheral Channel
IPC	Integrated Personal Computer
IPC	Integrated Protective Circuits
IPC	Intelligent Peripheral Controller
IPC	Inter-Process Communication(s)
IPC	Inter-Process Controller|Coupler
IPC	Inventory Process Control
IPCCC	International Performance, Computing & Communications Conference
IPCES	Intelligent Process Control by means of Expert Systems
IPCP	Internet Protocol Control Protocol
IPCS	Interactive Problem|Programming Control System
IPD	Inbound Processor Dialog
IPD	Information Processing Department
IPD	Insertion Phase Delay
IPD	Integrated Program Design
IPD	Internet Protocol Datagram
IPDF	Input Data Funnel
IPDF	Interactive Program Debugging Facility
IPDMUG	International Product Data Management Users Group
IPDPS	International Parallel & Distributed Processing Symposium

IPDS	Integrated Personnel Data System
IPDS	Intelligent Printer Data Stream
IPDT	Integrated Processing of Data and Text
IPDU	Internetwork Protocol Data Unit
IPE	In-band Parameter Exchange
IPE	Installation Performance Evaluation
IPE	Integrated Programming Environment
IPE	Interpret Parity Error
IPEC	Internet Protocol Exchange Communicator
IPEMS	Integrated Production Engineering Management System
IPF	Image Processing Facility|Format
IPF	Information|Interactive Productivity Facility
IPF	Information Presentation|Processing Facility
IPFC	Information Presentation Facility Compiler
IPFM	Integral Pulse Frequency Modulation
IPG	In-circuit Program Generator
IPG	Interactive Presentation Graphics
iph	impressions per hour
IPI	Industrial Products Information
IPI	Initial Protocol Identifier
IPI	Intelligent Peripheral|Printer Interface
IPIC	Initial Production and Inventory Control
IPIC	In-Process Inventory Control
IPICS	Initial Production and Information|Inventory Control System
IPIM	Integrated Product Information Model
IPL	Information Processing|Programming Language
IPL	Initial Program Load(er)
IPL	Internet Public Library
IPL	Interrupt Priority Level
IPL	Ion Projection Lithography
IPL	Iverson's Programming Language
IPLS	Instant Program Location System
ipm	impulses per minute
ipm	inches per minute
IPM	Incidental|Incremental Phase Modulation
IPM	Information Products Marketing
IPM	Institute of Applied Mathematics (DE)
IPM	Integrated Pull Manufacturing
IPM	Inter-Personal Message|Messaging
IPMF	In-Process Material File
IP/MP	In-Phase/Mid-Phase
IPMS	Interpersonal Messaging Service|System
IPM-UA	Interpersonal Messaging User Agent
IPN	Infopulse Private Network
IPN	Integrated|Intelligent Packet Network
IPN	Inter-Planetary Network
IPng	Internet Protocol next generation

IPNS	ISDN PABX Networking Specification (ISDN = Integrated Services Digital Network)
IPO	IGES/PDES Organization
IPO	Initial Public Offering
IPO	Input, Process, and Output
IPO	Inquiry Programmed Operations
IPO	Installation Planning Organization
IPO	Installation Productivity Option
IPO/E	Installation Productivity Option/Extended
IPOM	Installation Planning Operation Manual
IPOT	Inductive Potentiometer
IPP	In-Plant Printing
IPP	Internet Printing Protocol
IPPD	Integrated Product and Process Development
IPPF	Instruction Pre-Processing Function
IPPS/SPDP	International Parallel Processing Symposium & Symposium on Parallel and Distributed Processing
IPR	Impostor Pass Rate
IPR	Information Processing & Retrieval
IPR	Installation Planning Review
IPR	Institute for Process control and Robotics (DE)
IPR	Interrupt Priority Register
IPR	Isolated Pacing Response
IPRL	ISPICS Requirements List
IPS	Image Processing System
ips	inches per second
IPS	Index Participation
IPS	Information Performance Specification(s)|Standards
IPS	Information Processing System
IPS	Installation Performance Specification
IPS	Instruction Prescription System
IPS	Integrated Peripheral System
IPS	Intelligent Power Saver
IPS	Interactive Planning|Processing System
IPS	Interactive Product Simulation
IPS	Internet Publishing System
IPS	Interpretive Programming System
IPSA	International Planning System Architecture
IPSB	Inter-Processor Signal Bus
IPSC	Information Processing Standards for Computers
IPSE	Integrated Programming|Project Support Environment
IPSec	Internet Protocol Security
IPSE-IG	Integrated Project Support Environment Interest Group
IPSI	Integrated Publications and information Systems Institute (DE)
IPSIT	International Public Sector Information Technology
IPSJ	Information Processing Society of Japan
IPSS	Information Processing System Simulator
IPSS	Instant Program Search System

IPSX	Inter-Processor Switch Matrix
IPSY	Interactive Planning System
IPT	Improved Programming Technique(s)|Technology
IPT	Interrupt Priority Table
IPT	Inverse Path Table
IPTC	International Press Telecommunications Council
IP/TCP	Internet Protocol/Transmission Control Protocol
IPTO	Information Processing Techniques Office (US)
IPTS	In-Plant Terminal System
IPTS	Inter-Plant Transmission System
IPU	Immediate Pick-Up
IPU	Instruction Processing Unit
IPU	Integrated Processor Unit
IPU	Internal Processing Unit
IPU	Inter-Processor Unit
IPV4	Internet Protocol Version 4
IPX	Internetwork Packet Exchange
IPX/SPX	Internetwork Packet Exchange/Sequenced Packet Exchange
IQ	Information Quick
IQ	Intelligence Quotient
IQ	Interactive Query
IQ	Iraq
IQE	Interruption Queue Element
IQF	Interactive Query Facility
IQI	Image Quality Indicator
IQL	Interactive Query Language
IQMH	Input Queue Message Handler
IQ/O	Interactive Query/Objects
IQR	Inquiry
IQRP	Interactive Query and Report Processor
IR	Current multiplied with Resistance
IR	Improvement Request
IR	Incident Report
IR	Independent Release|Research
IR	Index Register
IR	Industrial Robot
IR	Industry Remarker
IR	Informal Report
IR	Information Record|Resource
IR	Information Retrieval
I&R	Information & Retrieval
IR	Infra-Red
IR	Inspection Report
IR	Installation Removal|Report
IR	Instruction Register|Repertory
IR	Insulation Resistance
IR	Integrated Research|Resource
IR	Interim Report

IR	Intermediate Representation
IR	Internet Registry
IR	Internet(work) Router
I-R	Interrogator-Responder
IR	Interrupt Register|Request
IR	Investor Relations
IR	Iran (Islamic Republic of)
IRA	Information Resource Administration
IRA	Instruction Register, Address portion
IRA	Interprocedural Register Allocation
IRAC	Interdepartmental Radio Advisory Committee
IRAF	Image Reduction and Analysis Facility
IRAM	Indexed Random Access Method
IRAM	Intelligent Random Access Memory
IRAP	Industrial Research Assistance Program
IRAR	Integrated Random Access Reservation
IRASER	Infra-Red Amplification by Stimulated Emission of Radiation
IRB	Institutional Review Board
IRB	Interruption Request Block
IRBM	Intelligent Repeater Bridge Module
IRC	Information Retrieval Center
IRC	Infra-Red Connection|Control
IRC	International Record Carrier
IRC	Internet Relay Chat
IRCSE	Integrated Real-time Control and Simulation Environment
IR&D	Independent Research & Development
IRD	Information Resource Dictionary
IRD	Initial Requirements Determination
IRD	Integrated Receiver/Descrambler
IRDA	Infra-Red Data|Device Association
IRDP	ICMP Router Discovery Protocol
IRDS	Information Resource Dictionary System
IRDS	Infra-Red Detection Set
IRE	Institute of Radio Engineers
IRED	Infrared Emitting Diode
IREP	Internal Representation
IRET	Interrupt Return
IRF	Inheritance Rights Filter
IRF	Input Register Full
IRF	Intermediate Routing Function
IRFITS	Infra-Red Fault Isolation Test System
IRG	Inter-Record Gap
IRH	Inductive Recording Head
IRI	Interference Interval
IRIG	Inter-Range Instrumentation Group
IRIS	Included Rank-Intervals Set
IRIS	Instant Response Information Service
IRIS	Institute for Research in Information and Scholarship (US)

IRJE	Interactive Remote Job Entry
IRL	Information Retrieval Language
IRL	In Real Life
IRL	Interactive Reader Language
IRL	Inter-divisional Records List
IRL	Inter-Repeater Link
IRL	Intuitive Robot Language
IRLED	Infra-Red Light Emitting Diode
IRLM	IMS/VS Resource Lock Manager
IRLS	Interrogation, Recording and Locating System
IRM	Information Resource(s) Management
IRM	Inherent|Inherited Rights Mask
IRM	Inspection, Repair and Maintenance
IRM	Intelligent Remote Multiplexer
IRM	Intelligent Repeater Module
IRMA	Illinois Reliable Multicast Architecture
IRMA	Integrated Resource Management Architecture
IRMR	Infra-Red Micro Radiometry
IRMS	Information Resources Management Service
IRN	Intermediate|Internal Routing Node
IROS	Increased Reliability of Operational Systems
IROS	Intelligent Robots and Systems (international conference)
IRP	Image Refinery Productions
IRP	Initial Receiving Point
IRP	Ink Ribbon Printer
IRP	Installation Readiness Plan
IRP	Installation Record Performance
IRP	Integrated Reference Package
IRP	Inventory and Requirements Planning
IRPG	Interactive Report Generator
IRPT	Interrupt
IRQ	Interrupt Request (signal)
IRQ	Intervention Required
IRQL	Interrupt Request Level|Line
IRR	Interrupt Request Register
IRRAD	Infra-Red Range And Direction (equipment)
IRS	Inertial Reference System
IRS	Information Record Separator
IRS	Information Reference|Retrieval System
IRS	Inquiry and Reporting System
IRS	Interchange Record Separator
IRS	Internetworking Routing Service
IRS	Inter-Record Separator (character)
IRSG	Internet Research Steering Group
IRSS	Intelligent Remote Station Support
IRT	Index Return (character)
IRTF	Industrial Real-Time Fortran
IRTF	Internet Research Task Force

IRTU	Intelligent Remote Terminal Unit
IRU	Inertial Reference Unit
IRU	Interface Repeater Unit
IRUC	Intermediate Resource Usage Condition
IRV	International Reference Version
IRV	Interrupt Request Vector
IRW	Indirect Reference Word
IRX	Information Retrieval Experiment
IRX	Interactive Resource Executive
IS	Iceland
IS	Index(ed) Sequential
IS	Information Science
IS	Information Separator (character)
IS	Information Services|System
IS	Interface Summary
IS	Intermediate System
IS	International Standard
IS	Internet Standard
IS	Interrupt Set|Status
IS	Interval Signal
ISA	Icon-based Structured Analysis
ISA	Informationssysteme für computerintegrierende Automation (DE)
ISA	Inspection Summary Analysis
ISA	Instruction-Set Architecture
ISA	Instrument Society of America
ISA	Instrument Systems Analysis
ISA	Integrated Systems Architecture
ISA	Interrupt Storage Area
ISA	Invalid Storage Address
ISAAC	International Symposium on Algorithms And Computation
ISAB	Inhibit Switch A Bit
ISAC	Information Systems work and Analysis of Changes
ISACA	Information Systems Audit and Control Association
ISADS	International Symposium on Autonomous Decentralized Systems
ISAI	International Society of Applied Intelligence
ISAL	Information System Access Line
ISAM	Indexed Sequential Access Management|Method
ISAM	Integrated Switching And Multiplexing
ISAP	Information Sort And Predict
ISAP	International STEP Automotive Project (STEP = Standard for The Exchange of Product model data)
ISAPI	Internet Server Application Programming Interface
ISAR	Indirect Scratchpad Address Register
ISAR	Information Storage And Retrieval
ISARL	Indirect Scratchpad Address Register Lower
ISARU	Indirect Scratchpad Address Register Upper

ISAT	Internal System Acceptance Testing
ISAT	Interrupt Storage Area Table
ISBB	Inhibit Switch B Bit
ISBD	International Standard for Bibliographic Description
ISBL	Information System Base Language
ISBN	International Standard Book Number
ISC	Ideal Standard Cost
ISC	Information Separator Character
ISC	Information Services Center
ISC	Installation Support Center
ISC	Instruction Set Computer
ISC	Integrated Semiconductor Circuit
ISC	Integrated Storage Control
ISC	Intelligent Systems Center
ISC	International Switching|Systems Center
ISC	Inter-System Communication|Connection
ISCA	Intelligent Synchronous Communications Adapter
ISCA	International Symposium (on) Computer Architecture
ISCAN	Inertialess Steerable Communications Antenna
ISCB	Inhibit Switch C Bit
ISCBL	Interrupt System Control Block List
ISCC	Intelligent System Control Console
ISCC	Inter-Society Colour Council
ISCF	Inter-System Control Facility
ISCIS	International Symposium on Computer & Information Science
ISCP	Improved Space Computer Program
ISCS	Information Systems Computing Services
ISD	Image Section Descriptor
ISD	Information Structure Design|Diagram
ISD	Instructional Systems Design
ISD	Intermediate Storage Device
ISD	Internal Symbol Dictionary
ISD	International Subscriber Dialling
ISDE	International Software Development Environment
ISDIS	Information System Development and Implementation Services
ISDN	Integrated Services|Subscriber Digital Network
ISDN2	ISDN basic connection 2bd (ISDN = Integrated Services Digital Network)
ISDN30	ISDN basic connection 30bd
ISDN-UP	ISDN – User Part
ISDOS	Information System Design and Optimization System
ISDS	Indexed Sequential Data Set
ISDS	Instruction Set Design System
ISDS	Integrated Software Development System
ISDT	Integrated Systems Development Tool
ISE	Institute for Software Engineering (US)
ISE	In-System Emulator
ISE	Interactive Software Engineering

ISE	International Society for Epistecybernetics
ISE	Interrupt System Enable
ISEE	Integrated Software Engineering Environment
ISEM	Improved Standard Electronic Module
ISEP	International Standard Equipment Practice
ISEQ	Indexed Sequential
ISESS	International symposium & forum on Software Engineering Standards
ISF	Individual Store and Forward
ISF	Information Systems Factory
ISF	International Supply Function
ISFD	Integrated Software Functional Design
ISFM	Indexed Sequential File Management|Manager
ISFMS	Indexed Sequential File Management System
ISFO	Indexed Sequential File Organization
ISFUG	Integrated Software Federal User Group
ISG	IETF Steering Group
ISG	Inter-Subblock Gap
ISH	Information Super Highway
ISH	Intermediate-System Hello
ISHM	International Society of Hybrid Microelectronics
ISI	Information Sciences Institute (US)
ISI	Information Society Index
ISI	Information Structure Implementation
ISI	Inter-Symbol Interference
ISIC	International Standard Industrial Classification
ISIC	International Standard Industry Code
ISIL	Information Structure Identification Language
ISIS	Integrated Software Invocation System
ISIS	Integrated Staff Information System
ISIS	Integrated Systems and Information Services
ISIS	Intelligent Scheduling and Information System
IS-IS	Intermediate System to Intermediate System
ISITC	Industry Standardization for Institutional Trade Communication
ISK	Insert Storage Key
ISK	Instruction Space Key
ISL	Information Search Language
ISL	Integrated Schottky Logic
ISL	Interactive Simulation|System Language
ISL	Interface Socket Listing
ISL	Inter-Satellite Link(s)
ISL	Inter-System Link(s)
ISLM	Integrated Services Line Module
ISLN	Integrated Services Local Network
ISLS	Integrated Set of Library Systems
ISLU	Integrated Services Line Unit
ISM	Image Support Module

ISM	Industrial, Scientific, and Medical
ISM	Information Systems Management
ISM	Instruction Stream Memory
ISMA	Information Systems Management Architecture
ISMEC	Information Service in Mechanical engineering
ISMF	Interactive Storage Management Facility
ISMH	Input Source Message Handler
ISML	Inter-Shop Markup Language
ISMM	International Society for Mini- and Microcomputers
ISMOD	Indexed Sequential Module
ISMS	IBM Subscriber Management System
ISMS	Image Store Management System
ISMVL	International Symposium on Multiple-Valued Logic
ISN	Information Services|Systems Network
ISN	Initial|Input Sequence Number
ISN	Integrated Systems Network
ISN	Internal Sequence|Statement Number
ISN	Internet Society News
ISNOT	Is Not equal to
ISNSA	Independent Software Nuclear Safety Analysis
ISO	Individual System Operation
ISO	Information Systems Office
ISO	In Search Of
ISO	International Organization for Standardization
ISO9000	International Organization for Standardization 9000-series quality guidelines
ISOC	Individual System/Organization Cost
ISOC	Internet Society
ISOO	Information Security Oversight Office
ISO/OSI	International Organization for Standardization/Open Systems Interconnect(ion)
ISORC	International Symposium on Object-oriented Real-time distributed Computing
ISORM	International Standardization Organization Reference Model
ISP	Index Sequential Processor
ISP	Information Search Processor
ISP	Instruction Set Processor
ISP	In-System Programmability
ISP	Internal Stored Program
ISP	International Standardized Profile
ISP	Internet Service Provider
ISP	Interrupt Stack Pointer
ISP	Interrupt Status Port
I-SPAN	International Symposium on Parallel Architectures, algorithms and Networks
ISPASS	International Symposium on Performance Analysis of Systems & Software
ISPATS	International Standardized Profile Abstract Test Suite

IS-PBX	Integrated Services – Private Branch Exchange
ISPC	International Sound Programme Center
ISPETS	International Standardized Profile Executable Test Suite
ISPF	Interactive System Productivity|Programming Facility
ISPICE	Interactive Simulation Program with Integrated Circuit Emphasis
ISPICS	International Standardized Profile Implementation Conformance Statement
ISPIXIT	Internet Service Provider protocol Implementation Extra Information for Testing
ISPN	International Standard Package Number
ISPO	Information Society Project Office (EU)
ISPT	Instituto Superiore Poste e Telecommunicazioni (IT)
ISPX	ISDN Private Branch Exchange (ISDN = Integrated Subscriber Digital Network)
ISQED	International Symposium on Quality of Electronic Design
ISQL	Informix Standard Query Language
ISR	Image Storage and Retrieval
IS&R	Information Storage & Retrieval
ISR	In-Service Register
ISR	Integrated Switch Router
ISR	Interrupt Service Routine
ISR	Intersecting Storage Ring
ISRAD	Integrated Software Research And Development
ISRC	International Standard Recording Code
ISRD	Information Storage, Retrieval and Dissemination
ISRM	Information Systems Resource Manager
ISRN	Indonesian Small Ruminant Network
ISRN	International Standard Record Number
ISS	Image Symbol Set
ISS	Information Storage System
ISS	Information Systems Security|Services
ISS	Integrated Sounding|Support System
ISS	Integrated Switching System
ISS	Intelligent SCSI Subsystem
ISS	Intelligent Support System
ISS	International Software Solutions
ISS	Internet Security Systems
ISS	Interrupt Service Subroutines
ISSA	Information Support Services for Agriculture
ISSB	Information Systems Standards Board
ISSC	Integrated Systems Solution(s) Corporation
ISSCC	International Solid-State Circuits Conference
ISSE	Internet Streaming SIMD Extensions
ISSG	International System Support Group
ISSMB	Information Systems Standards Management Board
ISSN	Integrated Special Services Network
ISSN	International Standard Serial Number

ISSO	Information Systems Security Organization
ISSP	Installation Specified Selection Parameters
ISSR	Information Storage, Selection, and Retrieval
ISSRE	International Symposium on Software Reliability Engineering
ISSS	Information Selection and Sampling System
ISSS	International Symposium on Systems Synthesis
ISSS	ISDN Supporting System (ISDN = Integrated Services Digital Network)
ISSUE	Information System and Software Update Environment
IST	Initial System Test
IST	Interrupt Service Task
ISTAB	Information Systems Technical Advisory Board
ISTDT	Issues To Date
ISTM	It Seems To Me
ISTR	Indexed Sequential Table Retrieval
ISU	Instruction Storage Unit
ISU	Integrated Service(s) Unit
ISU	Interface Sharing Unit
ISUP	Immediate Suppress
ISUP	Integrated Services User Part
ISUP	ISDN User Part (ISDN = Integrated Services Digital Network)
ISV	Independent Software Vendor
ISVD	Information System for Vocational Decision making
ISW	Initial Status Word
ISWC	International Symposium on Wearable Computers
ISWM	Inhibit Switch Word Mark
ISWYM	I See What You Mean
ISZ	Increment and Skip on Zero
IT	Image Technology
IT	Indent Tab (character)
I(t)	Indicial response
IT	Industrial Technology
IT	Information Technology|Type
IT	Input Terminal|Translator
IT	Inspection Tag
IT	Intelligent Terminal
IT	Interface Tape
IT	Internal Translator
IT	Italy
IT	Item Transfer
ITA	Initial Teaching Alphabet
ITA	Integrated Technical Architecture
ITA	Interface Test Adapter
ITA	International Telegraph Alphabet
ITAA	Information Technology Association of America
ITAC	Interprocessor Tasking And Communications
ITAEGC	Information Technology Advisory Expert Group on Certification (EU)

ITAEGM	Information Technology Advisory Expert Group on advanced Manufacturing technologies (EU)
ITAEGS	Information Technology Advisory Expert Group on Standardization (EU)
ITAEGT	Information Technology Advisory Expert Group on Telecommunications (EU)
ITAM	Interdata Telecommunications Access Method
ITAP	Information Technology Advisory Panel (report)
Itapac	The Italian packet-switched network
ITAVS	Integrated Testing, Analysis and Verification System
ITB	Information Technology Branch
ITB	Intermediate Text Block (character)
ITB	Intermediate Transmission Block
ITB	Internal Transfer Bus
ITBLT	Information Technology Based on Learning and Training
ITC	Individual Table of Contents
ITC	Installation Time and Cost
ITC	Integrated|Intelligent Transaction Controller
ITC	International Telemetering|Test Conference
ITC	International Teletraffic Congress (DK)
ITC	International Typeface Corporation
ITC	Inter-Task Communication
ITCA	International Tele-Conferencing Association
ITCC	Information Technology Consultative Committee
ITCLC	Item Class Code
ITDF	Interactive Transaction Dump Facility
ITDG	Information Technology Decision-Guru
ITDM	Intelligent Time-Division Multiplexer
ITDNS	Integrated Tour operating Digital Network Service
ITDS	Integrated Technical Data Systems
ITDSC	Item Description
ITE	Implement – Test – Edit
ITE	Independent Trade Exchange
ITE	Information Technology Equipment
ITEA	Information Technology for European Advancement
ITEM	Information Technology in Educational Management
ITEM	Institute for Technology Management (CH)
ITEM	Integrated Test and Maintenance
ITEM	Interactive Technique for Effective Management
ITEX	Interactive TTCN Editor and Executor
ITEX-DE	Interactive TTCN Editor and Executor – Development Environment
ITF	Integrated Test Facility
ITF	Interactive Terminal Facility
ITF	Interleave Two of Five
ITGEN	Input Tape Generator
ITI	Information Technology Industry (council)
ITI	Information Technology Interfaces (international conference)

ITI	Interactive Terminal Interface
ITIC	International Tsunami Information Center
ITIL	Information Technology Infrastructure Library
I-time	Instruction time
ITIMS	In-service Transmission Impairment Measurement Set
ITIRC	IBM Technical Information Retrieval Center
ITIS	Interactive Terminal Interface System
ITK	Integration Toolkit
ITL	Integrated Tape Controller
ITL	Integrated Transfer Launch
ITL	Inter-divisional Technical Liaison
ITL	Intermediate Transfer Language
ITM	Indirect Tag Memory
ITM	Information Technology Management
ITM	Instruction Trace Monitor
ITMS	International Transport Messages Scenario
ITN	Identification Tasking and Networking
ITN	Integrated Teleprocessing Network
ITN	International Transit Node
ITNBR	Item Number
ITNC	International Telecommunication Network Center
ITO	Input-Transformation-Output
ITON	Internet Transaction Operations Network
ITOS	Interactive Terminal Oriented Software
ITP	Integrated Transaction Processor
ITP	Interactive Terminal Protocol
ITP	Interactive Transactional Program
ITP	Interpretive Translation Program
ITPA	Ideal Transport Protocol (over) ATM (= Asynchronous Transfer Mode)
ITPM	Information Technology Process Model
ITPS	Interactive Text Preparation System
ITPS	Internal Tele-Processing System
ITR	Internet Talk Radio
ITRAC	Interdata Transaction Controller
ITRC	Information Technology Requirements Council
ITRM	Information Technology Reference Model
ITRN	Iteration
ITRON	Industrial TRON (= The Real-time Operating system Nucleus)
ITS	Industrial Translator System
ITS	Information Transmission System
ITS	Institute of Telecommunication Sciences (US)
ITS	Intelligent Terminal Service
ITS	Intelligent Transport|Tutoring System(s)
ITS	Interactive Terminal Support
ITS	Interactive Training System
ITS	Invitation To Send
ITSB	Image Technology Standards Board

ITSC	International Technical Support Center
ITSC	Inter-regional Telecommunications Standards Conference
ITSEC	Information Technology Security Evaluation Criteria
ITSO	Incoming Trunk Service Observation
ITSP	Information Technology and System Planning
ITSP	Internet Telephony Service Provider
ITSS	Interim Time Sharing System
ITSTC	Information Technology Steering Committee
ITSW	International Test Synthesis Workshop
ITT	Import Transit Time
ITT	International Telephone and Telegraph
ITT	Inter-Toll Trunk
ITT	Invitation To Transmit
ITU	International Telecommunications Union
ITUG	International Telecommunications User Group
ITU-T	International Telecommunications Union -Telecommunications (standardization section)
ITV	Interactive Television
ITVA	International Television Association
ITX	Intermediate Text (block)
ITXC	Internet Telephony Exchange Carrier
IU	Information|Input Unit
IU	Installable Unit
IU	Instruction|Instrument Unit
IU	Integer|Interface Unit
IUA	Interactive User Access
IUAP	Internet User Account Provider
IUB	Instruction Used Bit
IUCM	Input Unit Creation Module
IUCP	Interface Unit Creation Processor
IUE	Instruction Unit Execution
IUMA	Internet Underground Music Archive
IUP	Installed User Program
IUR	Irreducible Unitary Representation
IURP	Integrated Unit Record Processor
IUS	Inertial Upper Stage
IUS	Information|Intermediate Unit Separator
IUS/ITB	Interchange Unit Separator/Intermediate Transmission Block
IUT	Implementation Under Test
IUW	ISDN User's Workshop (ISDN = Integrated Services Data Network)
IV	Independent Validation|Variable
IV	Independent Verification
IV	Interactive Video
IV	Interface Vector
IV	Invalid
IV	Inverter
IVAN	International Value Added Network

IVAR	Internal Variable
IVC	Integrated Visual Computing
IVC	Internet Voice Chat
IVCP	Interoperable Virtual Connection Protocol
IVD	Integrated Voice and Data
IVD	Interactive Video Disk
IVD	Invalid Decimal
IVDMS	Integrated Voice and Data Multiplexers
IVDT	Integrated Voice/Data Terminal
IVDTE	Integrated Voice/Data Terminal Equipment
IVDTS	Integrated Voice/Data Terminal System
IVESS	Interactive Vehicle Scheduling System
IVG	Immediate Visual Gratification
IVG	Interrupt Vector Generator
IVIS	Interactive Video Information System
IVL	Intel Verification Lab
IVL	In Virtual Life
IVM	Initial Virtual Memory
IVM	Interface Virtual Machine
IVMO	Initial Value Managed Object
IVN	International Voice Network
IVP	Installation Verification Procedure|Program
IVR	Integrated Voltage Regulator
IVR	Interactive Voice Recognition|Response
IVRS	Interactive Voice Response System
IVRU	Interactive Voice Response Unit
IVS	Interactive Videodisc System
IVT	Integrated Video Terminal
IVT	Interrupt Vector Table
IVTS	International Video Teleconferencing Service
IV&V	Independent Verification & Validation
IW	Impulse Width
IW	Index|Instruction Word
IWA	Interpreter|Interrupt Work Area
IWAN	International Working conference on Active Networks
I-WAY	Information Highway
IWC	Inside Wire Cable
IWF	Inter-Working Function
IWG	Implementation Working Group
IWGDE	Inter-laboratory Working Group for Data Exchange
IWI	International Workload Index
IWPA	International Word Processing Association
IWPC	International Workshop on Program Comprehension
IWPS	Info-Window Presentation System
IWPTS	International Workshop on Protocol Test Systems
IWQ	Input Work Queue
IWS	Instruction Work Stack|Station
IWS	Interactive Work Station

IWSED	International Workshop on Software Engineering Data
IWT	Inhibit Word Trigger
IWTCS	International Workshop on Testing Communication Systems
IWU	Intermediate Working Unit
IWU	Inter-Working Unit
IX	Index (register)
IX	Interactive Executive
IXC	Inter-Exchange Carrier|Channel
IXC	Inter-Exchange Channel
IXCU	Integrated Transmission Control Unit
IXI	International X.25 Infrastructure
IXL	Introduction of Extremely Large databases
IXM	Index Manager
IXR	Index Register
IXU	Index translation Unit
IYB	Inhibit Y Bit

J

J	Jack
J	Joule
J	Jump (instruction)
J	Symbol for connector
J2EE	Java 2 Enterprise Edition
J2ME	Java 2 Micro Edition
J2SE	Java 2 Standard Edition
JA	Job Analysis
JA	Jump Address
JA	Jump if Above
JAB	Job Analysis and Billing
JACC	Joint Automatic Control Conference
JACI	Just Another CASE Implementation (CASE = Computer Aided Software Engineering)
JACM	Journal of the Association for Computing Machinery
JAD	Joint Application Design|Development
JADE	Jasmine Application Development Environment
JAE	Java Application Environment
JAE	Jump if Above or Equal
JAF	Job Accounting Facility
JAI	Job Accounting Interface
JAI	Journal of Artificial Intelligence
JAIN	Java Advanced Intelligent Network
JAIN	Java API's for Integrated Network (API = Application Programming Interface)
JAM	Just A Minute
JANET	Joint Academic Network (GB)
jar	java archive (file extension indicator)
JAR	Jump Address Register

JARS	Job Accounting Report System
JAS	Job Accounting|Analysis System
JASIS	Journal of the American Society for Information Science
JAVAC	Java Compiler
JAVS	Josephson Array Voltage Standards
JBE	Jump if Below or Equal
JBIG	Joint Bi-level Imaging Group
JBIG	Joint Binary Image Group
JBOD	Just a Bunch Of Disks
JC	Job Control
JC	Jump Condition
JC	Jump if Carry set
JCALS	Joint Computer-aided Acquisition and Logistics Support
JCB	Job Control Block
JCI	Job Control Information
JCIT	Jerusalem Conference on Information Technology
JCL	Job Control Language
JCLGEN	Job Control Language Generation
JCLPREP	Job Control Language Preprocessor
JCN	Jump on Condition
JCP	Java Community Process
JCP	Job Control Processor|Program
JCP	Joint Committee on Printing
JCS	Job Control Statement|System
JCSE	Joint Communications Support Element
JCT	Job Control Table
jctn	junction
JDA	Joint Development Agreement
JDBC	Java Data Base Connectivity
JDBD	Java-enabled Data Base Development
JDC	Java Developer Connection
JDEF	Joint Demonstration (and) Evaluation Facility
JDK	Java Development Kit
JDL	Job Description Language|Library
JDO	Java Data Object
JDS	Job Data Sheet
JDS	Joint Deployment System
JE	Job Entry
JE	Jump if Equal
JECC	Japan Electronic Computer Company
JECF	Java Electronic Commerce Framework
JECL	Job Entry Control Language
JECS	Job Entry Central Services
JED	Job Entry Definition
JEDEC	Joint Electronic Device Engineering Council
JEDI	Joint Electronic Data Interchange
JEEEA	Journal of Electrical and Electronics Engineering, Australia
JEF	Japanese processing Extended Feature

JEIDA	Japanese Electronics Industry Development Association
JEMA	Japanese Electric Machinery Association
JEP	Job Entry Program
JEPES	Joint Engineering Planning Execution System
JEPS	Job Entry Peripheral Services
JES	Job Entry Subsystem|System
JESI	Joint European Standards Institution
JESSI	Joint European Submicron Silicon Initiative
JESUS	Job Entry System of the University of Saskatuan
JETD	Joint Electronics Type Designation
JETEC	Joint Electron Tube Engineering Council
JETRO	Japan External Trade Organization
JETS	Job Executive and Transport Satellite
JFAST	Joint Flow and Analysis System for Transportation
JFC	Java Foundation Class(es)
JFC	Job Flow Control
JFCB	Job File Control Block
JFET	Junction Field Effect Transistor
JFIF	JPEG File Interchange Format
JFN	Job File Number
JFS	Journalized File System
JG	Jump if Greater
JGE	Jump if Greater or Equal
JGL	Java Generic Library
JGN	Junction Gate Number
JIB	Job Information Block
JIC	Joint Information|Intelligent Center
JICST	Japan Information Center for Science and Technology
JID	Job Input Device
JIDL	Java Interface Definition Language
JIEO	Joint Interoperability Engineering Organization
JIF	Job Input File
JIM	Job Information Memorandum
JIPS	JANET Internet Protocol Service (GB)
JIQ	Job Input Queue
JIRA	Japanese Industrial Robot Association
JIS	Japanese Institute for Standards
JIS	Job Information System
JIS	Job Input Station|Stream
JISC	Japanese Industrial Standards Committee
JISC	Joint Information Systems Committee (GB)
JIT	Job Instruction Training
JIT	Just-In-Time
JITC	Joint Interoperability Testing Center
JITEC	Joint Information Technology Experts Committee
JJ	Josephson Junction
JK	Inputs to a flip-flop (J and K have no specific meaning)
J/K	Joules per Kelvin

JKFF	JK Flip-Flop (J and K have no specific meaning)
JL	Jump if Less
JLE	Jump if Less than or Equal to
JLSC	Joint Logistics Systems Center
JM	Jamaica
JM	Job Memory|Mix
JMAPI	Java Management Application Programming Interface
JMEM	Job Memory
JMF	Java Media Framework
JMP	Jump
JMPR	Jumper
JMPS	Journal of the Mechanics and Physics of Solids
JMS	Java Messaging Service
JMSC	Japan Midi Standards Council
JMSX	Job Memory Switch Matrix
JMX	Jumbogroup Multiplex
JNA	Jump if Not Above
JNAE	Jump if Not Above or Equal
JNB	Jump if Not Below
JNBE	Jump if Not Below or Equal
JNDI	Java Naming and Directory Interface
JNF	Job Networking Facility
JNG	Jump if Not Greater
JNGE	Jump if Not Greater or Equal
JNI	Java Native Interface
jnk	junk (file extension indicator)
JNLE	Jump if Not Less or Equal
JNO	Jump if No Overflow
JNP	Jump if No Parity
JNS	Jump if No Sign
JNT	Joint Network Team
JNZ	Jump if Not Zero
JO	Jordan
JOBDOC	Job Documentation
JOBID	Job Identifier
JOBLIB	Job Library
JOBNO	Job Number
JOBQ	Job Queue
JOD	Job Output Device
JOD	Just Off Drum
JOE	Java Objects Everywhere
JOF	Job Output File
JOHNNIAC	John von Neumann Integrator and Automatic Computer
JOL	Job Organization|Oriented Language
JOMO	Job Mix Optimization
JOOP	Journal of Object-Oriented Programming
JOPS	Joint inter-Operability Planning System
JOS	Job Output Stream

JOSS	JOHNNIAC Open Shop System
JOT	Job Oriented Terminal
JOVIAL	Jules' Own Version of International Algorithmic Languages
JP	Japan
JP	Job Processing|Processor
JPA	Job Pack Area
JPACQ	Job Pack Area Control Queue
JPCS	Job Processing Control System
JPE	Jump if Parity Even
JPEG	Joint Photographic Experts Group
jpg	joint photographic group (file extension indicator)
JPMF	Job Processing Master File
JPNIC	Japan Network Information Center
JPO	Jump if Parity Odd
JPS	Job Processing System
JPT	Job Process Terminating
JPU	Job Processing Unit
JPW	Job Processing Word
JR	Job Rotation
JR	Joint Return
JRAG	Joint Registration Advisory Group
JRB	Java Relational Binding
JRE	Java Runtime Environment
JRMI	Java Remote Method Invocation
JRP	Joint Requirements Planning
JRST	Jump and Restore (flags)
JS	JavaScript
J/s	Joules per second
JS	Jump if Sign
JSA	Japan(ese) Standards Association
JSA	Job Start Address
JSCS	Job Shop Control System
JSD	Jackson System Development
JSF	Job Services File
JSI	Job Step Index
JSIA	Japan Software Industry Association
JSL	Job Specification Language
JSM	Job Stream Manager
JSN	Junction Switch Number
JSP	Jackson Structured Programming
JSP	Java Server Page
JSR	Jump to Subroutine
JSS	Javascript Style Sheet(s)
JSS	Job Segment Scheduler
JST	Job Step Task
JSTCB	Job Step Control Block
JSTEP	Japan STEP promotion center (STEP = Standard for The Exchange of Product model data)

JSTOR	Journal Storage
JSW	Junctor Switch
JSWAP	Job Swapping memory
JT	Jarno Taper
JT	Job Table
jt	joint
JTA	Java Transaction API (= Application Programming Interface)
JTAG	Joint Test Action Group
JTAM	Job Transfer And Manipulation
JTAPI	Java Telephony Application Program Interface
JTAS	Jydsk Telefon Aktieselskab (DK)
JTASC	Joint Training, Analysis and Simulation Center
JTC	Joint Technical Committee (of ISO and IEC) (ISO = International Organization for Standardization; IEC = International Electrotechnical Committee)
JTDS	Joint Track Data Storage
JTIDS	Joint Tactical Information Distribution System
JTM	Job Transfer and Manipulation
JTM	Josephson Tunnelling Memory
JTPS	Job and Tape Planning System
JTS	Java Transaction Service(s)
JUCFET	Junction Field Effect Transistor
JUDGE	Judged Utility Decision Generator
JUG	Joint Users Group
JUGHEAD	Jonzy's Universal Gopher Hierarchy Excavation And Display
JUGL	JANET User Group for Libraries (GB)
JULLS	Joint Universal Lessons Learned System database
JUN	Jump Unconditionally
JUNCFET	Junction Field Effect Transistor
JUNET	Japanese Universities and research Network
JUNET	Japanese Unix Network
JUSTIS	Joint Uniform Service Technical Information System
JUTCPS	Joint Uniform Telephone Communication Precedence System
JVM	Java Virtual Machine
JVNCC	John Von Neumann Computer Center
JVNCNET	John Von Neumann Center Network
JVTOS	Joint Viewing and Tele-Operation Service
JWG	Joint Working Group
JZ	Jump if Zero

K

K	Cathode
k	(dielectric) constant
K	Kelvin
K	Key
k	kilo (prefix 1000)
K	Kilo (prefix 1024 bytes)

K	Measured computer storage capacity
KA	Keyed Address
kA	kiloAmpere
KADS	Knowledge Acquisition and Documentation System
KAK	Key-Auto-Key
KAM	Keep Alive Memory
KAPES	Knowledge-Aided Process planning and Estimation
KASS	Keyboard A Scale Start
KAT	Key-to-Address Transformation
KAU	Key Station Adapter Unit
KB	Keyboard
Kb	Kilobit(s)
KB	Kilobyte(s)
KBD	Keyboard
KBD$	Keyboard of operating system 2 of IBM
KBE	Keyboard Entry
KBI	Key Buying Influence
Kb/i	Kilobits per inch
Kbit/s	Kilobits/second
KBL	Keyboard Listener
KBMEDM	Knowledge Based Model of the Experienced Decision Maker
KBMS	Knowledge Based Management System
KBP	Keyboard Processor
Kbps	Kilobits per second
KBps	Kilobytes per second
KBS	Keyboard System
Kb/s	Kilobits per second
KBS	Knowledge Based System
Kb/sec	Kilobits per second
KB/sec	Kilobytes per second
KBSR	Keyboard B Scale Reset
KBT	Keyboard Branch Table
kc	kilocycle(s)
KCC	Keyboard Common Contact
KCL	Key-station Control Language
kcs	1000 characters per second
kc/s	1000 cycles per second
kcs	kilo-characters per second
KCS/SO	Keyboard Class Select/Statistics Output
KCU	Keyboard Control Unit
KD	Keyboard and Display
KD	Key Definition
KD	Knock-Down (signal)
KDCI	Key Display Call Indicator
KDD	Knowledge Discovery from Databases
KDE	K Desktop Environment
KDE	Keyboard Data Entry
KDEM	Kurzweil Data Entry Machine

KDF	Knowledge Discovery Framework
KDOS	Key Display Operating System
KDP	Keyboard, Display, and Printer
KDS	Key(board) Display Station|System
KDS	Key-to-Diskette System
KDSI	Thousands of Delivered Source Instructions
KDSS	Key to Disk Sub-System
KDT	Keyboard Display Terminal
KDT	Key Data Terminal
KDT	Key Definition Table
KDU	Keyboard Display Unit
KE	Kenya
KEE	Knowledge Engineering Environment
KEG	Key Gap
KEP	Key Entry Processing
KEPROM	Keyed access Erasable Programmable Read-Only Memory
KER	Keying Error Rate
KEYBBE	Keyboard foreign language (BE)
KEYBBR	Keyboard foreign language (BR)
KEYBCF	Keyboard foreign language (CA-FR)
KEYBCZ	Keyboard foreign language (CZ)
KEYBDK	Keyboard foreign language (DK)
KEYBFR	Keyboard foreign language (FR)
KEYBGR	Keyboard foreign language (DE)
KEYBHU	Keyboard foreign language (HU)
KEYBIT	Keyboard foreign language (IT)
KEYBLA	Keyboard foreign language (ES-PT)
KEYBNL	Keyboard foreign language (NL)
KEYBNO	Keyboard foreign language (NO)
KEYBPL	Keyboard foreign language (PL)
KEYBPO	Keyboard foreign language (PT)
KEYBSF	Keyboard foreign language (CH-FR)
KEYBSG	Keyboard foreign language (CH-DE)
KEYBSL	Keyboard foreign language (SK)
KEYBSP	Keyboard foreign language (ES)
KEYBSU	Keyboard foreign language (FI)
KEYBSV	Keyboard foreign language (SE)
KEYBUK	Keyboard foreign language (GB)
KEYBUS	Keyboard foreign language (US)
KEYBYU	Keyboard foreign language (YU)
KF	Key Filed
KFAS	Keyed File Access System
KFD	Knowledge Flow Diagram
KFM	Kohonen' Feature Map
Kfr/cm	Kilo flux reversals per centimeter
KFT	Kilo-Feet
kg	kilogram
KG	Kyrgyzstan

kgm	kilogram-meter
kgps	kilograms per second
KH	Cambodia
Kh	Kilohour
KHz	Kilohertz
KI	Kiribati
KIC	Kernel Input Controller
KIC	Knowledge Intensive CAD (workshop) (CAD = Computer Aided Design)
KICU	Keyboard Interface Control Unit
KIF	Key Index File
KIF	Knowledge Interchange Format
KIL	Keyed Input Language
KIM	Keyboard Input Matrix
KIN	Keyboard Input
Kips	1000 instructions per second
KIPS	Knowledge Information Processing System(s)
kips	thousands of instructions per second
KIS	Keyboard Input Simulation
KIS	Knowbot Information Service
KISS	Keep It Safe and Simple
KISS	Keyed Indexed Sequential Search
kJ	kilojoule
KK	Key-encrypting Key
KK	Know that you Know
KL	Key Length|Lever
KLF	Kings (of the) Low Frequency
klips	1000 logical inferences per second
KLOC	1000 Lines Of Code
KLT	Karhunen-Loeve Transform(ation)
KM	Comoros
KM	Kilo-Mega
km	kilometer
KM	Knowledge Management
KMC	Kilomegacycle
KMON	Keyboard Monitor
KMP	Key Management Protocol
KMS	Knowledge Management System
KMUX	Synchronous Multiplexer
KN	Saint Kitts and Nevis
KNAW	Koninklijke Nederlandse Akademie van Wetenschappen
KNCSS	Thousands of Non-Commented Source Statements
KO	Knock-Out
KOBOL	Key station On-line Business Oriented Language
KOD	Knowledge Oriented Design
KOF	Knock-Off
KOGGE	Koblenzer Generator (für) Graphische Entwurfsumgebungen (DE)

kops	thousands of operations per second
KORSTIC	Korean Science and Technological Information Center
KOX	Keyboard Operated Transmission
KP	Key Pulse|Pulsing
KP	Keypunch
KP	Kick Plate
KP	Korea, Democratic People's Republic of
KPC	Keyboard Printer Control
KPDL	Kyocera Printer Description Language
kph	kilometers per hour
KPI	Kernel Programming Interface
KPI	Key Performance Indicator
KPO	Key Performance Objectives
KPO	Key Punch Operator
KPR	Key Punch Replacement
kps	kilometers per second
KPV	Keypunch Verifier
KQML	Knowledge Query and Management|Manipulation Language
K&R	Kernighan & Ritchie
KR	Key Register
KR	Knowledge Representation
KR	Korea, Republic of
KRA	Key Recovery Alliance
KRL	Knowledge Representation Language
KRS	Knowledge Retrieval System
KS	Kansas (US)
KSAM	Keyed Sequential Access Method
KSDS	Key(ed) Sequence(d) Data Set
KSF	Key Success Factor
ksh	key strokes per hour
KSH	Korn Shell (program) (Unix)
KSL	Keyslot
ksloc	1000 source lines of code
KSM	Kane Security Monitor
KSO	Keyboard Send Only
ksph	keystrokes per hour
KSR	Keyboard Send/Receive
KSR/T	Keyboard Send/Receive Terminal
KST	Known Segment Table
KSU	Key Service Unit
KSU	Key System Unit
KT	Keyboard Training
KT	Key Tape
KTAS	Kobenhavns Telefon Aktieselskab (DK)
KTD	Key To Disk
KTDS	Key-To-Disk Software
KTL	Key-edit Terminal Language
KTM	Key Transport Module

KTMS	Knowledge-based Tank Management System
KTR	Keyboard Typing Reperforator
KTR	Knowledge Template Repository
KTS	Key Telephone System
KTT	Key To Tape
KTU	Key Telephone Unit
KUIP	Kernel User Interface Package
KV	Kilovolt
KVA	Kilovolt-ampere
KVI	Known Value Item
KV/mm	Kilovolt per millimeter
KW	Kilowatt
KW	Kuwait
KWAC	Key-Word And Context
KWh	Kilowatt-hour
KWhr	Kilowatt-hour
KWIC	Key-Word In Context
KWIT	Key-Word In Title
KWOC	Key-Word Out of Context
KWY	Keyway
KXU	Keyword Transformation Unit
KY	Cayman Islands
KY	Kentucky (US)
KY	Keying (device)
KYC	Know Your Customer
KZ	Kazakhstan

L

L	Laplace-transformation
l	length
L	Lightness
L	Load (instruction)
L	Low
l	lumen
L	Symbol for coil, lambert, inductance, and mean life
L2TP	Layer Two Tunnelling Protocol
LA	Laboratory Automation
LA	Lao People's Democratic Republic
LA	Line Adapter|Analysis
LA	Link Allotter
LA	Load|Local Address
LA	Logical Address
LA	Logic Automation
LA	Look Ahead
LA	Louisiana (US)
LAA	Locally-Administered Address
lab	laboratory

LAB	Laboratory Automation Based (system)
LAB	Line Adapter Base
LAC	Large Area Coverage
LAC	Load Accumulator
LAC	Local Area Communications|Coverage
LAC	Loop Assignment Center
LACC	Local Area Control Center
LACN	Local Area Communications Network
LAD	Local Area Disk
LADAR	Laser Detection And Ranging
LADDER	Language Access to Distributed Data with Error Recovery
LADDR	Layered Device Driver Architecture
LADGA	Layered Acyclic Directed Graph with Attributes
LADS	Local Area Data Set
LADS	Logic Automation Documentation System
LADT	Local Area Data Transport
LAG	Local Address Group
LAGER	Layout Generating Routine
LAIS	Local Automatic Intercept System
LAL	Leased Access Line
LAL	Local Analog Loopback
LALR	Left After Left-Right
LALS	Logic Automation Layout System
LAM	Laboratory Monitoring System
LAM	Laminated
LAM	Link Assembly Module
LAM	Load Accumulator with Magnitude
LAM	Lobe Attachment Module
LAM	Local Area Multicomputer
LAM	Look-Alike Machine
LAMA	Local Automatic Message Accounting
LAMP	Logic Analysis for Maintenance Planning
LAMPS	Light Airborne Multi-Purpose System
LAN	Local Area Network
LANAC	Laminar Navigation Anti-Collision
LANACS	Local Area Network Asynchronous Connection Server
LANAO	Local Area Network Automation Option
LANC	Local Application Numerical Control
LANCE	Local Area Network Controller for Ethernet
LANCO	Language for Abstract Numerical Control
LAND	Local Area Network Driver
landsat	land satellite
LANE	Local Area Network Emulation
LANG	Language
LAN/RM	Local Area Networks Reference Model
LAO	Line Advancing Order
LAP	Lapping
LAP	Link Access Procedure|Protocol

LAP	Linux Application Platform
LAP	List Assembly Programming
LAP	Local Analysis and Prediction (branch)
LAP-A	Link Access Procedure (byte oriented)
LAP-B	Link Access Procedure (bit Oriented)
LAPB	Link Access Protocol, B channel
LAPD	Link Access Procedure Direct (protocol)
LAPD	Link Access Procedure on the D channel
LAPM	Link Access Procedure for Modems
LAPS	Local Analysis and Prediction System
LAPX	Link Access Procedure, half Duplex
LAR	Lagging|Limit Address Register
LAR	Load Access Rights
LAR	Local Acquisition Radar
LARAM	Line Addressable Random Access Memory
LARC	Library Automation Research and Consulting
LARC	Livermore Automatic Research Calculator
LARP	Local Approval Review Program
LARPS	Local And Remote Printing Station
LAS	Link Active Scheduler
LAS	Local Address Space
LAS	Location Addressable Storage
LASCR	Light Activated Silicon Controlled Rectifier
LASCS	Light Activated Silicon Controlled Switch
LASD	Large Area Screen Display
LASER	Light Amplification by Stimulated Emission on Radiation
LASIM	Laser Aiming Simulation
LASP	Local Attached Support Processor
LASP	Logistics Analysis Simulation Program
LASS	Local Area Signalling Service
LAST	Logic Analysis and Simulation Technique
LASTport	Local Area Storage Transport
LAT	Latch
LAT	Local Access Terminal
LATA	Local Access and Transport Area
LaTeX	Leslie Lamport's TeX extensions
LATIS	Loop Activity Tracking Information System
LATMIC	Lateral Microstructures (EU)
LATRIX	Light Accessible Transistor Matrix
LAU	Line Adapter Unit
LAU	Link Assembly Unit
LAVA	Look Ahead Variable Acceleration
LAWN	Local Area Wireless Network
LB	Lebanon
LB	Left Button
LB	Line Buffer
LB	Load|Logical Block
LB	Lower Bound

lb	pound
LBA	Linear Bound(ed) Automation
LBA	Local Bus Adapter
LBA	Logical Block Addressing
LBC	Left Bounded Context
LBC	Local Bus Controller
LBE	Language Based Editor
LBE	Lower Band Edge
LBEN	Low Byte Enable
LBG	Load Balancing Group
LBL	Label
LBMI	Lease Base Machine Inventory (file)
LBN	Logical Block Number
LBO	Leveraged Buy-Out
LBO	Line Build-Out (network)
LBOT	Logical Beginning Of Tape
LBP	Line Binder Post
LBP	Local Batch Processing
lbr	library (file extension indicator)
LBRV	Low Bit Rate Voice
LBS	LAN Bridge Server
LBS	Line Buffer System
LBS	Load Balance System
LBT	Listen Before Talk
LBX	Local Bus Accelerator
LC	Inductance-Capacitance
LC	Large Card
LC	Learning Curve
LC	Line Card|Character
LC	Line Concentrator|Connector
L/C	Line Control
LC	Link Circuit|Control
LC	Liquid Crystal
LC	Load Cell
LC	Local Channel
LC	Location Code|Counter
lc	loco citato (in the place cited)
LC	Low cost Colour
LC	Lower Case (character)
LC	Saint Lucia
LCA	Life Cycle Analysis
LCA	Life Cycle Architecture|Assessment
LCA	Line Control Adapter
LCA	Local Communications Adapter
LCA	Logic Cell Array
LCA	Lotus Communications Architecture
LCA	Low Cost Automation
LCA	Lower Case Alphabet

LCAMOS	Loop Cable Maintenance Operation System
LCARS	Library Computer Access (and) Retrieval System
LCB	Line Control Block
LCB	Line to Computer Buffer
LCB	Link|Logic Control Block
LCB	Low Cost Bipolar (integrated circuit)
LCBX	Large Computerized (private) Branch Exchange
LCC	Laboratory of Computer Chemistry
LCC	Leaded|Leadless Chip Carrier
LCC	Lost Calls Cleared
LCCA	Low Cost Calculator|Computer Attachment
LCCC	Leadless Ceramic Chip Carrier
LCCIS	Local Common Channel Interoffice Signaling
LCCL	Line Card Cable
LCCLN	Line Card Cable Narrative
LCCM	LAN Client Control Manager
LCCPMP	Life Cycle Computer Program Management Plan
LCD	Least Common Denominator
LCD	Line Control Definer
LCD	Line Current Disconnect
LCD	Liquid Crystal Diode|Display
LCD	LISP Code Directory (LISP = List Processor or Processing)
LCD	Lost Calls Delayed
LCD	Lowest Common Denominator
LCDN	Last Called Directory Number
LCDTL	Load-Compensated Diode-Transistor Logic
LCES	Least Cost Estimating and Scheduling
LCF	Least Common Factor
LCF	Log Control Function
LCF	Logical Channel Fill
LCF	Low Cost Fiber|File
LCF	Low Cycle Fatigue
LCFIMS	Loosely Coupled Federated Information Management System
LCFS	Last Come, First Served
LCGN	Logical Channel Group Number
LCH	Latch
LCH	Logical Channel (queue)
LCHILD	Logical Child
LCHT	Logical Channel Table
LCI	Learner Controlled Instruction
LCI	Logical Channel Identifier
LCID	Local Character set Identifier
LCIE	Lightguide Cable Interconnection Equipment
LCK	Lock
LCKO	Lock-Out
LCL	Limited Channel Logout
LCL	Local
LCL	Lower Control Limit

LCLOC	Line Card Location
LCLS-SLT	Low Cost Low Speed – Solid Logic Technology
LCM	Large Core Memory
LCM	Least Common Multiple
LCM	Life Cycle Management
LCM	Line Concentrating|Control Module
LCM	Line Control Module
LCM	Logical Connection Manager
LCM	Low Cost Memory|Module
LCN	Local Computer Network
LCN	Logical Channel Number(s)
LCN	Loosely Coupled Network
LCNTR	Location Counter
LCO	Latching Contact Operate
LCOR	Load Character with Offset Register (instruction)
LCP	Language Conversion Program
LCP	Left-handed Circular Polarization
LCP	Link Control Procedure|Protocol
LCP	Local Control Point
LCP	Logical Construction of Programs
LCPRD	Low Cost Page Reader
LCR	Inductance-Capacitance-Resistance
LCR	Length Count Register
LCR	Library Change Request
LCR	Line Control Register
LCR	Load Character Register
LCRMKR	Line Card Remarks, Retained
LCROS	Large Capacity Read-Only Storage
LCRR	Length Count Recall Register
LCS	Laboratory for Computer Sciences
LCS	Large Capacity|Core Storage
LCS	Linear Control System
LCSAJ	Linear Code Sequence And Jump
LCSC	Logic Controlled Sequential Computer
LCSE	Line Card Service and Equipment
LCSEN	Line Card Service and Equipment Narrative
LCSLT	Low Cost Solid Logic Technology
LCSP	Logical Channels Switching Program
LCT	Last Compliance|Conformance Time
LCT	Latest Completion Time
LCT	Level|Line Control Table
LCT	Linkage Control Table
LCT	Log Control Table
LCT	Logical Channel Termination
LCT	Low Cost Technology
LC/TC	Line Control/Task Control
LCTD	Located
LCTDL	Load Compensated Transistor Diode Logic

LCU	Last Cluster Used
LCU	Level Converter Unit
LCU	Line|Local Control Unit
LCU	Logical|Loop Control Unit
LCW	Line|Lock Control Word
LCZR	Localizer
LD	Label Definition
LD	Laser Diode
LD	Linear Decision
LD	Line Delete (character)
LD	Line Driver
LD	Load(er)
LD	Logical Design
LD	Logic Driver
LD	Long Distance
LD	Low Density
LDA	Local Data Administrator|Area
LDA	Local Display Adapter
LDA	Logical Device Address
LDAM	Logical Data Access Method
LDAP	Light Directory Access Protocol
LDAP	Lightweight Directory Access Protocol
LDB	Large|Logical Data Base
LDBS	Local Data Base System
LDC	Line Delete|Deletion Character
LDC	Local Display Controller
LDC	Location Dependent Code
LDC	Logical Device Coordinates
LDC	Long Distance Carrier
LDC	Low Density Center
LDC	Low-speed Data Channel
LD-CELP	Long Delay – Code Excited Linear Prediction
LDCS	Linear Direct Conversion System
LDCS	Long Distance Control System
LDD	Local Data Distribution|Description
LDDC	Long Distance Direct Current
LDDI	Local Distributed Data Interface
LDDS	Limited Distance Data Service
LDDS	Long Distance Dial Service
LDDS	Low Density Data System
LDG	Loading
LDIF	Lightweight Directory Interchange Format
LDL	Local Digital Loopback
LDL	Logical Database Level
LDL	Logic Design Language
LDLA	Limited Distance Line Adapter
LDM	Limited Distance Modem
LDM	Linear Delta Modulation

LDM	Local Data Manager|Memory
LDM	Logical Data Model
LDM	Long Distance Modem
LDMS	Laboratory Data Management System
LDMTS	Long Distance Message Telecommunications Service
LDMX	Local Digital Message Exchange
LDN	Listed Directory Number
LDO	Language Dependent Objects
LDO	Logical Device Order
LDP	Label Distribution Protocol
LDP	Laboratory Data Products
LDPT	Load Point
LDR	Light Dependent Resistor
LDR	Linear Decision Rule(s)
LDR	Link Loader
LDRI	Low Data Rate Input
LDS	Language for Description and functional Specification
LDS	Last Data Sample
LDS	Leads
LDS	Legal Document Scanner
LDS	Logic(al) Data Structure
LDSYS	Load System
LDT	Language Dependent Translator
LDT	Linear Differential Transformer
LDT	Loader Definition Table
LDT	Load Terminate
LDT	Local Descriptor Table
LDT	Logical Device Table
LDT	Logic Design Translator
LDU	Line Drive Unit
LDU	Link Diagnostic Unit
LDX	Long Distance Extender
LDX	Long Distance Xerography
LE	Lance Ethernet
LE	Language Environment
LE	Latch Enable
LE	Left End
LE	Less than or Equal (to)
LE	Linkage Editor
LE	Link Encryption
LE	Local Exchange
LEA	Load Effective Address
LEAD	Leadership and Excellence in the Application and Development of computer integrated manufacturing
LEAD	Learn, Execute, And Diagnose
LEADS	Law Enforcement Automated Data System
LEAF	LISP Extended Algebraic Facility (LISP = List Processing language)

LEAP	Lambda Efficiency Analysis Program
LEAP	Language for the Expression of Associative Procedures
LEAP	Low-power Embedded Application Processor
LEAP	Low-power Enhanced At Portable
LEAR	Low Energy Antiproton Ring
LEAS	Lata Equal Access System
LEC	LANE Client
LEC	LAN Emulation Client
LEC	Local Exchange Carrier
LEC	Locking Escape Character
LECL	Linkage Editor Control Language
LECS	LANE Configuration Server
LECS	LAN Emulation Client Server
LED	Light Emitting Diode
LEDA	Library of Efficient Data types and Algorithms
LEDFOS	Light-Emitting Devices From Organic Semiconductors (EU)
LEED	Low Energy Electron Diffraction
LEF	Linear Energy (spectrophoto) Fluorometry
LEF	Line Expansion Function
LEGO	Low End Graphics Option
LEL	Link, Embed and Launch-to-edit
LEM	Language Extension Module
LEM	Logical End of Media
LEM	Logic Enhanced Memory
LEN	Large Extension Node
LEN	Low Entry Networking
LENCL	Line Equipment Number Class
LEO	Laser and Opto-Electronics
LEO	Logic Elementary Operation
LEOS	Low Earth Orbit Satellite
LEPIU	Low Energy Process Interface Unit
LEPM	Linkage Editor Program Map
LEQ	Less than or Equal (to)
LERB	Line Error Block
LEROM	Light Erasable Read-Only Memory
LES	Land Earth Station
LES	Language Engineering Services
LES	Large Eddy Simulation
LES	Laser Emulsion Storage
LES/BUS	LAN Server/Broadcast and Unknown Server
let	letter (file extension indicator)
LET	Logical Equipment Table
LETAM	Low-End Telecommunications Access Method
LETR	Least Effort Text Retrieval
LEV	Lever
LEVTAB	Level Table
LEX	Lexicon
LEX	Line Exchanger

LF	Largest Frame
LF	Line Feed (character)
LF	Line Finder
LF	Low Frequency
LFA	Last Field|File Address
LFA	Local Feature Analysis
LFA	Local File Access
LFACS	Loop Facilities Assignment and Control System
LFB	Linear Frame Buffer
LFC	Line Feed Character
LFC	Local Form Control
LFC	Local Function Capabilities
LFD	Line Feed
LFD	Local Frequency Distribution
LFF	Limited Fanout Free
LFI	Last File Indicator
LF-ID	Logical Frame Identifier
LFM	Limited-area Fine-mesh Model
LFM	Linear Frequency Modulation
LFM	Link Framing Module
LFM	Local File Management|Manager
LFM	Logic Feasibility Model
LFN	Logical File Name
LFN	Long Fat Network
LFN	Long File Name
LFO	Low Frequency Oscillator
LFP	LISP and Functional Programming (LISP = List Processing)
LFP	Local Front-end Processor
LFP	Low-Frequency Prediction
LFS	Local Format Storage
LFS	Logic File Structure
LFS	Loopback File System
LFSR	Linear Feedback Shift Register
LFSUX	Load Floating-point Single-precision with Update and Indexed
LFT	Low Function Terminal
LFU	Least Frequently Used
LG	Line Generator|Group
LGA	Low Gain Antenna
LGC	Line Group Controller
LGDT	Load Global Descriptor Table
LGI	Linear Gate and Integrator
LGMR	Laser-Guided Magnetic Recording
LGN	Line Generator Number
LGN	Logical Group Number
LGP	Language Generator and Processor
LGW	Link Graphics Workstation
LH	Left Hand
LH	Low-noise High-output

L/H	Low to High
lha	lharc compression method (file extension indicator)
LHASA	Logic and Heuristic Applied to Synthetic Analysis
LHC	Left Hand Chain|Component(s)
LHC	Local Host Computer
LHF	List Handling Facility
LHL	Long Haul Link
LHN	Load Half Name
LHN	Long Haul Network
LHOR	Load Halfword with Offset Register (instruction)
LHOTS	Long Haul Optical Transmission Set
LHR	Load Halfword Register (instruction)
LHR	Lower Hybrid Resonance
LH-RH	Left Hand – Right Hand
LHS	Left Hand Side
LHT	Long Holding Time
LI	Lead In
LI	Left In
LI	Length Indicator
L/I	Letter of Intent
LI	Liechtenstein
Li	Lithium
LIA	Label Information Area
LIA	Loop Interface Address
LIA	Low-speed Input Adapter
LIAS	Library Information Access System
lib	library (file extension indicator)
LIB	Line Interface Base
LIBEDIT	Library Editor
LIBMAN	Library Management
LIBNAME	Library Name
LIB/OL	Librarian/On-Line
LIBRIS	Library Information System
LIC	Last In Chain
LIC	Licensed Internal Code
LIC	Linear Integrated Circuit
LIC	Line Interface Coupler
LICD	Large Information Content Display
licrystal	liquid crystal
LICS	Logic In Computer Science (symposium)
LICS	Lotus International Character Set
LID	Leadless Inverted Device
LID	Local Identifier
LID	Local Injection/Detection
LIDB	Line Information Data Base
LIDB	Line Interface Data Bus
LIDO	Logic In, Documents Out
LIDT	Load Interrupt Descriptor Table

LIED	Linkage Editor
LIEP	Large Internet Exchange Packet
LIF	Line Interface Feature
LIF	Low Insertion Force
LIFE	Logistics Interface for a Factory Environment
LIFER	Language Interface Facility with Ellipsis and Recursion
LIFIA	Labaratoire d'Informatique Fondamentale et d'Intelligence Artificielle (FR)
LIFO	Last In, First Out
LIGHT	Lifecycle Global HyperText (CH)
LIH	Line Interface Handler
LILO	Last In, Last Out
LIM	Language Interpretation Module
LIM	Limit(er)
LIM	Line Interface Module
LIMA	Light Induced Modulation of Absorption
LIMA	Logic-In-Memory Array
LIMB	Look In (your) Mail Box
LIMIT	Lot-size Inventory Management Interpolation Technique
LIML	Linear propagation time Immediate Language
LIMS	Laboratory Information Management Systems
LIMS	Library Information Management System
LIMS	Lotus Intel Microsoft Specifications
linasec	a line a second
LINC	Laboratory Instrument Computer
LINCS	Language Information Network and Clearinghouse System
LINDI	Line to Disk
LINED	Line Editor
LINGO	Linguistic Operation
LINKLIB	Linked Library
LINUS	Logical Inquiry and Update System
LINUX	Linus Torvalds' Unix
LIOCS	Logical Input/Output Control System
LIOI	Local In-Out Interface
LIOP	Logical Input/Output Processor
LIOP	Low-speed Input/Output Processor
LIP	Large Internet Packet
LIP	Line Interface Protocol
LIPL	Linear Information Processing Language
LIPS	Laboratory Interconnecting Programming System
LIPS	Laser Image Processing Scanner
lips	logical inferences per second
LIPX	Large Internetwork Packet Exchange
LIQ	Liquid
LIROC	Last Instruction Read Out Cycle
LIRS	Library Information Retrieval System
LIS	Laboratory|Land Information System
LIS	Large Interactive Surface

LIS	Library (and) Information Science
LIS	Logistic Information System
LISA	Large Installation Systems Administration
LISA	Linear Systems Analysis
LISA	Linked Indexed Sequential Access
LISA	Locally Integrated Software Architecture
LISARDS	Library Information Search And Retrieval Data System
LISD	Library and Information Services Division (US)
LISH	Last In, Still Here
LISN	Line Impedance Stabilization Network
LISP	List-oriented Processing (language)
LISP	List Processing|Processor
LISP	Local Initiatives Support Project
LISTSERV	List Server
LIT	Line Insulation Test
LIT	Literal
LIT	Load Initial Table
LIT	Logical Interface Tape
LITHP	Link Type description language for Hypertext Processing
Litprog	Literate programming
LIU	Line Interface Unit
LIU	Logical Installation Unit
LIU	Logically Installable Unit
LIU	Logic Input and Update
LIVE-NET	London Interactive Video Educational Network
LIW	Long Instruction Word
LIX	Liquid Crystal
LJE	Local Job Entry
LJP	Local Job Processing
LK	Sri Lanka
LKAD	Look Ahead
LKED	Linkage Editor
LKM	Low Key Maintenance
LKUP	Look Up
LL	Leased Line
LL	Line Length (identification)
LL	Local Line|Loopback
LL	Long Lines
LL	Loudness Level
LL	Lower Limit
LL	Low Level
LLA	Leased Line Adapter
LLA	Long Line Adapter
LLA	Low-speed Line Adapter
LLAP	Local-talk Link Access Protocol
LLBA	Language and Language Behaviour Abstracts
LLC	Logical Link Control
LLC	Logic Level Control

LLC	Low Level Code
LLCC	Leadless Chip Carrier
LLC/CC	Low Level Code/Continuity Check
LLCLINK	Logical Layer Control Link (protocol)
LLCn	Logical Link Control type n
LLCS	Logical Link Control Security
LLCSC	Lower-Level Computer Software Component
LLCTL	Last Line Control
LLDT	Load Local Descriptor Table
LLE	Load List Element
LLF	Low Level Format
LLG	Logical Line Group
LLI	Low Level Interface
LLIB	Load module Librarian
LLJ	Low-Level Jet
LLL	Last Look Logic
LLL	Low-Level Language|Logic
LLLTV	Low Light Level Television
LLM	Local Linear Mapping
LLM	Logical Link Module
LLM	Low Level Multiplexer
LLN	Line-Link Network
LLP	Laserbeam Line Printer
LLP	Line Link Pulsing
LLP	Linking Loader Program
LLP	Link Level Protocol
LLP	Lower Layer Protocol
LLP	Low-Level Protocol
LLPDU	Logical Link Protocol Data Unit
LLPI	Logical Low Power Inverter
LLPIU	Low Level Process Interface Unit
LLPN	Lumped Linear Paramagnetic Network
LLR	Line-Loop Resistance
LLR	Long Length Record
LLS	LAN-Like Switching
LLS	Low Level Software
LLSIG	Lower Layer Special Interest Group
LLU	Logical Link Unit
LLWANP	LAN-to-LAN Wide Area Network Program
LM	LAN Manager
LM	Layer Management
LM	Library Material
l/m	lines per minute
LM	Link Manager
LM	Load Multiple
LM	Logic Module
LM	Loop Multiplexer
lm	lumen

LMB	Left Most Bit
LMB	Left Mouse Button
LMBCS	Lotus Multi-Byte Character Set
LMBI	Local Memory Bank|Bus Interface
LMC	Logistical Maintenance Computer
LMCSS	Letter Mail Code Sort System
LMD	Library Macro-Definition
LMDL	Longitudinal Mode Delay Line
LMDS	Local Multipoint Distribution Service(s)
LME	Layer Management Entity
L/MF	Low and Medium Frequency
lm-hr	lumen-hour
LMI	Layer Management|Manufacturing Interface
LMI	Linkage Macro Instruction
LMI	Local Management Interface
LMI	Local Memory Image
LMIS	Logistics Management Information System
LML	Load Module Library
LML	Logical Memory Level
LMLR	Load Memory Lockout Register
LMMP	LAN/MAN Management Protocol (MAN = Metropolitan Area Network)
LMMS	LAN/MAN Management Service
LMMS	Local Message Metering System
LMOS	Line|Loop Maintenance Operations System
LMS	Laser Mass Storage
LMS	Least Mean Square(s)
LMS	Library Management System
LMS	List Management System
LMS	Lotus Messaging Switch
LMSW	Load Machine Status Word
LMT	Limit
LMT	Logical Mapping Table
LMT	Logic Master Tape
LMU	LAN Management Utilities
LMU	LAN Manager for Unix
LMU	Line Monitor Unit
lm/W	lumens per Watt
LM/X	LAN Manager for Unix
LMX	L Multiplex
LMX	Local Multiplexer
LMXRB	Low Mass X-Ray Binary
L-N	Lexis-Nexis
LN	Line
LN	Link Number
ln	natural logarithm
LNA	Layered Network Architecture
LNA	Low Noise Amplifier

LNB	Local Name Base
LNB	Low Noise Block (convertor)
LNC	Low Noise (block down) Convertor
LND	Local Number Dialling
LN-DI	Lotus Notes – Document Imaging
LNE	Local Network Emulator
LNK	Link
LNKEDT	Link(age) Edit(or)
LNM	LAN Network Manager
LNM	Logical Network Machine
LNN	Linear Nearest Neighbour
LNO	Local Network Operations
LNR	Last Number Redial
LNR	Line Ring
LNR	Low Noise Receiver
LNSTAT	Line Status
LNU	Last Name Unknown
LO	Lay-Out
LO	Lead|Left Out
LO	Leverage Out
LO	Line Occupancy|Office
LO	Local Oscillator
LO	Log Out
LO	Longitudinal Optical
lo	low
LO	Low Order
LOA	Leave Of Absence
LOA	Length of Output Area
LOA	Low-speed Output Adapter
LOAP	Length Of Adjacency Process
LOB	Line Of Balance
LOC	Large Optical Cavity diode
LOC	Last Of Chain
LOC	Line(s) Of Code
LOC	Load On Call
LOC	Local
LOC	Local Operating Company
LOC	Location
LOC	Location Counter
LOC	Loop On-line Control
LOCAL	Load On Call (basis)
LOCAP	Low Capacitance
LOCATS	Low-cost Card Test System
LOCD	Location Dependent
LOCI	Location Information
LOCIS	Library Of Congress Information System
LOCKD	Lock Daemon
LOCMOS	Local Oxidation Complementary Metal-Oxide Semiconductor

LOCO	Local Copy
LOCOS	Local Oxidation Semiconductor
LOCS	Logic and Control Simulator
LOD	Level Of Detail (control)
LODSB	Load String Byte
LOE	Level Of Effort
LOF	List Overflow
LOF	Local Oscillator Frequency
LOF	Lock Off-line
LOF	Look-ahead On Fault
LOFAR	Low Frequency Analysis Recording
log	logarithm
LOG	Logging
LOG	Logic(al)
LOG	Logistics
LOGCON	Logic Connection
LOGFED	Log File Editor
LOGIK	Logical Organizing and Gathering of Information Knowledge
LOGLAN	Logical Language
LOGMSG	Log Message
LOGOS	Language Of Generalized Operational Simulation
LOGSAFE	Logistics Sustainability Analysis (and) Feasibility Estimator
LOI	Letter Of Intent
LOIS	Law Office Information Systems
LOL	Laughing Out Loud
LOL	Lots Of Love|Luck
LOLA	Layman Oriented Language
LOLA	Library On-Line Acquisition (subsystem)
LOLA	Local Line Adapter(s)
LOLIST	Locator Listing
LOLITA	Language for the On-Line Investigation and Transformation of Abstractions
LOMAPS	Logical and Operational Methods in the Analysis of Programs and Systems (EU)
LOMAT	Language Oriented Machine Analysis Table
LON	Lock On-line
LONG	Longitudinal
LOOM	Line of Operator Oriented Machines
LOOPE	Loop while Equal
LOOPNE	Loop while Not Equal
LOOPNZ	Loop while Not Zero
LOOPS	LISP Object Oriented Programming System (LISP = List Processing language)
LOOPZ	Loop while Zero
LOP	Line Oriented Protocol
LOP	Lines Of Position
LOR	Level Of Repair
LOR	Load with Offset Register (instruction)

LOR	Look-ahead On Request
LORAN	Long Range Aid to Navigation
LORE	Line Oriented Editor
LORO	Lobe-On-Receive Only
LORPGAC	Long Range Proving Ground Automatic Computer
LOS	Level Of Support
LOS	Line Of Sight
LOS	Loss Of Signal|Station
LOSAT	Language Oriented System Analysis|Audit Table
LOSOS	Local Oxidation of Silicon On Sapphire
LOSR	Limit Of Stack Register
LOSS	LAPS Observing System Simulation
LOSTPED	Load, Orientation, Speed, Travel, Precision, Environment, and Duty cycle
LOT	Ley de Ordenación de las Telecomunicaciones (ES)
LOT	Light(spot) Operated Typewriter
LOT	Logic Optimization with Testability
LOTIS	Logic, Timing, and Sequencing
LOTOS	Language Of Temporal Ordering of observational behaviour of Systems
LOTP	Logical Operation Time Projection
LOTS	Low Overhead Time-sharing System
LOWL	Low level Language
LP	Latch Pick
LP	Lead Programmer
LP	Licensed Program
LP	Light Pen
LP	Linear Polarization|Processing
LP	Linear Programming
LP	Line Printer|Protocol
LP	Load Point
LP	Local Program|Process
LP	Logic Pin|Probe
LP	Long Play(ing)
LP	Loop
LP	Low Pass (filter)
LP	Low Power|Primary
LPA	Link Pack Area
LPA	Lower order Path Adaptation
LPACQ	Link Pack Area Control Queue
LPAGE	Logical Page
LPAR	Line Printer Address Register
LPB	Lighted Push-Button
LPB	Load Program Block
LPC	Language Conversion Processor
LPC	Light Pen Control
LPC	Linear Power Control(ler)
LPC	Linear Predictive Coding

lpc	lines per centimeter
LPC	Local Procedure Call
LPC	Longitudinal Parity Check
LPCL	Linear Programming Control Language
LPCM	Linear Phase Code Modulation
LPCOMP	Logical Physical Comparator
LPD	Legendary Pink Dots
LPD	Line Printer Daemon
LPD	Logical Physical Design
LPD	Log Periodic Dipole
LPDA	Link Problem Determination Aid
LPDC	Least Positive Down Count
LPDE	Leeds Product Data Editor
LPDU	Link layer Protocol Data Unit
LPE	Layer Primitive Equation
LPE	Local Peripheral Equipment
LPE	Logic Program Error
LPF	Low Pass Filter
LPFK	Lighted Program Function Keyboard
lph	lines per hour
LPI	Learner Paced Instruction
LPI	Leverage Point for Improvement
LPI	Line Program Impulse
lpi	lines per inch
LPI	Linux Professional Institute
LPI	Logic Parts Indicator
LPIA	Line Printer Image Address
LPID	Logical Page Identifier
LPL	List Processing Language
LPL	Local Processor Link
LPL	Location Programming List
LPL	Lotus Programming Language
LPM	Laleh's Pattern Matcher
lpm	lines per minute
LPM	Load Point Marker
LPM	Local Program Memory
LPM	Logical Port Multiplexer
LPMOSS	Linear Programming Mathematical Optimization Subroutine System
LPN	Local Packet Network
LPN	Logical Page Number
LPO	Logical-Physical Output
LPP	Latest Precedence Partition
LPP	Licensed Program Products
LPP	Lightweight Presentation Protocol
LPP	Location Programming Practice
LPR	Line Printer
LPRB	Load Program Request Block|Buffer

LPRD	Latch Program Register D
LPS	Language for Programming in-the-Small
LPS	Laser Printing|Projection System
LPS	Layered Protocol Structure
LPS	Licensed Program Support
LPS	Linear Programming System
LPS	Line Procedure Specification
LPS	Line Program Selector
lps	lines per second
LPS	Location Programming Standard
LPS	Low Power Schottky
LPS	Low Primary Sequence
LPSC	Local Program Support Charges
LPSTTL	Low Power Schottky Transistor-Transistor Logic
LPSW	Load Program Status Word
LPT	Largest Processing Time
LPT	Line Printer
LPT	Local Point
LPT	Longest Processing Time
LPT1	First parallel Line Printer port
LPT2	Second parallel Line Printer port
LPT3	Third parallel Line Printer port
LPTTL	Low Power Transistor-Transistor Logic
LPTV	Low Power Television (service)
LPU	Language Processor Unit
LPU	Line Printer|Processing Unit
LPU	Logic Physical Update
LPUL	Least Positive Up Level
LPVS	Link Packetized Voice Server
LPVT	Large Print Video Terminal
lpW	lumens per Watt
LQ	Letter Quality
LQA	Line Quality Analysis
LQG	Linear Quadratic Gaussian
LQM	Link Quality Monitoring (protocol)
L-R	difference between Left and Right stereo channels
LR	Label Reference
LR	Last Record (indicator)
LR	Left to Right
LR	Level Recorder
LR	Liberia
LR	Limited Response
LR	Limit Register
LR	Line Relay
LR	Link|Load Register
LR	Logical Record
LR	Long Range
L+R	Sum of Left and Right stereo channels

LRA	Load Real Address
LRA	Logical Record Access
LRAP	Long Route Analysis Program
LRB	Load(ed) Request Block
LRBC	Left-Right Bounded Context
LRC	Learning Resource Center
LRC	Local Register Cache
LRC	Longitudinal Redundancy Character|Check
LRCC	Longitudinal Redundancy Check Character
LRCR	Longitudinal Redundancy Check Register
LRD	Long Range Data
LRECL	Logical Record Length
LREP	Left-bracketed Representation
LRG	Large
LRG	Long Range
LRI	Load Register Immediate (instruction)
LRIM	Long Range Input Monitor
LRL	Least Recently Loaded
LRL	Linking Relocating Loader
LRL	Location Records List
LRL	Logical Record Length|Location
LRM	Language Reference Manual
LRM	LAN Reporting Mechanism
LRM	Least Recently-used Master
LRM	Line Replaceable Module
LRO	Long Range Output
LRP	Long Range Plan
LRPC	Lightweight Remote Procedure Call
LRR	Loop Regenerative Repeater
lrs	language resource (file extension indicator)
LRS	Line Repeater Station
LRS	Local Regional Server
LRS	Long Range Search
LRSP	Long Range Strategic Planning
LRSP	Long Range System Plan
LRSS	Long Range Switching Studies
LRU	Least Recently Used
LRU	Line Replaceable Unit(s)
LRUP	Least Recently Used Page
LS	Laser System
LS	Last Sent
LS	Least Significant
LS	Left Side
LS	Lesotho
LS	Level Switch
LS	Link Status
LS	Load Storage
LS	Local Single-layer

LS	Local Stor(ag)e
LS	Logic Synthesis
LS	Long Stub
LS	Low-power Schottky
LS	Low Secondary|Speed
LSA	LAN and SCSI Adapter
LSA	LAN Security Architecture
LSA	Limited Space-charge Accumulation
LSA	Line Sensing Amplifier
LSA	Line Sharing Adapter
LSA	Local Security Authority
LSA	Logarithmic Sense Amplifier
LSA	Logic Sequential Access
LSA	Logic State Analysis
LSA	Logistic Support Analysis
LSA	Low Speed Adapter
LSAP	Link|Logical Service Access Point
LSAP	Long Service Access Point
LSAPI	Licensing Service(s) Application Program Interface
LSAR	Local Storage Address Register
LSAR	Logistic Support Analysis Record
LSASS	Local Security Authority Sub-System
LSb	Least Significant bit
LSB	Least Significant Byte
LSB	Line Speed Buffer
LSB	Linux Standards Base
LSB	Lower Side-Band
LSBS	Line Scanning Before Sending
LSC	Least Significant Character
LSC	Least-Square mean Circle
LSC	Linear Sequential Circuit
LSC	Locking Shift Character
LSC	Loop Station Connector
LSC	Low Speed Concentrator
LSC	Low Speed interface Control
LSCB	Load System Control Block
LSCU	Link Service data Unit
LSD	Language for System(s) Development
LSD	Least Significant Difference|Digit
LSD	Line Sharing Device
LSD	Line Signal Detector
LSDB	Launch Support Data Base
LSDM	Lagrangian Stochastic Dispersion Model
LSDR	Local Store Data Register
LSE	Language Sensitive Editor
LSE	Local Single-layer Embedded
LSE	Local System Environment
LSE	Longitudinal Section Electric

LSEL	Link Selector
LSFR	Local Storage Function Register
LSHER	Load Sheet Reference
LSI	Large Scale Integration
LSIC	Large Scale Integrated Circuit(ry)
LSID	Local Session Identification
LSIG	Least Significant
LSL	Ladder Static Logic
LSL	Link and Selector Language
LSL	Link Support Layer
LSL	Load Segment Limit
LSL	Location Systems List
LSL	Logical Shift Left
LSL	Low Speed Logic
LS/LC	Line Stabilizer/Line Conditioner
LSLM	Low Speed Line Manager
LSM	Line Select Module
LSM	Logic Selection Module
LSMA	Low Speed Multiplexer Arrangement
LSMLC	Low Speed Multi-Line Controller
LSN	Logical Session Number
LSP	Least Significant Position
LSP	Link State Packet
LSP	Local Store Pointer
LSP	Loop Splice Plate
LSPD	Low Speed
LSPS	Local Stor(ag)e Protect Storage
LSQA	Local System Queue Area
LSQM	Large Systems Qualification Monitor
LSR	Label-Switching Router
LSR	Linear Shift Register
LSR	Local Shared Resources
LSR	Local Storage Register
LSR	Logical Shift Right
LSR	Low Speed Reader
LSRP	Local Switching Replacement Planning (system)
LSRR	Loose Source and Record Route
LSS	Language for Symbolic Simulation
LSS	Large Scale System(s)
LSS	Location Systems Standard
LSS	Loop Surge Suppresser
LSS	Loop Switching System
LSS	Low Speed Storage
LSSC	Lower Sideband Suppressed Carrier
LSSD	Level Sensitive Scan Design
LSSE	Life Support System Evaluator
LSSS	Local Store Swap Sequencer
LSSU	Link Status Signal Unit

LST	Large (capacity) Storage
LST	Line Schedule Terminal
lst	list (file extension indicator)
LST	Logic Service Terminal
LST	Low Speed Tape|Terminal
LSTOR	Local Storage
LSTTL	Low-power Schottky Transistor-Transistor Logic
LSU	Languages, Sorts, Utilities
LSU	Least Significant Unit
LSU	Library Storage Unit
LSU	Line Sharing|Switching Unit
LSU	Load Stor(ag)e Unit
LSU	Local|Logic Storage Unit
LSUP	Loader Storage Unit support Program
LSV	Line Status Verifier
LSVI	Large Size Visual Interface
LSWD	Linear System With Display
LSWS	Line Scanning While Sending
LSX	LSI-Unix system
L&T	Language & Terminal
LT	Language Translation|Transmission
LT	Latch Trip
LT	Lead Time
LT	Left
LT	Less Than
LT	Life Test
LT	Light
LT	Line Termination (Terminator)
LT	Lithuania
LT	Logic Theory
LT	Loop Termination|Terminator
LT	Lower Tester
LT	Low Tension
LTA	Line Turnaround
LTA	Logical Transient Area
LTAB	Line Test Access Bus
LTAS	Load Transit And Set
LTB	Last Trunk Busy
LTB	Logical Twin Backward pointer
LTBL	Level Table
LTC	Line Time Clock
LTC	Line Traffic Coordinator
LTC	Local Terminal Controller
LTC	Local Test Cabinet
LTD	Line Transfer Device
LTD	Local Test Desk
LTDS	Logic Test Data System
LTE	Linear Threshold Element

LTE	Line Terminating Entity|Equipment
LTE	Local Telephone Exchange
LTE	Local Truncation Error
LTERM	Logical Terminal
LTF	Lightwave Terminating Frame
LTF	Line Trunk Frame
LTF	Logical Twin Forward pointer
LTG	Line Trunk Group
LTH	Latch
LTH	Logical Track Header
LTI	LAN Telephony Integration
LTM	LAN Traffic Monitor
LTM	Leverage Transaction Merchant
LTM	Long Term Memory
LTMS	Laser Thread Measurement System
LTN	Line Terminating Network
LTO	Linear Tape Open
LTP	Long Term Planning
LTPC	Local Time Pseudo Clock
LTPD	Limited Temperature Power Dissipation
LTPD	Lot Tolerance Percent Defective
LTPLT	Low Temperature Power Life Test
LTPN	Lightpen
LTR	Left-To-Right
LTR	Letter
LTR	Load Task Register
LTR	Long Term Research
LTROM	Linear Transformer Read-Only Memory
LTRS	Letters
LTRS	Letter Shift
LTS	Line Task (control block)
LTS	Line Transient Suppression
LTS	Logical Technical Services (corporation)
LTS	Loss Test Set
LTT	Long Term Test
LTTL	Low-power Transistor Transistor Logic
LTU	Line Terminating Unit
LTWA	Log Tape Write Ahead
LU	Line Unit
LU	Logical Unit
LU	Loudness level Unit(s)
LU	Lower-Upper
LU	Luxembourg
LUA	Logical Unit Application
LUB	Logical Unit Block
LUCID	Language for Utility Checkout and Instrumentation Development
LUD	Logical Unit Description

LUE	Link Utilization Efficiency
LUF	Limiting system Utilization Factor
LUF	Lowest Usable Frequency
LUFO	Last Used, First Out
LUFO	Longest Unused, First Out
LUG	Local Users Group
LUHF	Lowest Useful High Frequency
LUI	Local User Input
LUIS	Library User Information Service
LULMP	Limited Use Licensed Maintenance Program(s)
LU-LU	Logical Unit to Logical Unit
lum	lumen
LUN	Logic(al) Unit Number
LUS	Logical Unit Services
LUSVC	Logical Unit Services (manager)
LUT	Look-Up Table
LUT-DAC	Look-Up Table – Digital to Analog Converter
LUW	Logical Units of Work
LV	Latvia
LV	Level
L/V	Loader Verifier
LV	Logical Volume
LV	Low Voltage
LVA	Line Voltage Analyzer
LVA	Local Virtual Address
LVCD	Least Voltage Coincidence Detection
LVD	Low Voltage Differential
LVDR	Low Volume Document Reader
LVDS	Low-Voltage Differential Signal
LVDT	Linear Variable Differential Transformer
LVM	Logical Volume Management
LVP	Low Volume Product
LVR	Longitudinal Video Recording
LW	Last Word
LW	Long Wave
l/W	lumens per Watt
LWA	Last Word Address
LWA	Level|Logical Work Area
LWB	Lower Bound
LWC	Loop Wiring Concentrator|Connector
LWD	Limit Word
LWKR	Least Work Remaining
LWP	Long Wavelength Pass (filter)
LWR	Lower
LWS	Library Work Space
LWS	Link Work-Station
LWSP	Linear|Logical White Space
LWT	Listen While Talk

lx	lux
LXE	Lightguide Express Entry
LXMAR	Load External Memory Address Register
LY	Libyan Arab Jamahiriya
LZ	Lazy (writer)
LZ	Left Zero
LZH	Lempel-Ziv-Huffman
LZH	Lzari Huffman (code)
LZP	Left Zero Print
LZS	Lempel-Ziv-Stac (compression technique)
LZW	Lempel-Ziv-Welch (compression method)

M

M	Magenta
M	Magnetic
m	mantissa
M	Medium
M	Mega (prefix)
M	Memory
M	Merge (order)
m	meter
M	Million
m	milli (prefix)
M	Modification
M	Monitor
M	Mutual inductance (symbol for)
M2FM	Modified Modified Frequency Modulation
MA	Maintenance Administrator|Agreement
MA	Massachusetts (US)
MA	Megampere
MA	Memory Address
mA	milli-ampere(s)
MA	Morocco
MAA	Mathematical Association of America
MAA	Maximum Acceptance Angle
MAAP	Management And Administration Panel(s)
MA-ASE	Multiple Association Application Service Element
MAB	Macro-Address Bus
MABO	Multiplier-Arithmetic Bolton
MAC	Machine Aided Cognition
MAC	Macintosh computer
mac	macpaint (file extension indicator)
mac	macro (file extension indicator)
MAC	Maintenance Agreement Charge
MAC	Maintenance Allocation Chart
MAC	Mandatory Access Control
MAC	Media|Medium Access Control

MAC	Memory Address Control(ler)|Counter
MAC	Message Authentication Code
MAC	Minimal Access Coding
MAC	Multi-Access Computer
MAC	Multi-Application Controller
MAC	Multiple Access|Address Computer
MAC	Multiple Analog Component(s)
MAC	Multiple Aperture Core
MAC	Multiplexed Analog Components
MAC	Multiply Accumulate
MACBS	Multi-Access Cable Billing System
MACC	Micro Asynchronous Communications Controller
MACDAC	Man Communication and Display to Automatic Computer(s)
MACE	Management Applications in a Computer Environment
MACH	Machine
MACH	Manual Assist Chip Handler
MACH	Multilayer Actuator Head
MACHAN	Machine Analysis (program)
MACLIB	Macro Library
MacOS	MacIntosh Operating System
MACRO	Macro-assembler
MACRO	Macroprocessor
MACRUG	Macintosh's Reflex User Group
MACS	Media Account Control System
MACS	Multiline Automatic Calling System
MACSTAR	Multiple Access Customer Station Rearrangement
MACSYM	Measurement And Control System
MACSYMA	Macintosh's Symbolic Manipulation (system)
MACT	MOS Automatic Capacitance Tester
MACTCP	Macintosh Transmission Control Protocol
MACU	Multidrop Auto Call Unit
MAD	Magnetic Anomaly Detection
MAD	Mean Absolute Deviation
MAD	Memory Access Director
MAD	Message Assembler and Distributor
MAD	Message Assembly Director
MAD	Model-based Application Development
MAD	Multi-Aperture Device
MADA	Multiple Access Discrete Address
MADALINE	Multiple Adaptive Linear Elements
MADAM	Moderately Advanced Data Management
MADAM	Multipurpose Automatic Data Analysis Machine
MADC	Multiplexed Analog-to-Digital Converter
MADCAP	Mosaic Array Data Compression And Projection
MADD	Multiply-Add
MADDAM	Macro-module And Digital Differential Analyzer Machine
MADE	Manufacturing Automation and Design Engineering
MADE	Minimal Airborne Digital Equipment

MADIC	Machinery Acoustic Data Information Center
MADICT	Modular Advanced Development IC-Tester
MADL	Maximum Allowable Defect Level
MADM	Multi-Attribute Decision Making
MADN	Multiple Access Directory Numbers
MADR	Microprogram Address Register
MADRE	Magnetic Drum Receiving Equipment
MADT	Micro-Alloy Diffused-base Transistor
MADYMO	Mathematical Dynamic Modelling
MAE	Macintosh Application Environment
MAE	Memory Address Extension
MAE	Metropolitan Area Ethernet
MAEP	Major Application Extension Program
MAFIA	Multi-Access executive with Fast Interrupt Acceptance
MAG	Magazine
MAG	Magnetic
MAG	Maximum Available Gain
MAGAMP	Magnetic Amplifier
MAGEN	Matrix (and report) Generator
MAGPI	Manufacturing Automatic General Packaging Interface
MAGPIE	Machine Automatically Generating Production Inventory Evaluation
mAh	milli-Amperehour
MAI	Machine Aided Indexing
mai	mail (file extension indicator)
MAI	Multiple Access Interface
MAID	Maintenance Aid
MAIDS	Multipurpose Automatic Inspection and Diagnostic System
MAINSTAR	Main Storage Address Register
MAINT	Maintenance
MAJ	Major
MAJSR	Major State Register
MAL	Memory Access Logic
MAL	Meta Assembly Language
MALF	Malfunction
MAM	Memory Allocation Manager
MAM	Memory Array Module
MAM	Monolithic Analysis Method
MAM	Monolithic Array Memory|Module
MAN	Manual(ly)
MAN	Metropolitan Area Network
MANDATE	Manufacturing Data Exchange
MANDATE	Multiline Automatic Network Diagnostic And Transmission Equipment
MANIAC	Mathematical Analyzer, Numerical Integrator, And Computer
MANIAC	Mechanical And Numerical Integrator And Computer
MANOP	Manual Operation
MANPAGE	Manual Page (Unix)

MAOS	Metal Alumina dielectric Oxide Semiconductor
MAOS2UG	Mid-Atlantic Operating System 2 User Group
map	(linker) map (file extension indicator)
MAP	Macro-Assembly Program
MAP	Maintenance Analysis Procedure|Program
MAP	Major Point
MAP	Management Analysis and Projection
MAP	Management Assessment Program
MAP	Manpower Analysis Procedure
MAP	Manufacturing Automation Protocol
MAP	Mathematical Analysis without Programming
MAP	Measurement Analysis Program
MAP	Mechanized Assignment Processing
MAP	Memory Allocation and Protection
MAP	Memory Allocation Map
MAP	Micro-programmed Array Processor
MAP	Modular Application Program
MAP	Modular Arithmetic Processor
MAPD	Maximum Allowable Percent Defective
MAPD	Mean Absolute Percent Deviation
MAPDU	Management Application Protocol Data Unit
MAPE	Mean Absolute Percentage Error
MAPGEN	Map Generator
maph	milli-ampere per hour
MAPI	Mail Applications Programming Interface
MAPI	Messaging Application Program Interface
MAPICS	Manufacturing, Accounting and Production Information (and) Control System
MAPPER	Maintaining, Preparing and Processing Executive Reports
MAPR	Manual Action Pre-Requisite
MAPR	Matched filter Processor
MAPS	Mail Abuse Prevention System
MAPS	Maintenance Analysis Procedure & System chart(s)
MAPS	Management|Marketing Analysis and Planning System
MAPS	Multiple Application Partition Supervisor
MAPS	Multivariate Analysis, Participation, and Structure
MAPT	Modular Automatic Panel Test
MAP/TOP	Manufacturing Automation Protocol/Technical Office Protocol
MAR	Macro-Address Register
MAR	Memory Address Register
MAR	Micro(program) Address Register
MARBI	Machine Readable form of Bibliographic Information
MARC	Machine-Readable Catalog
MARC	Manufacturing Activity Release and Control
MARC	Market Analysis of Revenue and Customers system
MARECS	Maritime European Communication Satellite
MARG	Margin(al)
MARGA	Market Game

MARITIME	Modeling And Reuse of Information over Time
MARLIS	Multi-Aspect Relevance Linkage Information System
MARS	Manufacturing Automated Records System
MARS	Memory Address Register Storage
MARS	Modular Access Random Storage
MARS	Multiple Access Retrieval System
MARS	Multiple Aperture Reluctance Switch
MARS	Multi-user Archival and Retrieval System
MART	Maintenance Analysis and Review Technique
MART	Mean Active Repair Time
MARVEL	Machine-Assisted Realization of the Virtual Electronic Library
MARVIN	Mobile Autonomous Robot with Video-based Navigation
MAS	Main Store
MAS	Managed Application System
MAS	Manufacturing Agility Server
MAS	Master
MAS	Material Application Service
MAS	Micro-Assembly System
MAS	Microprogram(ming) Automation System
MAS	Multi-Access Spool
MAS	Multiple Access System
MAS	Multiple Application Screen
MASA	Methods Analysis Standards Automation
MASA	Methods And Standards Automation
MASB	Main Store Bus
MASC	Main Store Controller
MASC	Master Cost(s)
MASC	Mobitex Asynchronous Communications (protocol)
MASCADA	Manufacturing control Systems Capable of managing production change and Disturbances (EU)
MASCOT	Model Analysis and Syntactic Complexity Utility
MASCOT	Modular Approach to Software Construction Operation and Test
MASCOT	Motorola Automatic Sequential Computer-Operated Tester
MASCOTS	Modelling, Analysis & Simulation of Computer Telecommunication Systems
MASER	Microwave Amplification by Stimulated Emission of Radiation
MASH	Multi-Stage noise Shaping
MASINT	Measurement And Signature Intelligence
MASIS	Management And Scientific Information System
MASK	Multimodal Automated Service Kiosk
MASK	Multiple Amplitude Shift Keying
MASM	Macro-Assembler
MASM	Main Store Memory
MASM	Meta-Assembler
MASM	Microsoft Assembler
MASR	Memory Address Select Register
MASS	Machine And System Scheduling

MASS	Maximum Availability and Support Subsystem
MASS	Multiple Access Switching System
MAST	Multivalued Advanced Simulation Techniques
MASTAP	Master Tape
MASTER	Microwave Amplification by Stimulated Emission of Radiation
MASTER	Multiple Access Shared Time Executive Routines
MASTOR	Main Storage
MAT	Machine Analysis|Audit Table
MAT	Macro-Alloy Transistor
MAT	Maintenance Access Terminal
MAT	Medial Axis Transform(ation)
MAT	Memory Address Test
MAT	Micro-Alloy Transistor
MAT	MilliAmpere Turns
MAT	Monolithic Array|Automatic Tester
MATE	Modular Automatic Test Equipment (system)
MATEX	Macro-Text editor
MATFAP	Metropolitan Area Transmission Facility Analysis Program
MATH	Mathematics
MATHMOD	Mathematical Modelling (international symposium)
MATIC	Multiple Area Technical Information Center
MATILDA	Multimedia Authoring Through Intelligent Linking and Directed Assistance
MATLAB	Materials Laboratory
MATLAN	Matrix (manipulation) Language
MATO	Management Tool
MATR	Management Access To Records
MATR	Minimum Average Time Requirement
MATS	Multi-Application Teleprocessing System
MATS	Multiple-Access Time-Sharing
MATSYS	Matrix System
MAU	Math Acceleration Unit
MAU	Media Access Unit
MAU	Medium Attachment Unit
MAU	Memory Access Unit
MAU	Modem Adapter Unit
MAU	Multiple Access Unit
MAU	Multistation Access Unit
MAUD	Memory Address Utilization Display
MAVDM	Multiple Application Virtual DOS Machine (DOS = Disk Operating System)
MAVIN	Mobile Adaptive Visual Navigator
MAX	Maximal
max	maximum
MAXA	Maximum Available
MAXBL	Maximum Balance
MAXI	Modular Architecture for the Exchange of Information
MB	Macro Block

MB	Magnetic Belt
MB	Master Block
Mb	Megabit(s)
MB	Megabyte(s)
MB	Memory Buffer|Bus
MB	Memoryless Behaviour
MB	Middle Button
mb	millibit(s)
mB	millibyte(s)
MBASE	Model-Based system Architecting and Software Engineering
MBBS	Managed Broad-Band Services
MBC	Memory Bus Controller
MBC	Multiple Base Channel
MBCD	Modified Binary Coded Decimal
MBCS	Multi-Byte Character Set
MBD	Magnetic Bubble Device|Domain
MBE	Management By Exception(s)
MBE	Modelling By Example
MBE	Molecular Beam Epitaxy
MBF	Monotonic Boolean Function
MBGA	Metal Ball Grid Array
MBI	Memory Bank Interface
MBIO	Microprogrammable Block Input/Output
Mbit	Megabit
Mbit/s	Megabits/second
MBK	Multiple Beam Klystron
MBLT	Mobile Bothway Line Trunk
MBM	Magnetic Bubble Memory
MBM	Main Bulk Memory
MBM	Monolithic Buffer Memory
MBMS	Model Base Management System
MBO	Monostable Blocking Oscillator
MBONE	Multicast Backbone
MBPC	Model Based Predictive Control
Mbps	Megabits per second
MBps	Megabytes per second
MBQ	Message-Based Queuing
MBQ	Modified Bi-Quinary (code)
MBR	Master Boot Record
MBR	Member
MBR	Memory Base|Buffer Register
MBR-E	Memory Buffer Register, Even
MBR-O	Memory Buffer Register, Odd
MBS	Magnetic Beam Switching
MBS	Magnetic Bubble Storage
MBS	Main Buffer Storage
MBS	Master Build Schedule
MBS	Maximum Batch Size

Mb/s	Megabits per second
MB/s	Megabytes per second
MBS	Multi-Body Systems analysis
MBS	Multiple Business Systems
MBU	Memory Buffer Unit
MBX	Mailbox
Mbyte	Megabyte
MBZS	Maximum Bandwidth Zero Suppression
MC	Machine Cycle
MC	Magnetic Card|Core
MC	Main line and Control
M/C	Maintenance Console
MC	Management Center
MC	Master Control
Mc	Megacurie
MC	Megacycles
MC	Memory Control(ler)
mc	meter-candle
MC	Micro-Computer
MC	Mission Computer
MC	Mode Control
MC	Momentary Contact
MC	Monaco
MC	Monitor Call
MCA	Material Control Area
MCA	Maximum Calling Area
MCA	Micro Channel Adapter|Architecture
MCA	Minimum Charge Area
MCA	Multi-Channel Access
MCA	Multiplexing Channel Adapter
MCA	Multiprocessor Communications Adapter
MCAD	Mechanical Computer Aided Design
MCAD	Monte Carlo Design
MCAE	Mechanical Computer Aided Engineering
MCAO	Manually Controlled Automatic Operation(al mode)
MCAR	Machine Check Analysis and Recording
MCAX	Monte Carlo Analysis
MCB	Master Circuit Breaker
MCB	Memory Control Block
MCB	Micro-Computer Board
MCB	Miniature Circuit Breaker
MCBC	Main Channel Byte Counter
MCBF	Mean Characters Between Failures
MCBF	Mean Cycles Between Failure
MCBP	Mini-Computer Business Package
MCC	Machine Control Character(s)
MCC	Magnetic Card Code
MCC	Maintenance Control Circuit

MCC	Master Control Code|Console
MCC	Micro-Computer Chip
MCC	Micro CPU Chip (CPU = Central Processing Unit)
MCC	Modem Controller Chip
MCC	Multi-Channel Controller|Communication(s)
MCC	Multi-Chip Carrier
MCC	Multi-Components Circuit
MCC	Multiple Column Control
MCCA	Mission Critical Custom Applications
MCCB	Molded Case Circuit Breaker
MCCCT	Manufacturing Computer-Controlled Circuit Tester
MCCCT	Modular Computer-Controlled Continuity Tester
MCCD	Message Cryptographic Check Digits
MCCS	Manual Closed-loop Control System
MCCS	Mechanized Calling Card Service
MCCU	Multiple Channel Control Unit
MCCX	Multimedia Communications Exchange
MCD	Manipulative Communication Deception
MCD	Marginal Check(ing) and Distribution (unit)
MCD	Master Copy Distribution
MCD	Missing Cycle Detector
MCD	Monitor Console routine Dispatcher
MCDBSU	Master Control and Data Buffer Storage Unit
MCDM	Multi-Criteria Decision Making
MCDS	Management Control Data System
MC/DSS	Multiple-Criteria Decision Support System
MCE	Magnetic Card Executive
MCE	Mode of Consistency Enforcement
MCEL	Machine Check Extended Logout
MCF	Magnetic Card File
MCF	Master Circuit File
MCF	Master Clock Frequency
MCF	Master Copy Flag
MCF	Medium-access Control convergence Function
MCF	Meta Content Framework
MCGA	Multi-Colour Graphics Array
MCH	Machine Check Handler
MCH	Maintenance Channel
MCHB	Maintenance Channel Buffer (register)
MCHC	Maintenance Channel Command (register)
Mchr	Megacharacter
MCHTR	Maintenance Channel Transmit Receiver register
MCI	Machine Check Interrupt(ion)
MCI	Media Control Interface
MCIAS	Multi-Channel Intelligent Announcement System
MCIC	Machine Check Interruption Code
MCIC	Multi-Channel Interface Controller
MCID	Malicious Call Identification

MCIS	Maintenance Control Information System
MCIS	Management Controlled Information System
MCIS	Map and Chart Information System
MCIS	Materials Control Information System
MCIS	Microsoft Commercial Internet Systems
MCIS	Multiple Corridor Identification System
MCL	Manufacturing Control Language
MCL	Maximum Contaminant Level
MCL	Memory Core Loader
MCL	Micro-Code Language
MCL	Microprogram Control Logic
MCL	Microsoft Compatibility Laboratories
MCL	Monitor Control Language
MCLA	Microcoded Communications Line Adapter
MCLK	Master Clock
MCM	Management Control Model
MCM	Media and Communication Management (CH)
MCM	Memory Control Module
MCM	Monte Carlo Method
MCM	Multi-Channel Multiplex
MCM	Multi-Chip Module
MCM	Multiple Constant Multiplication
MCMC	Multiple Channel/Multiple Choice
MCMS	Multi-Channel Memory System
MCN	Metropolitan Campus Network
MCNS	Multimedia Cable Network System
MCO	Multiplexer Control Option
MCOES	Manufacturing Cell Operator's Expert System
MCOS	Microprogrammable Computer Operating System
MCP	Main Control Program
MCP	Master Communications|Control Program
MCP	Message Control Program
MCP	Micro-Channel Plate
MCP	Microsoft Certified Professional
MCP	Multiport|Multiprotocol Communication Processor
MCP/AS	Master Control Program/Advanced System
MCPG	Media Conversion Program Generator
MC-PGA	Metallized Ceramic – Pin Grid Array
mcps	megacycles per second
MCPU	Multiple Central Processing Unit
MCPYF	Master Copy Flag
MCQ	Multiple Choice Question
MC-QFP	Metallized Ceramic – Quad Flat Pack
MCR	Machine Character Recognition
MCR	Magnetic Card|Character Reader
MCR	Master Control Register|Routine
MCR	Memory Controller
MCR	Memory Control Register

MCR	Micro Copier-Reproducer
MCR	Monitor Console Routine
MCR	Multi-Contact Relay
MCRR	Machine Check Recognition and Recording
MCRR	Machine Check Recording and Recovery
MCS	Machine Control System
MCS	Magnetic Card Store
MCS	Magnetic Character Sensing
MCS	Maintenance Control Subsystem
MCS	Management|Master Control System
MCS	Meeting Communications Service
MC/s	Megacycles per second
MCS	Mesoscale Convective System
MCS	Message Control Supervisor|System
MCS	Microcode Control Storage
MCS	Micro-Computer System
MCS	Microprogram Certification System
MCS	Modified Current Switch
MCS	Monitor Control|Converter System
MCS	Multichannel Communications Support
MCS	Multi-Computer System
MCS	Multiple Column Select(or)
MCS	Multiple Console Support
MCS	Multiple Copy Screen
MCS	Multipurpose Communications and Signalling
MCSD	Microsoft Certified Solutions Developer
MCSE	Microsoft Certified Systems Engineer
MCST	Magnetic Card Selectric Typewriter
MCT	Machine Cycle Time
MCT	MOS Controlled Thyristor
MCTD	Mean Cell Transfer Delay
MCTS	Memory Card Test System
MCU	Machine (tool) Control Unit
MCU	Maintenance|Management Control Unit
MCU	Master|Memory Control Unit
MCU	Medium Close Up
MCU	Micro-Computer Unit
MCU	Microprocessor|Microprogram Control Unit
MCU	Multi-Chip Unit
MCU	Multipoint Conference|Control Unit
MCU	Multiprocessor|Multisystem Communications Unit
MCUSR	Memory Control Unit Special Register
MCV	Mean Cell Volume
MCV	Mesoscale Convectively-generated Vortices
MCVF	Multi-Channel Voice Frequency
MCVFC	Multi-Channel Voice Frequency Cable
MCW	Maintenance Control Word
MCW	Modulated Continuous Wave

MD Machine Description|Dialogue
MD Macro Directory
MD Magnetic Driver|Drum
MD Maintenance Documentation
MD Make Directory
MD Management Domain
MD Manual Data
MD Maryland (US)
MD Mediation Device
MD Medium Duty
MD Memory Data register
MD Memory Decrement
MD Message Data
MD Micro Diagnostics
MD Mini Disk
MD Modify
M-D Modulation-Demodulation
MD Moldova
M&D Monitoring & Diagnostics
MD Monochrome Display
MD Multiply/Divide
MD2/HD Mini floppy Disk Double Sided High Density
MDA Manual Data Acquisition
MDA Mechanical Design Automation
MDA Mechanical Differential Analyzer
MDA Mesocyclone Detection Algorithm
MDA Monochrome Display Adapter
MDA Monolithic Design Automation
MDA Multi-Dimensional Access
MDAC Multiplying Digital to Analog Converter
MDACS Modular Digital Access Control System
MDAP Machining and Display Application Program
MDAS Magnetic Drum Auxiliary Sender
MDB Master Data Bank
MDBMS Micro Data Base Management System
MDBS Microcomputer Data Base System
MDBS Mobile Data Base System
MDC Magnetic Drum Calculator
MDC Main Digital Computer
MDC Manipulation Detecting Code
MDC Marker Distributor Control
MDC Memory Disk Controller
MDC Meridian Digital Centrex
MDC Message Display Control
MDC Meta Data Coalition
MDC Microcomputer Development Center
MDC Minimum Distance Code
MDC Modification Detection Code

MDC	Multiple Device Controller
MDC&R	Management Data Collection & Reporting
MDCR	Mini Digital Cassette Recorder
MDCU	Magnetic Disk Control Unit
MDCU	Multi-Display Control Unit
MDD	Magnetic Disk Drive
MDD	Meteorological Data Distribution
MDD	Multiple Disk Drive
MDDB	Multi-Dimensional Data Base
MDDBMS	Multi-Dimensional Data Base Management System
MDDPM	Magnetic Drum Data Processing Machine
MDDS	Media Documentation Distribution Set
MDDU	Manual Data Display Unit
MDE	Magnetic Decision Element
MDE	Modular|Multiple Design Environment
MDES	Multi-Data Entry System
MDF	Main Distribution Frame
MDF	Master Data File
MDF	Master Distribution Frame
mdf	menu definition file (file extension indicator)
MDF	Microcomputer Development Facility
MDF	Multi-Disk File
MDFM	Modified Double-Frequency Modulation
MDH	Macro Definition Header
MDH	Multidrop Design Heuristic
MDI	Magnetic Data Inscriber
MDI	Magnetic Direction Indicator
MDI	Manual Data Input
MDI	Medium Dependent Interface
MDI	Multiple Document Interface
MDIC	Manchester Decoder and Interface Chip
MDIS	Meta-Data Interchange Specification
MD-IS	Mobile Data, Intermediate System
MDIU	Manual Data Insertion Unit
MDK	Multimedia Development Kit
MDKU	Manual Data Keyboard Unit
MDL	Machine Dependent Language
MDL	Macro-Data Language
MDL	Magneto-restrictive Delay Line
MDL	Maintenance Diagnostic Logic
MDL	Mechanical Design Language
MDL	Mercury Delay Line
MDL	Method Definition Language
MDL	Microprocessor Development Laboratory
MDL	Microstation Development Language
MDL	Model
MDL	Modular Development Language
MDLC	Multiple Data Link Controller

MDLP	Module Data Link Protocol
MDM	Main Data Memory
MDM	Maintenance Diagnostic|Diagram Manual
MDM	Manufacturing Data Management
MDM	Marketing Data Management
MDM	Multidimensional Data Model
MDMA	Multithreaded Daemon for Multimedia Access
MDMFM	Miniature Digital Matched Filter Module
MDN	Managed Data Network
MDNF	Minimal Disjunctive Normal Form
MDNS	Managed Data Network Service
MDOS	Multiprocessor Disk Operating System
MDP	Main Data Path
MDP	Management|Manual Data Processing
MDP	Micro-Display Processor
MDPS	MICR Document Processing System
MDQS	Management Data Query System
MDQS	Multi-Device Queuing System
MDR	Machine Design Rate
MDR	Magnetic Disk Recorder
MDR	Master Data Record
MDR	Memory Data Register
MDR	Message Detail Recording
MDR	Minimum Design Requirement
MDR	Multichannel Data Recorder
MDRAM	Multibank Dynamic Random Access Memory
MD/ROM	Mini Disk Read Only Memory
MDRU	Manual Data Readout Unit
MDS	Magnetic Disk Store
MDS	Main Device Scheduler
MDS	Maintenance Data System
MDS	Malfunction Detection Subsystem
MDS	Management Decision|Display System
MDS	Marketing Design System
MDS	Master Data Set
MDS	Mechanical|Mechanism Design System
MDS	Meta-computing Directory Service
MDS	Microcode Development System
MDS	Microcomputer Development System
MDS	Minimum Data Set
MDS	Minimum Discernible Signal
MDS	Modular Data System
MDS	Multi-Dialogue System
MDS	Multiple Data Set
MDS	Multiple Dataset System
MDS	Multipoint Distribution Service|System
MDSE	Message Delivery Service Element
MDSS	Microprocessor Development Support System

MDT	Macro Definition Trailer
MDT	Mean Down Time
MDT	Mechanical Desktop
MDT	Merchant Deposit Transmittal
MDT	Micro Debugging Tool
MDT	Mobile Data Terminal
MDT	Modified Data Tag
MDT	Multi-Disciplinary Team
MDTS	Modem Diagnostic and Test System
MDU	Maintenance Diagnostic Unit
MDU	Marker Decoder Unit
MDU	Message Display Unit
MDU	Multiply/Divide Unit
MDUL	Module
MDX	Modular Digital Exchange
MDY	Month-Day-Year
ME	Maine (US)
ME	Memory Element
ME	Micro-Electronics
ME	Molecular Electronics
ME	Motion Estimation
ME	Multi-Edit
me	opening information (file extension indicator)
MEA	Memory inspection Ending Address
MEAN	Measurement and Analysis (program)
MEB	Memory Expansion Board
MEB	Modem Evaluation Board
MEB	Multiplexed one Electrode per Bit
MEBAS	Method Base System
MECCA	Memory Environmental Control of Circuits and Array
MECCA	Multi-Element Component Comparison and Analysis
MECH	Mechanism
MECHA	Mechanical Analysis (program)
MECL	Motorola Emitter-Coupled Logic
MECO	Measurement and Control
MECT	Module Electrical Crossover Tester
MED	Manipulative Electronic Deception
MED	Manual Entry Device
MED	Medium|Media
MED	Multi-Exit Discriminator
MEDAPE	Median Absolute Percentage Error
MEDEA	Measurements Description Evaluation and Analysis tool
MEDESS	Methodology for the Design of Expert Support System
MEDINFO	Medical Informatics (world conference)
MEDLAR	Mechanizing Deduction in Logic And practical Reasoning (EU)
MEDLARS	Medical Literature Analysis and Retrieval System
MEDS	Manufacturing Efficiency Decision Support
MEDS	Marine Ecological Database System

MEDS	Meteorological and Environmental Data Services
MEECES	Multi-Experimental Event-Controlled Entry System
MEF	Master Edit File
MEF	Microsoft Easy Fulfillment
MEG	Megabyte
meg	megohm
mega	millions (prefix)
megaflop	one million floating-point operations per second
MEGRIN	Multipurpose European Ground Related Information Network
MEI	Module Execution Interval
MEIS	Manufacturing Engineering Information System
MEKAS	Methodology for Knowledge Analysis
MELC	Manufacturing Equipment Level Control
MELCU	Multiple External Line Control Unit
MELISSA	Meta-Linguistic Syntax Specification Analyzer
MELO	Multiple Eigenvector Linear Orderings
MELODY	Management Environment for Large Open Distributed Systems
MEM	Memory
MEM	Memory Emulation Module
MEM	Micro-Electromechanical Machine
MEMA	Micro-Electronic Modulators Assembly
MEMC	Memory Controller
ME/ME	Multiple Entry/Multiple Exit
MEMIS	Maintenance and Engineering Management Information System
MEMO	Memorandum
MEMR	Memory Read
MEMS	Micro-Electro-Mechanical System
MEMSEL	Memory Select
MEMW	Memory Write
men	menu (file extension indicator)
MEP	Microfiche Enlarger Printer
MEPRO	Memory Protection
MER	Machine Experience Report
MER	Minimum Energy Requirements
MER	Multiple Ejector Rack
MERA	Molecular Electronics for Radar Applications
MERANDA	Multiple Events Recorder And Data Analysis
MERIE	Magnetically Enhanced Reactive Ion Etch
MERISE	Methode d'Etude et de Réalisation Informatique pour les Systèmes d'Enterprise (FR)
MERM	Multilateral Exchange Rate Model
MERS	Most Economic Route Selection
MES	Manufacturing Execution System(s)
MES	Message
MES	Metal Semiconductor
MES	Mobile End System
MES	Multiple Earning Statement

MESA	Model Experimental Systems Analysis
MESA	Modularized Equipment Storage Assembly
MESFET	Metal Silicon Field-Effect Transistor
MESH	Multimedia services on the Electronic Super Highway
MESI	Modified, Exclusive, Shared and Invalid (protocol)
MET	Memory Enhancement Technology (HP)
MET	Metafile
MET	Metal
MET	Module Evaluation Tester
MET	Multibutton Electronic Telephone
METAL	Meta-Language
METC	Metallic
meteosat	meteorological satellite
METL	Metallurgy
METRG	Metering
METRIC	Multi-Echelon Technique for Recoverable Item Control
METS	Multi-Executive Time-Sharing
MEU	Maximization of Expected Utility
MEU	Memory Expansion and protection Unit
MeV	Mega-electronvolt
MEV	Million Electron Volts
MEWERS	Message Writer and Errata Subsystem
MEWT	Matrix Electrostatic Writing Technique
M/F	Mainframe
MF	Master File
MF	Measurement Facility
MF	Mediation Function
MF	Medium Frequency
mf	microfarad
MF	Micro Film
mF	millifarad
MF	Modified Field
MF	Multi(ple) Frequency
MFB	Monochrome Frame Buffer
MFC	Magnetic tape Field Scan
MFC	Micro-Functional Circuit
MFC	Microsoft's Foundation Class
MFCA	Multi-Function Communications Adapter
MFCM	Multi-Function Card Machine
M&FCS	Management & Financial Control System
MFCS	Measurement Feedback Control System
MFCU	Multi-Function Card Unit
MFD	Master File Directory
mfd	microfarad
MFD	Multi-Function Display
MFDB	Micro-Fiche Data Bank
MFDSUL	Multi-Function Data Set Utility Language
MFE	Machine Functional Entity

MFE	Multi-Function Equipment|Executive
MFENET	Magnetic Fusion Energy Network
MFFC	Master Function Flowchart(s)
MFFS	Microsoft Flash File System
MFG	Mark Flow Graph
MFG	Message Flow Graph
MFG	More Friendly Garbage
MFI	Machine File Index
MFI	Mainframe Interactive
MFI	Multisensor Fusion and Integration
MFJ	Modified Final Judgment
MFKP	Multi-Frequency Key Pulsing
MFLD	Message Field
mflops	one million floating point operations per second
MFLP	Multi-File Linear Programming
MFM	Memory Feasibility Model
MFM	Modified Frequency Modulation
MFM	Multilevel Flow Modeling
MFO	Machine Functional Operation
MFOM	Multi-Function Office Machine
MFOTS	Military Fiber-Optic Transmission System
MFP	Mainframe Processor
MFP	Mathematic(al) Function(s) Program
MFP	Mobile Flux Platform
MFP	Multi-Form Printer
MFP	Multi-Frequency Pulsing
MFP	Multi-Function Peripheral
MFPC	Multi-Function Protocol Converter
MFPE	Minimum Final Prediction Error
MFPI	Multi-Function Peripheral Interface
MFR	Manufacturer
MFR	Modified Frequency Recording
MFR	Multi-Frequency Receiver
MFS	Macintosh File System
MFS	Magnetic Film Storage
MFS	Magnetic Flux Sensor
MFS	Magnetic tape Field Search
MFS	Message Format Service(s)
MFS	Modified Filing System
MFS	Multiple File Systems
MFSK	Multiple Frequency Shift Keying
MFSL	Mathematical and Functional Subroutine Library
MFT	Machine Function Test
MFT	Master File Table
MFT	Memory Final Test
MFT	Mixed Form Text
MFT	Monolithic Functional Tester
MFT	Multi-Function Tasking

MFT	Multiprogramming with a Fixed number of Tasks
MFTT	Multi-Function Telephone Terminal
MFTU	Multi-Function Tape Unit
MG	Madagascar
MG	Master Group
mg	milligram
MGA	Monochrome Graphics Adapter
MGAP	Magnetic Attitude Prediction
MGC	Manual Gain Control
MGCB	Master Gate Control Block
MGCP	Media Gateway Control Protocol
mgd	million gallons per day
MGDE	Minimum Geometric Data Element
MGE	Modular GIS Environment (GIS = Geographic Information System)
MGEN	Module Generation (system)
MGFY	Magnify
MGG	Matrix-Generator Generator
mgm	milligram
MGOS	Memory Graphics Operating System
MGP	Macro Generating Program
MGP	Magnetic Graphic Products
MGP	Memory Graphics Program
MGP	Monochrome Graphics Printer (port)
MGP	Multiple Goal Programming
MGP/PP	Memory Graphics Program/Post Processing
MGPS	Memory Graphic Processing System
MGR	Manager
MGRW	Matrix Generator and Report Writer
MGT	Management
MGT	Master-Group Translator
MH	Magnet(ic) Head
MH	Manufacturing Hardware (testing)
MH	Marshall Islands
MH	Message Handling
mh	millihenry
MHD	Magneto-Hydro-Dynamic(s)
MHD	Message Header
MHD	Moving Head Disk
MHEG	Multimedia/Hypermedia Experts Group
MHHZO	Move High-to-High Zone
MHL	Microprocessor Host Loader
MHLZO	Move High-to-Low Zone
MHMS	Modular Hydrologic Modeling System
MHOS	Multi-Hospital Operating System
MHOTY	My Hat's Off To You
MHP	Materials Handling Processor
MHP	Message Handling Protocol

MHPCC	Maui High Performance Computing Center
MHS	Message Handling Service|System
MHS	Modular Hardware System
MHS	Multiple Host Support
MHSDC	Multiple High Speed Data Channel
MHSS	Message Handling System Service(s)
MHTS	Message Handling Test System
MHz	Megahertz
MI	Machine Independent|Interface
MI	Maintenance|Management Interface
MI	Manual Input
MI	Manual of Instruction
MI	Maskable Interrupt
MI	Memory Input (register)
MI	Memory Interface
MI	Michigan (US)
MI	Mode Indicator
MIA	Minor Acknowledgment
MIA	Multiplex Interface Adapter
MIACS	Manufacturing Information And Control System
MIAMI	Multimodal Integration for Advanced Multimedia Interfaces (EU)
MIAP	Member (of the) Institution (of) Analysts (and) Programmers
MIAR	Micro-Address Register
MIARS	Maintenance Information Automated Retrieval System
MIAS	Management Information and Accounting System
MIAS	Multipoint Interactive Audiovisual System
MIB	Management Information Base
MIB	Manual Input Buffer
MIB	Micro Instruction Bus
MIB	Module Interconnection Bus
MIB	Multilayer Interconnection Board
MIB	Multilayer Internal plane Board
MIBA	Multiple Input Binary Adder
MIBN	Microsoft International Business Network
MIC	Magnetic Ink Character
MIC	Management Inventory Classification
MIC	Mathematics Information Center
MIC	Medium Interface Cable|Connector
MIC	Message Identification Code
mic	microphone
MIC	Microwave Integrated Circuit
MIC	Middle In Chain
MIC	Minimum Issue Control
MIC	Missing Interrupt Checker
MIC	Mobile Interface Control
MIC	Monolithic Integrated Circuit
MIC	Multicharacter Input Channel

MIC	Multiple Input Change
MICA	Macro-Instruction Compiler Assembler
MICALL	Microprocedure Call
MICC	Mineral Insulated Conductor Cable
MICE	Modular Integrated Communications Environment
MICE	Monitoring In Complex Environments
MICE	Multimedia Integrated Conferencing for Europe
MICIS	Midwestern Climate Information System
MICOT	Minimum Completion Time
MICR	Magnetic Ink Character Reader|Recognition
MICRO	Microcomputer
MICRO	Microprocessor
micro	millionths (extremely small; prefix)
MICROM	Micro-Instruction Read Only Memory
micron	micrometer
MICROSIM	Micro-instruction Simulator
MICS	Machine Inventory Control System
MICS	Macro Interpretive Commands
MICS	Magnetic Ink Character Set
MICS	Management Information and Control System
MICS	Management Information Conformance Statement
MICSP	Microscope
MID	Maintenance Information Department
MID	Message Identification
MID	Message Input Description|Descriptor
MID	Middle
MID	Multiplexing Identifier
MIDA	Message Interchange for Distributed Application
MIDA	Multiple Integrated Digital Access (GB)
MIDAR	Micro-Diagnostics for Analysis and Repair
MIDAS	Management Interactive Data Accounting System
MIDAS	Materials Inventory Data Acquisition System
MIDAS	Memory Implement Data Acquisition System(s)
MIDAS	Microprogrammable Integrated Data Acquisition System
MIDAS	Mixed Data Structure
MIDAS	Mixed-mode Digital and Analog Simulator
MIDAS	Modular Integrated Direct Access System
MIDAS	Monte Carlo Investigation of Data System
MIDAS	Multi-tier Distributed Application Services
MIDAS-2000	Mixed-mode Digital and Analog Simulator for the years 2000
MIDDLE	Microprogram Design and Description Language
MIDEF	Microprocedure Definition
MIDI	Medium Independent Digital Image
MIDI	Musical Instrument Digital Interface
MIDL	Microsoft Interface Definition Language
MIDL	Modular Interoperable Data Link
MIDMS	Machine Independent Data Management System
MID-PAC	Middle power Package

MIDS	Management Information and Decision System
MIDS	Mission Information Dispensing System
MIDS	Multimode Information Distribution System
MIE	Management Information Element
MIE	Mission-Independent Equipment
MIEC	Master Interrupt and Executive Control (program)
MIF	Machine Installed|Interface File
MIF	Management Information File|Format
MIF	Master Index File
MIF	Minimum Internetworking Functionality
MIFF	Management Information Format File
MIFR	Master International Frequency Register
MIG	Magnetic Injection Gun
MIGET	Miniature Interface General purpose Economy Terminal
MIH	Missing Interrupt(ion) Handler
MIH	Multiplex Interface Handler
MII	Microsoft/IBM/Intel
MIKADOS	Mini Instant Keyboard Assembler, Debug, and Operating System
mike	microphone
MIL	Machine Independent Language
MIL	Machine Interface Layer
MIL	Macro-Instruction Link
MIL	Management Information Library
MIL	Micro-Implementation Language
MIL	Module Interconnection Language
mil	one thousands of an inch
MILE	Matrix Inversion and Linear Equations
MILES	Mixed-Level mixed-mode Simulator
MILNET	Military Network
MILP	Mixed Integer Linear Programming
MIL-STD	Military Standard (US)
MIM	Management Information Model
MIM	Manufacturing Integration Model
MIM	Media|Modem Interface Module
MIM	Metal-Insulator-Metal (screen)
MIM	Morality In Media
MIMD	Multiple Instruction, Multiple Data
MIME	Multipurpose Internet Mail Extension
MIMI	Mini- and Microcomputers and their applications (international symposium and exhibition)
MIMO	Multiple Input, Multiple Output
MIMOLA	Machine Independent Microprogramming Language
MIMOSE	Machine Independent Model Oriented Simulation Environment
MIMS	Multiple Image Masking System
min	minimum
MIN	Minor

min	minute
MIN	Mobile Identification Number
MIN	Multipath Interconnection Network
MINBL	Minimum Balance
MINC	Modular Instrumentation Computer
MINCOS	Modular Inventory Control System
MIND	Management of Information through Natural Disclosure(s)
MIND	Modular Interactive Network Designer
MINI	Minicomputer
MINI	Minicomputer Industry National Interchange
MINIAC	Minimal Automatic Computer
MINIDOS	Mini Disk Operating System
MINI-IR	Minimum Incident Report
MINIPERT	Mini Program Evaluation and Review Technique
MINOS	Modular Input/Output System
MINSOP	Minimum Slack time per Operation
MINUET	Minnesota Internet User's Essential Tool
MINX	Multimedia Information Network Exchange
MIO	Multiple Input/Output
MIOCB	Master Input/Output Control Block
MIOP	Multiplexer Input/Output Processor
MIOS	Metal Insulator Oxide Silicon
MIOS	Modular Input/Output System
MIP	Machine Improvement Program
MIP	Machine Instruction Processor
MIP	Main Instructions Processor
MIP	Management Information Protocol
MIP	Maximum Intensity Projection
MIP	Memory In Pulse
MIP	Methods Improvement Program
MIP	Minor Point
MIP	Mixed Integer Programming
MIP	Multipurpose Information Processor
MIPAS	Management Information Planning and Accounting Service
MIPS	Meaningless Indications of Performance Statistics
MIPS	Microprocessor without Interlocked Piped Stages
mips	millions of instructions per second
MIPS	Modular Information Processing System
MIPS	Moving Interactive Planning System
MIR	Machine Incident Report
MIR	Master Interrupt Register
MIR	Memory Information|Input Register
MIR	Metered Incident Report
MIR	Micro-Instruction Register
MIRA	Microfilm Information Retrieval Access
MIRAC	Management Information Retrieval And Communication|Control
MIRAC	Master Index Remote Access Capability

MIRACL	Management Information Report Access without Computer Languages
MIRAGES	Materials Information Reference And General Enquiry System
MIRS	Management Information and Reporting System
MIS	Machine Instruction Set
MIS	Management Information Service|System
MIS	Manufacturing Information System
MIS	Medical Information System
MIS	Metal Insulated Semiconductor
MIS	Multimedia Information Sources
MIS	Multiple Interactive Screen(s)
MISAM	Multiple Index Sequential Access Method
MISAR	Microfilm Information Storage And Retrieval
MISC	Miscellaneous
MISD	Multiple Instruction, Single Data (stream)
MISE	Mean Integrated Square(d) Error
MISER	Minimum Size Executive Routine(s)
MISG	Missing
MISI	Management Information System Interface
MIS-MDS	Multiple Instruction Streams – Multiple Data Streams
MISP	Medical Information System Programs
MISP	Microelectronics Industry Support Program
MISR	Machine Supply Requisition
MISR	Multi Input Signature Register
MIS-SDS	Multiple Instruction Streams – Single Data Streams
MISSI	Multilevel Information System Security Initiative
MISSIL	Management Information System Symbolic Interpretive Language
MISSION	Manufacturing Information System Support Integrated On-line
MIT	Management Information Tree
MIT	Marked If Touched
MIT	Massachusetts Institute of Technology
MIT	Master Instruction|Interface Tape
MIT	Modular Intelligent Terminal
MITA	Microcomputer Industry Trade Association (US)
MITE	Microelectronic Integrated Test Equipment
MITE	Microprocessor Industrial Terminal
MITE	Miniaturized Integrated Telegraph Equipment
MITE	Multiple Input Terminal Equipment
MITI	Ministry of International Trade and Industry (JP)
MITOS	Metal-Insulator Trap-Oxide-Silicon
MITS	Management Information and Text System
MITS	Microcomputer Interactive Test System
MITS	Micro Instrumentation Telemetry Systems
MIU	Modem Interface Unit
MIW	Micro-Instruction Word
MIX	Member Information Exchange
MIX	Metropolitan Information Exchange

MIX Microprogram Index (register)
MJ Manual Jack
MJ MegaJoule
MJ Modular Jack
MJP Multiple Job Processing
MJPEG Motion Joint Photographic Experts Group
MJT Multi-Job Terminal
MK Mark
MKDIR Make Directory
MKDS Master Key Data Set
MKH Multiple Key Hashing
mks meter-kilogram-second
mksa meter, kilogram, second, ampere
MKTG Marketing
ML Machine Language|Learning
ML Macro Library
ML Maintenance Level
ML Mali
M&L Matched & Lost
ML Maximum Likelihood
ML Mega-Language
ML Memory Location
ML Meta-Language
ML Microprogramming Language
mL millilambert
ml milliliter
ML Missing Lower
ML Multi-Line
MLA Matching Logic and Adder
MLA Microwave Link Analyzer
MLA Multi-Line Adapter
MLA Multiple Line Adapter
MLAPI Multi-Lingual Application Programming Interface
MLB Multi-Layer Board
MLC Machine Level Control
MLC Magnetic Ledger Card
MLC Mesh Level Control
MLC Microprogram Location Counter
MLC Mini-Line Card
MLC Multi-Language Computer
MLC Multi-Layer Circuit
MLC Multi-Layer(ed) Ceramic (modules)
MLC Multilayer Laminated Ceramic
MLC Multi-Line Controller
MLCA Machine Level Control Address
MLCA Multi-Line Communications Adapter|Attachment
MLCD Multi-Line Call Detail
MLCH Machine Level Control History

MLCI	Multi-Link Channel Interface
MLCP	Machine Level Control Program
MLCP	Multi-Line Communications Processor
MLCSP	Multi-Level Continuous Sampling Plan
MLCU	Magnetic Ledger Card Unit
MLD	Machine Language Debugger
MLD	Masking Level Differences
MLD	Mixed Layer Depth
MLE	Maximum Likelihood Estimate|Estimation
MLE	Meta-Language Extension
MLEM	Multi-Language Environment
MLF	Modelling Language and Formalism
MLFN	Malfunction
MLFS	Master Library File System
MLHZO	Move Low-to-High Zone
MLI	Machine Language Instruction
MLI	Marker Light Indicator
MLI	Multi-Leaving Interface
MLI	Multiple Link Interface
MLIA	Multiplex Loop Interface Adapter
MLID	Multiple Link Interface Driver
MLIM	Matrix Log-In Memory
MLIN	Massachusetts Library and Information Network
MLL	Master Logic List
MLLZO	Move Low-to-Low Zone
MLM	Mailing List Manager
MLM	Maintenance Library Manual
MLM	Membrane Light Modulator
MLM	Multileaving Line Manager
MLM	Multi-Level Marketing
MLMA	Multi-Level Multi-Access
MLMP	Multi-Line/Multi-Point
MLN	Main Listed Number
MLO	Multi-Layer Organic
MLOC	Move Location
MLOCR	Multi-Line Optical Character Reader
MLOLT	Multiple Level On-Line Trace
MLP	Machine Language Program
MLP	Machine Layout Problem
MLP	Multi-Level Precedence
MLP	Multi-Link Procedures
MLP	Multiple Line Printer
MLPA	Modified Link Pack Area
MLPC	Multiplicand
MLPP	Multi-Level Precedence and Preemption
MLPPP	Multi-Link Point-to-Point Protocol
MLPR	Multiplier
MLPSC	Monthly Licensed Program Support Charge

MLR	Machine Location Report
MLR	Maximum Logical Records
MLR	Memory Lockout Register
MLR	Multichannel Linear Recording
MLR	Multiple Line Read(ing)
MLR	Multiply and Round
MLRTP	Multi-Leaving Remote Terminal Processor
MLS	Machine Literature Searching
MLS	Microprocessor Line Set
MLS	Microprogram List System
MLS	Multi-Language Support
MLS	Multi-Level Security
MLSE	Maximum Likelihood Sequence Estimation
MLSU	Multiple Listening Station Unit
MLT	Main Logic Table
MLT	Master Library Tape
MLT	Module Logic Technology
MLT	Monolithic (module) Technology
MLT	Multi-Level Technology
MLT	Multiple Logical Terminals
MLTA	Multiple Line Terminal Adapter
MLTP	Machine Language Test Program
MLU	Memory Logic Unit
MLU	Multiple Logical Unit
MM	Main|Mass Memory
MM	Maintenance Manual
MM	Master Machine|Monitor
MM	Material Manager|Manager
MM	Memory Module|Multiplexer
MM	Mesoscale Model
MM	Message Management
mm	millimeter
MM	Monitoring Module
MM	Monitor Mode
MM	Multi-Media
MM	Multiple Master
MM	Myanmar
MMA	Machine Maintenance Analysis
MMA	Main Memory Array
MMA	Major Maintenance Availability
MMA	Microcomputer Managers Association (US)
MMA	MIDI Manufacturers Association (MIDI = Musical Instrument Digital Interface)
MMA	Multiple Module Access
MMAC	Micro-Miniature Analog Circuit
MMAC	Multi-Media Access Center
MMAR	Main Memory Address Register
MMB	Multiport Memory bank

MMC	Magical Mystery Chip
MMC	Main Memory Controller
M/MC	Man/Machine Communication
MMC	Matched Memory Cycle
MMC	Microcomputer Marketing Council
MMC	Microsoft Management Console|Control
MMC	MIDI Machine Control (MIDI = Musical Instrument Digital Interface)
MMC	Minicomputer Maintenance Center
MMC	Monthly Maintenance Charge
MMC	Multiport Memory Controller
MMCA	Message Mode Communications Adapter
MMCC	Multi-Mini Computer Compiler
MMCD	Multi-Media Compact Disk
MMCF	Multi-Media Communications Forum
MMCS	Manual Monitored Control System
MMCS	Manufacturing Management Control System
MMCX	Multi-Media Communications Exchange
M/MD	Man/Machine Dialog
MMD	Multi-Media Document
mm-dd-yy	month-day-year
MMDF	Multichannel Memorandum Distribution Facility
MMDS	Multichannel Multipoint Distribution Service
MMDS	Multipoint Microwave Distribution System
MMDS	Multipoint Multidrop Distribution Service
MME	Multi-Media Extension(s)
MMF	Message Management Facility
mmf	micro-microfarad
MMF	Multi-Mode Fiber
MMFS	Manufacturing Message Format Standard
MMGT	Multi-Master-Group Translator
MMH	Maintenance Man-Hours
MMH	Monthly Maintenance Hours
MMH/OH	Maintenance Man-Hours per Operating Hour
MMI	Main Memory Interface
MMI	Man-Machine Interface
MMI	Modified Mercalli Intensity
MMI	Multi-Message Interface
MMI	Multiport Memory Interface
MMIC	Monolithic Microwave Integrated Circuit
MMIO	Memory Mapped Input/Output
MMIS	Maintenance|Manufacturing Management Information System
MMIS	Materials Manager Information System
MMIU	Multiport Memory Interface Unit
MMJ	Modified Modular Jack
MML	Major-Minor Loop
MML	Man-Machine Language
MML	Micro-programmed MOSFET Logic

MML	Monitor Meta-Language
MMLPSC	Monthly Multiple Licensed Program Support Charge
MMM	Main Memory Module
MMM	Man-Machine Model
MMM	Monolithic Main Memory
MMM	Multiport Memory Multiplexer
MMMC	Minimum Monthly Maintenance Charge
MMOC	Minicomputer Maintenance Operations Center
mmol	millimole
MMOS	Message Multiplexer Operating System
MMP	Machine Main Performance
MMP	Main Micro-Processor
MMP	Massively Multi-Processing
MMP	Modified Modular Plug
MMP	Multiple Machine Plan
MMP	Multiple Micro-Processors
MMPM	Multi-Media Presentation Manager
MMPS	Manpower and Machine Planning System
mmps	millimeters per second
MMPU	Memory Management and Protection Unit
MMR	Main Memory Register
MMR	Memory Management Register
MMR	Multi-Modular Redundancy
MMR	Multiple Match Resolver
MMS	Main Memory Status
MMS	Man-Machine Service|System
MMS	Manufacturing Message Service
MMS	Manufacturing Message Specification|Standard
MMS	Manufacturing Message Standard|Service
MMS	Manufacturing Monitoring System
MMS	Memory Management System
MMS	Micro-Memory System
mms	milli-microsecond
MMS	Monolithic Main Storage
MMS	Multimedia Manufacturing System
MMSE	Minimum Mean-Squared Error
MMSS	Material Management Standard System
MMT	Multimedia Multiparty Teleconferencing
MMT	Multiple Mirror Telescope
MMU	Main Memory Unit
MMU	Memory Management|Mapping Unit
MMU	Minimum Mapping Unit
MMV	Modified Microfiche Viewer
MMV	Monostable Multi-Vibrator
MMW	Multi-Mega Watt
MMX	Mastergroup Multiplex
MMX	Multi-Media Extensions
MN	Main

MN	Manual
MN	Message Number
MN	Minnesota (US)
MN	Minus
MN	Mnemonic(s)
MN	Mongolia
MNA	Modified Nodal Analysis
MNA	Multishare Network Architecture
MNCS	Master Net Control System
MNCS	Multipoint Network Control System
MNDP	Multi-National Data Processing
MNDS	Multi-Network Design System
mnem	mnemonic
MNET	Measuring Network
MNF	Minimum Normal Form
MNF	Multisystem Networking Facility
MNIC	Mnemonic Instruction Code
MNIN	Multi-Net Invariant Network
MNOS	Metal Nitride Oxide Semiconductor
MNP	Microcom Networking Protocol
MNP	Multiple Network Protocols
MNR	Maximum Number of Records
MNS	Metal Nitride Semiconductor
mnu	menu (file extension indicator)
MO	Macau
MO	Magneto-Optical
MO	Mail Order
MO	Managed Object
MO	Manually Operated
MO	Manual of Operation
MO	Master Oscillator
MO	Memory Output
MO	Method of Operation
MO	Missing Operand
MO	Missouri (US)
MO	Mode of Operation
MO	Multiple Operations
MOA	Memorandum Of Agreement
MOAC	Message Origin Authentication Check
MOARC	Motorola Open Architecture Robot Controller
MOAT	Methods Of Appraisal and Test
MOB	Memory-Order Buffer
MOBIDIC	Mobile Digital Computer
MOBL	Macro-Oriented Business Language
MOBO	Mother Board
MOBOT	Mobile Robot
MOC	Management Oriented Computing
MOC	Mask Order Control

MOC	Master Operational Controller
MOC	Microfilm Output Computer
MOC	Middle Of Chain
MOC	Minimum Operating Condition
MOCNESS	Multiple Opening-Closing Net Environmental Sampling System
MOCS	Managed-Object Conformance Statement
MOCS	Movement Order Control System
MOD	Magneto-Optical Disk
MOD	Masters Of Downloading
MOD	Message Output Description|Descriptor
MOD	Model
MOD	Modem
MOD	Modification
MOD	Modulation
MOD	Module
mod	modulus
MODAC	Modular Data Acquisition
MODACS	Modular Data Acquisition and Control Subsystem
MODAL	Modular Diagnostic Application Language
MO:DCA	Mixed Object: Document Content Architecture
MODE	Merchant Oriented Data Entry
MODEM	Modulator/Demodulator
MODI	Modular Optical Digital Interface
MODL	Modular
MODOS	Modular Distributed Operating System
MODS	Master Operational Data Set
MODSIM	Modular Simulator
MODUS	Modular One Dynamic User System
MOE	Measure Of Effectiveness
MOESI	Modified, Owned, Exclusive, Shared, Invalid
MOET	Microsoft Order Entry Tool
MOF	Managed Object Format
MOF	Master Order File
MOF	Meta Object Facility
MOG	Minicomputer Operations Group
MOHLL	Machine Oriented High Level Language
MOHOL	Machine Oriented Higher Order Language
MOIP	Manual On-line Improvement Process
MOKE	Magneto-Optic Kerr Effect
MOL	Machine Oriented Language
MOL	Mask Order Listing
MOL	Maximum Output Level
MOL	Memory Organization Language
mol	mole
MOLAP	Multidimensional On Line Analytical Processing
MOLDS	Managerial On-Line Data System
MOLE	Market Odd-Lot Execution system
MOLECOM	Molecularized digital Computer

MOLP	Multiple Objective Linear Programming
MOM	Message Oriented Middleware
MOM	Microsoft Office Manager
MOM	Multipurpose Office Machine
MOMACT	Monolithic Memory A-C Tester
MOMENTS	Mobile Media and Entertainment Services (EU)
MOMTY	Momentary
MON	Monitor
MONET	Multiwave Optical Networking
MONGEN	Monitor Generator
MONOS	Metal Oxide Nitride Oxide Semiconductor
MONS	Monitoring System
MOO	MUD, Object-Oriented (MUD = Multi-User Domain)
MOO	Multi-Object Oriented (environment)
MOO	Multi-user dungeon Object Oriented
MOODS	Music Object-Oriented Distributed System
MOOSE	Method for Object-Oriented Software Engineering
MOP	Machine Out Pulse
MOP	Maintenance Operations Protocol
MOP	Method Of Procedure(s)
MOP	Multi On-line Printer
MOP	Multiple Operation
MOPA	Master Oscillator Power Amplifier
MOPAC	Molecular Orbital Package
MOPS	Mainframe Office Processing System
MOPS	Microcomputer Operation System
MOPS	Micro-monitor Operating System
mops	million operations per second
MOPS	Mini Operating System
MOPSY	Matrix Operating System
MOR	Magneto-Optical Reading
MOR	Management by Objectives and Results
MOR	Master Operations Record
MOR	Memory Output Register
MOR	Monthly Operating Review
MORE	Management with Operations Research for Engineering
MOREPS	Monitor station Reports
MORIF	Microprogram Optimization technique considering Resource occupancy and Instruction Formats
MOS	Management|Manufacturing Operating System
MOS	Memory Oriented System
MOS	Metal Oxide Semiconductor|Substrate
MOS	Microprogram Operating System
MOS	Model Output Statistics
MOS	Modem|Modular Operating System
MOS	Multiprogramming Operating System
MOSAIC	Machine-tool Open System Advanced Intelligent Controller
MOSAIC	Metal Oxide Semiconductor Arrays of Integrated Circuits

MOSCOR	Modular System for Computation Of Requirements
MOSFET	Metal Oxide Semiconductor Field Effect Transistor
MOSIS	Metal Oxide Semiconductor Implementation System
MOSIS	Micro-Optical Silicon Systems (EU)
MOSIS	Microprocessor Operating System Interface Specification
MOSPF	Multicast Open Shortest Path First
MOSROM	Metal Oxide Semiconductor Read Only Memory
MOSS	Mathematical Optimization Subroutine System
MOSS	MIME Object Security Services
MOSS	Modelling Surfaces
MOST	Management Operation System Technique
MOST	Metal Oxide Semiconductor|Silicon Transistor
MOST	Modular Office System Terminal
MOT	Managed Object to Test
MOT	Means Of Testing
MOT	Multiple Observation Time
MOTCHK	Motion Check
MOTD	Message Of The Day
MOTI	Message Oriented Text Interchange
MOTIF	Message Oriented Text Information Function
MOTIS	Management Oriented Terminal Information System
MOTIS	Message Oriented Text Interchange System
MOTIS/MHS	Message Oriented Text Interchange System/Message Handling System
MOU	Memorandum Of Understanding
MOU	Months Of Usage
MOUTH	Modular Output Unit for Talking to Humans
MOV	Metal Oxide Varistor
MOV	Move (instruction)
MOVI	Mobile Visualization
MOVI	Move Immediate
MOVL	Moving Left
MOVR	Moving Right
MOVS	Move String
MP	Machine Processable
MP	Managing Process
MP	Massively Parallel (processing)
MP	Materials Planning
MP	Mathematical Programming
MP	Mechanical Part
MP	Medium Power
MP	Melting Point
MP	Message Processor
MP	Micro-Processor
MP	Mixed Projection
MP	Modem Port
MP	Modular Plug
MP	Multiple Processors

MP	Multiplex
MP	Multipoint
MP	Multi-Processor
MP	Multi-Programming
MP	Northern Mariana Islands
MPA	Multiple Peripheral Adapter
MPA	Multiple Precision Arithmetic
MPACS	Management Planning And Control System
MPACT	Micro-Processor Application to Control firmware Translator
MPAD	Multiple Adapter
MPAR	Micro-Program Address Register
MPAT	Mainframe Programmer/Analyst Track
MPB	Media Parameter Block
MPC	Manufacturing Process Control
MPC	Micro-Program Control|Counter
MPC	Modular Peripheral interface Converter
MPC	Multimedia Personal Computer
MPC	Multi-Path Channel
MPC	Multi-Processing
MPC	Multi-Process(or) Communications
MPCC	Micro-Processor Common Control
MPCC	Micro-Programmable Communications Controller
MPCC	Multi-Protocol Communications Controller
MPCCG	Maintenance and Program Control Console Groups
MPCD	Minimum Perceptible Chromaticity Difference
MPCD	Multiplicand
MPCF	Manufacturing Process Capability File
MPCH	Main Parallel Channel
MPCI	Multiport Programmable Communications Interface
MPCM	Micro-Program Control Memory
MPCN	Multiplication
MPCNC	Multi-Processor Computer Numerical Control(ler)
MPCP	Manufacturing Process Control Procedure
MPCR	Micro-Program Count Register
MPCS	Manufacturing Planning and Control System
MPCS	Manufacturing Process Control System
MPCS	Multi-Programming Control System
MPD	Magneto-Plasma Dynamics
MPD	Missing Pulse Detector
MPDC	Most Positive Down Current
MPDS	Message Processing and Distribution System
MPDT	Multi-Peer Data Transmission
MPDU	Message Protocol Data Unit
MP&E	Maintenance Planning & Execution
MPE	Maximum Permitted Error
MPE	Memory Parity Error
MPE	Monaural Phase Effects
MPE	Multi-Programming Executive

MPECS	Multi-Pass Expert Control System
MPEG	Motion Pictures Expert Group
MPES	Multi-Programming Executive System
MPF	Manufacturing Process Files
MPF	Master Part File
MPF	Master Process|Program File
MPG	Microwave Pulse Generator
MPG	Multimedia Presentation Generator
MPG	Multi-Programming
MPGS	Micro-Program Generating System
MPI	Message Passing Interface
MPI	Micro-Processor Interface
MPI	Multimedia & Photo-Imaging
MPI	Multiple Protocol Interface
MPI	Multi-Precision Integer
MPIF	Message Passing Interface Forum
MPIF	Multi-Processor Interface
MPIN	Mapping Parcel Identification Number
MPL	Machine Processable Language
MPL	Macro-Procedure Language
MPL	Message Processing Language
MPL	Micro-Programming Language
MPL	Multimedia Presentation Language
MPL	Multischedule Private Line
MPLE	Multiple
MPLR	Multiplier
MPLS	Multi-Protocol Label|Layer Switching
MPLT	Multiple Programmable Light Table
MPLXR	Multiplexer
MPLY	Multiply
MPM	Message Passing Method
MPM	Metra Potential Method
MPM	Micro-Program Memory
MP/M	Multi-Programming control system for Microcomputer
MPM	Multi-Programming for Microprocessors
MPM	Multi-Programming Monitor
MPM	Multi-user Program for Microcomputers
MPMC	Micro-Processor Memory Controller
MPMCU	Micro-Program Memory Control Unit
MPMI	Multi-Port Memory Interface
MPO	Maximum Power Output
MPO	Memory Protect Override
MPOA	Multi-Protocol Over ATM (= Asynchronous Transfer Mode)
MPOS	Micro-Processor Operating System
MPOS	Multi-Programming Operating System
MPOW	Multiple Purpose Operator Workstation
MPP	Massive(ly) Parallel Processor
MPP	Message Posting Protocol

MPP	Message Processing Program
MPP	Multi-Programmable Processor
MPP	Multiprogrammable Processor Port
MPPD	Multi-Purpose Peripheral Device
MPPL	Multi-Purpose Programming Language
MPPR	Monthly Performance and Progress Report
MPQP	Multi-Protocol Quad Port
MPR	Make Pages Resident
MPR	Marked Page Reader
MPR	Micro-Program Register
MPR	Model Parts Release
MPR	Monthly Program Review
MPR	Multi-Part Repeater
MPR	Multi-Protocol Router
MPRF	Master Parts Record File
MPRG	Microprogram
MPRN	Multi-Protocol Router Network
MPROM	Mask Programmed Read Only Memory
MPRS	Master Parts Record System
MPS	Macro-Processing System
MPS	Master Production Schedule
MPS	Mathematical Programming System
MPS	Medium Power Standard
MPS	Micro-Processor System
MPS	Micro Program Storage
MPS	Mid Pack System
MPS	Modular Processing System
MPS	Modulator Power Supply
MPS	Multimedia Presentation System
MPS	Multiple Partition Support
MPS	Multiple Processor System
MPS	Multi-Processing System
MPS	Multi-Processor Specification
MPS	Multi-Programming System
MPSA	Multi-Processor Server Architecture
MPSA	Multi-Protocol Server over ATM (= Asynchronous Transfer Mode)
MPSCC	Micro-Program Scan Count Check
MPSCL	Mathematical Programming System Control Language
MP/SCM	Multi-Port Semi-Conductor Memory
MPSM	Multi-Programming Set Manager
MPSR	Multi-Product, Single-Row
MPSS	Multi-Purpose System Simulator
MPSX	Mathematical Programming System Extended
MPT	Maximum Power Transfer
MPT	Memory Processing Time
MPT	Micro-Probe Tester
MPT	Monolithic Packaging Technology

MPT	Multi-Port Transceiver
MPTF	Micro-Program Temporary Fix
MPTM	Multi-Party Test Method
MPTN	Multi-Protocol Transport Network
MPTR	Message Pointer
MPTS	Multi-Protocol Transport Services
MPU	Main Power Unit
MPU	Maintenance Processor Unit
MPU	Memory Protection Unit
MPU	Micro-Processor Unit
MPU	Multi-Processor Unit
MPUL	Most Positive Up Level
MPVL	Matrix Padé Via a Lanczos-type process
MPW	Macintosh Programmer's Workshop
MPX	Multiplex(er)
MPX	Multi-Programming Executive
MPXG	Multiplexing
MPXR	Multiplexer
MPY	Multiply
MQ	Martinique
MQ	Message Queue
MQ	Multiplier/Quotient (register)
MQCR	Master Queue Control Record
MQE	Message Queue Element
MQEs	Managed Query Environments
MQFP	Metric Quad-Flat Package
MQG	Multithreaded Query Gate(way)
MQH	Message Queue Handler|Header
MQI	Message Queuing Interface
MQID	Message Queue Identification
MQL	Mean Queue Length
MQN	Message Queue Name
MQR	Multiplex(er) Quotient Register
MQRC	Master Queue control Record
MQS	Message Queue Server
MR	Magnetic Resonance
MR	Magneto-Resistance
M&R	Maintainability & Reliability
M&R	Maintenance & Repair
MR	Maintenance Report
MR	Map Reference
MR	Mark Reading|Recognition
MR	Mask Register
MR	Master Reset
MR	Matching Record
MR	Mauritania
MR	Memorandum Report
MR	Memory Read|Register

MR	Message Requester|Retrieval
MR	Miniatures Rules
MR	Minimal Reality (toolkit)
MR	Mixture Ratio
MR	Modem Ready (toolkit)
MR	Modular Redundancy
MR	Monthly Rental
MR	Multiple Register|Regression
MR	Multiple Requesting
MRA	Multiple Regression Analysis
MRAC	Model Reference Adaptive Control
MRACS	Model Reference Adaptive Control System
mrad	milliradian
MRB	Modification Review Board
MRC	Machine Readable Character|Code
MRC	Magnetic Recording Channel
MRC	Magnetic Rectifier Control
MRC	Master Resident Core
MRC	Memory Request Controller
MRCF	Microsoft Real-time Compression Format
MRCI	Microsoft Real-time Compression Interface
MRCS	Multiple Report Creation System
MRCT	Modified Residue Calculus Technique
MRCU	Multi-Remote Control Unit
MRD	Memory Read
MRDF	Machine Readable Data File
MRDOS	Mapped Real-time Disk Operating System
MRDY	Message Ready
MRE	Measuring Recording Equipment
MRE	Memory Register Exponent
MREGAD	Multiplexer Regenerator Address
MREN	Metropolitan Research and Education Network
MRF	Maintenance Reset Function
MRF	Master Record File
MRF	Medium-Range Forecast
MRF	Multipath Reduction Factor
MRG	Medium Range
MRG	Multi-Resolution Geometry
MRH	Magneto-resistive Recording Head
MRH	Maximum Rectangular Hierarchy
MRI	Machine-Readable Information
MRI	Magnetic Resonance Imaging
MRI	Memory Reference Instruction
MRIO	Multi-Regional Input/Output
MRIR	Medium Resolution Infra-Red (measuring)
MRJE	Multileaving Remote Job Entry
MRJE/WS	Multileaving Remote Job Entry/Work Station
MRL	Machine Representation Language

MRM	Machine-Readable Material|Medium
MRM	Maximum Right Mask
MRM	Most Recently-used Master
MRNet	Minnesota Regional Network
MRO	Maintenance, Repair, and Operating (supplies)
MRO	Maintenance, Replenishment, and Operating (materials)
MRO	Maintenance Replenishment Order
MRO	Multiple Rank Order
MRO	Multi-Regional Operation
MROM	Macro Read-Only Memory
MRP	Manufacturing Records Processing
MRP	Manufacturing Report Processor
MRPC	Multi-Regional Processing Center
MRP-II	Manufacturing Resource Planning
MRPL	Main Ring Path Length
MRPN	Multi-Protocol Router Network
MRQ	Master Request
MRR	Mode Request Register
MRR	Molecular Rotational Resonance
MRR	Multiple Response Resolver
MRS	Management Reporting System
MRS	Master Release Sequence
MRS	Media Recognition System
MRS	Message Routing System
MRS	Micro Records System
MRS	Modular Retrieval System
MRS	Music Reading Software
MRSAP	Magnetic Recording System Analysis Program
MRSE	Message Retrieval Service Element
MRST	Master Reset
MRT	Master Relocatable|Reply Tape
MRT	Maximum|Mean Repair Time
MRT	Multiple Requester Terminal
MRTMS	Mini Real-Time Monitor System
MRTS	Modification Request Tracking System
MRU	Minimum Resolvable Unit
MRU	Most Recently Used
MRV	Monthly Requirement Value
MR/W	Multiple Read/Write
MS	Machine Sketches
MS	Main Storage
MS	Maintenance and Service
MS	Maintenance State
MS	Management Science
MS	Margin of Safety
MS	Mark Scanning|Sensing
MS	Mass Storage
MS	Master Schedule(r)

MS	Master Slave|Slice
MS	Match Station
MS	Mean Square
MS	Medium Shot|Speed
MS	Memory System
MS	Message Store|Switching
MS	Messaging Services
MS	Metric System
ms	microsecond
MS	Microsoft (corporation)
ms	millisecond
MS	Mississippi (US)
MS	Mobile Station
M&S	Modelling & Simulation
MS	Molecular Stuffing (process)
MS	Montserrat
MS	More Segments
MS	Multiplexer Storage
MSA	Mass Storage Adapter
MSA	Measurement Systems Analysis
MSACM	Microsoft Audio Compression Manager
MSAM	Multiple Sequential Access Method
MSAP	MAC Service Access Point (MAC = Multi-Access Computer)
MSAP	Message Store Access Protocol
MSAU	Multi-Station Access Unit
MSAV	Microsoft Anti Virus
MSB	Machine Status Byte
MSB	Most Significant Bit
MSBF	Mean Swaps Between Failures
MSBR	Maximum Storage Bus Rate
MSBY	Most Significant Byte
MSC	Main Storage Capacity|Controller
MSC	Maintenance Service Contract
MSC	Management Service Center
MSC	Massachusetts Software Council
MSC	Mass Storage Control(ler)
MSC	Media Stimulated Calling
MSC	Memory System Controller
MSC	Message Sequence Chart
MSC	Message Switching Center|Computer
MSC	Message Switching Concentrator
MSC	Microsoft C
MSC	MIDI Show Control (MIDI = Musical Instrument Digital Interface)
MSC	Minnesota Supercomputer Center
MSC	Mobile Switching Center
MSC	Monolithic Storage Cell
MSC	Multimedia Super-Corridor

MSC	Multiple Systems Coupling
MSC	Multi-System Coupling
MSCDEX	Microsoft Compact Disc Extensions
MSCE	Main Storage Control Element
MSCE	Microsoft Site Commerce Edition
MSCM	Multi-System Configuration Manager
MSCP	Mass Storage Control Protocol
MSCS	Microsoft Cluster Server
MSCTC	Mass Storage Control Table Create
MSCU	Modular Store Control Unit
MSCW	Marked Stack Control Word
MSD	Mass Storage Device
MSD	Master Standard Data
MSD	Microsoft System Diagnostics
MSD	Microwave Semiconductor Device
MSD	Modem Sharing Device
MSD	Most Significant Digit
MSDB	Main Storage Data Base
MSDE	Microsoft Data Engine
MS/DOS	Microsoft Disk Oriented System
MSDR	Main Storage Data Register
MSDR	Multiplexed Streaming Data Request
MSDR	Multiplexer Storage Data Register
MSDS	Message Switching Data Service
MSDS	Microsoft Developer Support
MSDTR	Multi-Speed Digital Tape Recorder
MSE	Mean Square Error
MSE	Mobile Subscriber Equipment
msec	millisecond(s)
MSEL	Master Scenario Event List
MSERV	Macro library Service
MSF	Main Storage Frame
MSF	Mass Storage Facility
MSF	Master Schedule|Structure File
MSF	Master Supplier File
msf	minutes, seconds and frames
MSF	Multiservice Switching Forum
MSFR	Minimum Security Function Requirements
MSG	Maintenance Service Group
msg	program message (file extension indicator)
MSGFLG	Message Flag
MSG/WTG	Message Waiting
MSHER	Master Sheet Reference
MSHF	Matrix Switch Host Facility
MSHP	Maintain System History Program
MSHP	Maintenance and Service History Program
MSI	Medium Scale Integration
MSIG	Most Significant

MSIO	Mass Storage Input/Output
MSIR	Machine Survey and/or Installation Report
MSIS	Multi-State Information System
MSISDN	Mobile Station Integrated Services Digital Network
MSIT	Multi-Source Interconnect Tree
MSK	Mask
MSK	Minimal Shift Keying
MSL	Machine Specification Language
MSL	Map Specification Library
MSL	Mathematical Subprogram Library
MSL	Member Specification Library
MSL	Mirrored Server Link
MSL	Motor Simulation Laboratory
MSLC	Multi-Sub-Loop Control
MSLS	Medium Speed Local Store
MSLS	Multi-terminal Shared Logic System
MSLT	Monolithic Solid Logic Technology
MSM	Main Storage Management|Module
MSM	Matrix Switch Module
MSM	Memory Storage|Support Module
MSM	Monolithic Support Module
MSM	Multilevel Storage Model
MSM	Multiple Speed Modem
MSML	Microsoft Markup Language
MSMQ	Microsoft Message Queue server
MSN	Message Switching Network
MSN	Microsoft Network
MSN	Multiple Subscriber Number
MSNF	Multi-System Networking Facility
MSO	Manufacturers Statement of Ownership
MSO	Media Services Operator
MSO	Multimedia Services Operator
MSO	Multiple-Systems Operator
MSO	Multi-Stage Operation(s)
MSOS	Mass Storage Operating System
MSP	Maintenance Spare Parts
MSP	Management of Software Projects (international workshop)
MSP	Manual Switching Position
MSP	Manufacturing Software Package
MSP	Mass Storage Pedestal
MSP	Message Security Protocol
msp	microsoft paint (file extension indicator)
MSP	Modular System Program
MSP	Most Significant Position
MSP	Multiple Summation Processor
MSPISO	Member Service Provider, Independent Sales Organization
MSR	Machine State Register
MSR	Magnetic Send/Receive

MSR	Magnetic Stripe Reader
MSR	Main Status Register
MSR	Mark Sense Reading
MSR	Mark Sheet Reader
MSR	Mark-to-Space Ratio
MSR	Mechanized Storage and Retrieval
MSR	Mode Status Register
MSR	Multitrack Serpentine Recording
MSRJE	Multiple Session Remote Job Entry
MSRT	Master Standard Reference Tape
MSRT	Mini System Real-Time
MSS	Magnetic Strip Storage
MSS	Managed Security Services
MSS	Management Support Service|System
MSS	Mass Storage System
MSS	Maximum Segment Size
MSS	Message Switching System
MSS	Metal Silicone Silicon
MSS	Metropolitan Switching System
MSS	Mobile Satellite Service
MSS	Modem Substitution Switch
MSS	Modular Software System
MSS	Multiple Secondary and Selection
MSS	Multiservice Satellite System
MSSB	Multiple Supplier System Bulletin
MSSC	Mass Storage System Communicator
MSSC	Medium Speed Store and Compare
MSSE	Mass Storage System Extension
MSSE	Message Submission Service Element
MSSSE	Message Submission and Storage Service Element
MST	Macro Symbol Table
MST	Mark-to-Space Transition
MST	Mass Storage Tape
MST	Master Station
MST	Mean Service Time
MST	Medium Speed Terminal
MST	Micro-Systems Technology
MST	Minimum Spanning Tree
MST	Modular Semiconductor Technology
MST	Monolithic Solid Technology
MST	Monolithic System Technology
MSTDA	Monolithic System Technology Design Automation
MSTOR	Main Storage
MSTR	Master
MSU	Mass|Memory Storage Unit
MSU	Message Signal Unit
MSU	Modem-Sharing Unit
MSUS	Minnesota State University System

MSV	Mass Storage Volume
MSV	Multi-Service Vendor
MSVC	Mass Storage Volume Control
MSVC	Meta-Signalling Virtual Channel
MSVC	Microsoft Visual C++
MSW	Machine Status Word
MS/WG	Module Select/Write Gate
MSX	MicroSoft Extended
MSYN	Master Synchronization
MT	Machine Translation|Type
MT	Magnetic Tape
MT	Malta
MT	Maximum Total
MT	Measured Time
MT	Mechanical Translation
MT	Message Transfer
MT	Mode Transducer
MT	Montana (US)
MT	Motor Terminal
MT	Multiple Tracks|Transfer
MT	Multi-Tasking
MTA	Mail Transfer Agent
MTA	Message Transfer Agent|Architecture
MTA	Motion Time Analysis
MTA	Multiple Terminal Access
MTAC	Mathematical Tables and other Aids to Computation
MTAE	Message Transfer Agent Entity
MTAM	Multileaving Telecommunications Access Method
MTAP	Machine Timing Analysis Program
MTAU	Metallic Test Access Unit
MTB	Maximum Theoretical Bandwidth
MTBB	Mean Time Between Breakdowns
MTBCD	Mean Time Between Confirmed Defects
MTBCF	Mean Time Between Component Failure
MTBCF	Mean Time Between Critical Failures
MTBDD	Multi-Terminal Binary Decision Diagram
MTBE	Mean Time Between Errors
MTBF	Mean Time Before|Between Failure(s)
MTBI	Mean Time Between Interrupts
MTBJ	Mean Time Between Jams
MTBM	Mean Time Between Maintenance
MTBO	Mean Time Between Outages
MTBO	Mean Time Between Overhauls
MTBR	Mean Time Between Repair(s)
MTBS	Mean Time Between System breakdown
MTBSE	Mean Time Between Software Errors
MTBSF	Mean Time Between Significant Failures
MTBUM	Mean Time Between Unscheduled Maintenance

MTBUR	Mean Time Between Unscheduled Removal
MTC	Magnetic Tape Cassette|Channel
MTC	Magnetic Tape Command|Control
MTC	Magnetic Tape Controller
MTC	Maintenance Time Constraint
MTC	Master Table of Contents
MTC	Message Transmission Controller
MTC	MIDI Time Code (MIDI = Musical Instrument Digital Interface)
MTC	Minor Transaction Code
MTC	Monostable Trigger Circuit
MTCA	Multiple Terminal Communications Adapter
MTCB	Master Timer Control Block
MTCN	Minimum Throughput Class Negotiation
MTCS	Minimum Teleprocessing Communications System
MTCU	Magnetic Tape Control Unit
MTD	Mean Time Down
MTD	Month To Date
MTD	Mounted
MTD	Multimodal Transport Document
MTDB	Machine Type Data Base
MTDC	Modified Total Direct Costs
MTDS	Manufacturing Test Data System
MTDT	Memory Technology, Design & Testing (international workshop)
MTE	Machine Transaction Entry
MTE	Monostable Trigger Element
MTE	Multiple Terminal Emulator
MTEL	Macro-Time Event List
MTEL	Mobile Telecommunications technologies corporation
MTEX	Multi-Threading Executive
MTEXT	Multiline Text
MTF	Magnetic Thin Film
MTF	Master Test Frame
MTF	Mean Time to Failure
MTF	Message Transfer Facility
MTF	Microsoft Tape Format
MTF	Modulation Transfer Function
MTFBWY	May The Force Be With You
MTFP	Message Transfer Facility Program
MTG	Machine Type Group
MTG	Meeting
MTG	Mounting
MTH	Magnetic Tape Handler
MTH	Month
MTHD	Method
MTI	Mission Time Improvement
MTI	Moving Target Indicator

MTI	Multichannel Time Intervalometer
MTI	Multi-Terminal Interface
MTIAC	Manufacturing Technology Information Analysis Center
MTID	Machine Type Identification Data
MTIF	Mission Time Improvement Factor
MTL	Merged|Mixed Transistor Logic
MTL	Message Transfer Layer
MTM	Machine Type Model
MTM	Methods Time Management
MTM	Model Test-Model
MTM	Multiple Terminal Manager
MTM	Multiple Time Measurement
MTMOD	Magnetic Tape Module
MTMS	Machine Tool Management System
MTMT	Multiple Terminal Monitor Task
MTN	Message Transport Network
MTNS	Metal Thick-Nitride Semiconductor
MTO	Master Terminal Operator
MTO	Multi-Task Operation
MTOAD	More Than One Address Detected
mtops	million of theoretical operations per second
MTOS	Magnetic Tape Operating System
MTOS	Metal Thick-Oxide Semiconductor
MTOS	Multi-Tasking Operating System
MTP	Mass Transaction Processing
MTP	Mean Term Planning
MTP	Message Transfer Part
MTP	Message Transfer Protocol
MTPR	Miniature Temperature Pressure Recorder
MTPT	Minimal Total Processing Time
MTR	Magnetic Tape Reader|Recording
MTR	Message Trailer
MTR	Microwave Thermal Radiation
MTR	Miniature Temperature Recorder
MTR	Minimum Time Requirement
MTR	Modular Tree Representation
MTR	Monitor
MTR	Multiple Token Ring
MTR	Multiple Track Range
MTRCB	Master Timer Control Block
MTRON	Macro TRON (= The Real-Time Operating system Nucleus)
MTRR	Magnetic Tape Recorder/Reproducer
MTRR	Magnetic Tape Remote Record
MTRS	Magnetic Tape Reformatting System
MTS	Magnetic Tape Station|Storage
MTS	Message Telecommunication Service
MTS	Message Transfer Service|System
MTS	Microsoft Transaction Server

MTS	Mobile Telephone Service
MTS	Modular Training|Television System
MTS	Module Testing System
MTS	Multichannel Television Sound
MTS	Multiple Terminal|Transient System
MTS	Multipoint Terminal Software
MTSA	Mantissa
MTSC	Magnetic Tape Selectric Composer
MTSE	Message Transfer Service Element
MTSF	Mean Time to System Failure
MTSI	Mean Time to System Interrogation
MTSL	Message Transfer Sub-Layer
MTSN	Machine Type and Serial Number
MTSO	Mean Time to Switch-Over
MTSO	Mobile Telephone Switching Office
MTSO	Multiprogramming Time-Sharing Operating system
MTSR	Mean Time to System Restoration
MTST	Magnetic Tape Selectric Typewriter
MTT	Magnetic Tape Terminal|Typewriter
MTT	Maritime Telephone and Telegraph
MTT	Message Transfer Time
MTT	Microprogram Trace Tape
MTT	Microsoft Travel Technologies
MTTA	Multi-Tenant Telecommunications Association (US)
MTTD	Mean Time To Diagnose
MTTF	Mean Time To Failure
MTTFF	Mean Time To First Failure
MTTFSF	Mean Time To First System Failure
MTTFSR	Mean Time To First System Repair
MTTM	Mean Time To Maintenance
MTTR	Maximum Time To Repair
MTTR	Mean Time To Repair|Restore
MTTS	Multi-Task Terminal System
MTTSF	Mean Time To System Failure
MTTSR	Mean Time To Service Restoration
MTTU	Magnetic Tape Transmission Unit
MTU	Magnetic Tape Unit
MTU	Maintenance Termination Unit
MTU	Maximum Transfer|Transmission Unit
MTU	Maximum Transmission Unit
MTU	Mean Time Up
MTU	Media Tech Unit
MTU	Memory Transfer Unit
MTU	Metric Units
MTU	Modem Transfer Unit
MTU	Multiplex(er) Terminal Unit
MTV	Modem To Voice
MTW	Minimum Trimmed Width

MTX	Matrix
MTX	Mobile Telephone Exchange
MTX	Multi-Tasking Executive
MTX	Multi-Terminal Executive
MU	Machine Unit|Utilization
MU	Mauritius
MU	Memory|Message Unit
mu	micro(meter)
MU	Missing Upper
MU	Multiple Unit
MUA	Mail User Agent
MUBUS	Microprocessor Bus
MUCOM	Multi-sensory Control of Movement (EU)
MUD	Master User Directory
MUD	Multi-User Dialogue|Domain
MUD	Multi-User Dimension|Dragon
MUD	Multi-User Dungeon
MUDAID	Multivariate, Univariate and Discriminant Analysis of Irregular Data
MUDDC	Multi-Unit Direct Digital Control
MUEXEC	Multi-User Executive
MUF	Maximum Usable Frequency
MUFF	Multi-User File Format
MUI	Magic User Interface
MUL	Multiply
muldex	multiplexer/demultiplexer
MULS	Multi-Slide
MULTI	Multiple
MULTICS	Multiplexed Information and Computing Service
MUM	Multi-User Monitor
MUMBLE	Multiple User Multicast Basic Language Exchange
MUMPS	Massachusetts Utility Multi-Programming System
MUO	Machine Used On
MUP	Multiple Uniform naming convention Provider
MUP	Multi-Processor
MUPO	Maximum Undistorted Power Output
MUPS	Multi-Processing System
MUR	Multi-Use Register
MURS	Machine Utilization Reporting System
MURTS	Multiple User Remote Terminal Supervisor
MUS	Multiprogramming Utility System
MUS	Multi-User System
MUSA	Multiple Unit Steerable Antenna
MUSC	Multi-Unit Supervisory Control
MUSE	Machine User Symbolic Environment
MUSE	Meditech Users Software Exchange
MUSE	Multidimensional User-oriented Synthetic Environment
MUSE	Multiple (sub-Nyquist) Sampling Encoder

MUSE	Multi-User Shared Environment
MUSH	Mail Users Shell
MUSI	Multi User Shared Illusion
MUSIC	Machine Utilization and Statistical Information Collection
MUSIC	Multiple Signal Classification
MUSIC	Multi-User System for Integrated Control
MUSIL	Multiprogramming Utility System Interpretive Language
MUST	Multipurpose User-oriented Software Technology
MUSYK	Multilevel planning and control System for one-of-a-Kind production
MUT	Monitor Under Test
MUTEX	Multi-User Terminal Executive
MUTEX	Multi-User Transactional Executive (system)
MUTEX	Music TeX
MUTEX	Mutually Exclusive
MUTT	Multi-Use Terminal Translator
MUX	Multiplex(er)
MV	Maldives
MV	Mean Value|Variation
MV	Measured Value
MV	Megavolt
mv	millivolt
MV	Multi-Vibrator
MVA	Machine Vision Association
MVA	Megavolt-Ampere
MVB	Multimedia Viewer Book
MVC	Model View Controller
MVC	Move Character
MVC	Multimedia Viewer Compiler
MVCOM	Move to Communication (region)
MVCU	Multi-Variable Control Unit
MVDM	Multiple Virtual DOS Machines (DOS = Disk Operating System)
MVDS	Modular Video Data System
MVF	Multi-Volume File
MVFG	Multi-Variable Function Generator
MVGA	Monochrome Video Graphics Array
MVI	Microgravity Vestibular Investigation
MVID	Major Vector Identification
MVIP	Multi-Vendor Integration Protocol
MVM	Manager Virtual Machine
mV/m	millivolts per meter
MVM	Minimum|Multiple Virtual Memory
MVMUA	Metropolitan Virtual Machine Users Association
MVP	Millivolt Potentiometer
MVP	Multiline Variety Package
MVP	Multimedia Video Processor
MVS	Multiple Virtual Storage|System

MVS/ESA	Multiple Virtual Storage/Enterprise System Architecture
MVSPC	Multi-Variate Statistical Process Control
MVS/SE	Multiple Virtual Storage/System Extensions
MVS/SP	Multiple Virtual Storage/System Product
MVS/TSO	Multiple Virtual Storage/Time-Sharing Option
MVS/XA	Multiple Virtual Storage/Extended Architecture
MVT	Multiprogramming with a Variable number of Tasks
MVZ	Move Zones
MW	Machine Word
MW	Malawi
MW	Manual Word
MW	Man Week
MW	Medium Wave
MW	Megawatt
MW	Memory Write
MW	Micro-Wave
MW	Middle Ware
MW	Million Words
mW	milliwatt
MW	Multi-Wink
MW	Music Wire
MWC	Multi-Wire Cable
MWG	Maintenance Working Group
MWh	Megawatt-Hour
MWI	Message Waiting Indicator
MWM	Motif Window Manager
MWR	Magnetic tape Write
mWRTL	milliwatt Resistor-Resistor Logic
MWS	Multi-Work Station
MWSR	Microwave Water Substance Radiometer
MX	Mail Exchanger
MX	Matrix
MX	Mexico
MX	Multiplex
MXA	Main Exchange Area
MXC	Multiplexer Channel
MXM	Matrix Memory
MXR	Mark Index Register
MXU	Multiplexer Unit
MY	Malaysia
MYAP	Mask limited Yield Analysis Program
MYOB	Mind Your Own Business
MZ	Mozambique
MZC	Minimum Zone Circle
MZR	Multiple Zone Recording

N

n	index of refraction
n	nano (prefix)
N	Negative
N	Neutral
N	New (issue)
N	Node
N	Normal
n	number
NA	Namibia
NA	Next Address
NA	No Action
NA	Not Accurate
NA	Not And
NA	Not Applicable|Assigned
NA	Not Authorized|Available
NA	Numerical Aperture
NAA	Name And Address
NAASTOR	Non-Addressable Auxiliary Storage
NAAUG	North American Autocad User's Group
NAB	National Association of Broadcasters (US)
NABTS	North American Broadcast Teletext Standard
NAC	Negative Acknowledge Character
NAC	Network Access Control(ler)
NAC	Network Adapter Card
NAC	Network Administration Center
NAC	Network Applications Consortium
NACCB	National Association of Computer Consultant Businesses (US)
NACCIRN	North American Coordinating Committee for Intercontinental Research Network
NACD	National Association of Computer Dealers (US)
NACK	Negative Acknowledgment
NACN	North American Cellular Network
NACOMEX	National Computer Exchange (US)
NACS	National Advisory Committee on Semiconductors (US)
NACS	NetWare Asynchronous Communications Server
NACT	Neural Adaptive Control Technology (EU)
NAD	Network Access Device
NAD	No Apparent Defect
NADAC	Navigation Data Computer
NADUG	National Data manager Users' Group (US)
NAE	Not Above or Equal
NAEB	North American EDIFACT Board
NAEC	Novell Authorized Education Center
NAF	Network Access Facilities
NAFIN	Netherlands Armed Forces Integrated Network

NAG	Network Advisers|Architecture Group
NAG	Numerical Algorithms Group
NAI	No Action Indicated
NAICC	National Association of Independent Computer Companies (US)
NAK	Negative Acknowledgment (character)
NAL	NetWare Application Launcher
NAL	New Assembly Language
NALOPKT	Not A Lot Of People Know That
NAM	Name and Address Module
NAM	National Average Maintenance
NAM	Network Access Machine|Method
NAM	Number Assignment Module
NAMPS	Narrow-band Analog Mobile Phone Service
NAMS	Network Analysis and Management System
NAN	National Area Network (US)
NAN	Network Application Node
NANA	Novel parallel Algorithms and New real-time (very large scale integration) Architectural methodologies (EU)
NANCAR	Nano-lithography using Chemically Amplified Resists (EU)
NAND	Negative AND (function)
NAND	Not-AND (function)
nano	billionth
NANO	Nonlinear AND function, Nonlinear OR function
NANOPT	Nanometer structures for future Optoelectronic applications (EU)
NANP	North American Numbering Plan
NAP	Network Access Protocol|Point(s)
NAP	Noise Analysis Program
NAPLPS	North American Presentation Level Protocol Syntax (graphics)
NAPSS	Numerical Analysis Problem Solving System
NAPT	Network Address Port Translation
NAR	Name, Address, Residence
NAR	Narrow
NAR	Net Annual Requirement
NAR	No Action Required
NARA	National Archives and Records Administration (US)
NARTB	National Association of Radio and Television Broadcasters (US)
NARTE	National Association (of) Radio (and) Telecommunications Engineers
NAS	Narrow Angle Scan
NAS	National Aerospace Standards
NAS	Network Acronym Server
NAS	Network Application Support
NAS	Network-Attached Storage
NAS	Numerical Aerodynamic Simulation
NAS	Numerical and Atmospheric Sciences network
NASA	National Aeronautics and Space Administration (US)

NASAP	Network Analysis for Systems Application Program
NASCOM	NASA Communications (network)
NASDAC	National Aviation Safety Data Analysis Center
NASI	NetWare Asynchronous Services Interface
NASIRC	NASA Automated Systems Internet Response Capability
NASKER	NASA Ames Kernel (benchmark)
NASREM	NASA-NIST Standard Reference Model
NASTD	National Association of State Telecommunications Directors (US)
NASTRAN	NASA Structural Analysis (program)
nat	natural
NAT	Natural unit (information content)
NAT	Network Address Translation
NAT	No Action Taken
NAT	Nonlinear and Adaptive Techniques (EU)
NATA	National Association of Testing Authorities (US)
NATA	North American Telecommunications Association
NATA	Numerical Analysis Thermal Application
NATD	National Association of Telecommunication Dealers (US)
NATS	Negative Authorization Terminal System
NATURE	Novel Approaches to Theories Underlying Requirements Engineering (EU)
NAU	Network Access(ible) Unit
NAU	Network Addressable Unit
NAUN	Nearest Active|Addressable Upstream Neighbour
NAV	Norton Anti-Virus
NAVA	Non-Added Value Activity
NAVAIDS	Navigation Aids
NAVCOM	Naval Communications
NAVCOMSAT	Naval Communications Satellite
NAVNET	Naval Network
NAVSAT	Navigation Satellite
NAVSTAR	Navigational Satellite Timing|Tracking And Ranging
NAW	National Average Workload
NAW	Net Available Workload
NB	Narrow Band
NB	Negative Balance
NB	No Branch
NB	Noise Block
NB	Number of Bytes
NBA	Narrow Band Allocation
NBAC	Negative Balance All Cycles
NBCD	Natural Binary Coded Decimal
NBD	No Big Deal
NBDL	Narrow-Band Data Line
NBDP	Narrow-Band Direct Printing
NBE	Not Below or Equal
NBEC	Non-Bell Exchange Carrier

NBFM	Narrow-Band Frequency Modulation
NBFS	New Balanced File organization Scheme
NBH	Network Busy Hour
NBMA	Non-Broadcast Multiple Access
NBNC	Noted But Not Corrected
NBO	Network Build Out
NBO	Network Business Opportunity
NBP	Name Binding Protocol
NBPM	Narrow Band Phase Modulation
NBR	Number
NBRS	Next Basic Records System
NBS	National Bureau of Standards (US)
NBS	Numeric Backspace character
NBSP	Non-Breaking Space
NBT	Negative Balance Test
NBTDR	Narrow-Band Time-Domain Reflectometry
NBU	Natural Business Unit
NBVM	Narrow-Band Voice Modulation
nbw	noise bandwidth
NC	Narrow Coverage
NC	Network Computer|Control
NC	Network Connection
NC	New Caledonia
NC	No Card|Carry
NC	No Change|Charge
NC	No Connection|Cost
NC	Normally Closed
NC	North Carolina (US)
NC	Numerical Code|Control(led)
NCA	Network Communications Adapter
NCA	Network Computer Architecture
NCA	Network Control Analysis
NCA	Non-Contractual Authorization
NCAM	Network Communication Access Method
NCAPI	Netscape Client Application Programmers Protocol
NCB	National Computer Board (SG)
NCB	Network Control Block
NCB	Nickel Cadmium Battery
NCBI	National Center for Biological Information
NCC	National Computer|Computing Center (GB)
NCC	National Computer Conference
NCC	Network Control Center
NCC	New Common Carrier
NCC	Normally Closed Contact
NCCE	Northwest Council for Computer Education
NCCF	Network Communications Control Facility
NCCL	National Commission for Communications and Liberalization (US)

NCCP	National Concealed Carry Policy
NCD	Network Computing Device(s)
NCD	Network Cryptographic Device
NCDC	National Climatic Data Center (US)
NCDE	No Code
NCDS	Numerical Control Distribution System
NCE	Network Connection Element
NCET	National Council for Educational Technology (GB)
NCF	National Capital Free net (CA)
NCF	National Communications Forum (US)
NCF	NetWare Command File
NCG	Numerical Control Graphics
NCGA	National Computer Graphics Association (US)
NCGIA	National Center for Geographic Information and Analysis
NCH	Network Connection Handler
NCI	Non-Coded Information
NCIC	National Cartographic Information Center (US)
NCIC	National Crime Information Center (US)
NCIC	Network Control Interface Channel
NCITS	National Committee for Information Technology Standards (US)
NCL	Network Control Language
NCL	Node Control Logic
NCL	Numerical Control Language
NCM	Network Connection Management
NCM	Network Control Module
NCM	Non-Conforming Material
NCM	Normalized Correlation Method
NCM	Numerical(ly) Control(led) Machine
NCMC	Non-Conforming Material Control
NCMOS	N-Channel Metal Oxide Silicon
NCMS	Network Connection Management Sub-protocol
NCMS	Network Control and Management System
NCMT	Numerical Control for Machine Tools
NCN	Network Control Node
NCN	New Converged Network
NCND	Non Circumvent Non Disclosure
NCO	Number Controlled Oscillator
NCO	Numerically Controlled Oscillator
NCON	Name Constant
NCOP	Network Code Of Practice
NCOS	Non-Concurrent Operating System
NCP	NetWare Core Protocol
NCP	Network Control Point|Processor
NCP	Network Control Program|Protocol
NCP	Not Copy Protected
NCPAS	National Computer Program Abstract Service
NCPG	Numerical Control Part-program Generator

NCR	No Carbon Required (paper)
NCR-DNA	National Cash Registers corporation – Distributed Network Architecture
NCRI	Net Cash Rental Income
NCS	Name file Change Sheet
NCS	National Communications Systems
NCS	National Cryptologic School
NCS	Net Control Station
NCS	Network Computing|Control System
NCSA	National Center for Supercomputing Applications (US)
NCSA	National Computer Security Association (US)
NCSC	National Computer Security Center (US)
NCSI	Network Communications Services Interface
NCSL	National Computer Systems Laboratory (US)
NCSS	Non-Comment(ed) Source Statement
NCSTRL	Networked Computer Science Technical Reference Library
NCT	Network Control Terminal
NCTE	Network Channel Termination Equipment
NCTL	National Computer and Telecommunications Laboratory
NCTS	Non-Contacting Test System
NCU	Network Control Unit
NCU	Number Crunching Unit
NCUG	National Centrex Users' Group (US)
NCV	No Commercial|Core Value
ND	Network Development
ND	Network Digit|Disk
ND	Neutral Density
ND	No Date
ND	No Defects|Detect
ND	North Dakota (US)
ND	Not Dated|Desirable
NDA	Non-Disclosure Agreement
NDAC	Not Data Accepted
NDAM	New Disk Access Method
NDB	Node Data Base
NDB	Non-Directional Beacon(s)
NDBC	National Data Buoy Center (US)
NDBMS	Network Data Base Management System
NDBO	National Data Buoy Office (US)
NDC	National Data Center (US)
NDC	Negative Differential Conductivity
NDC	Network Diagnostic Control
NDC	Non-Destructive Cursor
NDC	Normalized Device Coordinates
NDCC	Network Data Collection Center
NDD	NetWare Directory Database
NDD	Non-Delivery Diagnostic
NDDE	Networking enabled Dynamic Data Exchange

NDDK	Network Device Development|Driver Kit
NDDL	Neutral Data Definition Language
NDE	Nonlinear Differential Equation
NDEF	Not to be Defined
NDF	No Defect Found
NDF	Non-Deterministic Fortran
NDGC	National Design Graphics Competition
NDHECN	North Dakota Higher Education Computer Network
NDI	Network Design & Installation
NDI	Non-Developmental Item
NDIR	Non-Dispersive Infra-Red
NDIS	National Document & Information Service (AU,NZ)
NDIS	Network Device|Driver Interface Specification
NDK	Network Development Kit
NDL	National|Network Database Language
NDLC	Network Data Link Control
NDM	Network Database Management
NDM	Normal(ly) Disconnected Mode
NDMS	NetWare Distributed Management Services
NDMS	Network Design and Management System
NDN	Non-Delivery Notification
NDOC	Network Design and Operations Center
NDP	Numeric Data Processor
NDR	Net Difference Report
NDR	Network Data Representation
NDR	Non-Destructive Read
NDRO	Non-Destructive Read-Out
NDROS	Non-Destructive Read-Only Storage
NDS	NetWare Directory Service
NDS	Network Data Series
NDS	Network Development System
NDS	Network Directory Services
NDS	New Display System
NDS	Non-temporary Data Set
NDS	Novell Directory Services
NDSTOR	Non-Destructive Storage
NDT	Net Data Throughput
NDT	Network Description Table
NDT	Normative Decision Theory
NDTS	Network Diagnostic and Test System
ndx	index (file extension indicator)
NE	Nebraska (US)
NE	Network Element
NE	Niger
NE	Non Exempt
NE	Not Editable
NE	Not Equal to
NEA	National Electronics Association (US)

NEAP	Novell Education Academic Partner
NEAR	National Electronic Accounting and Reporting system (US)
NEAT	National Electronic Autocoding Technique (US)
NEAT	New Eindhoven Architectural Toolbox (NL)
NEB	Noise Equivalent Bandwidth
NEBS	Network|New Equipment-Building System
NEBULA	Natural Electronic Business Language
NEC	National Electric(al) Code
NEC	Necessary
NEC	Never Ending Conflict
NEC	Non-Error Check
NEC	Not Elsewhere Classified
NECA	National Exchange Carrier Association (US)
NED	NASA Extragalactic Database
NED	No Expiration Date
NEDA	National Electronic Distributors' Association
NEDC	National Engineering Design Challenge
NEF	Network Element Function
NEF	Noise Equivalent Flux
NEF	Noise Exposure Forecast(s)
NEFS	Network Extensible File System
NEG	Negative
NEGT	Negate
NEI	Noise Equivalent Input
NEI	Not Elsewhere Indicated
NEM	Non-Erasable Medium
NEMA	National Electrical Manufacturers Association (US)
NEMI	National Electronics Manufacturing Initiative (US)
NEO	Networked Objects
NEON	New Era Of Networks
NEP	Network Entry Point
NEP	Noise Equivalent Power
NEPC	Neon Photo-Conductor
NEPD	Noise Equivalent Power Density
NEQ	Not Equal to
NERC	New En-Route Centre
NERD	Network Event Recording Device
NERO	Next Reader Optics
NERP	New Equipment Rental Plan
NERSC	National Energy Research Supercomputer Center
NES	National Education Supercomputer (US)
NESAC	National Electronic Switching Assistance Center
NESC	National Electrical Safety Code
NESC	National Energy Software Center
NESDIS	National Environmental Satellite, Data, (and) Information Service (US)
NESDRES	National Environmental Satellite Data Referral Service (US)

NEST	Novell Embedded Systems Technology
NET	Network
NET	Network-Entity Title
NET	Network Equipment Technologies
NET	Noise Equivalent Temperature
NET	Norme Européenne de Télécommunications
NET	Not Earlier Than
NETA	New England Telecommunications Association
NetBEUI	NetBIOS Extended User Interface
NetBIOS	Network Basic Input/Output System
NETBLT	Network Block Transfer
netCDF	network Common Data Form
NETCON	Network Control
NETCONSTA	Net Control Station
NETDA	Network Design and Analysis
NetDDE	Network Dynamic Data Exchange
NETGEN	Network Generation
netiquette	net + etiquette
NETOP	Network Operator Process
NETPARS	Network Performance Analysis Reporting System
NetPC	Net Personal Computer
NETPLAN	Network Planning
NETSET	Network Synthesis and Evaluation Technique
NETT	Network for Environmental Technology Transfer
NEU	Network|Numeric Extension Unit
NEUTRABAS	Neutral product definition data Base for large multifunctional Systems
new	new (information) (file extension indicator)
NEWS	Network Error Warning System
NEWS	Network Extensible Window(ing) System
NEWT	News Terminal
NEWVOL	New Volume
NEXPERT	Neuron data Expert system
NEXT	Near End (differential) Crosstalk
NEXT	NLR Engineering Expert system Toolkit (NLR = Nationaal (Nederlands) Lucht- en Ruimtevaartlaboratorium)
NF	Next without For
NF	No Funds
NF	Noise Factor|Figure
NF	Norfolk Island
NF	Normal Form
NF	Not Finished
NFAM	Network File Access Method
NFAP	Network File Access Protocol
NFB	Negative Feedback
NFB	No Flat Band (charge)
NFC	Not Favourably Considered
NFD	Negative Feature Decomposition

NFD	No Fixed Date
NFD	Non-Fatal Defect
NFE	Network Front End
NFE	No First Error
NFERO	Non Ferrous
NFET	N-channel junction Field-Effect Transistor
NFF	Neutral File Format
NFF	No Fault Found
NFF	No Form Feed
NFG	Network Flow Graph
NFIS	Non-Formatted Information System
NFM	Narrowband Frequency Modulation
NFM	Network File Manager
NFMR	Nonlinear Ferro-Magnetic Resonance
nfo	info (file extension indicator)
NFP	No File Protect
NFPR	Normalized Floating Point Representation
NFR	Not a Functional Requirement
NFS	Network File Server|System
NFS	Number Field Sieve
NFT	Network File Transfer
NFU	Not For Us
NG	Negative Glow
NG	Netgram
NG	Nigeria
NG	No Good
NGC	New General Catalog
NGC	Next Generation Controller(s)
NGDA	New Generation Design Automation
NGDC	National Geophysical Data Center (US)
NGDLC	Next Generation Digital Loop Carrier
NGE	Not Greater or Equal
NGEN	New Generation Enterprise Network
NGI	Nederlands Genootschap voor Informatica
NGI	Next Generation Internet
NGIO	Next Generation Input/Output
NGIS	Next Generation Information System
NGL	Next Generation Learning (technology)
NGLC	Next Generation Level Control
NGM	Nested Grid Model
NGMRD	New Generation Mark Reader
NGP	Network Graphics Protocol
NGP	Next Generation Processing
NGT	Next Generation Technology
NGT	Noise Generator Tube
NGT	Nominal Group Technique
NGT	Not Greater Than
NGWS	New Group (of) World Servers

NGWS	Next Generation Windows Services
NH	New Hampshire (US)
NH	Non-busy Hour
NH	Not Held
NHA	Next Higher Assembly
NHOB	Nonspecific Hierarchical Operational Binding
NHR	National Handwriting Recognition
NHR	Non-Hierarchial Routing
NHRP	Next Hop Resolution Protocol
NI	Natural Intelligence
NI	Near Instantaneous
NI	Network Interface
NI	Nicaragua
NI	Non Indicate
NI	Non-inhabitable Interrupt
N/I	Non-Interlaced
NI	Normal Information
NIA	Next Instruction Address
NIA	No Input Acknowledge
NIAM	Natural language Information Analysis Method
NIAM	Nijsens Information Analysis Method
NIAP	National Information Assurance Partnership
NIAS	Novell Internet Access Server
NIB	Nickel Iron Battery
NIB	Node Initialization Block
NIB	Not In Budget
NIBL	National Industrial Basic Language
NIBS	National Inventory Billing System
NIC	Navigation Information Connection
NIC	Negative Impedance Converter
NIC	Network Information Center|Card
NIC	Network Interface Card|Control
NIC	Not In Contact
NIC	Numeric Intensive Computing
NICAD	Nickel-Cadmium
NICAM	Near Instantaneous(ly) Companded Audio Multiplex
NICE	National Information Conference and Exhibition
NICE	Network Information and Control Exchange
NICI	Novell International Cryptographic Infrastructure
NICOLAS	Network Information Center On-Line Aid System
NICROSP	Neural networks for Identification, Control, Robotics and Signal/image Processing (international workshop)
NICS	Network Integrity Control System
NID	Network In-Dialling
NID	Network Interface Device
NID	New Interactive Display
NID	Next Identification
NIDA	Numerically Integrating Differential Analyzer

NIDDESC	Navy-Industry Digital Data Exchange Standards Committee
NIDE	Numerical Integration of Differential Equation(s)
NIEMR	Non-Ionizing Electro-Magnetic Radiation
NIF	Network Information File
NIF	Noise Improvement Factor
NIFF	Notation Interchange File Format
NIFO	Next In, First Out
NIFs	Not In Files
NIFTP	Network Independent File Transfer Protocol
NIH	Not Invented Here
NII	National Information Infrastructure (US)
NIIIP	National Industrial Information Infrastructure Protocol(s)
NIIT	National Information Infrastructure Testbed
NIL	Network Interface Layer
nil	nothing
NILE	Number of Inverters along any Loop is Even
NIM	Net Intersected with an Image
NIM	Network Installation Manager
NIM	Network Interface Machine|Manager
NIM	Network Interface Module|Monitor
NIM	New Inside Macintosh
NIM	Nuclear Instrumentation Module
NIMA	National Imagery and Mapping Agency (US)
NIMBUS	Network Information Management client-Based User Service
NIMBY	Not In My Back Yard
NIMIS	Networked Inventory Management Information System
NIMMS	Nineteen hundred Integrated Modular Management System
NIMPA	Newly Installed Machine Performance Analysis
NIMS	Near-Infrared Mapping Spectrometer
NIMS	Non-Invasive Monitoring Systems
NIMS	Novell Internet Messaging System
NINO	Nothing In Nothing Out
NIO	Native Input/Output
NIO	Non-Interruptible Operation
NIOD	Network Inward/Outward Dialling
NIOPSWL	New Input/Output Program Status Word Location
NIOS	Network In-Out System
NIP	Non-Impact Printer|Printing
NIP	Non-Internal Plane
NIP	Normal Investment Practice
NIP	Nucleus Initialization Procedure|Program
NIPDE	National Initiative for Product Data Exchange
NIPO	Negative Input/Positive Output
NIPRNET	Non-classified Internet Protocol Router Network
NIPRNET	Non-secure Internet Protocol Router Network
nips	network inputs/outputs per second
NIPTS	Noise-Induced Permanent Threshold Shift
NIR	Near Infra-Red

NIR	Net Installed Revenue
NIR	Network Information Registry|Retrieval
NIRI	Net Installed Record Increase
NIS	Names Information Socket
NIS	Network Information Service|System
NIS	Network Interface System
NISC	Network Information and Support Center
N-ISDN	Narrowband|National Integrated Services Digital Network
NISO	National Information Standards Organization (US)
NISP	Networked Information Services Project
NISPC	New Information Systems Professional Council
NIST	National Institute for Standards and Technology (US)
NIT	Native Interface Tester
NIT	Network Interface Task
NIT	Non-Intelligent Terminal
NITC	National Information Technology Center (US)
NITF	National Imagery Transmission Format
NITFS	National Imagery Transmission Format Standard
NIU	Network Interface Unit
NIU	North-American ISDN Users (ISDN = Integrated Services Digital Network)
NIUF	North-American ISDN Users' Forum
NJ	New Jersey (US)
NJB	Network Job Processor
NJCL	Network Job Control Language
NJE	Network Job Entry (protocol)
NJI	Network Job Interface
NJOBS	Number of Jobs
NJP	Network Job Process(ing)
NJSZT	Neumann János Számítógéptudományi Társaság (HU)
NKS	Network Knowledge Server
NL	Natural Language
NL	Netherlands
NL	Network Layer
NL	New Line (character)
NL	Noiseless
NL	Noise Level
NL	No Label|Load
NLA	Next Level Aggregator
NLA	Normalized Local Address
NLC	Network Language Center
NLC	New Line Character
NLC	Number of Lower Classes
NLD	Network Logical Disk
NLDM	Network Logical Data Manager
NLE	NetList Extractor
NLE	Non-Linear (video) Editor
NLE	Not Less or Equal

NLIN	NOAA Library and Information Network (NOAA = National Oceanic and Atmospheric Administration)
NLM	Network Loadable Module
NLOS	Natural Language Operating System
NLP	Natural Language Processing|Processor
NLP	No Longer Pre-competitive
NLP	Non-Linear Programming
NLPID	Network Layer Protocol Identifier
NLQ	Near Letter Quality
NLR	No Load Ratio
NLR	Non-Linear Resistor
NLRI	Network Layer Reachability Information
NLS	National Language Support
NLS	Native Language System
NLS	Network License Server
NLS	Node Logic Shelf
NLS	No-Load Speed
NLSC	National Language Services Center
NLSC	Non-Locking Shift Character
NLSDAP	Non-Linear System Data Presentation
NLSFUNC	National Language Support Function
NLSP	NetWare Link Service Protocol
NLSP	Novell Link State Protocol
NLT	New Logic Technology
NLT	Not Later|Less Then
NLT-HP	New Logic Technology – High Performance
NLT-S	New Logic Technology – Slow
NLUS	Network Logical Unit Services
NLV	National Language Version
NM	Network Management|Manager
NM	Network Module
NM	New Mexico (US)
NM	Night Message
NM	No Match
NM	No(t) Mark(ed)
NM	Not Matched|Measured
NMA	National Micrographics Association
NMA	NetWare Management Agent
NMA	Network Management Architecture
NMAP	Network Management Application Program
NMC	National Management Committee
NMC	Network Management Center|Console
NMC	Network Management Controller
NMC	Non-Marginal Check
NMC	Numerical Machine Control
NMCC	Network Management Control Center
NMDA	New Media Development Association
NME	Network Management Entity

NME	Noise Measuring Equipment
NMF	Network Management Forum
NMF	New Master File
NMG	Numerical Master Geometry
NMHS	Not-Made-Here Syndrome
NMI	National Maintenance Index
NMI	Native Method Invocation
NMI	New Model Introduction
NMI	Non-Maskable Interrupt
NML	Network Management Layer
NMM	NetWare Management Map
NMO	National Member Organization
NMO	Network Management Offering
NMO	Number of critical Micro-Operations
NMOS	Negative-channel Metal Oxide Semiconductor
NMP	Network Management Protocol
NMPAP	Noise Minimization Pad Assignment Problem
NMPF	Network Management Productivity Facility
NMPIS	National Marine Pollution Information Systems (US)
NMR	NetWare Management Response
NMR	N-Modular Redundancy
NMR	Normal Mode Rejection
NMR	Nuclear Magnetic Resonance
NMRR	Normal Mode Rejection Ratio
NMS	Network Management Services|Station
NMS	Network Management System
NMS	Network Monitoring Station
NMSE	Normalized Mean Square(d) Error
NMSI	Non-Multiplexed Serial Interface
NMSIG	Network Management Special Interest Group
NMSS	Network Management Support Services
NMT	Nordic Mobile Telephone
NMTS	Nordic Mobile Telephone System
NMVT	Network Management Vector Transport
NN	Negative Notification
NN	Network Node
NN	Neural Network
NN	No News
NNA	New Name Account
NNA	New Network Architecture
NND	National Number Dialling
NNDP	Non-Numeric Data Processing
NNI	Network-Network Interface
NNI	Network-Node Interface
NNI	Next Node Index
NNN	Non-Normalized Number
NNP	Net National Product
NNP	Network Node Processor

nnq	news.newusers.questions (file extension indicator)
NNS	NetWare Name Service
NNSC	NSF Network Service Center (NSF = National Science Foundation)
NNT	Nearest Neighbour Tool
NNTP	NetNews Transfer Protocol
NNTP	Network News Transfer Protocol
NNX	Network Numbering Exchange
NO	National Organization
NO	Network Operator
NO	Normally Open
NO	Norway
No	Number
NOA	Network Oriented Analysis
NOA	Not Available
NOAE	No Observed Adverse Effect
NOAEL	No Observed Adverse Effect Level
NOC	Network Operating|Operations Center
NOC	Normally Open Contact
NOC	Not Carry
NOC	Not Otherwise Classified
NOC	Number Of Classes
NOCN	National Ocean Communications Network (US)
NOCP	Network Operator Control Program
NOCS	Network Operations Center System
NOCS	Number Of Control Systems
NOD	Navigation Oriented Display
NOD	Network Out-Dialling
NODAL	Network Oriented Data Acquisition Language
NODAS	Network Oriented Data Acquisition System
NODC	National Oceanographic Data Center (US)
NODES	Network Operation (and) Design Engineering System
NODES	Nonlinear and active Optical Devices on Electronic Substrates (EU)
NOE	Network Of Excellence
NOF	National Optical Font
NOIBN	Not Otherwise Indexed By Name
NOISE	Near Optimal Iso-Surface Extraction
NOISE	Netscape, Oracle, IBM, Sun, and Everyone else
NOLCOS	Non-Linear Control Systems (symposium)
NOLOG	No Logging
NOM	Nominal
NOMC	Network Operation and Management Center
NOMDA	National Office Machine Dealers Association (US)
NOMEN	Nomenclature
NONCHK	Non Check
NONIND	Non Indicate
NONP	Non Print

NOOP	Network OSI Operations (OSI = Open Systems Interconnect)
NO-OP	No Operation
NOP	Not Otherwise Provided for
NOPAC	Network Online Public Access Catalog
NOPE	Not On Planet Earth
NOPS	Null Operations (instruction)
NOR	Negative OR
NOR	No Record
NOR	Notice Of Revision
NOR	Not OR (logical operation)
NORD	Nondestructive Readout
NORGEN	Network Operations Report Generator
NORM	Normal
NORM	Not Operationally Ready (due to) Maintenance
NORMZ	Normalize
NORTEL	Northern Telecom
NOS	Nederlandse Omroep Stichting
NOS	Network Operating System
NOS	Network Operational Schema
NOS	Node Operating System
NOS	Not Otherwise Specified
NOS/BE	Network Operating System/Batch Environment
NOSC	Network Operations and Service Centre
NOSP	Network Operating Support Program
NOS/VE	Network Operating System/Virtual Environment
NOTE	Not Over There Either
NOTIS	Network Operator Trouble Information System
NOVOLSTOR	Non-Volatile Storage
NOVON	Non Von Neumann
NOVRAM	Non-Volatile Random Access Memory
NOW	Negotiable Order of Withdrawal
NOW	Network Of Workstations
NOW	Number Of Windows
N-P	Negative-Positive
NP	Nepal
NP	Netgram Protocol
NP	Network Provider
NP	New Page|Project
NP	Non-deterministic Polynomial
NP	Non Print
NP	No Parity
NP	Noun Phrase
NP	Number of steps, Polynomial time
NP/1	New Program Language
NPA	Network Performance Analyzer
NPA	Network Printer Alliance
NPA	Network Programmers Association
NPA	New Products Announcement

NPA	No Power Alarm
NPA	Numbering Plan Area
NP/ADE	Non-Procedural Application Development Environment
NPAI	Network Protocol Address(ing) Information
NPAP	Network Printing Alliance Protocol
NPAS	New Products Analysis System
NPB	Net Plan Band
NPC	Nano-Program Counter
NPC	Non-Player Character
NPC	North Pacific Cable
NPD	Network Protective Device
NPDA	Network Problem Determination Application
NPDN	Nordic Public Data Network
NPDU	Network Protocol Data Unit(s)
NPEL	Noise Power Emission Level
NPERMTD	Not Permitted
NPEX	Normal Priority Exit
NPF	Network Partitioning Facility
NPG	New Power Generation
NPG	New Product Group
NPI	Network Printer Interface
NPI	New Product Introduction
NPII	Net Points Installed Increase
NPIRI	Net Product Installed Record Increase
NPIU	Numerical Processing and Interface Unit
NPL	New Process(or)|Product Line
NPL	New Programming Language
NPL	Non-Procedural Language
NPLA	New Product Line Audit
npm	counts per minute
NPM	Network Performance Monitor
npn	negative-positive-negative
NPN	New Public Network
NPNA	No Protest Non-Acceptance
npnp	negative-positive-negative-positive
NPO	Negative-Positive-Zero
NPP	Net Primary Productivity
NPP	Network Protocol Processor(s)
NPP	New Product Planning|Products
NPR	Neoprene Photoresist
NPR	New Product Release
NPR	Noise Power Ratio
NPR	Non-Procedural Reference
NPR	Non-Processor Request
NPRI	Net Product Record Increase
NPRL	Non-Procedural Referencing Language
NPRM	Notice of Proposed Rule-Making
NPRO	Non-Process Run-Out

NPROCL	Non-Procedural Language
NPRZ	Non-Polarized Return-to-Zero recording
nps	counts per second
NPS	Network Processing Supervisor
NPS	Network Product Support
NPS	New Products Support|System
NPS	Novell Productivity Specialist
NPS	Numerical Plotting System
NPSI	Network Packet Switch Interface
NPSI	Network Protocol Service Interface
NPSM	Non-Productive Standard Minute
NPSTN	National Public Switched Telecommunications Network (US)
NPSWL	New Program Status Word Location
NPT	National Pipe Tapered
NPT	Network Planning Tool
NPT	Non-Packet mode Terminal
NPTN	National Public Telecomputing Network (US)
NPTR	Net Process Throughput Rate
NPU	Natural|Network Processing Unit
NPU	Node Processor Unit
NPU	Non-elementary Physical Unit
NPU	Numeric Processing Unit
NPUI	Net Product Unit Increase
NPUS	Network Physical Unit Services
NPV	Net Present Value
NPV	No Par Value
NPX	Numeric Processor Extension
NQ	No Quote
NQ	Not Qualified
NQCAR	No Q Carry
NQR	Non-Quadratic Residues
NQR	Nuclear Quadrupole Resonance
NQS	Network Queuing System
NR	Nauru
NR	Negative Resistance
NR	New Release
NR	Noise Ratio|Reduction
NR	Noise Reduction
N/R	Non Recoverable
NR	No Resume
NR	Not Rated
NR	Not Responsible for
NR	Number Received
NRB	New Record Build
NRC	Networking Routing Center
NRCC	Network Resource Consultants and Company
NRD	Negative Resistance Diode
NR/D	Not Required, but Desired

NRE	Negative Resistance Element
NRE	Non-Recurring Engineering
NREN	National Research and Education Network (US)
NRFD	Not Ready For Data
NRFI	Not Ready For Issue
NRI	Net Revenue Installed
NRIC	Network Reliability and Interoperability Council
NRL	Network Restructuring Language
NRL	Normal Rated Load
NRM	Natural Remnant Magnetism
NRM	Network Resource Manager
NRM	Normalize
NRM	Normal Response Mode
NRN	National Research Network (IE)
NRN	Next Record Number
NRN	No Reply Necessary
NRP	Narrowband Random Process(es)
NRRR	Non-Return-to-Reference Recording
NRS	Name Registration Scheme
NRT	Non Real-Time
NRTC	Normalized Re-instrumented Terrain Computer
NRTZ	Non-Return To Zero
NRU	Network Repeater|Resource Unit
NRV	Nodal Route Vector
NRZ	Non-Return-to-Zero (coding)
NRZ0	Non-Return-to-Zero change on zeros
NRZ1	Non-Return-to-Zero change on ones
NRZC	Non-Return-to-Zero Change
NRZI	Non-Return-to-Zero Indiscrete
NRZI	Non-Return-to-Zero Invert(ed)
NRZL	Non-Return-to-Zero Level
NRZM	Non-Return-to-Zero Mark
NRZS	Non-Return-to-Zero Space
NS	Name Server
ns	nanosecond
NS	Network Service(s)|Supervisor
NS	New Signal|System(s)
NS	Non Stop
NS	Normal Stress
NS	Not Specified|Sufficient
NSA	Network Service Amendment
NSA	Network Standard Architecture
NSA	Next Station Addressing
NSA	Non-Sequenced Acknowledgment
NSAI	National Standards Authority of Ireland
NSAP	Network Service(s) Access Point
NSAPI	Netscape Server Application Programming Interface
NSC	Native Sub-Channel

NSC	Network Service|Switching Center
NSC	Nodal Switching Center
NSC	Noise Suppression Circuit
NSC	Non-Sequential Computer
NSCS	Network Service Center System
NSD	Nonlinear Sampled Data
NSDI	National Spatial Data Infrastructure
NSDRC	National Standard Reference Data Center
NSDU	Network protocol Service and Distribution Unit
NSDU	N-Service Data Unit
NSE	Network Software Environment
NSE	Network Support Encyclopedia
nsec	nanosecond
NSEC	Network Switching Engineering Center
NSEL	Network (service) Selector
NSEM	Network Software Environment
NSF	National Science Foundation (US)
NSF	Negotiated Search Facility
NSF	Not Sufficient Funds
NSFL	New Strip File
NSFnet	National Science Foundation network (US)
NSI	NASA Science Internet
NSI	Next Sequential Instruction
NSI	Non-Sequenced Information
NSI	Non-SNA Interconnection (NSA = System Network Architecture)
NSI	Non-Standard Item
NSIDC	National Snow and Ice Data Center (US)
NSIL	Non-Saturating Inverter Logic
NSK	Not Specified by Kind
NSL	New Simulation Language
NSL	Non-Standard Label
NSLOOKUP	Name Server Lookup (Unix)
NSM	Network Security Module
NSM	Network Services Manager
NSM	Node Supervision Module
NSML	Netscape Markup Language
NSNP	No Space, No Print
NSO	National Statistical Office (MW)
NSOS	N-channel Sapphire On Silicon
NSP	Native Signal Processing
NSP	Network Service Part|Provider
NSP	Network Service(s) Protocol
NSP	Numeric Space (character)
NSP	Numeric Subroutine Package
NSPE	Network Services Procedure Error
NSPF	Not Specifically Provided For
NSPMP	Network Switching Performance Measurement Plan

NSR	Normal Service Request
NSR	No Slot Release
NSR	Numerical Supplier Rating
NSRD	National Software Reuse Directory
NSRDS	National Standard Reference Data System
NSRS	National Spatial Reference System (US)
NSS	Network Supervisor System
NSS	New Simulation System
NSS	New Software Support
NSS	New System Services
NSS	Nodal Switching System
NSSC	NASA Standard Spacecraft Computer
NSSDC	National Space Science Data Center (US)
NSSDU	Normal Session Service Data Unit
NSSII	Network Supervisory System II
NSSN	National Standards Systems Network
NSSR	Non-Specific Subordinate Reference
NSTC	National Science and Technology Council (US)
NSTC	Not Subject To Call
NSTL	National Software Testing Laboratory
NSU	Networking Support Utilities
NSUG	Nihon Sun User's Group
NSUI	Net Sales Unit Increase
NSV	Non-automatic Self Verification
NSW	National Software Works
NT	Narrower Term
NT	Network Technology|Terminator
NT	Neutral zone
NT	New Technology
NT	Northern Telecom (CA)
NT	Norwegian Telecom
NT	No Text|Transmission
NT	Not True
NT	Numbering Transmitter
NT	Number of Tracks
NT1	Network Termination 1
NTA	National Telecommunications Agency
NTA	Noise Transmission Attenuation
NTAS	New Technology Advanced Server
NTC	National Telecommunications Conference
NTC	Negative Temperature Coefficient
NTCA	National Telephone Cooperative Association (US)
NTCA	Non-Tutorial Computer Application
NTCH	Notch
NTD	Network Tools for Design
NTE	Network Terminating Equipment
NTE	Not To Exceed
NTEC	Network Technical Equipment Center

NTF National Transfer Format
NTF No Trouble Found
NTFS New Technology File System
NTI Noise Transmission Impairment
NTIA National Telecommunications and Information Agency (US)
NTIBOA Not That I'm Bitter Or Anything
NTK Newton Tool Kit
NTLM New Technology LAN Manager
NTM Network Traffic Management
NTN National Trends Network (US)
NTN Network Terminal Number
NTN Neutralized Twisted Nematic
NTO Network Terminal Operator|Option
NTP Network Terminal Protocol|Point
NTP Network Termination Processor
NTP Network Time Protocol
NTP Network Transaction Processing
NTP Normal Temperature and Pressure
NTPD Network Time Protocol Daemon
ntpf number of terminals per failure
NTR Net Total Requirement
NTR Next Task Register
NTRAS NT Remote Access Services (NT = New Technology)
NTRS National Technology Roadmap for Semiconductors
NTRT Nine-Tracks Tape
NTS Network Technical Support
NTS Network Test|Tracking System
NTS Note To Scale
NTSA NetWare Telephony Services Architecture
NTSA Networking Technical Support Alliance
NTSC National Television Standard Code
NTSC National Television Standards Committee
NTSR Noise-To-Signal Ratio
NTT Net-To-Time
NTT Nine-Track Tape
NTT Number Theoretic Transform
NTU Network Terminating Unit
NTV National Televoice
NTZ Neutral Zone
NU Name Unknown
NU Net Utilization
NU Niue
NU Nothing Unsatisfactory
NU Number Unobtainable
NU0B Numeric 0 Bit
NU1B Numeric 1 Bit
NUA Network User Address
NUA Network Users' Association

NUBLU	New Basic Logic Unit
NUC	Number of Upper Classes
NUCFS	NetWare Unix Client File System
NUDOR	Numerical Data Processor
NUF	Noise Ulterior Flux
NUG	Net Unit Group
NUI	Network User Identification|Interface
NUI	Notebook User Interface
NUL	No Upper Limit
NUL	Null (character)
NULF	Nullify
NULS	Net Unit Load Size
NUM	Number
NUM	Numeric
NUMA	Non-Uniform Memory Access
NUMA-Q	Non-Uniform Memory Access from Sequent
NUMA-RC	Non-Uniform Memory Access – Remote Cache
NURBS	Non-Uniform Relational B-Spline
NUTL	Non-Uniform Transmission Lines
nV	nanovolt
NV	Nevada (US)
NV	No Overflow
NVDM	NetView Distribution Manager
NVE	Network Visible Entity
NVEB	Non-Vacuum Electron Beam
NVH	Noise, Vibration, Harshness
NVI	Node Value Incidence
NVLAP	National Validation Laboratory Program (US)
NVM	Non-Volatile Memory
NVOD	Near-Video On Demand
NVP	Nominal Velocity of Propagation
NVR	Nonspecific Volume Request
NVRAM	Non-Volatile Random Access Memory
NVT	Network|Novell Virtual Terminal
NVTS	Network Virtual Terminal Service
nW	nanowatt
NW	Needle Wire
NWAdmin	NetWare Administrator
NWCS	NetWare Cluster Services
NWD	Network Wide Directory
NWDS	Network Wide Directory System
NWH	Normal Working Hours
NWI	National Workload Index
NWI	New Work Item (proposal)
NWNET	North-West Network
NWSTG	National Weather Service Telecommunications Gateway
NWTC	National Wind Tunnel Complex
NXA	Nodal Exchange Area

NXM	Non-Existent Memory
NY	Net Yield
NY	New York (US)
NYAP	New York Assembly Program
NYSE	New York Stock Exchange
NYSERNet	New York State Education and Research Network
NZ	New Zealand
NZ	Non-Zero
NZLIA	New Zealand Library and Information Association
NZS	Near-Zero Stamping
NZSG	Non-Zero-Sum Game
NZT	Non-Zero Test|Transfer
NZUSUGI	New Zealand Unix System User Group, Inc.

O

o	origin
o	out
O	Output
O	Overflow
OA	Office Automation
OA	Operand Address register
OA	Operating Assemblies|Authorization
OA	Operational Analysis
OA	Operator Assistance|Availability
OA	Order Action
OAAC	Objects and Attributes for Access Control
OAAU	Orthogonal Array Arithmetic Unit
OAB	One-to-All Broadcast
OABETA	Office Appliance and Business Equipment Trades Association
OAC	Office Automation Conference
OAC	Office of Academic Computing
OAC	Operation Activity Control
OAC	Operational Amplifier Characteristics
OAC	Operations Analysis Center
OACB	Output Area Control Block
OACDT	Order Acknowledge Date
OAD	Open Architecture Driver
OA/DDP	Office Automation/Distributed Data Processing
OADG	Open Architecture Development Group
OADW	Oracle Applications Data Warehouse
OAF	Origin Address Field
OAFP	Origin Address Field Prime
OAG	On-line Air Guide
OAG	Operand Address Generator
OAI	Open Application Interface
OAIDE	Operational Assistance and Instructive Data Equipment
OAM	Operand Addressing Mode

OAM	Operation And Maintenance
OAM	Operations, Administration, and Maintenance (functions)
OAMC	Operation And Maintenance Center
OAM&P	Operations, Administration, Maintenance, and Provisioning
OAMP	Optical Analog Matrix Processing
OAN	Optical Access Networking
OAO	Over And Out
OAP	Orthogonal Array Processor
OAPM	Optimal Amplitude and Phase Modulation
OAR	Operand Address Register
OAR	Operations Activity Recorder
OAR	Operations Analysis Report
OAR	Operator Authorization Record
OAR	Optional Address Register
OAR	Overhaul And Repair
OARS	Opening Automated Report Service
OAS	One-to-All Scatter
OAS	Oracle Application Server
OAS	Order Allocation System
OAS	Output Amplitude Stability
OASIS	Office Automation Services Informatics Sector
OASIS	Okuma America Supervisory Information System
OASIS	On-line Access to the Standards Information Service
OASIS	On order Activity Simulation System
OASIS	Open Architecture System Integration Strategy
OASIS	Operational Analyses Strategic Interaction Simulator
OASIS	Organization for the Advancement of Structured Information Standards
OASYS	Office Automation System(s)
OASYS	Operational Amplifier Synthesis
OASYS	Organization and Analysis System
OAT	Operating Acceptance Test
OAT	Operating Ambient Temperature
OATS	Office Automation Technology and Services
OAUG	Oracle Applications User Group
OAUS	On An Unrelated Subject
O-A-V	Object-Attribute-Value
OAV	Original Animation Video
OB	Obligatory
OB	Or Better
OB	Ordered Back
OB	Output Buffer|Bus
OBAC	One-Bit Adder Computer
OBB	Operation Better Block
OBC	On-Board Computer
OBC	Outside Back Cover
OBD	On Board Diagnostics
OBD	Online Bugs Database

OBDC	Open Data Base Connectivity
OBDD	Ordered Binary Decision Diagram
OBDICS	Order Backlog Delivery and Installation Control System
OBE	Output Buffer Empty
OBEX	Object Exchange
OBF	Operational Base Facility
OBI	Open Buying on Internet
OBJ	Object
OBLIST	Object program Listing
OBO	Official Business Only
OBO	Order Book Official
OBP	On-Board Processor(s)
OBR	Optical Barcode Reader
OBR	Outboard Recorder(s)
OBS	Obsolete
OBS	Omni-Bearing Selector
OBS	On-line Business Systems
OBS	Optical Beam Scanner
OBS	Optimum Blending System
OBTN	Obtain
OBTW	Oh, By The Way
OBV	On-Balance Volume
OBWO	O-type Backward-Wave Oscillator
OC	Official Classification
OC	Open Circuit|Collector
OC	Operating Characteristic(s)
OC	Operation Check|Control
OC	Operator Centralization
OC	Optical Carrier
OC	Output Check|Computer
OC	Output Controller
OC1	Optical Carrier (level) 1
OC-12	Optical Carrier 12
OC-192	Optical Carrier 192
OC3	Optical Carrier (level) 3
OCA	Open Communication Architecture
OCA	Output Communications Adapter
OCAI	Open CAD Architecture Initiative (CAD = Computer Aided Design)
OCAL	On-line Cryptanalytic Aid Language
OCB	Outgoing Calls Barred
OCB	Override Control Bits
OCB	Over-the-Counter Batch
OCBP	Output Control Block Pointer
OCC	Office Communications Cabinet
OCC	Operating Characteristic Curve
OCC	Operations Control Center
OCC	Operator Control Command

OCC	Order Control Card
OCC	Other Common Carrier(s)
OCCB	Operational Configuration Control Board
OCCF	Operator Communication Control Facility
OCD	Off Chip Driver
OCD	On-line Communication Driver
OCDMS	Onboard Checkout and Data Management System
OCDS	Output Command Data Set
OCE	Open Collaborative Environment
OCE	Other common Carrier channel Equipment
OCF	Objects Components Framework
OCF	Onboard Computational Facility
OCF	Open Channel Flow
OCF	Operator Console Facility
OCFP	Operator Command Function Processor
OCG	Optimal Code Generation
OCG	Overall Conflict Graph
OCIS	Operational Control Information System
OCL	Object Constraint Language
OCL	Operating Control Language
OCL	Operational Check List
OCL	Operator(s) Control|Command Language
OCLC	Online Computer Library Center
OCM	Ocular Connection Machine
OCM	One-Chip Module
OCM	Operator Console Monitor
OCM	Oscillator and Check Module
OCMODL	Operating Cost Model
OCMV	Output Common Mode Voltage
OCO	Object Code Only
OCO	One Cancels the Other
OCO	Open-Close-Open (contact)
OCO	Operations Control Operator
OCP	Obligatory Contour Principle
OCP	One-Chip Processor
OCP	Operational Control Panel
OCP	Operator Console Panel
OCP	Output Control Program|Pulse(s)
OCR	Optical Card|Character Reader
OCR	Optical Character Reader|Recognition
OCR	Output Control Register
OCR-A	Optical Character Recognition – ANSI standard
OCRA	Order Change Record Access
OCR-B	Optical Character Recognition – International standard
OCRE	Optical Character Recognition Equipment
OCRUA	Optical Character Recognition Users Association
OCS	Object Compatibility Standard
OCS	Office Computing System

OCS	On-Card Sequencer
OCS	Operating Control System
OCS	Operational Computer Software
OCS	Operation(al) Control System
OCS	Operator Console Services
OCS	Optical Character Scanner
OCS	Order Communications Systems
OCS	Order|Output Control System
OCSP	On-line Certificate Status Protocol
OCT	Octal
OCT	Operation(al) Cycle Time
octree	octal tree
OCU	Office Channel Unit
OCU	Operational Control Unit
OCW	Operation Command Word
OCX	Open Compact Exchange
OD	Octal Dump
OD	Original Design
OD	Out of Data
OD	Output Disable
ODA	Octal Debugging Aid
ODA	Office Document Architecture
ODA	Open Document Architecture
ODA	Open DWG Alliance (DWG = Drawing)
ODA	Operational Data Analysis
ODAC	Operations Distribution Administration Center
ODA/ODIF	Office Document Architecture/Open Document Interchange Format
ODAPI	Open Database Application Programming Interface
ODAS	Ocean Data Acquisition System
ODBC	Open Data Base Connectivity
ODBMS	Object-oriented Database Management System
ODBR	Output Data Buffer Register
ODC	On-line Data Capture
ODC	Operational Document Control
ODC	Output Data Control
ODCS	Open Distributed Computing Structure
ODCS	Operational Data Collection System
ODD	Operator Distance Dialling
ODD	Optical Data Digitizer
ODDD	Operator Direct Distance Dialing
ODDH	Onboard Digital Data Handling
ODE	Ordinary Differential Equation
ODES	Office for Data Exchange Standards
ODESY	On-line Data Entry System
ODETTE	Organization for Data Exchange by Tele-Transmission in Europe
ODG	Off-line Data Generator

ODI	Open Datalink Interface
ODI	Open Device Interconnect
ODI	Optical Digital Image
ODIF	Office|Open Document Interchange Format
ODINSUP	Open Datalink Interface/Network driver interface specification Support
ODISS	Optical Digital Image Storage System
ODL	Overlay Description Language
ODM	Object Data Manager
ODM	Outboard Data Manager
ODM	Output Decomposition Method
ODMA	Open Document Management API (= Application Program Interface)
ODMA	Open Document Management Association
ODMG	Object Database Management Group
ODN	Optical Data Network
ODP	Open Data Path
ODP	Open Distributed Processing
ODP	Optical Data Processing
ODP	Original Document Processing
ODPCS	Oceanographic Data Processing and Control System
ODR	Optical Data Recognition
ODR	Optical Document Reader
ODR	Original Data Record
ODR	Output Definition Register
ODS	Open Data Services
ODS	Operational Data Store|Summary
ODS	Operational Data System
ODS	Optical Discrimination System
ODS	Optical Document Sorter
ODS	Output Data Strobe
ODS	Overhead Data Stream
ODSI	Open Directory Services Interface
ODSI	Optical Domain Service Interconnect
ODT	Object Definition Table
ODT	Octal Debugging Technique
ODT	On-line Debugging Technique(s)
ODT	Open Desktop
ODU	Output Display Unit
ODW	Object Development Workbench
ODW	Optional Data Warehouse
OE	Order Entry
OE	Or-Exclusive
OE	Output Enable
OEA	Operating Expense Analysis
OEAP	Operational Error Analysis Program
OEC	Odd-Even Check
OEC	Open Environment Corporation

OECD	Organization for Economic Cooperation and Development
OEDIPE	OSI Electronic Data Interchange for Energy Providers (OSI = Open System Interconnect)
OEDIT	Octal Editor
OEF	Online Education Facility
OEF	Origin Element Field
OEI	Order and Equipment Installed
OEI	Own Equipment Inventory
OEM	Original Equipment Manufacturer|Market
OEM	Other Equipment Manufacturer
OEMI	Original Equipment Manufacturers Information
OEO	Operational Equipment Objective
OE/OE	One Entry/One Exit
OEP	Operand Execution Pipeline
OEP	Original Element Processor
OER	Operating Equipment Requirements
OER	Original Equipment Replacement
OES	Output Enable Serial
OET	Objective End Time
OEV	Observer's Eye View
OEXP	Office of Exploration
OF	On File
OF	Operational Fixed
OF	Order Feedback
OF	Original Finish
OF	Overflow Flag
OFB	Output Feedback
OFC	Open Financial Connectivity
OFC	Orthonormal Function Coding
OFC	Outside Front Cover
OFDM	Orthogonal Frequency Division Multiplexing
OFEX	Office automation Equipment Exhibition
OFHC	Oxygen-Free High Conductivity
OFI	On-line Free format Input
OFI	Originating Financial Institution
OFL	Off-Line
OFLDT	Off-Line Data Transmission
OFLO	Overflow
OFLSTOR	Off-Line Storage
OFMT	Output Format (for numbers)
OFN	Open File Number
OFNP	Optical Fiber, Nonconductive Plenum
OFNPS	Outstate Facility Network Planning System
OFNR	Optical Fiber, Nonconductive Riser
OFR	Open File Report
OFR	Optical Form Reader
OFR	Ordering Function Register
OFR	Over Frequency delay

OFS	Object File System
OFS	Offset
OFS	Operating Functional Summary
OFS	Output Field Separator
OFSO	Oracle Field Sales Online
OFT	Optical Fiber Tube
OFTEL	Office For Telecommunications (GB)
OFTF	Optical Fiber Transfer Function
OG	Operations Guide
OG	OR Gate
OG	Outgoing
OGC	Open GIS Consortium (GIS = Graphical Information System)
OGF	Online Guidance Facility
OGIS	Open Geodata Interoperability Specification
OGL	Outgoing Line
OGL	Overlay Generation Language
OGT	Out-Going Trunk
OH	Off Hook
OH	Office Hours
OH	Ohio (US)
OHL	Over High Limit
OHP	Overhead Projector
OHQ	Off-Hook Queue
OHR	OR-Halfword Register (instruction)
OHV	Overhead Valve
OI	Operating Instructions
OI	OR Inverter
OI	Output Impedance
OIA	Operator Information Area
OIB	Operation Instruction Book
OIC	Oh, I See
OIC	Only In Chain
OIC	Operations Instrumentation Coordinator
OIC	Operator's Instruction Chart
OIC	Optimized Image Compression
OICC	Operations Interface Control Chart
OICR	Operator Identification Card Reader
OID	Object Identification Diagram
OID	Object|Octal Identifier
OID	Oracle Internet Directory
OIDI	Optically Isolated Digital Input
OIDL	Object Interface Definition Language
OIE	Optical Incremental Encoder
OIF	Output Interface
OIG	Operational Interface Group
OIL	Only Input Line
OIL	Operation Inspection Log
OIL	Our Intermediate Language

OIM	Open Information Model
OIM	Optical Index Modulation
OIM	OSI Internet Management (OSI = Open Systems Interconnect)
OIOPSWL	Old Input/Output Program Status Word Location
OIP	Operational Improvement Plan|Program
OIP	Optical Image Processor
OIPS	Optical Image Processing System
OIR	Online Information Retrieval
OIRA	Office of Information and Regulatory Affairs
OIS	Office Information System
OIS	Operating Information System
OIS	Operational Instrumentation System
OIS	Optical Image Sensor
OIT	Optical Image Terminal
OITDA	Optoelectronics Industry and Technology Development Association (JP)
OIU	Office Interface Unit
OIU	Optical Image Unit
OIW	Open Information Warehouse
OIW	OSI Implementors Workshop (OSI = Open System Interconnect)
OJS	Output Job Stream
OJT	On-the-Job Training
OK	Oklahoma (US)
OL	On Line
OL	Open Loop
OL	Operating Location|Log
OL	Operations/Logistics
OL	Or Less
OL	Outgoing Line
OL	Output Latch
OL	Outside Location
OL	Overlap
OL	Oxide Layer
OLAC	Off-Line Adaptive Computer
OLAP	On-Line Analytical Processing
OLB	Online Banking (system)
OLC	On-Line Computer
OLC	Open Loop Control
OLC	Operation Load Code
OLC	Outgoing Line Circuit
OLCA	On-Line Circuit Analysis
OLCC	Optimum Life Cycle Costing
OLCP	On-Line Complex Processing
old	old (version) (file extension indicator)
OLDAS	On-Line Digital-Analog Simulator
OLDB	On-Line Data Base
OLDC	On-Line Data Collection

OLDE	On-Line Data Entry
OLDS	On-Line Display System
OLDT	On-Line Data Transmission
OLE	Object Linking and Embedding
OLE	On-Line Edit|Equipment
OLE4D&M	Object Linking and Embedding For Design and Modelling
OLED	Organic Light Emitting Display
OLERT	On-Line Executive for Real-Time
OLFREFL	Outline Font Reflection
OLFSHAD	Outline Font Shadow
OLG	Open Level Generation
OLG	Open Loop Gain
OLI	Optical Line Interface
OLIFLM	On-Line Image Forming Light Modulator
OLIR	On-Line Information Retrieval
OLIT	Open Look Interface Toolkit
OLIVER	On-Line Interactive Vicarious Expediter and Responder
OLL	Output Logic Level
OLLS	On-Line Logical Simulation system
OLM	Overlay Load Module
OLMC	On-Line Machine Control
OLMC	Output Logic Macrocell
OLO	On-Line Operation
OLOE	On-Line Order Entry
OLP	Object Language Program
OLP	On-Line Processing|Programming
OLP	Order Load Point
OLPARS	On-Line Pattern Analysis and Recognition System
OLPC	Odd Longitudinal Parity Check
OLPS	On-Line Programming System
OLQ	On-Line Query
OLR	Open Loop Receiver
OLR	Open Loop Response
OLRT	On-Line Real-Time
OLRTP	On-Line Real-Time Processing
OLS	Operational Linescan System
OLSA	Off-Line Selectric Analyzer
OLSAT	Off-Line Selectric Analyzer – Transistorized
OLSC	On-Line Scientific Computer
OLSF	On-Line Subsystem Facility
OLT	On-Line Teller
OLT	On-Line Test(ing)
OLTD	On-Line Test(ing)/Debug(ing)
OLTEP	On-Line Test Executive Program
OLTL	On-Line Tape Library
OLTM	Optical Line Terminating Multiplexer
OLTP	On-Line Tele-Processing
OLTP	On-Line Transaction(al) Processing

OLTS	On-Line Tape System
OLTS	On-Line Test Section|System
OLTS	On-Line Time-Share
OLTSEP	On-Line Test Stand-alone Executive Program
OLTT	On-Line Teller Terminal
OLTT	On-Line Terminal Test
OLVWM	Open Look Virtual Window Manager
OLWM	Open Look Window Manager
OLX	On-Line Executive
OM	Object Manager|Model
OM	Oman
OM	Operating Memorandum|Monitor
OM	Operation Manual
O&M	Operations and Maintenance
OM	Operations Manager|Manual
OM	Operators' Manual
OM	Optical Modulator
O&M	Organization and Methods
OM	Orthogonal Memory
OM	Out of Memory
OM	Output Module
OMA	Object Management Architecture
OMA	Operations Monitor Alarm
OMAC	On-line Manufacturing, Accounting, and Control system
OMAC	Open Modular Architecture Control(ler)
OMAP	Object Module Assembly Program
OMAR	Optical Mark Reader
OMAR	Order Maintenance Analysis Report
OMAR	Order Management And Routing
OMC	Operation and Maintenance Center
OMC	Operations Monitoring Computer
OMD	Open Macro Definition
OME	Open Messaging Environment (protocol)
OMEC	Optimized Microminiature Electronic Circuit
OMEF	Office Machines and Equipment Federation
OMEGA	Open-ended Modular Electronic Graphics Art
OMEN	Orthogonal Mini-Embedment
OMF	Object Management Facility|Function
OMF	Object Module Format
OMF	Old Master File
OMF	Open Media Framework
OMF	Open Message Format
OMFI	Open Media Framework Interchange format
OMG	Object Management Group
OMI	Open Messaging Interface
OMI	Operations Maintenance Instructions
OMI	Organization for Micro-Information
OMIS	Operational Management Information System

OML	Object Module Library
OML	Open Modelling Language
OMM	Output Message Manual
OMNI	Organizing Medical Networked Information
OMNS	Open Network Management System
OMPF	Operation and Maintenance Processor Frame
OMPR	Optical Mark Printer
OMR	Optical Mark Reader|Recognition
OMRC	Optical Mark Reader Card
OMRS	Optical Mark Reader Sheet
OMS	Object Management System
OMS	Office Mail System
OMS	Optical Modulation System
OMS	Oracle Media Server
OMS	Organization Management System
OMT	Object Modelling Technique
OMTool	Object Management Tool
OMX	Ordinal Mapped Crossover
ONA	Open Network Architecture
ONAC	Operations Network Administration Center
ONAL	Off Network Access Line
ONC	Open Network Computing
OND	Operator Number Display
ONDS	Open Network Distribution Services
ONE	Open Network Environment
ONI	Operator Number Identification
ONITA	Of No Interest To Anybody
ONLP	On-Line Programming
ONLY	On-Line Yield
ONMS	Open Network Management System
ONN	Open Network Node
ONNA	Oh No, Not Again!
ONP	Open Network Provision
ONS	Off-Normal Switch
ONT	Operational Navigation Tool
OO	Object Oriented
OO	Over and Out
O&O	Owned and Operated
O&O	Owner and Operator
OOA	Object Oriented Analysis
OOAD	Object Oriented Analysis (and) Design
OOB	Out Of Band|Bounds
OOBE	Out-Of-Box Experience
OOBF	Out-Of-Box Failure
OOCC	Object Oriented Cell Control
OOD	Object Oriented Design
OODB	Object Oriented Data Base
OODBMS	Object Oriented Data Base Management System

OODMS	Object Oriented Database Management System
OOK	On-Off Keying
OOL	Object Oriented Language
OOL	Operator Oriented Language
OOOS	Object Oriented Operating System
OOP	Object Oriented Programming
OOPL	Object Oriented Programming Language
OOPS	Object Oriented Programming System
OOPS	Off-line Operating Simulator
OOPS	Operator Oriented Problem Source
OOPSLA	Object-Oriented Programming Systems, Languages and Applications (conference)
OOPSTAD	Object Oriented Programming for Smalltalk Application Developers (association)
OOR	Out Of Range
OOS	Object Oriented Software|System(s)
OOS	Off-line Operating Simulator
OOS	Out Of Service|Sync
OOSD	Object Oriented Structured Design
OOSH	Object Oriented Shell
OOSP	Object Oriented Solutions Practice
OOT	Object Oriented Technology
OOT	Out Of Territory
OOTB	Out Of The Box
OOUI	Object Oriented User Interface
OOW	Object Oriented Workflow
OP	Open Position|Potential
OP	Operand
OP	Operate
OP	Operating Plan|Point
OP	Operating Procedure(s)
OP	Operation(al)
OP	Operation Part|Procedure
O/P	Operations Planning
OP	Operator
OP	Operator Performance
OP	Optical
OP	Order Profile
OP	Outboard Processor
OP	Out of Print
OP	Output
OPAC	On-line Public Access Catalog
OPACK	Operation Acknowledge
OPAL	Operational Performance Analysis (program)
OP-AMP	Operational Amplifier
OPASYN	Operational Amplifier Synthesis
OPBLE	Operable
OPC	Odd Parity Check

OPC	OLE for Process Control (OLE = Object Linking and Embedding)
OPC	OpenGL Performance Characterization
OPC	Operation Code
OPC	Operations Planning and Control
OPC	Optical Photo-Conductor
OPC	Optimum Power Calibration
OPC	Organic Photo-conducting Cartridge
OPC	Organic Photosensitive Cartridge
OPC	Originating Point Codes
OPCE	Operator Control Element
OP-COD	Operating Code
OPCODE	Operation Code
OPCOM	Operator Communications
OPCTR	Operations Center
OPD	Opening Delayed
OPD	Operand
OPDAC	Optical Data Converter
OPDATS	Operational Performance Data System
OPDESC	Operation Description
OPDF	Output Data Funnel
OPDIN	Ocean Pollution Data and Information (center)
OPDU	Operation Protocol Data Unit
OPE	Optimized Processing Element
OPECO	Operations Coordinator
OPEDS	Outside Plant Engineering Design System
OPEN	Object-oriented Product model Engineering Network
OPEN	Open Protocol Enhanced Network
OpenGL	Open Graphics Library
OpenPIC	Open Programmable Interrupt Controller
OPEVAL	Operational Evaluation
OPEX	Operational Executive
OPF	Official Personnel Folder
OPI	Old Paper Information
OPI	Open Prepress Interface
OPICT	Operator Interface Control (block)
OPIM	Order Processing and Inventory Monitoring
OPIS	Operational Priority Indicating System
OPIS	Opportunistic Intelligent Scheduler
OPL	Operational
OPL	Organizer Programming Language
OPLIB	Object Program Library
opm	operations per minute
OPM	Operator Master
OPM	Operator Programming Method
OPN	Open
OPNCPLT	Operation Not Complete
OPND	Operand

OPNML	Operations Normal
OPO	One Point Operation
OPO	Online Process Optimization
OPO	Other Programmed Operations
OPOL	Optimization Oriented Language
OPOP	Operator/Operation
OPP	Opposite
OPP	Optical Printer Projector
OPPP	Order Point and Peak Point
OPR	Operator
OPR	Optical Page Reader
OPRAD	Operations Research And Development management
OPREG	Operation Register
OPREP	Operational Reporting (system)
OPREQ	Operation Request
OPRFLT	Operator Fault
OPRND	Operand
OPS	Office Procedure Specification
OPS	Off Premises Station
OPS	On-line Process Synthesis
OPS	Open Profiling Standard
OPS	Operating Plans Summary
OPS	Operations
ops	operations per second
OPS	Operations Planning System
OPSADT	Optically Programmable Semi-Automatic Direct current Tester
OPSCON	Operations Control
OPSCOP	Operations Control (monitoring) Program
OPSEC	Operations Security
OPSER	Operator Service
OPSM	Outside Plant Subscriber Module
OPSREP	Operations Report
OPSWL	Old Program Status Word Location
OPS-X	Operational teletype message
OPSYS	Operating System
OPT	Open Protocol Technology
OPT	Optimization
OPT	Optimized Production Technology
OPT	Optimum
OPT	Option(al)
opt	options (file extension indicator)
OPT	Output Transformer
OPTIC	Optical Procedural Task Instruction Compiler
OPTS	On-line Peripheral Test System
OPTUL	Optical Pulse Transmitter Using Laser
OPU	Operating Processing Unit
OPU	Operations Priority Unit
OPU	Other Plant Usage

OPU	Out-Plant Usage
OPUR	Object Program Utility Routine
OPUS	Obvious Password Utility System
OPUS	Octal Program Updating System
OPW	Orthogonalized Plane Wave
OPX	Off-Premises Extension
OQ	Output (job) Queue
OQL	Object Query Language
O/R	On Request
OR	On Return
OR	Operand Register
OR	Operational Requirement
OR	Operation Record
OR	Operations Research
OR	Optical Reader
OR	Ordered Recorded
OR	Oregon (US)
OR	Orientation Ratio
OR	Origin
OR	Originating Register
O/R	Originator/Recipient
O&R	Overhaul and Repair
OR	Over Run
OR	Owner's Risk
ORA	Option Revision Agreement
ORA	Output Register Address
ORAP	Originator/Recipient Address Prefix
ORB	Object Request Broker
ORB	Office Repeater Bay
ORB	Operations Request Block
ORBIT	Online Real-time Branch Information Technique
ORBIT	Online Reduced Bandwidth Information Transfer
ORBIT	Online Retrieval of Bibliographic Text
ORC	Operations Research Center
ORC	Orthogonal Row Computer
ORCA	Optimizer for Receiver Architectures
ORD	Order
ORDBMS	Object Relational Data Base Management System
ORDD	Ordered
ORDER	online real-time Order (entry) system
ORDER	Organization and Retrieval of Data for Efficient Research
ORE	Output Register Empty
OR&F	Operations, Research and Facilities
org	organization (Internet domain name)
ORG	Origin
ORG	Out of Range
ORG	Outside Range
ORI	On-line Retrieval Interface

ORI	Operational Readiness Inspection
ori	original (file extension indicator)
ORIS	Object Recognition and Identification System
ORIS	Overlapped Rank-Intervals Set
ORKID	Open Real-time Kernel Interface Definition
ORM	Object Role Modelling
ORM	Optical Remote Module
ORMAC	Oral Response Machine
ORMS	Operating Resource Management System
ORMS	Operations Research and Management Science
ORNAME	Originator/Recipient Name
ORO	Operations Research Office
OROM	Optical Read Only Memory
OROS	Optical Read Only Storage
ORP	Optional Response Poll
ORPC	Object Remote Procedure Call
ORR	Operational Ready Rate
ORRAS	Optical Research Radiometrical Analysis System
ORS	Optimum Real Storage
ORS	Output Record Separator
ORSA	Operations Research Society of America
ORT	Ongoing Reliability Test
ORT	Operational Readiness Test
ORT	Optimal Response Time
ORT	Overload Recovery Time
ORTHO	Orthochromatic
ORU	On-line Replacement Unit
ORU	Operational Readiness Unit
ORWG	Open Routing Working Group
OS	Odd Symmetric(al)
OS	On Schedule
OS	Open System(s)
OS	Operating Statement|System
OS	Operational Safety|Sequence
OS	Operator Service|System
OS	Optical Scanning
OS	Optimum Size
OS	Out of String (space)
OS	Oversize(d)
OS/2	Operating System/2
OS/400	Operating System for IBM's AS/400 computer
OSA	Object System Adapter
OSA	Open Scripting|Systems Architecture
OSAC	Operator Services Assistance Center
OSACA	Open System Architecture for Control within Automation systems
OSAF	Origin Sub-Area Field
OSAK	OSI Applications Kernel (OSI = Open Systems Interconnect)

OSAM	Overflow Sequential Access Method
OSB	Operational Status Bit
OSC	Operating System Control
OSC	Operator Services Center
OSC	Oscillator
OSCAR	Object management System Clausthal And Rostock
OSCAR	On Screen Configuration & Activity Reporting
OSCAR	Optically Scanned Character Automatic Reader
OSCL	Operations System Control Language
OSCP	Oscilloscope
OSCR	Online Service Center Response
OSCRL	Operating System Command (and) Response Language
OSD	On-line System Driver
OSD	On Screen Display
OSD	Open Software Distribution
OSD	Operational Service Date
OSD	Optical Scanning Device
OSDIT	Office of Software Development and Information Technology
OSDM	Optical Space-Division Multiplexing
OSDP	On-Site Data Processor
OSDP	Operations System Development Program
OSDS	Operating System for Distributed Switching
OSE	Open Systems Environment
OS/E	Operating System/Environment
OSE	Operational Support Equipment
OSEDA	Office of Social and Economic Data Analysis
OSEXT	Operating System Extensions
OSF	Online Standard Interface
OSF	Open Software Foundation
OSF	Operation(al) Support Facility
OSF	Operations System Function
OSG	Open Service Gateway
OSG	Operand Select Gate
OSI	Open System Interconnect(ion)
OSI	Operating System Interface
OSI	Overhead Supply Inventory
OSIE	Open Systems Interconnect Environment
OSIL	Operating System Implementation Language
OSINet	Open Systems Interconnect Network
OSIRIS	Online Search Information Retrieval Information Storage
OSIRM	Open Systems Interconnection Reference Model
OSIS	Organization Structure Information System
OSI-TP	Open Systems Interconnect Transaction Processing
OSL	Operand Specification List
OSL	Operating System Language
OSLANS	Open System Local Area Network Standard
OSLU	Open Systems Link Unit
OSM	Off-Screen Model

OSM	Operating System Manual|Monitor
OSM	Operating system Specific Module
OSME	Open Systems Message Exchange
OSML	Operating System Machine Level
OSN	Office Systems Node
OSN	Open Systems Network
OSN	Output Sequence Number
OSNS	Open Systems Network Support
OSO	Office of Systems Operations
OSO	Originating Signalling Office
OSP	On-site Service Provider
OS/P	Operating Systems for People
OSP	Operator Station Processor
OSP	Optical Storage Processor
OSP	Outside Purchased
OSPF	Open Shortest Path First
OSPFIGP	Open Shortest-Path First Internal Gateway Protocol
OSPS	Operator Service Position System
OSQL	Object Structured Query Language
OSR	OEM Service Release (OEM = Original Equipment Manufacturers)
OSR	Operand Storage Register
OSR	Optical Scanning Recognition
OSR	Original Scheduled date for Removal
OSS	Observing Simulation System
OSS	Open-Source Software
OSS	Operating System Supervisor
OSS	Operational Sub-System
OSS	Operation(s) Support System
OSS	Operator Service System
OSSE	Observing System Simulation Experiment
OSSF	Operating System Storage|Support Facility
OSSL	Operating Systems Simulation Language
OSST	Operating System Symbol Table
OSSWG	Office System Standards Work Group
OST	Objectives, Strategies, and Tactics
OST	Operator Station Task
OSTC	Open Systems Testing Consortium
OSTC	Open System Transition Center
OSTF	Operational Suitability Test Facility
OSTN	Object State Transition Network
OSU	Operational Switching Unit
OS/VS	Operating System/Virtual Storage
OSWS	Operating System Workstation
OT	Object Technology
OT	Object Time
OT	Office of Telecommunications
OT	On Time

OT	Ontological Theory
OT	On Track
OT	Operating Time
OT	OR inverter Terminate
OT	Other
OT	Out of Territory
OT	Output Terminal
OT	Overtime
OTA	Operational Transconductance Amplifier
OTAF	Operating Time At Failure
OTANZ	Output Tape Analyzer
OTBS	One True Bracketing Style
OTC	Objective, Time and Cost
OTC	One-Time Charge
OTC	Operational Test Center
OTC	Output Technology Corporation
OTCC	Operator Test Control Console
OTCS	Operational Teletype Communications Subsystem
OTDC	Optical Target Designation Computer
OTE	Hellenic Telecommunications Organization SA (GR)
OT&E	Operational Test(ing) and Evaluation
OTF	Open Token Foundation
OTF	Optical Transfer Function
OTF	Optimum Traffic Frequency
OTG	Option Table Generator
OTH-B	Over-The-Horizon Backscatter
OTI	Object Technology International
OTI	Open Tool Interface
OTIS	Operation, Transport, Inspection, Storage
OTIS	Order Trend Information System
OTL	On-line Task Loader
OTL	Operating Time Log
OTL	OSI Testing Liaison (OSI = Open Systems Interconnect)
OTL	Output Transformerless
OTLN	Outline
OTM	Object Transaction Monitor
OTM	Office of Telecommunications Management
OTM	On Time Marker
OTN	Other Than New
OTO-D	Object Type Oriented – Data model
OTOH	On The Other Hand
OTP	Office of Telecommunications Policy
OTP	One-Time Programmable
OTQ	Outgoing Trunk Queuing
OTR	One Touch timer Recording
OTRAC	Oscillogram Trace reader
OTRT	Operating Time Record Tag
OTS	Object Time System

OTS	Object Transaction Service
OTS	Off The Shelf
OTS	On-line Terminal System
OTS	On The Spot
OTS	Operational Time Synchronization
OTSF	Open Telephony Server Forum
OTSS	Off-The-Shelf System
OTSS	Open systems Transport and Session Support
OTSS	Operational Test Support System
OTT	Over The Top
OTTOMH	Off The Top Of My Head
OTU	Operational Test Unit
OTW	On The Whole
OU	Operation(al) Unit
OU	Outlook Unusual
OUCL	Oxford University Computer Laboratory (GB)
OUD	Operational Unit Data
OUI	Organizationally Unique Identifier
OURS	Open Users Recommended Solutions
OUT	Outgoing
out	outlines (file extension indicator)
OUT	Output
OUTCWAR	Outbound Control Word Address Register
OUTLIM	Output Limiting
OUTRAN	Output Translator
OUTREG	Output Register
OUTS	Output String
OUTST	Outstanding
OUTWATS	Outgoing Wide Area Telephone Service
OV	Overflow
OVA	Original Video Animation
OVD	Optical Video Disk
OVD	Optional Valuation Date
OVFL	Overflow
OVG	Overage
OVI	Overvoltage Interruption
OVID	Object, View and Interaction Design
ovl	overlay (file extension indicator)
OVLD	Overload
OVLP	Overlap
OVLY	Overlay
OV/OC	Over Voltage/Over Current
OVPC	Odd Vertical Parity Check
ovr	overlay (file extension indicator)
OVRN	Overrun
OVT	Optical Voice Transmission
OW	Open Web
OW	Over-Write

OWA	Outlook Web Access
OWF	One-Way Function
OWF	Optimum Working Frequency
OWL	Object Window(s) Library
OWL	Open Windows Library
OWRTS	Open-Wire Radio Transmission System
OWS	One-Way Simultaneous
OWTL	Open-Wire Transmission Line
OWTTE	Or Words To That Effect
OX	Order Crossover
OY	Optimum Yield
oz	ounce(s)

P

p	page
P	Parity
P	Partitioned
p	per
P	Permeability
p	peta (prefix)
p	pico (prefix)
P	Point
P	Positive
P	Power
P	Primary
P	Principal
P	Printer
P	Probability
P	Program
P	Pulse
P2C2E	Processes Too Complicated To Explain
P3P	Platform for Privacy Preferences Project
PA	Package Application
PA	Panama
PA	Paper Advance
PA	Particular Average
PA	Pass
PA	Pending Availability
PA	Performance Analysis
PA	Physical Address
PA	Pin Assembly
PA	Power Alarm|Amplifier
PA	Pre-Assembler
PA	Precision Architecture
PA	Predictive Analyzer
PA	Problem Analysis
PA	Process Alert|Automation

PA	Product Analyst|Assurance
PA	Program Access|Address
PA	Program Analysis|Attention
P/A	Programmer/Analyst
PA	Project Analysis
PA	Pulse Amplifier
PA	Purchase Agreement
PAA	Peer Access Approval
PAAC	Program Analysis Adaptable Control
PAAD	Private Automatic Answering Device
PAAM	Practical Application of intelligent Agents and Multi-agent technology (international exhibition and conference)
PAAR	Product Assurance Analysis Report
PAB	Primary Application Block
PABD	Precise Access Block Diagram
PABX	Private Automatic Branch Exchange
PAC	Paging Area Controller
PAC	Perceptual Audio Coding
PAC	Performance Analysis and Control
PAC	Peripheral Autonomous Control
PAC	Personal Authentication Code
PAC	Perturbed Angular Correlations
PAC	Phoneme Access Controller
PAC	Pneumatic Analog Computer
PAC	Polled Access Circuit
PAC	Primary Address Code
PAC	Product Availability Code
PAC	Production Activity Control
PAC	Program Action Code
PAC	Program Authorized Credentials
PAC	Project Analysis and Control
PAC	Pulse Amplitude Modulation
PACCEPT	Presentation Accept
PACCOM	Pacific Computer Communications (network)
PACE	Packaged Cram Executive
PACE	Parts Automated Control through Electronics
PACE	Performance And Cost Evaluation
PACE	Planned Action with Constant Evaluation
PACE	Precision Analog Computing Equipment
PACE	Priority Access Control Enabled
PACE	Process and Assembly Computerized Environment
PACE	Processing And Control Element
PACE	Program Action Code Extension
PACE	Program for Arrangement of Cables and Equipment
PACE	Programmable Analog Computing Equipment
PACE	Programmed Automatic Communications Equipment
PACER	Process Assembly Case Evaluator Routine
PACI	Partnership for Advanced Computational Infrastructure

PACIT	Process Automation for Cable Interface Tape
PACM	Pulse Amplitude Code Modulation
PACOS	Process Automation Control Operating System
PACR	Performance And Compatibility Requirements
PACS	Personal Access Communications System(s)
PACS	Picture Archiving and Communication System
PACS	Program Authorization Control System
PACS-L	Public Access Computer Systems List
PACT	Pay Actual Computer Time
PACT	Prefix Access Code Translator
PACT	Print Active Computer Tables
PACT	Program for Automatic Coding Techniques
PACT	Programmable Asynchronous Clustered Teleprocessing
PACT	Programmed Analysis Computer Transfer
PACT	Programmed Automatic Circuit Tester
PACT	Public Access Cordless Telephony
PACUIT	Packet plus Circuit
PACX	Private Automatic Computer Exchange
PAD	Packet Assembler/Disassembler
PAD	Positioning Arm Disk
PAD	Program Analysis for Documentation
PAD	Propellant Actuated Device
PAD	Public Access Device
PADD	Personal Access Display Device
PADEL	Pattern Description Language
PADL	Part and Assembly Description Language
PADLA	Programmable Asynchronous Dual Line Adapter
PADRE	Portable Automatic Data Recording Equipment
PADS	Pen Application Development System
PADS	Performance Analysis Display System
PADS	Personnel Automated Data System
PADS	Program Allocator to Drum Storage
PADS	Programmer's Advanced Debugging System
PADT	Post Allow Diffused Transistor
PADU	Packet Assembling/Disassembling Unit
PAE	Peer Access Enforcement
PAEB	Pan-American EDIFACT Board
PAF	Page Address Field
PAF	Peripheral Address Field
PAF	Postcode Activated File
PAFC	Phase Automated Frequency Control
PAGAN	Pattern Generation Language
PAGCH	Paging and Access Grant Channel
PAGE	Preview And Graphics Editor
PAGEDEF	Page Definition
PAGES	Program Affinity Grouping and Evaluation System
PAGODA	Profile Alignment Group for Office Document Architecture
PAGSTOR	Page Storage

PAI	Parts Application Information
PAI	Pre-Arrival Inspection
PAI	Precise Angle Indicator
PAI	Process Automation Interface
PAI	Protocol Addressing Information
PAID	Programmers Aid In Debugging
PAIH	Public-Access Internet Host
PAINS	Patient Information System
PAIR	Performance And Improved Reliability
PAIR	Procurement Automated Integrated Requirements
PAIR	Product Analysis Incident Report
PAIS	Public-Access Internet Site
pak	packed (compressed file name extension)
PAK	Product Authorization Key
PAK	Program Attention Key
PAKDD	Pacific-Asia conference on Knowledge Discovery & Data mining
PAL	Pallet
PAL	Paradox Applications Language
PAL	Passive Activity Loss
PAL	Pedagogic Algorithmic Language
PAL	Personal Answer Line
PAL	Phase Alternating Line
PAL	Platform Abstraction Layer
PAL	Precision Artwork Language
PAL	Procedure Abstraction Language
PAL	Process Assembler Language
PAL	Process Audit List
PAL	Program Abstract Library
PAL	Programmable Array Logic
PAL	Programmed Application Library
PAL	Programmed Array Logic
PAL	Programming Assembly Language
PALASM	Programmable Array Logic Assembler
PALC	Plasma-Addressed Liquid Crystal
PALCD	Plasma-Addressed Liquid Crystal Display
PAL-DL	Phase Alternating Line with Delay Line
PALE	Phase Alternating Line Encoding
PALIS	Property And Liability Information System
PALS	Pre-Announcement Level System
PALS	Principles of the Alphabet Literacy System
PALS	Programmable Adapter Logic Sequence
PALS	Project for Automated Library Systems
PAM	Page Allocation Map
PAM	Panel Monitor
PAM	Parallel Arithmetic Mode
PAM	Partitioned Access Method
PAM	Peripheral Adapter Module
PAM	Primary Access Method

PAM	Printer Authorization Matrix
PAM	Programmable Active Memory
PAM	Programmed Associative Memory
PAM	Pulse Address Modem
PAM	Pulse Amplification|Amplitude Modulation
PAMA	Pulse-Address Multiple Access
PAMD	Payload Assist Module, Delta-class
PAM/D	Process Automation Monitor/Disk
PAMM	Precision Automatic Measuring Machine
PAMR	Public Access Mobile Radio
PAMS	Plan Analysis and Modelling System
PAN	Peripheral Area Network
PAN	Personal Account Number
PAN	Phase Advance Network
PAN	Practical Active Network
PAN	Primary Access Network
PAN	Primary Account Number
PANACEA	Philips Analysis program for Circuit Engineering Applications
PANDA	Passive Non-Destructive Assay
PANDA	Prestel Advanced Network Design Architecture
PANDORA	Preserving and Accessing Networked Documentary Resources of Australia
PANE	Performance Analysis of Electrical circuits
PANIC	Parameter Analysis of Integrated Circuits
PANTES	Panel Tester
PAP	Packet-level Procedure
PAP	Password Authentication Protocol
PAP	Phase Advance Pulse
PAP	Photonic Array Processor
PAP	Printer Access Protocol
PAP	Program Analysis Procedure
PAPA	Probabilistic Automatic Pattern Analyzer
PAPI	Precision Approach Path Indicator
PAPS	Performance Analysis and Prediction Study
PAQ	Process Average Quality
PAQ	Production, Accounting, Quality
PAR	Page Address Register
PAR	Paragraph
PAR	Parallel
PAR	Parameter
PAR	Peak-to-Average Ratio
PAR	Performance And Reliability
PAR	Personal Animation Recorder
PAR	Positive Acknowledgment with Retransmission
PAR	Program Access Request
PAR	Program Address Register
PAR	Program Analysis|Appraisal and Review

PAR	Project Authorization Request
PAR	Proposal Analysis Report
PAR	Pulse Address Register
PARA	Paragraph
PARADISE	Piloting A Researcher's Directory Service in Europe
PARAM	Parameter
paramp	parametric amplifier
PARASYN	Parametric Synthesis
PARC	Palo Alto Research Center
PARCC	Precision, Accuracy, Representativeness, Completeness, and Comparability
PARD	Precision Annotated Retrieval Display
PARDAC	Parallel Digital to Analog Converter
PARE	Program for Analytical Reliability Estimation
PAREX	Programmed Accounts Receivable Extra service
PARFORCE	Parallel Formal Computing Environment (EU)
PA-RISC	Precision Architecture – Reduced Instruction Set Computer
PARMA	Program for Analysis, Reporting and Maintenance
PARMLIB	Parameter Library
PARS	Performance Analysis and Reporting System
PARS	Programmed Airline Reservation System
PARSEC	Parallel State Event Condition
PARSTOR	Parallel Storage
PART	Participation
PARTNER	Proof of Analog Results Through a Numerical Equivalent Routine
PARTNERS	Physics and Application of Resonant Tunnelling for Novel Electronic, infra-Red and optical devices and Systems (EU)
pas	pascal source code (file extension indicator)
PAS	Passive
PAS	PDES Application protocol Suite for composites (PDES = Product Data Exchange using STEP)
PAS	Performance Animation System
PAS	Phase Address|Array System
PAS	Power Application Software
PAS	Process Automation System
PAS	Processed Array Signal
PAS	Publicly Available Specifications
PASC	Phase Accurate Sub-band Coding
PASC	Precision Adaptive Sub-band Coding
PASG	Pulse Amplifier Symbol Generator
PA/SI	Preliminary Assessment/Site Investigation
PASLA	Programmable Asynchronous Line Adapter
PASP	Public Answering Service Point
PASS	Private Automatic Switching System
PASS	Production Automated Scheduling System
PAST	Process Accessible Segment Table
pat	hatch pattern (file extension indicator)

pat	patch (file extension indicator)
PAT	Patent(ed)
PAT	Pattern
PAT	Person Activity Tracker
PAT	Personalized Array Translator
PAT	Planar Assembly and Test system
PAT	Portable Audio Terminal
PAT	Port Address Translation
PAT	Prediction Analysis Technique
PAT	Procedure for Automatic Testing
PAT	Process Analysis Technique
PAT	Product Adaption Tool
PAT	Program Analysis Table
PAT	Program Analyzer Tool
PAT	Programmer Aptitude Test(er)
PAT	Pseudo-Adder Tree
PATBX	Private Automated Telegraph Branch Exchange
PATE	Programmed Automatic Test Equipment
PATIO	Program Addressable Table Index Operation
PATN	Pattern
PATRIC	Pattern Recognition Interpretation and Correlation
PATS	Parameterized Abstract Test Suite
PATS	Preauthorized Automatic Transfer Scheme
PATS	Programmable Automatic Testing System
PATSY	Pulse Amplitude Transmission System
PATT	Partial Automatic Translation Technique
PATX	Private Automatic Telegraph Exchange
PAU	Pattern Articulation Unit
PAU	Picture Acquisition Unit
PAV	Phase-Angle Voltmeter
PAV	Program Activation Vector
PAVAT	Process Area Variables Tester
PAWS	Portfolio Accounting WWW Security
PAWS	Protect Against Wrapped Sequence numbers
PA-WW	Precision Architecture – Wide Word
PAX	Physical Address Extension
PAX	Pixel Addressing Extension
PAX	Place Address in Index
PAX	Portable Archive Exchange
PAX	Private Automatic Exchange
PB	Page Buffer
PB	Peripheral Buffer
PB	Phonetically Balanced
PB	Power Box
PB	Program Base
PB	Proportional Band
PB	Push Button
PBA	Pre-Boot Authentication

PBA	Purchase Burden Applied
PBAM	Problem Billing Analysis Module
PBAR	Print Buffer Address Register
PBC	Period Batch Control
PBC	Periodic Binary Convolutional
PBC	Peripheral Bus Computer
PBC	Personal Business Computer
PBC	Processor Bus Controller
PBC	Pure Binary Code
PBCD	Packed Binary Coded Decimal
PBD	Precise Block Diagram
PBE	Prompt By Example
PBGA	Plastic Ball Grid Array
PBI	Post-Bonding Inspection
PBI	Process Branch Indicator
PBIB	Partially Balanced Incomplete Block
PBIC	Programmable Buffer Interface Card
PBN	Physical Block Number
PBN	Policy Based Networking
PBO	Plotting Board Operator
PBO	Push Button Operation
PBOT	Physical Beginning Of Tape
PBP	Point By Point
PBP	Push Button Panel
PBR	Perspective-Based Reading
PBS	Process Batch Size
PBS	Program Buffer Storage
PBS	Punch Barrier Strip
PBS	Push Button Switch
PBSW	Push Button Switch
PBT	Pass Band Tuning
PBT	Personal Business Terminal
PBT	Push Button Telephone
PBX	Private Branch|Business Exchange
PC	Parity Check
pc	parsec
PC	Patch Conversion
PC	Path Control(ler)
PC	Per Cent
PC	Personal Computer
PC	Phase Corrector
PC	Photo Cell|Conductor
PC	Photo Copy
PC	Picture
PC	Piece
PC	Plant Communication(s)
PC	Plug Compatible
PC	Pluggable Connector

PC	Point Contact
PC	Portable Computer
PC	Positive Column
PC	Primary Center
PC	Print Check|Complement
PC	Printed Circuit
PC	Priority Code
PC	Private Code
PC	Privileged Character
PC	Procedure Coordinator
PC	Process Center|Change
PC	Process Control
PC	Product Category|Code
PC	Product Costing
PC	Production Computer|Control
PC	Professional Computer
PC	Program Counter
PC	Programmable Controller
PC	Programmed Check|Console
PC	Programming Change
PC	Project Control
PC	Pulse Counter|Control(ler)
PCA	Performance and Code Analysis
PCA	Photon-Coupled Amplifier
PCA	Physical Configuration Audit
PCA	Printed Circuit Assembly
PCA	Process Communications Architecture
PCA	Process Control Analyzer
PCA	Protective Connecting Arrangement
PCA	Pulse Counter Adapter
PCACIAS	Personal Computer Automated Calibration Interval Analysis System
PCAM	Partitioned Content Addressable Memory
PCAMI	Personal Computing Asset Management Institute (US)
PCAP	Process Characterization Analysis Package
PCAR	Project Completion Analysis Report
PCAS	Programming Components Announcement Summary
PC-AT	Personal Computer – Advanced Technology
PCAT	Product Category
PCB	Page Control Block
PCB	Printed Circuit Board
PCB	Process(or) Control Block
PCB	Program Communication Block
PCB	Program|Protocol Control Block
PCB	Punch Circuit Breaker
PCBA	Printed Circuit Board Assembly
PCBC	Plain Cipher Block Chaining
PCBE	Page Control Block Element

PCBS	Printed Circuit Board Socket
PCC	Personal Communication Computer
PCC	Portable C Compiler
PCC	Port Controller Chip
PCC	Print Character|Compare Counter
PCC	Print Control Character
PCC	Printed Circuit Card
PCC	Process Change Control
PCC	Process Control Computer|Card
PCC	Program Control Counter
PCC	Program Controlled Computer
PCCH	Program Control Channel
PCCO	Production Control Close Out
PCCP	Product Cost Curve Picture
PCCS	Ported Coaxial Cable System
PCCS	Process Control Computer System
PCCS	Processor Common Communications System
PCD	Partition Control Descriptor
pcd	photo (image) compact disc (file extension indicator)
PCD	Poly-Crystalline Diamond
PCD	Power Current Device
PCD	Pre-Configured Definition
PCD	Program Control Document
PCDA	Program Controlled Data Acquisition
PC/DOS	Personal Computer Disk Operating System
PCE	Photo-Cell Emitter
PCE	Presentation Connection Endpoint
PCE	Procedure Control Expression
PCE	Process Control Equipment
PCE	Processing and Control Element
PCE	Program Cost Estimate
PCEB	PCI to EISA Bridge (PCI = Programmable Communications Interface)
PCEH	Project Customer Engineering Hours
PCEI	Presentation Connection Endpoint Identifier
PCEO	Personal Computer Enhancement Operation
PCEP	Presentation Connection End-Point
PCES	Product Cost Estimate System
PCEX	Print Cycle Exit
PCF	Phase Crossover Frequency
PCF	Portable Compiled Font
PCF	Primary Control Field
PCF	Program Complex File|Facility
PCF	Programmed Cryptographic Facility
PCFF	Personal Computer File Finder
PCG	Printed Circuit Generator
PCG	Programmable Character Generator
PCGL	Printed Circuit Generated Level

PCGRIDS	Personal Computer Gridded Interactive Display and diagnostic System
PCH	Parallel Channel
PCH	Punch
PCHAR	Printing Character
PCI	Panel Call Indicator
PCI	Pattern Correspondence Index
PCI	Peripheral Command Indicator
PCI	Peripheral Component Interconnect|Interface
PCI	Peripheral Computer Interconnect
PCI	Peripheral Connection|Controller Interface
PCI	Personal Computer Interconnect
PCI	Presentation Context Identifier
PCI	Process Control Interface
PCI	Program Check Interrupt(ion)
PCI	Program Control(led) Interrupt(ion)
PCI	Programmable Communications Interface
PCI	Project|Protocol Control Information
PCIA	Personal Communications Industry Association
PCIC	Program Control Information Card
PCIL	Private Core Image Library
PC-I/O	Program Controlled Input/Output
PCIOS	Processor Common Input/Output System
PCIS	Process Control Information System
PCIU	Parts Controlled by Identifiable Unit
PCI/X	Personal Computer Interactive Executive
PCK	Phase|Printed Control Keyboard
PCKB	Phase Control Keyboard
PCKB	Printed Control Keyboard
PCKT	Printed Circuit
PCL	Print(er) Command|Control Language
PCL	Process Control Language
PCL	Programmable Command Language
PCLA	Process Control Language
PCLK	Program Clock
PCLR	Parallel Communications Link Receiver
PCLX	Parallel Communications Link Transmitter
PCM	Personal Computer Manufacturer
PCM	Plug Compatible Mainframe|Manufacturer
PCM	Plug Compatible Memory
PCM	Primary Control Module
PCM	Printed Circuit Motor
PCM	Process Communication|Control Monitor
PCM	Project Control Methods
PCM	Pulse Code Modulation
PCMC	Personal Computer Memory Card
PCMCIA	Personal Computer Memory Card International Association

PCMCIA	Personal Computer Miniature Communications Interface Adapter
PCMD	Pulse Code Modulation Digital
PCME	Pulse Code Modulation Event
PCMI	Photo-Chromic Micro-Image
PCMIA	Personal Computer Manufacturer Interface Adapter
PCMIM	Personal Computer Media Interface Module
PCMV	Plug Compatible Mainframe Vendor
PCN	Personal Communication(s) Networks
PCN	Personal Computer Network
PCN	Process Change Notice
PCNFS	Personal Computer Network File System
PCO	Process Change Order
PCO	Program Controlled Output
P-CODE	Pseudo-Code
PCONNECT	Presentation Connect
PCOS	Process|Production Control Operating System
PCP	Peripheral Control Pulse
PCP	Primary|Process Control Program
PCP	Program Change Proposal
PC/PA	Process Control/Product Acceptance
PCPC	Personal Computers Peripheral Corporation
PCPM	Programmable Call Progress Monitoring
PCR	Page Control Register
PCR	Partial Carriage Return
PCR	Peak Cell Rate
PCR	Per Call Rate
PCR	Polychromatic Colour Removal
PCR	Polymerism Chain Reaction
PCR	Print Command Register
PCR	Print Contrast Ratio
PCR	Process Control Rack
PCR	Production Control Record
PCR	Program Control Register
PCR	Program|Project Change Request
PCS	Patchable Control Store
PCS	Permanent Change of Station
PCS	Personal Communication(s) Services|System
PCS	Personal Computing System
PCS	Personal Conferencing Specification
pcs	pieces
PCS	Plastic Clad Silica
PCS	Preferred Character Set
PCS	Print Contrast Signal
PCS	Process Communication Supervisor|System
PCS	Process Control System(s)
PCS	Product Customization System
PCS	Program Center Store

PCS	Program Checkout|Control System
PCS	Program Counter Store
PCS	Programmable Character Set
PCS	Programmable Communications Subsystem|System
PCS	Project Control System
PCSA	Personal Computer System Architecture
PCSFSK	Phase Comparison Sinusoidal Frequency Shift Keying
PCSL	Procurement Component Supplier List
PCSN	Private Circuit-Switching Network
PCS/REAL	Project Control System/Resource Allocation
PCSS	Personal Computer Support Service
PCSS	Photo-Conductive Semiconductor Switch
PCSU	Peripheral Control Switching Unit
PCT	Partition Control Table
PCT	Percent(age)
PCT	Peripheral Control Terminal
pct	picture (file extension indicator)
PCT	Printed Circuit Tester
PCT	Processing Control Table
PCT	Program Call Table
PCT	Program Concept Trainer
PCT	Program Control Table
PCTE	Portable Common Tool Environment
PCTG	Programmable Channel Termination Group
PCTR	Program Counter
PCTR	Protocol Conformance Test Report
PCTS	POSIX Conformance Test Suite
PCTV	Printed Circuit Test Vehicle
PCTV	Program Controlled Transverters
PCU	Packet Control Unit
PCU	Page Clean-Up
PCU	Peripheral|Power Control Unit
PCU	Port Contention Unit
PCU	Process(or)|Primary Control Unit
PCU	Program(mable) Control Unit
PCUG	Private Closed User Group
PCUR	Procurement
PCV	Printed Circuit (test) Vehicle
PCW	Program Control Word
PCWG	Personal Conference Working Group
PCX	Personal Computer (format) Extension
pcx	picture (image) (file extension indicator)
PCX	Process Control Executive
PD	Panel Display
PD	Peripheral Decoder
PD	Phase Distortion
PD	Photo-Detector
PD	Physical Design

PD	Plasma Display
PD	Plate Dissipation
PD	Position Description
PD	Potential Difference
PD	Power Dissipation
PD	Probability Density
PD	Procedure Division
PD	Product Design
PD	Projected Display
PD	Proximity Detection
PD	Public Domain
PD	Pulse Driver|Duration
PDA	Parallel Data Adapter
PDA	Percent Defective Allowable
PDA	Personal Data|Digital Assistant
PDA	Photo Diode Array
PDA	Physical Device Address
PDA	Probability Distribution Analyzer
PDA	Problem Determination Application
PDA	Production Data Acquisition
PDA	Push Down Automat(i)on
PDAID	Problem Determination Aid
PDAS	Process Data Acquisition Server
PDAS	Process Design Analysis System
PDAS	Programmable Data Acquisition System
PDAU	Physical Delivery Access Unit
PDB	Physical Data Base
PDBIN	Processor Data Bus In
PDBM	Pulse Delay Binary Modulation
PDBR	Physical Data Base Record
PDC	Parallel Data Communicator|Controller
PDC	Peripheral Device Controller
PDC	Photo Data Card
PDC	Primary Domain Controller
PDC	Problem Domain Component
PDC	Product Display Center
PDC	Production Decision Criteria
PDC	Professional Developers Conference
PDC	Programmable Data Controller
PDC	Programmable Desk Calculator
PDC	Programmable Digital Clock
PDC	Programme Delivery Control
PDC	Project Data Card
PDCA	Plan, Do, Check, Act
PDCAA	Plan, Do, Check, Act, Analyze
PDCS	Parallel Digital Computer System
PDCS	Performance Data Computer System
PDCS	Predictably Dependable Computing Systems (EU)

PDCS	Processing Distribution and Control System
PDCU	Plotting Display Control Unit
PDD	Physical Data Description
PDD	Physical Device Driver
PDD	Processor Description Database
PDD	Product Definition Data
PDD	Program Description Document
PDD	Projected Data Display
PDDB	Product Definition Data Base
PDDI	Product Definition Data Interface
PDDT	Page Device Description Table
PDE	Partial Differential Equation(s)
PDE	Portable Development Environment
PDE	Product Data Exchange
PD&E	Product Design & Engineering
PDED	Partial Double Error Detection
PDEL	Partial Differential Equation Language
PDES	Product Data Exchange Standard
PDES	Product Data Exchange using STEP (Standard for the Exchange of Product model data)
PDF	Package Definition File
PDF	Parallel Data Field
PDF	Parallel Disk File
PDF	Portable Document File
pdf	portable document format (file extension indicator)
PDF	Power Distribution Frame
PDF	Printer Definition|Description File
pdf	printer description format (file extension indicator)
PDF	Probability Density|Distribution Function
PDF	Problem Determination Facility
PDF	Processor Defined Function
PDF	Program Data File
PDF	Program Development Facility
PDF	Pulse Duty Factor
PDFG	Planar Distributed Function Generator(s)
PDG	Program Documentation Generator
PDGL	Part Design Graphic Language
PDH	Plesiochronous Digital Hierarchy
PDI	Picture Definition|Description Instruction(s)
PDI	Power and Data Interface
PDI	Power Dissipation Index
PDI	Product Data Interchange
PD&I	Product Development & Introduction
PDIAL	Public Dial-up Internet Access List
PDIF	Product Definition Interchange Format
PDIO	Parallel Digital Input/Output
PDIP	Plastic Dual-In-line Package
PDIR	Peripheral Dataset Information Record

PDL	Page Description|Design Language
PDL	Picture Description Language
PDL	Positive Diode Logic
PDL	Preliminary Design Language
PDL	Procedure Definition Language
PDL	Process Design Language
PDL	Program Description|Design Language
PDL	Programmable Data Logger
PDL	Programmed Digital Logic
PDL	Push Down List
PDM	Practical Data Manager
PDM	Precedence Diagramming Method
PDM	Print Down Module
PDM	Process Decision Model
PDM	Product Data Management
PDM	Program Debugging Mode
PDM	Program Design Manual
PDM	Programmer Defined Macro
PDM	Pulse Deviation|Duration Modulation
PDM	Push Down Memory
PDMA	Polarization Division Multiple Access
PDMM	Push Down Memory Modem
PDMU	Passive Data Memory Unit
PDN	Portable Data Network
PDN	Public Data Network
PDNC	Parallel Digital Network Computer
PDO	Portable Distributed Objects
PDOSTOR	Push Down Storage
PDP	Parallel Data Processor
PDP	Partial Drive Pulse
PDP	Plasma Display Panel
PDP	Pneumatic Data Processing
PDP	Procedure Definition Processor
PDP	Program Development Plan
PDP	Programmed Data|Digital Processor
PD&P	Project Definition & Planning
PDPCS	Product Development Planning and Control System
PDPS	Program Data Processing System
PDPTA	Parallel and Distributed Processing Techniques & Applications
PDQ	Passed Dataset Queue
PDQ	Program for Descriptor Query
PDQ	Programmed Data Quantizer
PDR	Page Data|Description Register
PDR	Pre-Defined Report(s)
PDR	Preliminary Data Report
PDR	Preliminary Design Review
PDR	Processing Data Rate
PDRF	Parts Data Record File

PDRL	Procurement Data Requirements List
PDS	Packet Data Satellite
PDS	Packet Driver Specification
PDS	Page Data Set
PDS	Parallel Data Structure
PDS	Partitioned Data Set
PDS	Performance Display System
PDS	Personnel Data System
PDS	Photo-Digital Store
PDS	Photo Document Sensor
PDS	Planetary Data System
PDS	Portable Display Shell
PDS	Portable Document Software
PDS	Problem Definition Solution
PDS	Process Diagnosis System
PDS	Processor Direct Slot
PDS	Procurement Data Sheet
PDS	Product Development System
PDS	Program Data Sheet
PDS	Program Development System
PDS	Programmable Data Station
PDS	Programmable Devices and Systems
PDS	Programmed Decision System
PDS	Programming Development System
PDS	Protected Dynamic Storage
PDS	Public Domain Software
PDSA	Plan, Do, See, Approve
PDSM	Particular Design Specification Model
PDSMAN	Partitioned Data Set Management system
PDSP	Peripheral Data Storage Processor
PDSS	Post Development and Software Support
PDSS	Product Definition Support System
PDT	Parallel Data Transmission
PDT	Parameter Descriptor Table
PDT	Part Design Tree
PDT	Performance Diagnostic Tool
PDT	Product Data Technology
PDT	Product Design Tree
PDT	Program Distribution Transmittal
PDT	Programmable Data Terminal
PDT	Programmable Drive Table
PDT	Push Down Transducer
PDTS	Program Development Tracking System
PDU	Plasma Display Unit
PDU	Plug|Power Distribution Unit
PDU	Product Development Unit
PDU	Programmable Delay Unit
PDU	Protocol Data Unit(s)

pdx	paradox files (file extension indicator)
PDX	Program Development Executive
PE	Page End (character)
PE	Parity Error|Even
PE	Partial Evaluation
PE	Performance Enhancement
PE	Period Ending|Entry
PE	Peripheral Equipment
PE	Personal Electronics
PE	Peru
PE	Phase Encoded
PE	Portable Executable
PE	Print End
PE	Printer's Error
PE	Probable Error
PE	Processing Element
PE	Protect Enable
PE	Pulse Encoding
PEA	Pocket Ethernet Adapter
PEAC	Program Establishment And Control
PEAR	Program Error Analysis Report
PEARL	Process, Experiment, and Automation Real-time Language
PEB	PCM Expansion Bus (PCM = Primary Control Module)
PEBB	Power Electronic Building Blocks
PEBCAK	Problem Exists Between Chair And Keyboard
PEBS	Pulsed Electron Beam Source
PEBX	Private Electronic Branch Exchange
PEC	Photo-Electric Cell
PEC	Print Error Check
PEC	Program Element|Execution Code
PEC	Program Event Counter
PEC	Program Execution Control
PECA	Preliminary Engineering Change Analysis
PECA	Proposed Engineering Change Analysis
PECN	Process Equipment Change Notification
PECOS	Personal Electronic Catalog and Ordering System
PECOS	Program Environment Check-Out System
PECOS	Project Evaluation and Cost Optimization System
PECU	Print Edit Control Unit
PED	Period End Date
PED	Priority Error Dump
PEDI	Protocol for Electronic Data Interchange
P-EDIT	Parametric Editor
PEDS	Packaging Engineering Data System
PEE	Photoferro-Electric Effect
PEEP	Production Electronic Equipment Procurement
PEF	Periodic Focusing
PEF	Prototyper Extract File

PEG	Prime Event Generation
PEGS	Parametric Evaluation Geometric System
PEI	Parity Error Interrupt
PEI	Preferred Equipment Identifier
PEIC	Periodic Error Integrating Controller
PEL	Permissible Exposure Level|Limit
PEL	Picture Element
PEM	Performance Enhancement Module
PEM	Photo-Electro-Magnetic
PEM	Privacy-Enhanced Mail
PEM	Processing Element|Enhanced Memory
PEM	Program Element Monitor
PEN	Pacific Exchange Network
PENCIL	Pictorial Encoding Language
PEND	Pending
PENIS	Pan-European Network Information System
PENSDK	Pen computing Software Development Kit
PEOT	Physical End Of Tape
PEP	Packet Exchange Protocol
PEP	Packetized Ensemble Protocol
PEP	Paperless Electronic Payment
PEP	Partitioned Emulation Program(ming)
PEP	Peak Envelope Power
PEP	Performance Evaluation Program
PEP	Peripheral Event Processor
PEP	Planar Epitaxial Passivated
PEP	Platform Environment Profile
PEP	Process Evaluation Program
PEP	Program Evaluation Procedure
PEP	Programmed Emulation Partition
PEPC	Polynomial Error Protection Code
PEPE	Parallel Element Processing Ensemble
PEPPI	Projected Evaluation of Product Performance Indexes
PEPR	Precision Encoding and Pattern Recognition (device)
PEPY	Presentation Element Parser based on YACC
PER	Packed Encoding Rules
PER	Post-Execution Reporting
PER	Program Error Report
PER	Program Event Recording
PER	Program Execution Request
PERA	Purdue Enterprise Architecture
PERCOM	Peripheral Communications
PERCOMP	Personal Computing conference
PERCOS	Performance Coding System
PerDiS	Persistent Distributed Store (EU)
PERFECT	Performance evaluation For cost-Effective Transformations
PERL	Practical Extraction and Report Language
PERM	Permanent

PERM	Programmed Evaluation for Repetitive Manufacture
PERMACAPS	Personnel Management and Accounting Card Processor System
PERMT	Permit
PERP	Perpendicular
PERS	Performance Evaluation Reporting System
PERS	Program for Evaluation of Rejects and Substitutions
PERSIS	Personal Information System
PERT	Program Evaluation Research Test
PERT	Project Evaluation Review Technique
PERVAL	Performance/Valuation
PES	Paper End Signal
PES	Photo-Electric Scanning
PES	Positioning Error Signal
PES	Processor Enhancement Socket
PES	Program Execution System
PESD	Program Element Summary Data
PESDS	Program Element Summary Data Sheet
PESS	Product Evaluation and Selection System
PEST	Parameter Estimation by Sequential Testing
PESY	Peripheral Exchange Synchronization
PET	Peripheral Equipment Test(s)
PET	Personal Electronic Transaction|Translator
PET	Physical Equipment Table
PET	Position Event Time
PET	Print Enhancement Technology
PET	Privacy Enhancing Technology
PET	Process Evaluation Tester|Technique
PET	Program Evaluator and Tester
PET	Program Execution Time
PETS	Parameterized Executable Test Suite
PEU	Port Expander Unit
PEX	Packet Exchange protocol
PEX	PHIGS Extension
PF	French Polynesia
PF	Packet Fraction
PF	Page Footing|Format
PF	Permanent File
P/F	Poll/Final Bit
PF	Power Factor
PF	Program Function
PF	Programmable Format
PF	Programmed Function
PF	Pulse Frequency
PF	Punch off (character)
PFA	Predictive Failure Analysis
PFA	Product and Field Activity
P&FA	Program & File Analysis
PFAM	Programmed Frequency Amplitude Modulation

PFAR	Power Fail Automatic Restart
PFB	Pre-Fetch Buffer
PFB	Printer Font Binary
P-FBM	Parameterized Feature Based Modeller
PFC	Program Flow Chart
PFCB	Page Frame Control Block
PFCU	Parallel File Control Unit
PFD	Phase Frequency Distortion
PFDA	Precision Frequency Distribution Amplifier
PFDA	Pulse Frequency Distortion Analyzer
PFDE	Page File Description Entry
PFEP	Programmable Front End Processor
PFET	P-channel junction Field Effect Transistor
PFF	Page Fault Frequency
PFI	Physical Fault Insertion
PFK	Programmed Function Key(board)
PFKB	Program Function Keyboard
PFL	Pre Fade Listening
PFM	Package Feasibility Model
PFM	Performance Monitor
PFM	Power Frame
pfm	printer font metrics (file extension indicator)
PFM	Pulse Frequency Modulation
PFN	Permanent File Name
PFN	Prime Fanout Node
PFO	Partitioned File Organization
PFOR	Parallel Fortran
PFORM	Print Format
PFP	Power Failure Protection
PFP	Pre-Fetch Processor
PFP	Program File Processor
PFP	Programmable Function Panel
PFPU	Processor Frame Power Unit
PFR	Power Fail Recovery|Restart
PFR	Programmed Film Reader
PFR	Pulse Frequency Rate
PFS	Path Fault Secure
PFS	Programmable Frequency Standard
PFSK	Pulse Frequency Shift Keying
PFT	Page Frame Table
PFT	Paper Flat Tape
PFT	Program Fix Temporary
PFT	Pulse Fall Time
PG	Page
PG	Papua New Guinea
PG	Parity Generate
PG	Power Gain
PG	Program Generator|Graph

PG	Pulse Generation
PGA	Pin Grid Array
PGA	Professional Graphics Adapter
PGA	Programmable Gain Amplifier
PGA	Programmable Gate Array
PGBD	Plugboard
PGBL	Pluggable
PGC	Program Counter
PGC	Program Group Control
PGC	Programmed Gain Control
PGD	Planar Gas Discharge display
PGDN	Page Down
PGF	Presentation Graphics Feature
PGI	Parameter Group Identifier
pgl	graphics (file extension indicator)
PGL	Phosphating Granulating Liquid
PGLIN	Page and Line
PGM	Program(mer)
PGML	Precision Graphics Markup Language
PGNR	Page Number
PGP	Pretty Good Privacy
PGP	Programmable Graphics Processor
pgp	program parameter (file extension indicator)
PGR	Precision Graph Recorder
PGR	Presentation Graphics Routines
PGS	Program Generation Subsystem
PGSE	Pulsed Gradient Spin Echo
PGT	Page Table
PGT	Program Global Table
PGTN	Propagation
PgUp/PgDn	Page Up and Page Down (key)
PG/ZD	Propagated Group/Zero Defect
PH	Packet Handler|Header
PH	Page Heading
PH	Parity High bit
PH	Phase
PH	Philippines
PH	Physical Header
PH	Polarity|Process Hold
PHA	Pulse Height Analysis
PHALSE	Phreakers, Hackers, And Laundry Service Employers
phant	phantastron
PHB	Program Header Block
PHC	Power Handling Capability
PHD	Parallel Head Disk
PHD	Professional Help Desk
PHF	Packet Handling Function
PHF	Power Handling Factor

PHIGS	Programmer Hierarchical Interactive Graphics System
PHIGS+	Programmer Hierarchical Interactive Graphics System Extended
PHINTS	Phase-locked Interferometric Tracking System
PHM	Phase Modulation
PHNBR	Phone Number
pho	phone list (file extension indicator)
PHODEC	Photometric Determination of Equilibrium Constants
PHONO	Phonograph
PHOTOCD	Photographic Compact Disk
PHQ	Phase Qualification
PHR	Physical Record
PHR	Process Hazards Review
PHR	Pulse Height Resolution
PHRF	Performance Handicap Rating Formula
PHS	Personal Handyphone System
PHSS	Public Health Software Systems
PHT	Phototube
PI	Parameter Identifier
PI	Performance Index|Indicator
PI	Peripheral Inventory
PI	Peripherals Interface
PI	Photo Interpretation
PI	Planet Internet
PI	Power Indicator|Input
PI	Preliminary Input
P/I	Pressure to Current
PI	Primary Input
PI	Process Interrupt
PI	Processor Interface
PI	Program Identification|Indicator
PI	Program Interruption|Interrupt(or)
PI	Program Isolation
PI	Programmed Information|Instruction
PI	Proportional-plus Integral
PI	Protocol Identification|Interpreter
PI	Pulse Input
PIA	Peripheral Interface Adapter
PIA	Plug-In Administrator
PIA	Product Information Archive
PIB	Partition Information Block
PIB	Policy Information Base
PIB	Programmable Input Buffer
PIC	Peripheral Interface Channel|Controller
PIC	Personal Identification Code
PIC	Personal Information Carrier
PIC	Personal Intelligent Communicator
pic	picture file (file extension indicator)

PIC	Picture Image Compatibility
PIC	Plastic-Insulated Cable
PIC	Plug-In Card
PIC	Polyethylene-Insulated Conductor
PIC	Position Independent Code
PIC	Primary Independent Carrier
PIC	Primary Interexchange Carrier
PIC	Priority Interrupt Controller
PIC	Professional Image Computer
PIC	Program Information Code
PIC	Program(mable) Interrupt(ion) Control(ler)
PIC	Programmed Instruction Counter
PICA	Picture-Coding Algorithm
PICA	Program Interrupt Control Area
PICA	Project Integrated Catalogue Automation
PICASSO	Parts Inventory Control And Shipping System On-line
PICE	Program Instruction Control Element
PICE	Programmable Integrated Control Equipment
PICO	Program In, Chip Out
PICOS	Parts Inventory Control On-line System
PICRS	Program Information Control and Retrieval System
PICS	Personnel Information Communication System
PICS	Platform for Internet Content Selection
PICS	Platform Independent Content Selection
PICS	Protocol Implementation Conformance Statement
PICT	Picture (file)
PICU	Parallel Instruction Control Unit
PICU	Priority Interrupt Control Unit
PID	Personal Identification Device
PID	Pictorial Information Digitizer
PID	Process Identifier
P&ID	Process & Instrumentation Diagram
PID	Program Information Department
PID	Project Initiation Document
PID	Proportional, Integral, Derivative
PID	Protocol Identifier
PID	Pseudo Interrupt Device
PIDB	Peripheral Interface Data Bus
PIDC	Photo-Induced Discharge Characteristic
PIDCOM	Process Instruments Digital Communication system
PIDENT	Program Identification
PIDL	Personalized Information Description Language
PIDM	Particular Implementation Description Model
PIDS	Public Investment Data System
PIE	Parallel|Peripheral Interface Element
PIE	Plug-In Electronics|Equipment
PIE	Program Interrupt(ion) Element
PIE	Program Investment Evaluation

PIE | Pulse Interference Eliminator
PIECE | Productivity, Information, Education, Creativity, Entertainment
PIEEE | Posix IEEE
PIF | Phase Interface Fading
PIF | Picture Interchange File|Format
PIF | Process Interchange Format
PIF | Program Information File
PIF | Programmable Interface
PIFT | Protocol Interbank File Transfer
PIG | Passive Income Generator
PII | Program Integrated Information
PIL | Precision In-Line
PIL | Priority Interrupt Level(s)
PIL | Processing Information List
PILOT | Programmed Inquiry Learning Or Teaching
PilotACE | Pilot Automatic Computing Engine
PILP | Parametric Integer Linear Program
PIM | Personal Information Manager
PIM | Port|Process Interface Module
PIM | Processor-In-Memory
PIM | Processor Interface Module
PIM | Protocol Independent Multicast
PIM | Pulse Interval Modulation
PIML | Polynomial propagation time Intermediate Language
PIN | Personal Identification Number
pin | positive, intrinsic, negative
PIN | Privileged Instruction
PIN | Process Identification Number
PIN | Processor Independent (software)
PIN | Program Information
PINE | Pine Is Not Elm
PING | Packet Internet Groper
PINO | Positive Input/Negative Output
PINT | Power Integrated Transistors
PINT | Processor Interrupt
PINTE | Processor Interrupts Enable
PIO | Parallel|Peripheral Input/Output
PIO | Precision Interactive Operation
PIO | Process(or) Input/Output
PIO | Programmable|Programmed Input/Output
PIOA | Programmed Input/Output Address
PIOC | Programmed Input/Output Command
PIOCS | Physical Input/Output Control Subroutine|System
PIOM | Physical Input/Output Manager
PIOO | Programmed Input/Output Operation
PIOSP | Process Input/Output Subroutine Package
PIOU | Parallel Input/Output Unit
PIP | Packet Interface Port

PIP	Page Image Processor
PIP	Partner Interface Process
PIP	Path Independent Protocol
PIP	Peripheral Interchange Program
PIP	Personal Interaction Panel
PIP	Picture In Picture
PIP	P Internet Protocol
PIP	Polycoated Insert (data) Pack
PIP	Probabilistic Information Processing
PIP	Problem Isolation Procedure
PIP	Process Interface Processor
PIP	Programmable Interconnect Point
PIP	Programmed Interconnection Pattern(s)
PIPE	Plug-In Processing Element
PIPO	Parallel In/Parallel Out
PIPS	Paperless Item Processing System
PIPS	Pattern Information Processing System
PIPS	Product Information Pipeline System
PIQ	Parallel Instruction Queue
PIR	Passive Infra-Red
PIR	Program Incident Report
PIR	Program Instruction Register
PIR	Protocol Independent Routing
PIR	Pulse Input Register
PIRN	Preliminary Interface Revision Notice
PIRS	Project Information Retrieval System
PIRT	Placement by an Interchange and Rate Technique
PIRT	Programmed Instruction in Real Time
PIRV	Programmed Interrupt Request Vector
PISAB	Pulse Interference Suppression And Blanking
PISAM	Photon Induced Scanning Auger Microscope
PISO	Parallel In/Serial Out
PISW	Process Interrupt Status Word
PIT	Peripheral Input Tape
PIT	Print Illegal and Trace
PIT	Programmable Interval Timer
PIT	Projected Inactive Time
PITR	Product Inter-operation Test Report
PIU	Path Information Unit
PIU	Plug-In Unit
PIU	Process Interface Unit
PIV	Peak Inverse Voltage
PIV	Pivot
PIVOT	Programmer's Interface Verification and Organizational Tool
PIXEL	Picture Element
PIXIT	Protocol Implementation Extra Information for Testing
Pjava	Personal java
PJL	Printer Job Language

pk	packed (font file) (file extension indicator)
PK	Pakistan
PK	Protection Key
PK	Public Key
PKA	Public Key Algorithm
PKB	Public Key Block
PKC	Position Keeping Computer
PKC	Public Key Cryptography
PKCS	Public Key Cryptographic System
PKD	Packed
PKD	Programmable Keyboard/Display
PKE	Public Key Encryption
pkg	package
PKI	Private|Public Key Infrastructure
PKP	Public Key Partners
PK/PK	Peak to Peak
PKT	Packet
PKZIP	Phil Katz's Zip
PL	Parity Low bit
PL	Physical Layer
PL	Pluggable
PL	Poland
PL	Presentation Layer
PL	Private Line
PL	Procedure Library
PL	Program Level|Library
PL	Program Limit|Listing
PL	Programming Language
PL/1	Programming Language One
PLA	Print Load Analyzer
PLA	Programmable Line Adapter
PLA	Programmable Logic Array
PLAB	Party-Line Adapter Board
PLACE	Positioning, Layout, And Cell Evaluation
PLACE	Programming Language for Automatic Checkout Equipment
PLAMO	Planar Approximation Modelling system
PLAN	Problem Language Analyzer
PLAN	Public Library Automation Network
PLANES	Programmed Language-based Enquiry System
PLANET	Private Local Area Network
PLANIT	Programming Language for Interactive Teaching
PLANS	Programming Language for Allocating and Network Scheduling
PLAR	Private Line Auto-Ringdown
PLAS	Program Logical Address Space
PLATO	Programmed Logic for Automatic Teaching Operations
PLATON	Programming Language for Tree Operation
PLAW	Programming Language for Arc Welding
PLB	Personal Location Beacon

PLB	Print Line Buffer
PLC	Power Line Communication
PLC	Power Line Conditioner|Cycles
PLC	Print Line Complete
PLC	Process Line Control
PLC	Program Level Change
PLC	Programmable Logic Controller
PLCB	Program List Control Block
PLCB	Pseudo-Line Control Block
PLCC	Plastic Leadless Chip Carrier
PLCP	Physical Layer Convergence Procedure
PLD	Package Level Detail
PLD	Phase Lock(ed) Demodulator
PLD	Physical Logical Description
PLD	Power Line Disturbance
PLD	Programmable Logic Device
PLDI	Programming Language Design and Implementation
PLDS	Pilot Land Data System
PL/E	Programming Language/Edit
PLENG	Physical record Length
PLEX	Programming Language Extension
PLF	Page Length Field
PLF	Phone Line Formatter
PLF	Pixel Liberation Format|Front
PLG	Process Line Generator
PLI	Private Line Interface
PLIB	Program Library
PLIC	Procedural Language for Integrity Constraints
PLIMS	Programming Language for Information Management System
PLL	Phase-Lock(ed) Loop
pll	pre-linked library (file extension indicator)
PLLT	Program Load Library Tape
PLM	Passive Line Monitor
PLM	Planned Maintenance
PL/M	Programming Language for Microprocessors
PLM	Program(ming) Logic Manual
PLM	Pulse-Length Modulation
PLMATH	Procedure Library for Mathematics
PLMN	Public Land Mobile Network (Pan-European)
PLMS	Partitioned Libraries Management System
PLN	Platen
PLO	Phased Locked Oscillator
PLOD	Periodic List Of Data
PLOKTA	Press Lots Of Keys To Abort
PLOP	Pilot Line Operating Procedure
PLOS	Process Level Of Support
PLP	Packet-Layer Procedure|Protocol
PLP	Packet|Presentation Level Protocol

PLP	Procedural Language Processor
PLPA	Pageable Link Pack Area
PLR	Program Length Register
PLR	Program Library Release
PLR	Programming Language for Robots
PLRS	Phase Lock Receiving Station
PLRS	Position Location Reporting System
PLS	Personal Learning System
PLS	Physical Layer Signalling
PLS	Please (acknowledge)
PLS	Preliminary Location Summary
PLS	Primary Link Station
PLS	Private Line Service
PL/S	Programming Language/Systems
PLSAP	Physical Layer Service Access Point
PLT	Plot(ting language)
PLT	Power Line Transient
PLT	Private Line Teletypewriter
PLT	Program Library Tape
PLT	Program List Table
PLT	Programmed Light Table
PLU	Pluggable Unit
PLU	Plural
PLU	Power Logic Unit
PLU	Primary Logical Unit
PLUMS	Program Library Update and Maintenance System
PLUS	Program Library Update System
PM	Peripheral Module
PM	Permanent Magnet
PM	Phase Modulation
PM	Photo-Magnetic
PM	Photo Multiplier
PM	Physical Medium
PM	Planned Maintenance
PM	Post Meridian|Mortem
PM	Potentiometer
PM	Precedence Method
PM	Predictive Maintenance
PM	Premium Memory
PM	Prepared Message
PM	Presentation Manager
PM	Preventive Maintenance
PM	Process(ing) Module
PM	Program Manager|Memory
PM	Programming Manual
PM	Program Mode|Monitor
PM	Protocol Machine
PM	Pulse(d) Modulation

P/M	Put More
PM	St. Pierre et Miquelon
PMA	Performance Measurement Analysis
PMA	Physical Medium Attachment
PMA	Physical Memory Address
PMA	Preamplifier Module Assembly
PMA	Priority Memory Access
PMA	Program Memory Area
PMA	Protected Memory Address
PMAC	Packet Media Access Controller
PMAC	Parallel Memory Address Counter
PMAC	Peripheral Module Access Controller
PMAP	Procedure Map
PMAR	Page Map Address Register
PMB	Performance Measurement Baseline
PMB	Pilot Make Busy (circuit)
PMB	PROM Memory Board
PMBX	Private Manual Branch Exchange
PMC	Performance Management Computer
PMC	Permanently Manned Capability
PM&C	Process Monitor & Control
PMCD	Post Mortem Core Dump
PMCD	Program Module Connection Diagram
PMCS	Preventive Maintenance Checks and Services
PMCT	Program Management Control Table
PMCVG	Photometric Modelling for Computer Vision and Graphics
PMD	Packet Mode Data
PMD	Physical Media Dependent
PMD	Post Mortem Dump
PMD	Program Management Directive
PMD	Program Module Dictionary
PME	Phase Modulation Encoding
PME	Photo-Magnetic Effect
PME	Preventive Maintenance Effectiveness
PME	Privileged Mode Extensions
PME	Processor Memory Enhancement
PME	Program Memory Emulator
PMEG	Page Map Entry Group
PMEM	Processor Memory
PMF	Performance Monitor Function
PMF	Print Management Facility
PMF	Product Master File
PMF	Program Mode Field
PMF	Pulsed Magnetic Field
PMFJI	Pardon Me For Jumping In
PMG	Phase Modulation Generator
PMI	Personnel Management Information (system)
PMI	Programmable Machine Interface

PMI	Program Management Instruction
PMIC	Parallel Multiple Incremental Computer
PMIG	Programmer's Minimal Interface to Graphics
PMIRR	Pressure Modulated Infra-Red Radiometer
PMIS	Printing|Project Management Information System
PML	Physical Memory Loss
PML	Probable Maximum Loss
PML	Process Modelling Language
PMLC	Previous Micro-Location Counter
PMLC	Programmed Multi-Line Controller
PMM	Programmable Microcomputer Module
PMMB	Parallel Memory-to-Memory Bus
PMMU	Paged Memory Management Unit
PMN	Performance Monitoring
PMN	Project Management Network
PMO	Program Management Office
PMON	Performance Management Operations Network
PMOS	Positive channel Metal Oxide Semiconductor
PMP	Parallel Microprogrammed Processor
PMP	Performance Management Package
PMP	Physical Media Parameter
PMP	Post Mortem Program
PMP	Pre-Modulation Processor
PMP	Program Management Plan
PMPM	Phase Margin Performance Measure
PMR	Patient Master Record
PMR	Performance Measurement Report
PMR	Performance Monitoring Receiver
PMR	Permissive Make Relay
PMR	Phase Modulation Recording
PMR	Power Master Request
PMR	Private Mobile Radio
PMR	Problem Management Report
PMRS	Performance Management and Recognition System
PMS	Pantone Matching System
PMS	Performance Management|Measurement System
PMS	Process Messaging Service
PMS	Processor Memory Switch
PMS	Product|Program Management System
PMS	Programmed Mode Switch
PMS	Projected Moving Scale
PMS	Project Management System
PMS	Public Message Service
PMSD	Program Module Sequence Diagram
PMSS	Preventive Maintenance Scheduling System
PMSX	Processor Memory Switch Matrix
PMT	Packet-Mode Terminal
PMT	Photo-Multiplier Tube

PMT	Physical Master Tape
PMT	Printed Mechanical Transfer
PMT	Program Master Tape
PMTD	Post Mortem Tape Dump
PMTS	Predetermined Motion Time System
PMU	Parametric Measurement Unit
PMU	Performance Monitor Unit
PMU	Portable Memory Unit
PMUX	Programmable Multiplexer
PMX	Packet Multiplexer
PMX	Partially Matched Crossover
PMX	Presentation Manager for the X window system
PMX	Private Manual|Message Exchange
PMX	Protected Message Exchange
PN	Packet|Page Number
PN	Performance Number
PN	Petri Net
PN	Polish Notation
pn	positive/negative
PN	Positive Notification
PN	Processing Node
PN	Processor Number
PN	Programmable Network
PN	Pseudo Noise
PNA	Private Network Adapter
PNA	Programmable Network Access
PNA	Project Network Analysis
PNAF	Potential Network Access Facility
PNC	Paging Network Controller
PNC	Personal Number Calling
PNC	Programmed Numerical Control
PNCC	Partial Network Control Center
PNCH	Punch
PND	Present Next Digit
PNDG	Pending
PNET	Power Networking
PNEU	Pneumatic
PNEUC	Physical Non-Elementary Unit Control
PNEUC	Physical Non-Eliminated Unit Check
PNG	Portable Network Graphic(s)
PNI	Participate but do Not Initiate
PNIC	Private Network Identification Code
pnip	positive-negative-intrinsic-positive
PNL	Panel
PNM	Path Number Matrix
PNM	Pulse Number Modulation
PNNI	Private Network to Network Interface
PNO	Petri Net with Objects

pnp	positive-negative-positive
PNX	Private Network Exchange
PO	Parity Odd
PO	Patent Office
PO	Planning Objectives
P&O	Planning & Operations
PO	Polarity
PO	Power On
PO	Power Oscillator|Output
PO	Primary Output
PO	Privileged Operation
PO	Pulse Output
PO	Punch On (character)
POA	Portable Object Adapter
POA	Problem Oriented Assembler
POAC	Probe Origin Authentication Check
POB	Peripheral Order Buffer
POC	Particulate Organic Carbon
POC	Point Of Contact
POC	Power On Clear
POC	Power Oscillator
POC	Process Operator Console
POCIS	Physical Operational Control Information System
POCM	Portal Object Component Model
POCR	Processor Oriented Character Recognition
POCS	Patent Office Classification System
POCS	Process Operator Console Support
POCV	Perceptual Organization in Computer Vision (international workshop)
POD	Printing On Demand
POD	Probability Of Detection
POD	Program Operation Description
POD	Project Objectives Document
POD	Proton Omnidirectional Detector
PODA	Priority Oriented Demand Assignment
PODAF	Post Operation Data Analysis Facility
PODAPS	Portable Data Processing System
PODAS	Portable Data Acquisition System
PODR	Pixel Order
POE	Power Open Environment
POEM	Program Oriented External Monitor
POF	Point Of Failure
POF	Programmable|Programmed Operator Facility
POF	Program Order Form
POGO	Program(mer) Oriented Graphics Operation
POH	Path Overhead
POH	Power-On Hours
POI	Plan Of Instruction

POI	Point Of Information|Interconnect
POI	Program Of Instruction
POI	Program Operator Interface
POI	Purchase Of Installed (equipment)
POINT	Performance Oriented Integrated Technology
POK	Power on Okay signal
POKE	Processor Oriented Key Entry
POL	Problem|Procedure Oriented Language
POL	Produce Objectives – Logical/physical design
POL	Program Oriented Language
POLAR	Production Order Locating And Reporting
POLGEN	Problem Oriented Language Generator
POLIS	Polynomial Interpreting System
POLLG	Polling
POLMI	Problem Oriented Language – Machine-Independent
POM	Program Operation Mode
POMM	Preliminary Operating and Maintenance Manual
POMO	Production Oriented Maintenance Organization
POMS	Process Operations Management System
PON	Passive Optical Network
PON	Power On
PON	Purchase Of New (equipment)
PONA	Person Of No Account
POOL	Parallel Object Oriented Language
POOP	Principles Of Operation
POP	Perceived Outcome Potential
POP	Point Of Presence
POP	Pop (from stack)
POP	Post Office Protocol
POP	Power On/off Protection
POP	Processor On Plug-in
POP	Programmed Operators and Primitives
POPA	Pop All (registers)
POPAM	Parallel Optical Processors And Memories (EU)
POPF	Pop Flags
POPIN	Population Information Network
POPL	Principles Of Programming Languages
POPMail	Post Office Protocol Mail
POPO	Push On, Pull Off
POPS	Paperless Order Processing System
POPS	Parallel Optical Scanner
POPS	Predicted Order Programs
POPS	Process Operating System
POPS	Program for Operator Scheduling
POR	Plan Of Record
POR	Portion
POR	Power On Request|Reset
POR	Problem Oriented Routine

PORT Partnership Order Retrieval Tool
PORTS Physical Oceanographic Real-Time System
PORTS Port Objective for Real-Time Systems
POS Point Of Sale
POS Position
POS Positive
POS Primary Operating System
POS Processor|Professional Operating System
POS Programmable Object Select
POS Program Order Sequence
POSD Programs for Optical Systems Design
POSE Portable Operating System Extension
POSI Promoting the Open Systems Interconnect
POSIX Portable Operating System based on UNIX
POSIX Portable Operating System Interface Exchange|Extension
POSSO Polynomial System Solving (EU)
POST Point Of Sale Terminal
POST Power-On Self-Test
POST Production Oriented Scheduling Techniques
POSTNET Postal Numeric Encoding Technique
POSTOR Photo-Optical Storage
POT Picture Object Table
POT Point Of Termination|Train
POT Potential
POT Potentiometer
POT Program for Operational Trajectories
potmeter potentiometer
POTP Physical Operation Time Projection
POTS Photo-Optical Terrain Simulator
POTS Plain Old Telephone Service|System
POTS Plug-On Terminator System
POV Point Of View
POW Peripheral Output Writer
POWER Performance Optimization With Enhanced RISC (= Reduced Instruction Set Computer)
POWER Priority Output Writers, Execution processors and input Readers
POWERPC Performance Optimization With Enhanced RISC-Performance Computing
pp pages
PP Painting Processor
PP Parallel Processing|Processor
PP Partial Program
PP Past Participle
pp peak-to-peak
PP Peripheral Processor
PP Plasma Panel
PP Point to Point

PP	Post-Processor
PP	Pre-Printed
PP	Pre-Processor
PP	Primary Point|Processor
PP	Print Position
PP	Problem Program
PP	Program Product
PPA	Parts Performance Analysis
PPA	Photo-Peak Analysis
PPA	Pixel Processing Accelerator
PPA	Post Pack Audit
PPA	Pre-Planned Applications
PPA	Product Performance Analysis
PPA	Professional Photographers of America
PPA	Program Product Announcement
ppb	part(icle)s per billion
PPB	Part Period Balancing
PPB	Planning, Programming, Budgeting
PPB	PROM Programmer Board
PPB	Provisioning Parts Breakdown
PPBAS	Planning, Programming, Budgeting, Accounting System
PPBES	Program Planning-Budgeting-Evaluating System
PPBM	Pulse Polarization Binary Modulation
PPBS	Planning, Programming and Budgeting System
PPBS	Program Planning and Budgeting System
PPC	Personal Personal Computer
PPC	Personal Programmable Calculator
PPC	Platform Position Computer
PPC	Point to Point Connection
PPC	Portable Personal Computer
PPC	Power Personal Computer
PPC	Print Position Counter
PPC	Process-to-Process Communication
PPC	Program, Planning and Control
PPC	Program Product Center
PPC	Program-to-Program Communication(s)
PPCE	Print Position Control Exit
PPCE+E	Print Position Counter Entry and Exit
PPCI	Presentation Protocol Control Information
PPCM	Predictive Pulse Code Modulation
PPCS	Person to Person, Collect, Special
PPCU	Parallel Process Control Unit
PPD	Peripheral Pulse Distributor
PPD	Postscript Printer Description
PPD	Primary Paging Device
PPDD	Plan Position Data Display
PPDS	Personal Printer Data Stream
PPDS	Primary Processor and Data Storage

PPDU	Presentation Protocol Data Unit
PP/E	Parallel Print/Exact
PPE	Peripheral Processor Element
PPE	Pre-modulation Processing Equipment
PPE	Problem Program Efficiency|Evaluator
PPEP	Pen Plotter Emulation Program
PPF	Program Preparation Facilities
PPF	Program Production Fee
PPF	Program Protect Flags
PPFA	Page Printer Formatting Aid
PPGA	Plastic Pin Grade Array
pph	pages per hour
pph	prints per hour
P-PH-M	Pulse Phase Modulation
PPI	Plan(ned) Position Indicator
ppi	points per inch
PPI	Power Prime Implicant
PPI	Process Planning Interface
PPI	Programmable Peripheral Interface
PPI	Program Position Indicator
PPI	Project Performance Index
ppi	pulses per inch
PPIB	Programmable Protocol Interface Board
PPIU	Programmable Peripheral Interface Unit
PPL	Passed Parameter List
PPL	Plain Position Indicator
PPL	Point-to-Point Line
PPL	Polymorphic Programming Language
ppl	print positions per line
PPL	Program Production Library
PPLLT	Provisional Program Load Library Tape
PPM	Packet Processing Module
ppm	pages per minute
ppm	pages per month
PPM	Periodic Permanent Magnet
PPM	Planned Preventive Maintenance
PPM	Presentation Protocol Machine
PPM	Previous Processor Mode
PPM	Principal Period Maintenance
PPM	Pulse Position Modulation
ppm	pulses per minute
PPMS	Product Procedure Maintenance System
ppmv	parts per million by volume
PPMX	Primary Pulse code Multiplex
PPN	Parameterized Post-Newtonian (formalism)
PPN	Private Packet Network
PPN	Project Programmer Number
PPO	Point-to-Point Operation

PPP	Parallel Pattern Processor
PPP	Point-to-Point Protocol
PPP	Public-Private-Partnership
PPP	Purchasing Power Parity
PPPD	Point to Point Protocol Daemon
PPPOA	Point-to-Point Protocol Over ATM (= Asynchronous Transfer Mode
PPPOE	Point-to-Point Protocol Over Ethernet
PPQE	Position Page Queue Element
PPR	Partial Product Read
PPR	Price/Performance Ratio
PPR	Product Performance Report
PPR	Program Planning Report(ing)
PPRC	Peer (to) Peer Remote Copy
pps	packets per second
PPS	Page Printing System
PPS	Parallel Processing System
PPS	Partitioned Priority System
PPS	Patchboard Programming System
PPS	Post-Post-Scriptum
PPS	Potential Planning System
PPS	Power Personal Systems
PPS	Product Performance Specification(s)
PPS	Product Programming Services
PPS	Programmed Processor System
PPS	Program Preparation Subsystem
PPS	Program Product Specification
PPS	Project Profile System
PPS	Project Proposal Summary
PPS	Public Packet Switching (network)
pps	pulses per second
PPSAT	Peripheral Processor Saturation
PPSC	Parallel Processing System Compiler
PPSDN	Public Packet-Switched Data Network
PPSM	Programming Practices (and) Standards Manual
PPSN	Public Packet Switched Network
PPSS	Product and Process Status System
ppt	parts per thousand
ppt	parts per trillion
PPT	Parts Procurement Time
PPT	Periodic Program Termination
PPT	Point-to-Point Transmission
PPT	Primary Program operator interface Task
PPT	Processing Program Table
PPT	Programmer Productivity Techniques
PPT	Pulse Pair Timing
PPTM	Protocol Profile Testing Methodology
PPTP	Point-to-Point Tunnelling Protocol

PPTP	Purchase Pilot Test Plan
PPU	Peripheral Processing Unit
PPV	Pay Per View
PPW	Partial Product Write
PPX	Packet Protocol Extension
PPX	Power Parallel X connector
PPX	Private Packet Exchange
PQA	Palm Query Application
PQA	Protected Queue Area
PQEL	Partition Queue Element
PQFP	Plastic Quad Flat Pack
PQT	Preliminary Qualification Test
PR	Pattern Recognition
PR	Performance Report
PR	Photo Resist
PR	Physical Record
PR	Preliminary Report
PR	Prepare
PR	Price Record
PR	Primary Runout
PR	Principal Register
PR	Printer
PR	Print Restore
PR	Processor
PR	Product Reference|Register
PR	Program Register
PR	Progress Report
PR	Project Report
PR	Pseudo Random
PR	Public Relations
PR	Puerto Rico
PRA	Page Replacement Algorithm
PRA	Parabolic Reflector Antenna
PRA	Primary Rate Access
PRA	Print Alphanumerically
PRA	Probabilistic Risk Assessment
PRA	Program Reader Assembly
PRAAD	Photo-Resist Apply And Dry
PRACE	Photo-Resist Align, Contact and Expose
PRACSA	Public Remote Access Computer Standards Association
PRAM	Page Replacement Analysis Model(s)
PRAM	Parallel Random Access Machine
PRAM	Parameter Random Access Memory
PRAM	Pattern Random Access Memory
PRAM	Processing Rate Analytic Model
PRAM	Programmable Amplifier
PRAM	Programmable Random Access Memory
PRAM	Program Requirements Analysis Method

PRAN	Pre-Analysis
PRB	Packet Receive Buffer
PRB	Program Request Block
PRBAL	Previous Balance
PRBS	Printer Barrier Strip
PRBS	Pseudo-Random Binary Sequence
PRC	Primary Return Code
PRC	Printer Control
PRC	Procession Register Clock
PRC	Programmed Route Control
PRC	Program Range Change
PRC	Program Required Credentials
PRCA	Problem Reporting and Corrective Action
PRCD	Proceed
PRCPM	Print Complement
PRCTL	Print Control
PRD	Personnel Resources Data
PRD	Preliminary Review of the Design
prd	printer driver (file extension indicator)
PRD	Printer Dump
PRD	Product Reference Data
PRDM	Particular Requirements Definition Model
PRDMD	Private Directory-Management Domain
PRDY	Processor Ready
PRE	Picture Response Equipment
PRE	Prefix
PRE	Preformatted (hyper text markup language)
PRE	Print End
PRE	Pulse Rebalance Electronics
PREAMP	Pre-Amplifier
PREC	Precision
PRECIS	Preserved Contexed Indexing System
PRECON	Pre-Conditioning
PREF	Prefix
prefab	prefabricated
PREFUSE	Presentation Refuse
PREFX	Prefix
PREL	Programmable Rotary Encoded Logic
PREMIS	Premises Information System
PREMO	Presentation Environment for Multimedia Objects
PREP	Power-PC Reference Platform (PC = Personal Computer)
PREP	Preparation Program
PREP	Programmed Electronics Pattern
PREPAK	Prepackaging
PRE-REQ	Pre-Requisite
PRES	Pressure
PRESS	Product Records Engineering Support System
PREST4	Preprocessor for Structured Fortran

PREV	Prevent
PREV	Previous
PRF	Permanent Requirements File
PRF	Place, Route and Fold
PRF	Potential Risk Factor
prf	preferences (file extension indicator)
PRF	Pulse Repetition Frequency
PRFM	Pseudo-Random Frequency Modulated
prg	program (file extension indicator)
PRG	Purge
PRGCHK	Program Check
PRGEND	Program End
PRGEX	Program Exit|Expander
PRGGTG	Program Gating
PRGM	Program
PRGREG	Program Register
PRGSH	Program Shift
PRGSTP	Program Stop
PRGSTR	Program Start
PRGSUP	Program Suppress
PRI	Primary
PRI	Primary Rate Interface
PRI	Printer Interface
PRI	Priority
PRI	Priority Requirement for Information
PRI	Program Interrupt
PRI	Pulse Repetition Interval
PRID	Planning Record Identifier
PRID	Protocol Identifier
PRIDE	Profitable Information by Design
PRIDE	Programmed Reliability In Design Engineering
PRIDE	Projecting and Information system for Digital Exchanges
PRIME	Planning through Retrieval of Information for Management Extrapolation
PRIMECA	Pôle de Ressources Informatiques pour la Mécanique (FR)
PRIN	Principal
PRINCE	Programmed International Computer Environment
PRINSYS	Production Information System
PRINT	Pre-edited Interpretive system
PRINT	Print Recognition Input Terminal
PRINTF	Print with Formatting (C)
PRISEQ	Primary Sequence
PRISM	Parallel Reduced Instruction Set Multiprocessing
PRISM	Personnel Record Information System for Management
PRISM	Personnel Requirements Information System Methodology
PRISM	Production Requirements for Industrial Scheduling Manpower
PrISM	Program in Integrated Sustainable Manufacturing
PRISM	Program Integrated System Maintenance

PRISM	Programmed Integrated System Maintenance
PRISM	Program Reliability Information System for Management
PRISM	Progressive Refinement Integrated Supply Management
PRISM	Pulse Repetition Interval Sorting Matrix
PRITMR	Primary Timer
PRIV	Privilege(d)
PRIX	Primary X
PRIXPU	Primary X Pick-Up
PRK	Phase Reversal Keying
PRKB	Printer Keyboard
PRL	Print Lister
PRL	Processor Level
PRL	Program Reference Library
PRLM	Preliminary
PRM	Parameter
PRM	Program Reference Manual
PRM	Protocol Reference Model
PRMA	Packet Reservation Multiple Access
PRMD	Private Mail Domain
PRMD	Private Managed|Management Domain
PRMD	Program Release, Maintenance and Distribution
PRML	Partial-Response Maximum-Likelihood
PRMOD	Printer Module
PRMR	Primer
PRN	Printer
PRN	Print Numerically
PRN	Pseudo-Random Noise|Number
PRNET	Packet Radio Network
PRNMO	Performance Related Network Management Offering
PRO	Precision RISC Organization (RISC = Reduced Instruction Set Computer)
PRO	Print Octal
pro	profile (file extension indicator)
PROBE	Pre-Recognition Of Baleful Errors
PROBFOR	Probability Forecasting
PROC	Procedure
PROC	Processing
PROC	Processor
PROC	Programming Computer
PROC	Proposed Required Operational Capability
PROCAS	Process Calculation System
PROCD	Procedure
PROCL	Procedural Language
PROCLIB	Procedure Library
PROCNAME	Procedure Name
PROCOMP	Process Compiler
PROCOPT	Processing Option
PROCOS	Provably Correct Systems (EU)

PROCR	Processor
PROCSEQ	Processing Sequence
PROCSTOR	Program Storage
PROCTOT	Priority Routine Organizer for Computer Transfers and Operation Transfers
PROCU	Processing Unit
PRODAC	Programmed Digital Automatic Control
PRODEX	Project for Data Exchange
PRODIP	Program Distribution Program
PRODOC	Procedure Documentation
PRODOS	Professional Disk Operating System
PROF	Prediction of the Rate of Optimization Failures
PROFACTS	Production Formulation, Accounting and Cost System
PROFILE	Program Overview and File
PROFIT	Programmed Receiving, Ordering and Forecasting Inventory Technique
PROFS	Professional Office System
PROG	Programmer
PROG	Program(ming)
PROGDEV	Program Device
PROGMAN	Program Manager
PROGOFOP	Program Of Operation
PROGR	Programmer
PROGTOT	Progressive Total
PROJ	Project
PROJACS	Project Analysis and Control System
PROLAMAT	Programming Languages for numerically controlled Machine Tools
PROLAN	Processed Language
PROLOG	Programming in Logic
PROM	Programmable Read Only Memory
PROMATS	Programmable Magnetic Tape System
PROMIS	Project Management Information System
PROMIS	Project Oriented Management Information System
PROMISE	Process Oriented Machine Independent Simulation Environment
PROMISE	Programming Managers Information System Environment
PROMOTION	Planning Robot Motion (EU)
PROMPT	Project Management and Production team Technique
PROMS	Program Modelling and Simulation
PRONANO	Processing on a Nanometer scale (EU)
PRONTO	Program for Numerical (machine) Tool Operation
PRONTO	Programmable Network Telecommunications Operating system
PROOF	Projected Return on Open Office Facilities
PROP	Proposal On-line Preparation
PROP	Proprietor
PROPSIM	Propagation Simulator
PROSA	Product-Resource-Order-Staff Architecture
PROSA	Programming system with Symbolic Addresses

PROSIM	Production System Simulator
PROSPRO	Process Supervisory Program
PROT	Protect(ion)
PROTEC	Protection against Environmental Conditions
PROTEL	Procedure Oriented Type Enforcing Language
PROTIOS	Proton and Ion Switches (EU)
PROUT	Progressive Utilization Theory
PROWORD	Procedure Word
PRP	Partial Read Pulse
PRP	Prepare
PRP	Problem Recovery Procedure
PRP	Pseudo-Random Pulse
PRP	Public Review Period
PRPG	Pseudo-Random Pattern Generator
PRPQ	Programming Request for Price Quotation
PRPS	Program Requirement Process Specification
PRPT	Proportional
PRR	Pseudo-Resident Reader
PRR	Pulse Repetition Rate
PRS	Pattern Recognition System
PRS	Personal Response System
PRS	Polynomial Remainder Sequence
PRS	Primary Register Set
prs	printers (file extension indicator)
PRS	Program Requirements Summary
PRS	Pseudo-Random Sequence|Signal
PrSc	Print Screen
PRSC	Program Release Support Center
PR/SM	Processor Resource/Systems Manager
PRSMP	Print Sample
PRSS	Proprietary RAID Storage Solution (RAID = Redundant Array of Independent Disks)
PRST	Probability Ratio Sequential Test
PRSUP	Print Suppression
PRT	Print(er)
PRT	Process Response Time
PRT	Production Run Tape
PRT	Program|Provisional Reference Table
PRTG	Printing
PRTM	Printing Response Time Monitor
PRTN	Partition
PRTOT	Prototype Real-Time Optical Tracker
PRTR	Printer
PRTSC	Print Screen
PRTY	Priority
PRU	Packet Radio Unit
PRU	Printer Unit
PRUNCATS	Program Relay Universal Card Analysis Test System

PRV	Peak Reverse Voltage
PRV	Permanently Resident Volume
PRV	Pseudo Register Vector
PRVS	Previous
PRWP	Programmed Read/Write Protection
PRX	Processor Reed Exchange
PRX-A	Processor Reed Exchange – Analog version
PRZ	Polarized Return-to-Zero
PRZR	Polarized Return-to-Zero Recording
PS	Pace Setter
PS	Packet Switch(ing)
PS	Palestina (provisional)
P/S	Parallel to Serial
PS	Parity Switch
PS	Pentagrid Switch
PS	Perfect Shuffle
PS	Performance Specification
PS	Personal Services
PS	Physical Sequential
ps	picosecond
PS	Picture System
P&S	Planning & Scheduling
PS	Polling Sequence
ps	postscript (file extension indicator)
PS	Power Supply
PS	Preliminary Study
PS	Presentation Service(s)
PS	Primary Sequence
PS	Printer Start
PS	Print Server
PS	Process(or) Status
PS	Process Specification|Storage
PS	Product Safety
PS	Programmed Symbols
PS	Programming Services|System
PS	Program Shift|Start
PS	Program Specification(s)
PS	Program Store|Summary
PS	Proportional Spacing
PS	Proposed Standard
PS	Protect Status
PS	Pulse Shaping
PSA	Permanent Storage Area
PSA	Personal Satellite Assistant
PSA	Physical Sequential Access
PSA	Polycrystalline Silicon self-Aligned
PSA	Precipitation Series Algorithm
PSA	Prefix Save Area

PSA	Pressure Sensitive Adhesive
PSA	Problem Statement Analyzer|Analysis
PSA	Protected Storage Address|Area
PSAD	Prediction, Simulation, Adaptation, Decision
PSAF	Print Services Access Facility
PSAG	Program System Analysis Guide
PSAL	Programming Systems Activity Log
PSAM	Partitioned Sequential Access Method
PSAP	Presentation Service Access Point
PSAR	Process Storage Address Register
PSAS	Programming System Announcement Summary
PSB	Program Specification Block
PSBGEN	Program Specification Block Generator
PSBNAME	Program Specification Block Name
PSC	Packet Switching Center
PSC	Paper Skip Character
PSC	Parallel/Serial Converter
PSC	Personal Super Computer
PSC	Pittsburgh Supercomputer Center
PSC	Power System Control
PSC	Print Scan Counter
PSC	Print Server Command|Control
PSC	Process Schedule Control (program)
PSC	Program Schedule|Status Chart
PSC	Program Sequence Control
PSC	Program Switching Center
PSC	Project Systems Control
PSC	Protection Switching Circuit
PSC	Pulse Shape Control (circuit)
PSCE	Peripheral Storage Control Element
PSCE	Program Support Customer Engineer
PSCF	Primary System Control Facility
PSCF	Processor Storage Control Function
PSCL	Programmed Sequential Control Logic
PSCNET	Pittsburgh Supercomputing Center Network
PSCP	Presentation Services Command Processor
PSCS	Program Support Control System
PSD	Packed Switched Data
PSD	Phase Sensitive Detector
PSD	Power Spectral Density
PSD	Program Status Document(s)
PSD	Program Status Double-word
PSD	Program Structure Diagram
PSD	Protection Switching Duration
PSDC	Public Switched Digital Capability
PSDM	Presentation Services for Data Management
PSDN	Packet|Public Switched Data Network
PSDR	Process Storage Data Register

PSDR	Program Status Double-word Register
PSDS	Packet-Switched Data Service
PS&DS	Program Statistics & Data Systems
PSDS	Public Switched Digital Service
PSDU	Presentation Service Data Unit
PSDW	Program Status Double Word
PSE	Packet Switch(ing) Exchange
PSE	Print Scan Emitter
PSE	Product Structure Editor
PSE	Programming Support Environment
psec	picosecond
PSECT	Program Section
PSEL	Presentation Selector
PSERVER	Print Server
PSF	Permanent Swap File
PSF	Point Spread Function
PSF	Print Services Facility
PSF	Product Structure File
PSF	Program Support Facility
PSFO	Partitioned Sequential File Organization
PSG	Planning Systems Generator
PSG	Program Systems Guide
PSGEN	Program Specification (block) Generation
PSHF	Push Flag
PSHRPQ	Program Support for Hardware Request for Price Quotation
psi	angle
psi	flux
PSI	Packetnet System Interface
PSI	Packet Switch(ing) Interface
PSI	Parallel Sequential Inference
PSI	Parameter Signature Identification
PSI	Performance Summary Interval
PSI	Personal Security Identifier
PSI	Personal Sequential Inference
PSI	Problem Source Identification
PSI	Process to Support Interoperability
PSI	Programmable Serial Interface
PSI	Programmed Self-Instruction
PSI	Program Status Information
PSIA	Page Supervisor Information Area
PSID	PostScript Image Data
PSIS	Programming System Information Systems
PSIU	Packet Switch Interface Unit
PSK	Phase Shift Keying
PSKM	Phase Shift Keying Modem
PSK-PCM	Phase Shift Keying/Pulse Code Modulation
PS+L	Power Switching and Logic
PSL	Problem Statement Language

PSL	Processor Status Longword
PSL	Program Support Library
PSLAP	Power Supply Load Analysis Program
PSLI	Packet Switch Level Interface
PSL/PSA	Problem Statement Language/Problem Statement Analyzer
PSM	Packet Service Module
PSM	Packet Switched signalling Message
PSM	Peak Selector Memory
PSM	Persistent Storage Model
PSM	Persistent Stored Module(s)
PSM	Phase Shift Modulation
PSM	Position Switching Module
PSM	Power Supply Module
PSM	Printing Systems Manager
PSM	Problem Solving Module
PSM	Programmable State Machine
PSM	Programming System Memorandum
PSM	Programming Systems Manual
PSM	Program Support Material|Monitor
PSM	Pseudo-Symbolic Machine language
PSM	Pulse-Spacing Modulation
PSMI	Phase Shift Modal Interference
PSML	Processor System Modelling Language
PSML	Pseudo-Symbolic Machine Language
PSN	Packet Switched Network|Node
PSN	Planet Search Networks
PSN	Position
PSN	Print Sequence Number
PSN	Private Switching Network
PSN	Processor Serial Number
PSN	Program Summary Network
PSN	Public Switched Network
PSO	Programming Support Option
PSO	Provider Service Organization
PSOP	Power System Optimization Program
PSOS	Probably Secure Operating System
PSP	Packet Switching Processor
PSP	Personal Software Products
PSP	Planned Standard Programming
PSP	Power System Planning
PSP	Presentation Services Process
PSP	Programmable Signal Processor
PSP	Program Segment Prefix
PSPDN	Packet Switched Public Data Network
PSPN	Public Switched Packet Network
PSPS	Planar Silicon Photo-Switch
PSR	Page Send/Receive
PSR	Performance Summary Report

PSR	Processor Status Register
PSR	Product Specific Release
PSR	Program Start Register
PSR	Program Status Report
PSR	Program Support Representative
PSR	Protected Service Routine
PSRAM	Pseudo Static Random Access Memory
PSRT	PostScript Round Table
PSS	Packet Switched Services|System
PSS	Packet Switch Stream
PSS	Parallel Search Storage
PSS	Patent Search System
PSS	Planned Systems Schedule
PSS	Power System Stabilizer
PSS	Principle Support System
PSS	Printer Storage System
PSS	Print Sub-Scan
PSS	Process Support System
PSS	Programmable Store System
PSS	Programmable Symbol Set
PSS	Programming Systems Support
PSSHS	Programmable Store System Host Support
PSSP	Parallel System Support Programs
PST	Pair-Selected Ternary
PST	Partition Specification Table
PST	Periodic Self-Test
PST	Permanent Symbol Table
PST	Physical Storage Table
PST	Positive Sign Trigger
PST	Priority Selection Table
PST	Program Structure Technology
PSTN	Public Service Telecommunication Network
PSTN	Public Switched Telephone Network
PSTP	Program Stop
PSTT	Program Start
PSTU	Power Supply Test Unit
PSTV	Protocol Specification, Testing and Verification
PSU	Packet Switch(ing) Unit
PSU	Peripheral Switching Unit
PSU	Power Supply Unit
PSU	Problem Statement Unit
PSU	Processor Service|Storage Unit
PSU	Processor Speed Up
PSU	Program Storage Unit
PSV	Probabilistic State Variable
PSW	Processor|Program Status Word
PSWM	PostScript Window Manager
PSWR	Power Standing-Wave Ratio

PSYCE	Power Supply Current Estimate
PSYCHE	Programming Systems Yearly Cost/Headcount Estimate
PSYCO	Peripheral System Check-Out
PSYNC	Processor Synchronous
PT	Page Table
PT	Pass Through
PT	Performance Test|Tool
PT	Pilot Test
PT	Please Token
PT	Point
PT	Portugal
P&T	Postes et Télécommunications (LU)
P&T	Posti ja Tele (FI)
PT	Primary Timer
PT	Processing Time
PT	Processor Terminal
PT	Process Time
PT	Product Test
PT	Programmable Terminal
PT	Program Tab|Test
PT	Program Timer
PTA	Page Table Address
PTA	Post and Telecom Authority
PTA	Procrustes Target Analysis
PTA	Programmable Translation Array
PTA	Pulse Torquing Assembly
PTAN	Performance Testing Alliance for Networks
PTAP	Profiler Triangle Analysis Package
PTAT	Private Trans-Atlantic Telecommunication system
PTB	Page Table Base
PTB	Prohibitive To Bus
PTBX	Private Telegraph Branch Exchange
PTC	Packet Terminal Customer
PTC	Paper Throw Character
PTC	Positive Temperature Coefficient
PTC	Process and Test Control
PTC	Programmed Transmission Control
PTC	Program Test Controller
PTC	Public Telephone Companies
PTCC	Problem Tracking and Change Control
PTD	Parallel Transfer Disk|Drive
PTD	Pseudo-Terminal Driver
PTDOS	Processor Technology Disk Operating System
PTE	Page Table Entry
PTE	Path Terminating Entity
PTE	Process and Test Equipment
PTE	Protect Error
PTEN	Prime Time Entertainment Network

PTERM–PTS

PTERM	Physical Terminal
PTF	Pass-Through Facility
PTF	Problem Trouble Fix
PTF	Program Temporary Fix
PTH	Plated Through Hole
PTH	Project Team Head
PTI	Plant Transfer In
PTI	Portable Test Instrument
PTI	Program Transfer Interface
PTK	Protection Check
PTL	Parameter Table Load
PTL	Process and Test Language
PTM	Phase Time Modulation
PTM	Photo Tracing Machine
PTM	Programmable Terminal Multiplexer
PTM	Programmable Timer Module
PTM	Program Trouble Memorandum
PTM	Proof Test Model
PTM	Pulse Time Modulation
PTM	Pulse Transmission Mode
PTMS	Pattern Transformation Memory System
PTN	Partition
PTN	Personal Telecommunications Number
PTNX	Private Telecommunications Network Exchange
PTO	Pattern Trigger Output
PTO	Permeability Tuned Oscillator
PTO	Please Turn Over
PTO	Power Take-Off
PTO	Practical Theoretical Optimum
PTO	Public Telecommunications Operator
PTOCA	Presentation Text Object Content Architecture
PTOP	Program-To-Program
PTOPC	Program-To-Program Communications
PTP	Pilot Test Plan
PTP	Point-To-Point
PTP	Processor-To-Processor
PTPP	Point-To-Point Path
PTR	Photoelectric Tape Reader
ptr	pointer
PTR	Printer
PTR	Program Trouble Report
PTS	Personal Time-Sharing
PTS	Photo Type Setting
PTS	Points
PTS	Proceed To Send
PTS	Production Time-Share
PTS	Profile Test Specification
PTS	Programmable Terminal System

PTS Program Test System
PTS Public Telephone Service
PTT Post, Telephone and Telegraph
PTT Press To Talk
PTT Printing Teletypewriter Telegraphy
PTT Program Test Tape
PTT Push To Talk
PTTC Paper Tape (and) Transmission Code
PTTC Public Telephone and Telegraph Codes
PTTXAU Public Teletex Access Unit
PTU Package Transfer Unit
PTW Primary Translation Word
PTX Plus Teletypewriter Exchange
PTXAU Public Telex Access Unit
PTY Parity
PU Peripheral Unit
PU Physical Unit
PU Pick-Up
PU Pluggable|Processing Unit
PUB Physical Unit Block
pub publication (file extension indicator)
PUB Public (directory)
PUB Publish(er)
PUBL Publication
PUC Peripheral Unit Controller
PUC Personal Use Computer
PUC Processing Unit Cabinet
PUCP Physical Unit Control Point
PUCP Process Unit Control Panel
PUD Physical Unit Directory
PUD Planned Unit Development
PUDL Push-Down List (memory)
PUF Percent Unaccounted For
PUF Physical Update File
PUFI Pair-Usage-Frequency Indicator
PUIMP Pickup Impulse
PUL Program Update Library
PUMA Programmable Universal Machine for Assembly
PUMA Programmable Universal Micro Accelerator
PUMT Programmable Universal Module Tester
PUN Physical Unit Number
PUNC Practical, Unpretentious, Nomographic Computer
PUP PARC Universal Packet (protocol)
PUP Performance Units Plan
PUP Peripheral Unit|Universal Processor
PUS Performance Upgrade Socket
PUS Physical Unit Services
PUS Processor Upgrade Socket

P-use	Predicate use
PUSHA	Push All (registers)
PUSHF	Push Flags
PUSTA	Pushdown Stack
PUSVC	Physical Unit Services
PUT	Programmable Uni-junction Transistor
PUT	Program Update Tape
PV	Parameter Value
PV	Path Verification
PV	Photo-Voltaic
PV	Physical Volume
pV	picovolt
PV	Pore Volume
PV	Potential Vorticity
PV	Process Variable
PVAD	Position Velocity and Attitude Display
PVAN	Private Value-Added Networks
PVASCII	Plain Vanilla American Standard Code for Information Interchange
PVC	Permanent Virtual Call|Channel
PVC	Permanent Virtual Circuit
PVC	Photo-Voltaic Cell
PVC	Program and Velocity Computer
PVCC	Permanent Virtual Channel Connection
PVCC	Potential Valued Clause Combination
PVCS	Portable Voice Communication System
PVD	Plan View Display
PVG	Parallel Visualization and Graphics (symposium)
PVGA	Paradise Video Graphics Array
PVI	Programmable Video Interface
PVM	Parallel Virtual Machine
PVM	Pass-through Virtual Machine (protocol)
PVN	Private Virtual Network
PVP	Parallel Vector Processing
PVP	Program Verification Package
PVR	Precision Voltage Reference
PVR	Prefix Value Register
PVR	Process Variable Record
PVS	Parallel Visualization Server
PVS	Private Videotext System
PVS	Private Viewdata System
PVS	Program Validation Services
PVT	Parameter Variable Table
PVT	Performance Validation Test
PVT	Permanent Virtual Terminal
PVTR	Portable Video Tape Recorder
PVX	Packet Voice Exchange
PW	Palau

PW	Password
PW	Personal Workstation
pW	picowatt
PW	Printed Wiring
PW	Private Wire
PW	Processor Write
PW	Pulse Width
PWA	Pirates With Attitudes
PWA	Printed Wire Assembly
PWA	Private Write Area
PWAIT	Processor Wait (acknowledge)
PWB	Printed Wiring Board
PWB	Programmer's Workbench
PWBA	Printed Wiring Board Assembly
PWB/MM	Programmer's Workbench Memorandum Macro's
PWC	Pulse Width Coded
PWCM	Pulse Width with Carrier Modulation
PWD	Print Working Directory
PWD	Process Word
PWD	Pulse Width Discriminator
PWE	Pulse Width Encoder
PWEA	Printed Wiring and Electronic Assemblies
PWF	Print Work File
PWG	Permanent Working Group
PWG	Printer Working Group
PWI	Power Indicator
PWI	Public Windows Interface
PWID	Pair-Wise Interference Detection
PWM	Plated Wire Memory
PWM	Printed Wiring Master
PWM	Pulse Width Modulation
PWR	Power
PWR	Processor Write
PWR	Pulse Width Recording
PWRNO	Power No (failure)
PWRU	Power Unit
PWS	Private Wire System
PWS	Programmer Work Station
PWS	Program Work Statement
PWSCS	Programmable Work Station Communication Services
px	primary index (file extension indicator)
PX	Private Exchange
PXE	Preboot Execution Environment
pxl	pixel (file extension indicator)
PY	Paraguay
PZC	Point of Zero Charge
PZM	Pressure Zone Microphone
PZT	Piezoelectric Transducer

Q

Q	Electrical charge
Q	Q-band
Q	Q-output
Q	Quality (measure of quality)
Q	Quantity (of electric charge)
q	quantum (value)
Q	Query
Q	Queue
Q	Quintal
Q	Quotient
QA	Qatar
QA	Quality Analysis|Assurance
Q&A	Questions & Answers
QA	Queue Arbitrated
QAM	Quadrature Amplitude Modulation
QAM	Queue(d) Access Method
QAP	Quadratic Assignment Problem
QAPP	Quality Assurance Project Plan
QART	Quality Assurance Review Technique
QAS	Quasi-Associated Signalling
QAS	Question and Answering System
QASK	Quadrature Amplitude Shift Keying
QAT	Quality Assurance Test
QB	Quick Batch
QBD	Quality By Design
QBD	Quasi Bi-Directional
QBE	Query By Example
QBF	Query By Form
QBIC	Query By Image Content
QBO	Quasi-Biennial Oscillation
QC	Quality Check|Control
QC	Quantum Count
QC	Quiescent-Completed
QCAM	Queued Communications Access Method
Q-CAP	Quality Control activity Analysis Program
QCB	Queue Control Block
QCBE	Queue Control Block Extension
QCD	Query Complexity Degree
QCF	Quarterly Control Figures
QCIF	Quarter Common Interchange Format
QCM	Quantitative Computer Management
QCM	Quote Commission Master (file)
QCONF	Quantity Conversion Factor
QCR	Queue Control Record
QCRT	Quick Change Real-Time

QCT	Quiescent Carrier Telephony
QCW	Q-phase Continuous Wave (signal)
Q+D	Quick and Dirty
QD3D	Quick Draw 3 Dimensional
QDC	Quick Die Change
QDCS	Quality, Delivery, Cost, Service
QDEBUG	Quick diagnostic Debugging program
qdi	quicken dictionary (file extension indicator)
QDL	Quartz Delay Line
QDOS	Quick and Dirty Operating System
qdt	quicken data (file extension indicator)
QE	Quality Extract
QE	Queue Empty
QE	Quick Estimation
QEC	Quiescent at End of Chain
QECB	Queue Element Control Block
QECI	Quiescent at End of Chain Indicator
QEL	Queue Element
QEMM	Quarterdeck Expanded Memory Manager
QET	Quantum Effect Transistor
QF	Queue Full
QFA	Quick File Access
QFD	Quality Function Deployment
QFM	Quadratic Field Magnetic
QFM	Quantized Frequency Modulation
QFM	Quasi Frequency-Modulated
QFP	Quad Flat Pack
QFT	Queued File Transfer
QI	Quarterly Index
QIA	Quality In Automation
QIC	Quality Information using Cycle time
QIC	Quarter-Inch Cartridge
QIF	Quicken Import File
qif	quicken interchange format (file extension indicator)
QIL	Quad-In-Line
QIN	Quinary
QIO	Queue Input/Output
QIS	Quality Information System
QISAM	Queued Indexed Sequential Access Method
QL	Query Language
QL/1	Query Language One
QLI	Query Language Interpreter
QLLC	Qualified Link Level Control
QLLC	Qualified Logical Link Control
QLM	Quasi-Lagrangian Model
QLP	Query Language Processor
QLR	Queued Log-on Request
QLS	Quasi Linear Suppressor

QLS	Quick Lockup Sub-pool
QLSA	Queuing Line-Sharing Adapter
QM	Quadrature Modulation
QM	Quality|Queue Management
QMD	Quotient Multiplicand
QMF	Query Management Facility
QMIPS	Quantitative Modelling In Parallel Systems (EU)
QML	Quick Memorized List
QMN	Quotient Multiplication
QMP	Quarterly Machine Performance
QMPA	Queue Manager Parameter Area
QMRP	Qantel Manufacturing Resource Planning
QMS	Quality Micro Systems
qmt	quicken memorized list (file extension indicator)
QN	Query Normalization
QNS	Quantity Not Sufficient
QNT	Quantizer
qnx	quicken indexes (to data) (file extension indicator)
QoS	Quality of Service
QP	Quadratic Programming
QP	Quality Procedure
QPA	Quality Performance Analysis
QPAM	Quadrature Phase and Amplitude Modulation
QPG	Quantum Phase Gate
QPL	Qualified Products List
QPN	Queuing Petri Net
QPPD	Queue Processor Printing Dispatcher
QPRI	Qualitative Personnel Requirements Information
QPS	Query Property Similarity
QPS	Queued Printing Services
QPSK	Quadrature|Quaternary Phase Shift Keying
QPSX	Queued Packet and Synchronous (circuit) Exchange
QQP	Quick Query Program
QQS	Quick Query System
Q+R	Quality and Reliability
Q&RA	Quality & Reliability Assurance
QRA	Quality Reliability Assurance
QRC	Quick Reactions Communications
QRC	Quick Reference Card
QRCD	Quantity Received
QRG	Quick Reference Guide
QRGT	Quick Reaction Guidance and Targeting
QRL	Quick Relocate and Link
QRM	Quiet Recording Mode
QRP	Query and Reporting Processor
QRSS	Quasi Random Signal Source
QRT	Queue Run-Time
QS	Query Similarity

QS	Query System
QS	Queue Select
QSAM	Quadrature Sideband Amplitude Modulation
QSAM	Queued Sequential Access Method
QSAR	Quantitative Structure Activity Relationships
QSIF	Quarter Source Input Format
QSIG	Q-Signalling
QSIM	Qualitative Simulation
QSL	Queue Search Limit
QSM	Quantitative Software Management
QSPR	Quantitative Structure Property Relationships
qt	quart
QT	Queuing Theory
QT	Quick Time
QTAM	Queued Telecommunications|Teleprocessing Access Method
QTAM	Queued Terminal|Transmission Access Method
QTH	Queued Transaction Handling
QTL	Quantitative Trait Locus
QTM	Quadratic Texture Map
QTO	Quantity Take-Off
QTP	Quality Test Plan
QTR	Quality Technical Requirement
QTRCD	Quarter Code
QTS	Quantized Threshold Spacing
QTVR	Quick Time Virtual Reality
QTY	Quantity
QUAC	Quadratic Arc Computer
QUAL	Quality
QUALTA	Quad Asynchronous Local Terminal Adapter
quam	quadrature modulation
QUAM	Queued Access Method
QUANSY	Question and Answering System
QUART	Quadrature Ambiance with Reference Tone
quasar	quasi stellar
QUASTOR	Quick Access Storage
QUEL	Query English Language
QUEL	Query Language
QUEN	Quenching
QUEST	Query Evaluation and Search Technique
QUEST	Queuing Event Simulation Tool
QUICKTRAN	Quick Translation
QUIDAC	Quiet Design Aided by Computer
QUINTEC	Quantum optics for Information Technology (EU)
QUIP	Quad In-line Package
QUIP	Query Interactive Processor
QUIP	Quick Inquiry Processor
QUIP	Quota Input Processor
QUISAM	Queued Indexed Sequential Access Method

QUOT	Quotient
QVC	Quality, Value, Convenience
QVOD	Quick Video On Demand
QVSTP	Quick Video Streaming Protocol
QWERTY	standard English typewriter keyboard
QWEST	Quantum-Well Envelope State Transition
QWIP	Quantum-Well Infrared Photodetector
QXI	Queue Executive Interface

R

R	Radical
R	Read
R	Record
R	Red
R	Register
R	Relay
R	Request
R	Reset
R	Resistance
R	Reverse
R	Right
R	Ring
R	Roger (message received)
RA	Radium
RA	Random Access
RA	Rate Adapter
RA	Ratio Actuator
RA	Rational number
RA	Read Amplifier|Audit
RA	Recognition Arrangement
RA	Record Address|Automation
RA	Refer to Acceptor
RA	Relative Address
RA	Reliability Assessment
RA	Relocation Address
RA	Repair Action
RA	Repeat to Address
R&A	Research & Analysis
RA	Return Address|Authorization
RA	Right Angle
RAA	Relational Access Administrator
RAA	Remote Access Audio
RAC	Radio Adaptive Communications
RAC	Rapid Action Change
RAC	Read Address Counter
RACE	Random Access Card|Computer Equipment
RACE	Rapid Automatic Checkout Equipment

RACE	Rapid Automatic Core Evaluator
RACE	Research and development of Advanced Communication in Europe
RACE	Results, Analysis, Computation and Evaluation
RACEP	Random Access and Correlation for Extended Performance
RACER	Rapid Card Embedding and Routing
RACER	Runner Administration and Computerized Entry Routine
RACF	Resource Access Control Facility
RACH	Random Access Channel
RACKET	Routines for Arithmetic Computation of Key set Evaluation Tables
RACS	Random Access Communications|Control System
RACS	Remote Access Computing System
RACS	Remote Automatic Calibration|Control System
RACT	Remote Access Computing Technique
RAD	Radial
rad	radio
rad	radix
RAD	Random Access Device|Disk
RAD	Rapid Access Device|Disk
RAD	Rapid Application Development
RAD	Record Assembler and Distributor
RADA	Random Access Discrete Address
RADAC	Rapid Digital Automatic Computing
RADACS	Random Access Discrete Address Communications System
RADAR	Radio Detection And Ranging
RADAR	Real-time Aid to Diagnosis And Recovery
RADAR	Receivable Accounts Data-entry And Retrieval
RADAS	Random Access Discrete Address System
RADB	Routing Arbiter Data Base
RADEM	Random Access Delta Modulation
RADIAC	Radiation Detection, Identification And Computation
RADIC	Research And Development Information Center
radionics	radio-electronics
RADIR	Random Access Document Indexing and Retrieval
RADIUS	Remote Authentication Dial-In User Service
RADOC	Remote Automatic Detection Of Contingencies
RADOT	Real-time Automatic Digital Optical Tracker
RADS	Radar and Algorithm Display Model
RADS	Real-time Analysis and Display System
RADSL	Rate-adaptive Asymmetric Digital Subscriber Line
RAEBNC	Read And Enjoyed, But No Comment
RAES	Remote Access Editing System
RAF	Removal Adjustment Factor
RAF	Requirements Analysis Form
RAF	Resource Allocation Facility
RAFT	Recompilable Algebraic Formula Translator
RAG	ROM Address Gate (ROM = Read-Only Memory)

RAG	Row Address Generator
RAI	Random Access and Inquiry
RAI	Return After Interrupt
RAID	Rapid Automatic Inscribing Device
RAID	Redundant Arrays of Independent Disks
RAID	Retrieval And Information Database
RAIL	Robot Arm Instruction Language
RAIN	Relational Algebraic Interpreter
RAIR	Recorded Automated Information Retrieval
RAIR	Remote Access, Immediate Response
RAIS	Redundant Arrays of Inexpensive Systems
RAISE	Rigorous Approach to Industrial Software Engineering
RAK	Read Access Key (station)
RAL	Rapid Access Loop
RAL	Read And Lock
RALF	Relocatable Assembly Language Floating point
RALU	Register (and) Arithmetic Logic Unit
RALU	Register-equipped Arithmetic and Logic Unit
RAM	Random Access Machine|Memory
RAM	Random Access Measurement
RAM	Read And Modify
RAM	Real Audio Markup
RAM	Relational Access Manager
RAM	Remote Access Monitor
RAM	Remote Area Monitoring
RAM	Resident Access Method
RAMAC	RAID Architecture with Multilevel Adaptive Cache (RAID = Redundant Arrays of Independent Disks)
RAMAC	Random Access Memory Accounting and Control
RAMAC	Random Access Method of Accounting and Control
RAMAC	Random Access (storage)
RAMB	Random Access Memory Buffer
RAMCEASE	Reliability, Availability, Maintainability, Cost Effectiveness And Systems Effectiveness
RAMD	Random Access Memory Device
RAMDAC	Random Access Memory Digital-to-Analog Converter
RAMIS	Rapid Access Management Information System
RAMM	Random Access Memory Module
RAMP	Random Access Memory Process|Programs
RAMP	Remote Access Maintenance Protocol
RAMPAC	Random Access Memory Package
RAMPS	Resource Allocation and Multi-Project Scheduling
RAMS	Random Access Measurement System
RAMS	Remote Automatic Multipurpose Station
RAN	Random (selection)
RAN	Raw Area Normalization
RAN	Read Around Number
RANCID	Real And Not Corrected Input Data

RANCOM	Random Communication satellite
RAND	Random
RAND	Research And Development
RAND	Rural Area Network Design
RANDAM	Random Access Non-Destructive Advanced|Associative Memory
RANDID	Rapid Alpha-Numeric Digital Indicating Device
RAO	Related Application Object
RAP	Random Access Processing|Program
RAP	Rapid Action Pushbutton
RAP	Rapid Application Prototyping
RAP	Register Access Panel
RAP	Relational Analysis Planning
RAP	Relational Associative Processor
RAP	Remote Access Point
RAP	Resident Assembler Program
RAP	Resident Assistant Programmer
RAP	Resource Allocation Processor
RAP	Response Analysis Program
RAP	Review and Analysis Program
RAPCOE	Random Access Programming and Check-Out Equipment
RAPID	Rapid and Accurate Polygon Interference Detection
RAPID	Reactor And Plant Integrated Dynamics
RAPID	Relative Address Programming Implementation Device
RAPID	Retrieval And Processing Information for Display
RAPID	Retrieval And Production for Integrated Data
RAPID	Retrieval through Automated Publication and Information Digest
RAPIDS	Rapid Automated Problem Identification System
RAPPI	Random Access Plan Position Indicator
RAPS	Retrieval Analysis and Presentation System
RAPTAP	Random Access Parallel Tape
RAR	Reader Address Register
RAR	Read-only memory Address Register
RAR	Return Address Register
RARES	Rotating Associative Relational Store
RARP	Reverse Address Resolution Protocol
RAS	Random Access Storage
RAS	Raw Address Strobe
RAS	Regional Automated System
RAS	Regular Associated Solution
RAS	Reliability, Availability and Scalability
RAS	Remote Access Service(s)
RAS	Row Address Select|Strobe
RASAPI	Remote Access Service Application Programming Interface
RASDAMAN	Raster Data Management in databases (EU)
RASER	Radio Amplification by Stimulated Emission of Radiation
RASER	Random-to-Serial converter

RASM	Remote Analog Sub-Multiplexer
RASP	Random Access Stored Program
RASP	Remote Access Switching and Patching
RASP	Retrieval And Sort Processor
RASS	Radio Acoustic Sounding System
RASSP	Rapid prototyping of Application Specific Signal Processors
RAST	Receive-And-Send Terminal
RAST	Reliability And System Test
RASTAC	Random Access Storage And Control
RASTAD	Random Access Storage And Display
RAT	Random Adaptive Test
RAT	Ratio
RAT	Remote Area Terminal
RATC	Rate-Aided Tracking Computer
RATE	Remote Automatic Telemetry Equipment
RATFOR	Rational Fortran
RATR	Reliability Abstracts and Technical Review
RATS	Radio Amateur Telecommunications Society
RATS	Random Access Tape Store
RATTY	Radio Tele-Type
RAVAN	Random Vibration Analysis (program)
RAVE	Random Access Video Editing
RAVE	Random Access Viewing Equipment
RAVE	Rapid Automatic Variable Evaluator
RAVE	Reconfigurable Advanced Visualization Environment
RAVEN	Real-time Administrative Visual Environment
RAW	Read After Write
RAx	Rate Adapter x
RAX	Remote Access Execution
RAX	Rural Automatic Exchange
RB	Read Backward|Buffer
RB	Relay|Request Block
RB	Resource Broker
RB	Return to Bias
RB	Right Button (of 2 or 3 button mouse)
RBA	Relative Block|Byte Address
RBBP	Remote Batch Business Package
RBBS	Remote Bulletin Board System
RBC	Reflected Binary Code
RBC	Remote Balance Control
RBC	Remote Batch Computing
RBC	Right Bounded Context
RBCS	Remote Bar Code System
RBCS	Retail Batch Communications Subsystem
RBD	Reliability Block Diagram
RBD	Reliable Block Design
RBE	Remote Batch Entry
RBF	Remote Batch Facility

RBIM	Report-Based Information Management
RBL	Residual Byte Length
RBM	Real-time Batch Monitor
RBM	Relative Batch Monitor
RBM	Remote Batch Module
RBOC	Regional Bell Operating Company
RBOR	Request Basic Output Report
RBP	Remote Batch Processing
RBP	Robust Back Propagation
RBQ	Request Block Queue
RBS	Readiness-Based Sparing
RBS	Recovered Batch Storage
RBS	Remote Batch System
RBS	Resistor Barrier Strip
RBS	Robbed Bit Signalling
RBT	Relative Batch Throughput
RBT	Reliable Broadcast Toolkit
RBT	Remote Batch Terminal
RBTE	Remote Batch Terminal Emulator
RBTM	Remote Batch Terminal Module
RBTS	Raw Board Test System
R-BUN	Restricted (software) Bundle
RBV	Return Beam Vidicon
RC	Radio Controlled
RC	Raw Card
RC	Read Check
RC	Reader Code
RC	Real Circuit
RC	Receive Common
RC	Receiver Card|Clock
RC	Recode
RC	Record Count
RC	Reference Code
RC	Regional Center
RC	Remote Cache|Computer
RC	Remote Concentrator|Control
RC	Requirement Computation
RC	Resistance-Coupled
RC	Resistive Capacitive (circuit)
RC	Resistor-Capacitor
RC	Return Code
RC	Rewritable Consumer
RC	Routing Control
RC	Runtime Configuration
RCA	Remote Control Adapter
RCAC	Remote Computer Access Communications service
RCAR	Return Code Analysis Routine
RCAS	Reserve Component Automation System

RCB	Record Control Byte
RCB	Request Control Block
RCB	Residual Circuit Breaker
RCB	Resource Control Block
RCC	Radio Common Carrier
RCC	Read Channel Continue
RCC	Read Control Channel
RCC	Reader Common Contact
RCC	Real-time Computer Complex
RCC	Redundancy Check Character
RCC	Remote Cluster Controller
RCC	Remote Communications Complex
RCC	Remote Computing Capability
RCC	Reset Control Center
RCC	Reverse Command Channel
RCC	Routing Control Center
RCCAM	Remote Computer Communications Access Method
RCD	Receiver-Carrier Detector
RCD	Record
RCD	Registered Connective Device
RCD	Residual Current Device
RCDCD	Record Code
RCDD	Registered Communications Distribution Designer
RCDE	Recode
RCDG	Recording
RCDR	Recorder
RCE	Relay Communications Electronics
RCE	Remote Control Equipment
RCE	Restricted Coulomb Energy (network)
RCF	Reader's Comment From
RCF	Recall Finder
RCF	Remote Call Forwarding
RCF	Remote Cluster Facility
RCF	Retail Computer Facilities
RCHM	Remote Computer-controlled Hardware Monitor
RCI	Read Channel Initialize
RCI	Remote Control Interface
RCI	Rodent Cage Interface
RCIS	Remote Computer Interface Subsystem
RCIU	Remote Computer Interface Unit
RCL	Ramp Control Logic
RCL	Recalculate
RCL	Recall
RCL	Reliability Control Level
RCL	Resistance-Capacitance-Inductance
RCL	Rotate Carry Left
RCL	Runtime Control Library
RCLDN	Retrieval of Calling Line Directory Number

RCM	Read Clutch Magnet
RCM	Remote Carrier Module
RCMAC	Recent Change Memory Administration Center
RCN	Record|Report Control Number
RCO	Ramp Control Oscillator
RCO	Remote Control Oscillator
RCO	Representative Calculating Operation
RCOMP	Recomplement
RCOND	Reset Conditional
RCP	Receive Clock Pulse
RCP	Recognition and Control Processor
RCP	Remote Computer Pool
RCP	Remote Control Panel|Process
RCP	Remote Copy
RCP	Restore Cursor Position
RCP	Right-handed Circular Polarization
RCPN	Raw Card Part Number
RCQ	Record Correction and Quality
RCR	Reader Control Relay
RCR	Removal Card Request
RCR	Required Carrier Return (character)
RCR	Return Code Register
RCR	Rotate Carry Right
RCS	Reaction Control System
RCS	Rearward Communications System
RCS	Recode Selector
RCS	Records Communications Switching (system)
RCS	Reloadable Control|Core Storage
RCS	Remote Computing Service|System
RCS	Remote Control Support|Switch
RCS	Remote Control System
RCS	Requirements Computation System
RCS	Revision Control System (Unix)
RCSC	Remote Spooling Communications Subsystem
RCSDF	Reconfigurable Computer System Design Facility
RCSS	Random Communication Satellite System
RCT	Register Cycle Time
RCT	Representative Calculating Time
RCT	Resistor-Capacitor-Transistor
RCT	Resource Control Table
RCT	Reversible Counter
RCTL	Resistor-Capacitor-Transistor Logic
RCTL	Resistor-Coupled Transistor Logic
RCU	Radio Channel Unit
RCU	Remote Concentrator|Control Unit
RCV	Receive(r)
RCVD	Received
RCVD	Recovered

RCVR	Receiver
RCVY	Recovery
RCW	Read/Compute/Write
RCW	Return Control Word
RCY	Read Cycle
RCYC	Recycle
RD	Read Data
RD	Receive(d) Data
RD	Reference Document
RD	Register Drive
RD	Remove Directory
RD	Required Data
RD	Requirements Definition|Determination
R&D	Research & Development
R-D	Resolver to Digital
RD	Retention in Days
RD	Rotational Delay
RD	Route Descriptor
RD	Routing Domain
RDA	Read Data
RDA	Register Display Assembly
RDA	Remote Data(base) Access
RDA	Run-time Debugging Aid
RDAL	Representation Dependent Accessing Language
R-DAT	Rotary Digital Audio Tape
RDAU	Remote Data Access|Acquisition Unit
RDAV	Reset Data Available
RDB	Receive Data Buffer
RDB	Relational Data Base
RDBA	Remote Data Base Access
RDBK	Read Back
RDBL	Readable
RDBMS	Relational Data Base Management System
RDBMS	Remote Data Base Management System
RDC	Read Data Channel
RDC	Remote Data Collection|Concentrator
RDC	Remote Distribution Center
RDC	Resolver to Digital Converter
RDC	Routing Domain Confederation
RDCALL	Read Call
RD/CHK	Read Check
RDCI	Routing Domain Confederation Identifier
RDCLK	Received timing Clock
RDCM	Reduced Delta Code Modulation
RDCPL	Read Couple
RDCR	Reducer
RDDEL	Read Delay
RDE	Read End

RDE	Reader Emitter
RDE	Receive(d) Data Enable
RDE	Reliability Data Extractor
RDE	Remote Data Entry
RDES	Remote Data Entry System
RDEVQ	Reset Device Queues
RDF	Radial Distribution Function
RDF	Radio Direction Finding
RDF	Record Definition Field
RDF	Resource Definition Format|Framework
RDF	Resource Description Framework
RDG	Rounding
RDI	Remote Data Input
RDI	Remote Display Interface
RDI	Routing Domain Identifier
RDIN	Read In
RDIU	Read Interface Unit
RDIU	Remote Device Interface Unit
RDL	Random Dynamic Load
RDL	Remote Digital Loopback
RDL	Report Definition Language
RDL	Resistor-Diode Logic
RDL-SQL	Relational Database Language-Structured Query Language
RDM	Random
RDM	Real-time Database Manager
RDM	Reference Data Model
RDM	Remote Digital Multiplexer
RDMA	Random-Division Multiple Access
RDMOD	Read Modified
RDMS	Relational Data Management System
RDMS	Remote Data Management System
RDN	Redundancy
RDNBIT	Redundancy Bit
RDNS	Reverse Domain Name Service
RDO	Regular Data Organization
RDO	Resource Definition Online
RDOEX	Read Out Exit
RDORES	Read Out and Reset
RDOS	Real-time Disk Operating System
RDOUT	Read Out
RDP	Receipt Data at Plant
RDP	Reliable Data(gram) Protocol
RDPM	Random Data Processing Machine(s)
RDPUL	Read Pulse
RDR	Reader
RDR	Receive|Remote Data Register
RDR	Remote Digital Readout
RDRAM	Rambus Dynamic Random Access Memory

RDROM	Rambus Dynamic Read Only Memory
RDS	Radio Data|Digital System
RDS	Remote Data Scope
RDS	Remote Disk Station
RDS	Requirements Data System
RDS	Running Digital Sum
RDSEM	Real Data System Element Model
RDSM	Remote Digital Sub-Multiplexer
RDSR	Receiver Data Service Request
RDSTAT	Read Status
RDT	Radio Digital Terminal
RDT	Recall Dial Tone
RDT	Referenced Data Transfer
RDT	Remote Data Transmitter
RDT	Rework Data Tape
RDTO	Receive Data Transfer Offset
RDU	Raster Display Unit
RDU	Remote Disk Unit
RDVSTAT	Request Device Statistics
RDW	Record Descriptor Word
RDY	Ready
RE	Read Emitter
RE	Read End
RE	Real (number)
RE	Reference
RE	Reference Equivalent
re	regarding
RE	Request Element
RE	Reset
RE	Reunion
RE	Right End
RE	Routing Element
REACT	Recirculating A/C Tester
REACT	Register Enforced Automated Control Technique
REACT	Resource Allocation Control Tool
READ	Real-time Electronic Access and Display
READCOMM	Read Communications
READR	Re-Address
REAL	Realistic, Equal, Active (and for) Life
REAL	Relocatable Assembly Language
REAL	Resource Allocation
REALISE	Reality reconstructing from Image Sequences (EU)
REAP	Real Environment Application Program
REASM	Re-Assemble
REB	Rebound
REBOL	Relational Expression-Based Object Language
REC	Receive
REC	Recondition

REC	Record
rec	recorder (file extension indicator)
REC.	Recreation (USENET newsgroup category)
REC	Regional Engineering Center
REC	Remote Console
RECAL	Recalculate
RECAPS	Read/Encode/Capture/Proof/Sort
RECCE	Reconnaissance
RECD	Received
RECFM	Record Format
RECFMS	Record Formatted Maintenance Statistics
RECG	Recognition
RECMD	Reset at End of Command
RECMF	Radio and Electronic Component Manufacturers association
RECMK	Record Mark
RECMS	Record Maintenance Statistics
RECNUM	Record Number
RECO	Receive Only
RECOMP	Re-Complement
RECON	Re-Condition
RECON	Reconnaissance
RECON	Remote Console
RECONF	Re-Configure
RECOV	Recovery
RECOVER	Realize Effective Continuous Operation Via Error Response
RECP	Receptable
RECS	Reconnaissance Sensors
RECSTORMK	Record Store Mark
RECSYS	Recreation Systems analysis
rect	rectified
RECVD	Received
RECY	Rectifier
RECYC	Recycle
RED	Random Early Detection
RED	Reduce
RED	Reflection Electron Diffraction
REDAC	Real-time Data Acquisition
REDARS	Reference Engineering Data Automated Retrieval System
REDC	Read Control
REDE	Receiving Decoding
REDI	Review of Early Delivery Installations
REDN	Reduction
REDU	Reduce
REED	Restricted Edge Emitting Diode
REENG	Re-Engineering
ref	reference (file extension indicator)
REFS	Remote Entry Flexible Security
REFSYS	Reference System

REG	Range Extender with Gain
REG	Register
REG	Regulation
REGAD	Regenerate Address
REGAL	Rigid Epoxy Glass Acrylic Laminate
REGEN	Regenerative (repeater)
REGIO	Regional statistical data bank (EU)
REGIS	Relational General Information System
REGIS	Remote Graphics Instruction Set
REIF	Restructured Engineering Interface File
REINF	Reinforced
REINIT	Recovery Initialization
REINS	Requirements Electronic Input System
rej	reject(ion)
REJEN	Remote Job Entry
REL	Rapid Extensible Language
REL	Relational
rel	relative
REL	Release
REL	Reliability
REL	Relocatable
RELA	Relative
RELDR	Relocating (program) Loader
RELOC	Relocation
RELOCD	Relocated
RELQ	Release-Quiesce
RELSECT	Relative Sector
REM	Raster Entity Manipulation
REM	Rat Enclosure Module
REM	Recognition Memory
REM	Remark
REM	Remote
REM	Remote Equipment Module
REM	Ring Error Monitor
REMAC	Remote data Acquisition (subsystem)
REMAD	Remote Magnetic Anomaly Detection
REMAP	Record Extraction, Manipulation And Print
REMF	Reverse Electro-Magnetic Force
REMICS	Real-time Manufacturing Information Control System
REMS	Rohm Electronic Message System
REMV	Remove
REN	Remote Enable
REN	Rename
REN	Ring(er) Equivalence Number
RENT	Re-Enterable (program)
REO	Removable, Erasable, Optical
REP	Re-Entrant Processor
REP	Re-Entry Point

REP	Repeat
REP	Reply
REP	Representative
REP	Reproduce
REP	Request for Proposal
REP	Resolution Enhancement Program
REPE	Repeat while Equal
REPNE	Repeat while Not Equal
REPNZ	Repeat while Not Zero
REPOP	Repetitive Operation
REPRO	Reproduce
REPROC	Recovery Procedure
REPROM	Re-Programmable Read Only Memory
REPZ	Repeat while Zero
REQ	Request
REQD	Required
REQEX	Request Execute
REQSPEC	Requirements Specification
REQT	Requirement
RER	Residual Error Rate
RES	Remote Entry|Execution Service(s)
RES	Remote Entry Subsystem|System
RES	Reserve
RES	Reset
RES	Reset Signal
RES	Resistor
res	resolution
res	resource (file extension indicator)
RES	Response
RES	Restore (character)
RESB	Reverse Erased Second Breakdown
RESD	Reserved
ResEdit	Resource Editor
RESLOAD	Resident Loader
RESP	Response
RESPOND	Retrieval, Entry, Storage, and Processing of On-line Network Data
REST	Restore
RESTOR	Real Storage
RET	Resolution Enhancement Technology
RET	Retain
RET	Return
RETAIN	Remote Technical Assistance and Information Network
RETMA	Radio, Electronics, and Television Manufacturers Association
RETN	Retain
REU	Ready Extension Unit
REUS	Re-Usable
REV	Reverse

REV	Revision
REV	Revolution(s)
REW	Rewind
REWR	Rewrite
REX	Real-time Executive
REX	Regression Expert
REX	Relocatable Executable
REX	Route Extension
REXEC	Remote Execution
REXX	Restructured Extended Executor (language)
RF	Radio Frequency
RF	Rating Factor
RF	Read Forward
RF	Register File
RF	Reliability Factor
RF	Remote File
RF	Report Footing
RF	Reporting File
RFA	Reference File Administration
RFA	Remote File Access
RFA	Request For Announcement
RFA	Resource-For-Assignment
RFAM	Remote File Access Monitor
RFB	Reliability Functional Block
RFB	Request For Bid
RFC	Radio Frequency Choke
RFC	Report Format Control
RFC	Request For Comments
RFCP	Request For Computer Program
RFD	Ready For Data
RFD	Regional Frequency Divider
RFD	Request For Discussion
RFDC	Radio Frequency Data Collection
RFE	Request For Enhancement
RFG	Report Format Generator
RFI	Radio Frequency Interface|Interference
RFI	Ready For Issue
RFI	Request For Information
RFID	Radio Frequency Identification
RFLP	Restriction Fragment Length Polymorphism
RFM	Radio Frequency Modulation
RFM	Remote File Management
RFMS	Remote File Management System
RFMSS	Range Facility Management Support System
RFNM	Ready For Next Message
RFO	Radio Frequency Oscillator
RFO	Random File Organization
RFP	Request For Programming|Price

RFP	Request For Proposal(s)
RFR	Ready For Rework
RFR	Resource-For-Release
RFS	Random Filing System
RFS	Ready For Sending|Service
RFS	Real File Store
RFS	Remote File Server|System
RFS	Remote File Sharing
RFS	Report Forwarding System
RFSP	Request For System Proposal
RFSP	Rigid Frame Selection Program
RFT	Request For Technology|Test
RFT	Request Functional Transmission
RFT	Revisable Form(at) Text
RFTDCA	Revisable Form Text Document Content Architecture
RFU	Reserved for Future Use
RG	Radio Group
RG	Records Group
RG	Register
RG	Report Generator
RG	Reserve Gate
RG	Return without GOSUB
RG	Right
RG	Ring
RGA	Remote Gain Amplifier
RGB	Red, Green, Blue
RGBI	Red Green Blue Intensity
RGCAS	Remote Global Computer Access Service
RGE	Range
RGEN	Regenerate
RGLR	Regulator
RGM	Record Group Maintenance
RGM	Regeneration Module
RGP	Raster|Remote Graphics Processor
RGP	Report Generator Program
RGP	Return Good Parts
RGS	Radio Guidance System
RGT	Resonant Gate Transistor
RH	Relative Humidity
RH	Report Heading
RH	Request|Response Header
RH	Right Hand
RHA	Records Holding Area
RHC	Right-Hand Components
RHCE	Red Hat Certified Engineer
RHCS	Release History Control System
RHD	Round Head
RHEO	Rheostat

RHOB	Relevant Hierarchical Operational Binding
RHPM	Red Hat Package Manager
RHR	Radio Horizon Range
RHR	Receiver Holding Register
RHS	Right Hand Side
RHT	Release Hardware Test
RI	Radio Interface|Interference
RI	Read In
RI	Receiving & Inspection
RI	Reference Implementation
RI	Referential Integrity
RI	Register Immediate
RI	Reliability Index
RI	Rename Inhibit
RI	Rhode Island (US)
RI	Right In
RI	Ring Indication
RI	Routing Indicator|Information
RIA	Robotics Industries Association
RIAA	Recording Industry Association of America
RIACS	Research Institute for Advanced Computer Science (US)
RIB	Ribbon
RIB	Routing Information Base
RIC	Radio Identity Code
RIC	Read-In Counter
RIC	Relocation Instruction Counter
RICS	Range Instrumentation and Control System
RICS	Reports Index Control System
RICS	Requirements planning and Inventory Control System
RICS	Return Inventory Control System
RID	Remote Isolation Device
RID	Repair Identification
RID	Reputed Interactive Debugger
RIDE	Research Issues in Data Engineering (international workshop)
RIDF	Random Input Describing Function
RIDS	Reset Information Data Set
RIE	Radio Interference Elimination
RIF	Record Identifier
RIF	Relative Importance Factor
RIF	Reliability Improvement Factor
RIF	Resource Interchange Format
RIFF	Resource Interchange File Format
RIFI	Radio Interference Field Intensity
RIG	Reporters Internet Guide
RIGFET	Resistive Insulated Gate Field Effect Transistor
RIH	Read Inhibit
RII	Route Information Indicator
RIL	Representation Independent Language

RIM	Read-In Mode
RIM	Read Interrupt Mask
RIM	Records and Information Management
RIM	Remote Installation and Maintenance
RIM	Resource Information|Interface Model
RIME	RelayNet International Message Exchange
RIMM	Reset Immediate
RIMS	Remote Information Management System
RIMS	Requester-oriented Information Management System
RINT	Reverse Interrupt
RIO	Reconfiguration of Input/Output
RIO	Relocatable Input/Output
RIOC	Remote Input/Output Controller
RIOS	Remote Input and Output System
RIOS	Rotating Image Optical Scanner
RIOT	RAM Input/Output Timer (RAM = Random Access Memory)
RIOT	Real-time Input/Output Transducer
RIOT	Remote Input/Output Terminal
RIOT	Resolution of Initial Operational Techniques
RIOT	Retrieval of Information via an On-line Terminal
RIP	Random Input (sampling)
RIP	Random Inspection Program
RIP	Raster Image Processor
RIP	Real-time Integrated (control) Processor
RIP	Remote Imaging Protocol
RIP	Ring Index Pointer
RIP	Routing Information Protocol
RIPE	Réseaux Internet Provider Européens (EU)
RIPEM	Riordan's Internet Privacy Enhanced Mail
RIPFCOMTF	Rapid Item Processor to Facilitate Complex Operations on Magnetic Tape Files
RIPL	Representation Independent Programming Language
RIPL	Robot Independent Programming Language
RIQS	Remote Information Query System
RIR	Request Immediate Reply
RIR	ROM Instruction Register (ROM = Read-Only Memory)
RIRMS	Remote Information Retrieval and Management System
RI/RO	Roll In/Roll Out
RIS	Record Input Subroutine
RIS	Remote Information System
RIS	Rotating Image Scanner
RISC	Reduced Instruction Set Computer
RISC	Remote Information Systems Center
RISCOS	Reduced Instruction Set Computer Operating System
RISLU	Remote Integrated Services Line Unit
RISOS	Research In Secured Operating Systems
RISQ	Regional Internet Services for Quebec
RISS	Relational Inquiry and Storage System

RIT	Rapid Intelligent Tooling
RIT	Receiver Incremental Tuning
RIT	Release Information|Interface Tape
RITA	Recognition of Information Technology (achievement) Award
RITREAD	Rapid Iterative Re-analysis for Automated Design
RITS	Remote Input Terminal System
RIV	Radio Influence Voltage
RJ	Registered Jack
RJ	Reject
RJE	Remote Job Entry
RJEC	Remote Job Entry Communications
RJEF	Remote Job Entry Facility|Function
RJET	Remote Job Entry Terminal
RJEX	Remote Job Entry Executive
RJF	Remote Job entry Facility
RJO	Remote Job Output
RJP	Remote Job Processor
RJPA	Region Job Pack Area
RJU	Remote Job Update
RKB	Restoration Knowledge-Based system
RKM	Radar Keyboard Multiplexer
RKP	Relative Key Position
RKR	Rack Register
RKTS	Robot Keyboard Testing System
RL	Record Length
RL	Relay Logic
RL	Relocatable Library
RL	Remote Loopback
RL	Resistance-Inductance
RL	Return Loss
R-L	Right to Left
R/L	Rotate/Length
RLA	Remote Loop Adapter
RLC	Resistance Inductance Capacitance
RLC	ROM Location Counter (ROM = Read-Only Memory)
RLC	Run Length Coding
RLCM	Remote Line Concentrating Module
RLD	Relocation Dictionary
RLE	Reactivation List Element
RLE	Receiver Latch Enable
RLE	Request Loading Entry
RLE	Run Length Encoding
RLF	Recirculating Loop Frequency
RLF	Reuse Library Framework
RLIN	Research Library Information Network
RLL	Radio Lock Loop
RLL	Real Logic List
RLL	Relay Ladder Logic

RLL	Relocatable Library
RLL	Relocating Linking Loader
RLM	Random Logic Macro
RLM	Reentrant Load Module
RLM	Remote Line Module
RLM	Resident Load Module
RLN	Relation
RLN	Remote LAN Node
RLOGIN	Remote Login
RLP	Radio Link Protocol
RLR	Record Length Register
RLS	Record Level Security
RLS	Reels
RLS	Release
RLSD	Received Line Signal Detector
RLT	Remote Line Test
RLU	Recovery Log Unit
RLY	Relay
RM	Reactive Maintenance
RM	Record Mark(ing)
RM	Reference Manual
RM	Reference Mark|Model
RM	Register Memory
R/M	Reliability/Maintainability
RM	Remote Manipulator|Multiplexer
RM	Repository Manager
RM	Reset Mode
RM	Resource Management|Model
rm	revolutions per minute
RM	Routine Maintenance
RMA	Random Multiple Access
RMA	Rate Monotonic Analysis
RMA	Remote Maintenance Analysis
RMA	Restricted Manual Access
RMAS	Remote Memory Administration System
RMATS	Remote Maintenance And Testing System
RMAX	Range Maximum
RMB	Right-Most Bit
RMB	Right Mouse Button
RMC	Rack-Mount Control
RMC	Rack-Mounted Computer
RMC	Reset Must Complete
RMC	Rod Memory Computer
RMC	Role Membership Certificate
RMDIR	Remove Directory
RMDM	Reference Model of Data Management
RMDR	Remainder
RME	Rack-Mount Extender

RME Request Monitor Entry
RMF Remote Management Facility
RMF Resource Management|Measurement Facility
RMF Routing Master File
RMFC Resolved Motion Forced Control
RMI Radio Magnetic Indicator
RMI Remote Method Interface|Invocation
RMI Roll Mode Interrogation
RML Radar Microwave Link
RML Radio Microwave Links
RML Relational Machine Language
RML Remote Maintenance Line
RMM Read Mostly Memory|Mode
RMM Remote Maintenance Monitor
RMMC Real-time Multicast and Memory-replication Channel
RMMU Removable Media Memory Unit
RMNAME Randomizing Module Name
RM-ODE Reference Model for Open Distributed Environments
RMON Remote (network) Monitor(ing)
RMON Resident Monitor
RMON-MIB Remote network Monitoring Management Information Base
RMOS Refractory metal gate Metal Oxide Semiconductor
RMP Recovery Management Program
RMP Remote Maintenance Processor
RMPI Remote Memory Port Interface
RMR Remote Message Registers
RMS Random Mass Storage
RMS Record Management Service(s)|System
RMS Recovery Management Support
RMS Remote Manipulator System
RMS Resource Management System
RMS Risk Management Solutions
RMS Root Mean Square
RMS Rules Maintenance Subsystem
RMSE Root Mean Square Error
RMSR Recovery Management Support and Recording
RMSV Root Mean Square Value
RMT Remote
RMT Routine Maintenance Time
RMTB Reconfiguration Maximum Theoretical Bandwidth
RMU Remote Multiplexer Unit
RMU Resource Management Unit
RMV Remove
RMW Read-Modify-Write
RMX Remote Multiplexer
RN Read News
RN Reference Noise
RN Removal Number

RN	Requisition Number
RNAA	Radiochemical Neutron Activation Analysis
RNAC	Remote Network Access Controller
RNB	Received – Not Billed
RNC	Request Next Character
RND	Random
RND	Round
RNET	Remote Network
RNG	Random Number Generator
RNMC	Regional Network Measurement Center
RNOC	Regional Network Operations Center
RNP	Regional Network Provider
RNP	Remote Network Processor
RNR	Receive(r) Not Ready
R-NRZ-L	Randomized Non Return to Zero Level
RNSC	Reference Number Status Code
RNV	Radio Noise Voltage
RO	Radio Operator
RO	Read Only|Out
RO	Register Output
RO	Remote Operations
RO	Ring Out
RO	Romania
RO	Round Off
ROA	Recognized Operating Agency
ROAC	Report Origin Authentication Check
ROAR	Read Only Address Register
ROAR	Royal Optimizing Assembly Routine
ROB	Remote Order Buffer
ROB	Re-Order Buffer
ROBAR	Read Only Backup Address Register
ROBCAD	Robotics Computer Aided Design
ROC	Read-Out Clock
ROC	Receiver Operating Characteristics
ROC	Recovery Operations Center
ROC	Relative|Reliability Operating Characteristic
ROC	Remote Operation Call|Control
ROC	Remote Operator's Console
ROC	Re Our Cable
ROC	Required Operational Capability
ROCC	Reactive Organic Conversion Coating
ROCF	Remote Operator Console Facility
ROCH	Read Out and Check
ROCH	Routed Chain
ROCOMP	Read-Out Complete
ROCOND	Robust Control Design (conference)
ROCP	Remote Operator Control Panel
ROCR	Remote Optical Character Recognition

ROD	Read-Out Device
ROD	Reorder On Demand
ROD	Report Of Discrepancy
ROD	Reusable Object Domain
ROD	Rewritable Optical Disk
RODIAC	Rotary Dual Input for Analog Computation
ROER	Remote Operations Error
ROF	Remote Operator Facility
ROFF	Run-Off
ROFL	Rolling On the Floor Laughing
ROH	Receiver Off-Hook
ROHSC	Read-Out High-Speed Count
ROIN	Reorganization Of the Interconnection Network
ROIV	Remote Operations Invoke
ROK	Received Okay
ROL	Re Our Letter
ROL	Request On-Line
ROL	Rotate Left
ROLAP	Relational On-Line Analytical Processing
ROLS	Remote On-Line Subsystem
ROM	Read-Only Memory
ROM	Regional Oxidant Model
ROMAD	Read Only Memory Automatic Design
ROMBIOS	Read Only Memory Basic Input/Output System
ROMM	Read Only Memory Module
ROMON	Receiving Only Monitor
ROM/RAM	Read Only Memory/Random Access Memory
RONS	Read Only Nano Store
ROOM	Real-time Object-Oriented Modeling
ROP	Raster Operation
ROP	Real Optimal control Problem
ROP	Receive Only Printer
ROP	RISC Operation (RISC = Reduced Instruction Set Computer)
ROPES	Remote On-line Print Executive System
ROPM	Remote Operations Protocol Machine
ROPP	Receive Only Page Printer
ROPSE	Removal Order Processing Search Extractor
ROR	Rotate Right
RORE	Remote Operations Return Error
RORI	Roll Out, Roll In
RORJ	Remote Operations Reject
RoRo	Roll-on Roll-off
RORS	Remote Operations Response
ROS	Read Only Storage|Subpool
ROS	Real-time Operating System
ROS	Remote Operations Service
ROS	Resident Operating System
ROSAR	Read Only Storage Address Register

ROSCAR	Read Only Storage Channel Address Register
ROSCOE	Remote Operating System Conventional Operating Environment
ROSDR	Read Only Storage Data Register
ROSE	Real-time Object-oriented Simulation Environment
ROSE	Remote Operations Service Element
ROSE	Rensselaer Object Store for Engineering
ROSE	Research Open Systems for Europe
ROSE	Retrieval by On-line Search
ROSF	Remote Operation/Support Facility
ROSS	Route Oriented Simulation System
ROT	Receive-Only Terminal
ROT	Remaining Operating Time
ROT	Rotate
ROT	Running Object Table
ROTF	Rolling On The Floor
ROTFL	Roll(ing) On The Floor Laughing
ROTH	Read Only Tape Handler
ROTL	Remote Office Test Line
ROTR	Receive Only Typing Reperforator
ROTS	Rotary Out Trunks Selectors
ROTSAL	Rotate and Scale
ROTY	Rotary
ROUT	Retrieval Of Unformatted Text
ROW	Rest Of the World
ROYGBIV	Red, Orange, Yellow, Green, Blue, Indigo, Violet
RP	Random Processing
RP	Rapid Prototyping
RP	Reader-Printer
RP	Reader/Punch
RP	Read Path
RP	Real Processor
RP	Receive Processor
rp	record/playback
RP	Record Processor
RP	Reentry Point
RP	Relative Progress
RP	Repeater
RP	Rolling Plan
RPA	Release Process Automation
RPB	Remote Programming Box
RPC	Real Procedure Call
RPC	Regional Processing Center
RPC	Registered Protective Circuitry
RPC	Remote Position Control
RPC	Remote Procedure Call
RPCS	Reject Processing and Control System
RPE	Regular Pulse Excitation

RPE Remote Peripheral Equipment
RPE Required Page End (character)
RPG Raster Pattern Generator
RPG Remote Password Generator
RPG Report Generator
RPG Report Process|Program Generator
RPG Rotary Pulse Generator
RPI Read, Punch and Interpret
RPI Requirements Planning Interface
RPI Revenue Points Installed
RPI Rework Pictorial Instructions
RPI Rework Print Image
RPI Rockwell Protocol Interface
rpi rows per inch
RPL Radio-Photo-Luminescence
RPL Recognition of Prior Learning
RPL Remote Procedure Load
RPL Remote Program Load(er)
RPL Requested Privilege Level
RPL Request Parameter List
RPL Resident Programming Language
RPL Robot Programming Language
RPL Running Program Language
RPM Read Program Memory
RPM Remote Process Management
RPM Removable Peripheral Module
rpm repeats per minute
rpm revolutions per minute
rpm rotations per minute
rpm runs per minute
RPMC Remote Performance Monitoring and Control
RPN Real Page Number
RPN Regular Processor Network
RPN Reverse Polish Notation
RPOA Recognized Private Operating Agency
RPPROM Reprogrammable Programmable Read-Only Memory
RPQ Reorder Point Quantity
RPQ Request for Price Quotation
RPR Redundant Phase Recording
RPR Reflected Purchase Report
RPR Rejected Purchase Report
RPRINTER Remote Printer
RPROM Reprogrammable Read Only Memory
RPS Rapid Prototyping System
RPS Real-time Programming System
rps records per sector
RPS Remote Printing System
rps revolutions per second

RPS	Ring Parameter Service
RPS	Rotation(al) Position(ing) Sensing
rpt	records per track
RPT	Remaining Processing Time
RPT	Repeat (character)
RPT	Report
RPT	Request Process|Program(s) Termination
RPTC	Relative Priority Test Circuit
RPTN	Repetition
RPU	Regional|Remote Processing Unit
RPW	Record Processor Writer
RQ	Ready Queue
RQ	Repeat Request
RQ	Respiratory Quotient
RQBE	Relational Query By Example
RQD	Required
RQE	Reply Queue Element
RQE	Request Queue Element
RQI	Request Quality Investigation
RQL	Rejectable Quality Level
RQL	Relational Query Language
RQRD	Required
RQRMNT	Requirement
RQS	Rate/Quote System
RR	Rate & Route
RR	Read Record(s)
RR	Ready to Receive
RR	Real Reality
RR	Receive(r) Ready
RR	Record Removal
RR	Recurrence Rate
RR	Reed Relay
RR	Register to Register
R&R	Reliability & Response
RR	Relocatable Rectory
RR	Remove and Replace
R&R	Request & Reply
RR	Rest (and) Recreation
RR	Return Register
RR	Route Relay
RR	Running Reverse
RRA	Remote Record Address
RRAR	ROM Return Address Register (ROM = Read Only Memory)
RRAS	Routing and Remote Access Service
RRC	Remote Readable Counter
RRD	Re-Read
RRDS	Relative Record Data Set
RRE	Receiving Reference Equivalent

RRF	Routing Reference File
RRG	Resource Request Generator
RRIN	Readiness Risk Index Number
RRIP	Rock Ridge Interchange Protocol
RRN	Relative Record Number
RRN	Remote Request Number
RRNS	Redundant Residue Number System
RRO	Rate and Route Operator
RROS	Resistive Read Only Storage
RRPTN	Receiving Report Number
RRR	Return to Reference Recording
RRR	Run-time Reduction Ratio
RRS	Request Repeat System
RRT	Relative Retention Time
RRT	Reverse Recovery Time
RRVP	Resource Reservation Protocol
RS	Radio Shack
RS	Reader Stop
RS	Real Storage
RS	Recommended Standard
RS	Record Separator (character)
RS	Register and Storage
RS	Register Select
RS	Reliability and Serviceability
RS	Remote Single-layer
RS	Remote Site|Station
RS	Request to Send
R+S	Reset and Start
RS	Reset (key)
R-S	Reset-Set
RS	Resume Session
RS	Retrospective Search
RS	Return to Stream (indicator)
RS	Revised Status
RS	Right Side
RS	Ring Station
RS	Robot System
R-S	Rotate-Shift
R-S	Run-Stop
RS/6000	RISC System/6000 (RISC = Reduced Instruction Set Computer)
RSA	Read Signal Amplifier
RSA	Remote Station Alarm
RSA	Remote Storage Activities
RSA	Requirements Statement Analyzer
RSA	Resume Acknowledgment
RSA	Reusable Software Asset
RSA	Rivest, Shamir, Adelman (algorithm)

RSA129	129 digit cryptographic security number (named after Rivest, Shamir and Adelman)
RSAM	Relative Sequential Access Method
RSA-PPDU	Resynchronize Acknowledge Presentation Protocol Data Unit
RSAREF	Rivest, Shamir, Adelman Reference
RSB	Remote System Base
RSC	Record Separator Character
RSC	Remote Scientific Computing
RSC	Remote Store Controller
RSC	Remote Switching Center
RSC	Remote System Components
RSC	Reusable Software Component
RSCC	Remote Site Computer Complex
RSCS	Remote Source Control System
RSCS	Remote Spooling and Control Subsystem
RSCS	Remote Spooling Communications Subsystem
RSDP	Remote Site Data Processor
RSDS	Relative Sequential Data Set
RSDT	Remote Station Data Terminal
RSE	Record Selection Expression
RSE	Remote Single-layer Embedded
RSE	Request Select Entry
RSERV	Relocatable library Service
RSET	Register Set
RSEU	Remote Scanner-Encoder Unit
RSEXEC	Resource Sharing Executive
RSF	Remote Support Facility
RSFQ	Rapid Single-Flux Quantum
RSG	Reference Signal Generator
RSH	Remote Shell
RSH	Restricted Shell
RSH	Right Shift
RSI	Request for Shipping Instructions
RSID	Resource Identification (table)
RSIP	Realm-Specific Internet Protocol
RSIS	Relocatable Screen Interface Specification
RSL	RAISE Specification Language
RSL	Received Signal Level
RSL	Request-and-Status Link
RSL	Requirements Statement Language
RSLE	Remote Subscriber Line Equipment
RSLM	Remote Subscriber Line Module
RSLT	Result
RSLV	Resolve
RSM	Rapid Search Machine
RSM	Real Storage Management
RSM	Remote Switching Module
RSN	Real Soon Now

RSN	Rearrangeable Switching Network
RSN	Record Serial Number
RSOU	Read Sign Over Units
RSP	Rapid System Prototyping
RSP	Reader/Sorter Processor
RSP	Record Select Program
RSP	Reliable Stream Protocol
RSP	Required Space (character)
RS-PPDU	Resynchronize Presentation Protocol Data Unit
RSPT	Real Storage Page Table
RSPX	Remote Sequenced Packet Exchange
RSR	Refracted Surface Reflected
RSR	Re-Store
RSRC	Request for Special Review and Comment
RSRD	Restricted
RSS	Range Safety System
RSS	Real-time Switching System
RSS	Redundant Switch Selector
RSS	Relational Storage System
RSS	Remote Switching System
RSS	Remote Systems Scanner
RSS	Resident Support System
RSS	Resource Security System
RSS	Rework Support System
RSS	Routing and Switching System
RST	Readability Strength Tone
RST	Read Symbol Table
RST	Remotely Submitted Transaction
RST	Remote Station
RST	Representative Support Team
RST	Reset (flag)
RST	Restart
RSTC	Remote Sites Telemetry Computer
RSTCP	Remote Synchronous Terminal Control Program
RSTS	Resource-Sharing Time-Sharing (system)
RSTSE	Resource System Time-Sharing/Enhanced
RSU	Register|Relay Storage Unit
RSU	Remote Service|Switching Unit
RSV	Reserve
RSVD	Rapid Sequential Visual Display
RSVD	Reserved
RSVP	Rapid Serial Visual Presentation
RSVP	Répondez S'il Vous Plait
RSVP	Reservation Protocol
RSVP	Respond Soon (for) Verification Please
RSVR	Reservoir
RSW	Resistance Spot Welding
RSX	Realistic Sound Experience

RSX	Real-time resource-Sharing Executive
RSX	Resource Sharing Extension
RT	Radio Telephone|Transformer
RT	Real-Time
R/T	Receive/Transmit
RT	Receiving Time
RT	Record Type
RT	Register|Reliable Transfer
RT	Remote Terminal
RT	Reperforator/Transmitter
RT	Report
RT	Resistor-Transistor
RT	Right
RT	RISC Technology (RISC = Reduced Instruction Set Computer)
RT	Room Temperature
RT	Routing Table|Type
RT	Run Time
RTA	Ready To Assemble
RTA	Real-Time Accumulator|Analyzer
RTA	Reliability Test Assembly
RTA	Remote Trunk Arrangement
RTA	Resident Transient Area
RTAB	Reliable Transfer Abort
RTAC	Real-Time Adaptive Control
RTAC	Regional Technical Assistance Center
RTAM	Remote|Resident Terminal Access Method
RTAM	Remote Telecommunications|Teleprocessing Access Method
RTAS	Rapid Telephone Access System
RTAS	Real-time Technology and Applications Symposium
RTAW	Real-Time Applications Workshop
RTB	Response Throughput Bias
RTBM	Real-Time Bit Mapping
RTC	Reader Tape Contact
RTC	Real-Time Channel|Clock
RTC	Real-Time Command|Composition
RTC	Real-Time Computer|Conference
RTC	Reasonableness Test Constant
RTC	Relative Time Clock
RTC	Remote Terminal Controller
RTC	Right To Copy
RTC	Routing Control
RTCAD	Register Transfer Computer Aided Design
RTCC	Real-Time Computer Complex
RTCF	Real-Time Computer Facility
RTCG	Run-Time Code Generation
RTCM	Real-Time Control Memory
RTCP	Real-Time Control Program
RTCS	Real-Time Communication|Computer System

RTD	Resistance Temperature Detector
RTD	Retard(ed)
RTDBUG	Real-Time Debug(ging)
RTDM	Real-Time Data Migration
RT/DSS	Real-Time/Decision Support System
RTE	Real-Time Executive
RTE	Remote Terminal Emulator
RTE	Request To Expedite
RTE	Route
RTE	Run Time Environment
RTEC	Real-Time Error Corrector
RTEE	Real-Time Engineering Environment
RTEK	Real-Time Embedded Kernel
RTES	Real-Time Executive System
RTF	Real-Time Fortran
rtf	rich text format (file extension indicator)
RTFAQ	Read The Frequently Asked Questions
RTFM	Real-time Traffic Flow Measurement
RTG	Routing
RTG	Routing Table Generator
RTHCR	Real-Time Hand-print Character Recognition
RTI	Radiation Transfer Index
RTI	Real-Time Interface
RTI	Referred To Input
RTI	Removal Tie-In
RTIO	Real-Time Input/Output
RTIO	Remote Terminal Input/Output
RTIP	Remote Terminal Interface Package
RTIRS	Real-Time Information Retrieval System
RTIS	Real-Time Information System
RTJ	Return Jump
RTJEG	Real-Time for Java Experts Group
RTK	Real-Time Kernel
RTL	Real-Time Language|Link
RTL	Receive/Transmit Leader
RTL	Register Transfer Language|Level
RTL	Resistor-Transistor Logic
RTL	Right-To-Left
RTL	Run Time Library
RTM	Read The Manual
RTM	Real-Time Management|Mode
RTM	Real-Time Monitor
RTM	Register Transfer Module
RTM	Remote Test Module
RTM	Response Time Monitor
RTM	Run Time Manager
RTMBEP	Real-Time Minimal Byte Error Probability
RTMP	Routing Table Maintenance Protocol

RTMS	Real-Time Memory|Monitor System
RTMS	Real-Time Multiprogramming System
RTMTR	Remote Transmitter
RTN	Recursive Transition Network
RTN	Remote Terminal Network
RTNE	Routine
RTNR	Ringing Tone No Reply
RTO	Real-Time Operation
RTO	Referred To Output
RTO	Rejected Take Off
RTOAC	Reliable Transfer Open Accept
RTOP	Real-Time Optical Processing
RTORJ	Reliable Transfer Open Reject
RTORQ	Reliable Transfer Open Request
RTOS	Real-Time Operating System
RTP	Rapid Thermal Processing
RTP	Rapid Transport Protocol
RTP	Real-Time Processor|Protocol
RTP	Real-time Transport Protocol
RTP	Remote Terminal|Test Processor
RTP	Run Time Package
RTPA	Real-Time Prototype Analyzer
RTPC	RISC Technology Personal Computer (RISC = Reduced Instruction Set Computer)
RTPL	Real-Time Procedural Language
RTPM	Reliable Transfer Protocol Machine
RTPS	Real-Time Programming System
RTR	Ready To Receive
RTR	Real-Time Recognizer
RTR	Reel-To-Reel
RTR	Response Time Reporting
RTRL	Real-Time Recurrent Learning
RTS	Reactive Terminal Service
RTS	Ready To Send
RTS	Real-Time System
RTS	Reliable Transfer Service
RTS	Remote Take-over System
RTS	Remote Terminal Supervisor
RTS	Remote Terminal|Testing System
RTS	Request To Send
RTSA	Real-Time Structured Analysis
RTSE	Reliable Transfer Service Element
RTSP	Real-Time Streaming Protocol
RTSRS	Real-Time Simulation Research Systems
RTSS	Real-Time Specification System
RTSW	Real-Time Software
RTT	Régie des Télégraphes et des Téléphones (BE)
RTT	Request To Talk

RTT	Round-Trip Time
RTTI	Run Time Type Information
RTTR	Reliable Transfer Token Response
RTTY	Radio Teletype(writer)
RTU	Real-Time Unit(s)
RTU	Real-Time Unix
RTU	Remote Terminal|Trunking Unit
RTU	Right To Use
RTV	Real-Time Video
RTVS	Run Time Variable Stack
RTW	Right To Work
RTX	Real-Time Executive|Extension
RTXC	Real-Time Executive (in) C
RTYPE	Relation Type
RTZ	Return To Zero
RU	Are You?
RU	Receive|Request Unit
RU	Response Unit
RU	Restricted Usage
RU	Russia
RUA	Remote User Agent
RUC	Rapid Update Cycle
RUC	Reporting Unit Code
RUC	Restricted Usage Code
RUC	Rub-out Character
RUD	Recently Used Directory
RUF	Resource Utilization Factor
RUF	Revolving Underwriting Facility
RUG	Resource Utilization Graph
RUI	Restricted User Interface
RUIP	Remote User Information Program
RUIT	Rules release Interface Tape
RUM	Remote User Multiplex
RUM	Resource Utilization Monitor
RUN	Rewind and Unload
RUNIT	Route Unit
RUOK	Are You Okay?
RUP	Rational Unified Process
RUP	Remote Unit Processor
RUP	Routing Update Protocol
RUS	Routing Update Subroutine
RUSH	Remote Use of Shared Hardware
RUT	Resource Utilization Time
RVA	Relative Virtual Address
RVC	Relative Velocity Computer
RVD	Remote Virtual Disk
RVDT	Rotary Variable Differential Transformer
RVI	Reverse Interrupt (character)

RVN	Requirements Verification Network
RVRS	Reverse
RVT	Reliability Verification Test(s)
RVT	Resource Vector Table
RW	Read-Write
RW	Resume Without (error)
RW	Right Worthy
RW	Rwanda
RWC	Read, Write and Compare
RWC	Read/Write Calibration
RWC	Remote Work Center
RWD	Rewind
RWED	Read, Write, Extend and Delete
RWI	Radio Wire Interface
RWI	Read-Write-Initialize
RWM	Random Walk Method
RWM	Read/Write Memory
RWND	Rewind
RWO	Rights, Wrongs, Omits
RWOC	Right-Wrong-Omit Counter
RWOD	Re-Writable Optical Disk
RWR	Read/Write Register
RWS	Read/Write Storage
RWSCC	Regional Workshop Coordinating Committee
RWU	Rewind-Unload
RWW	Read While Writing
RWWF	Read While Write Feature
RWX	Read/Write Execute
RX	Receiver
RX	Register and Indexed (storage)
R&X	Register & Indexed
RX	Remote Exchange
RXD	Receive(d) Data
RXM	Read/write Expandable Memory
RY	Relay
RYC	Re Your Cable
RYL	Re Your Letter
RZ	Reset to Zero
RZ	Return to Zero (level)
RZI	Return to Zero, Inverted
RZM	Return to Zero Mark
RZ(NP)	Non-Polarized Return to Zero recording
RZ(P)	Polarized Return to Zero recording

S

S	Screen
s	second

S	Secondary
S	Sector
S	Select
S	Sender
S	Sensitivity (of deflection)
S	Sequential
S	Server
S	Shareable
S	Shell
S	Sine
S	Spool
S	Start
S	Strobe
S	Switch
S	Synchronous
S12	System 12
SA	Sample Array
SA	Saudi Arabia
SA	Scaling Amplifier
SA	Selection Addressing
SA	Selective Availability
SA	Semi-Automatic
SA	Sense Amplifier
SA	Sent Ahead
SA	Service Agreement
SA	Shift Advance
SA	Signal Analyzer|Attenuation
SA	Signature Analysis
SA	Signs Alike
SA	Simple-Adjoint (method)
SA	Situational Awareness
SA	Source Address
SA	Spaced Antenna
SA	Stack Access
SA	Staging Adapter
SA	Stand Alone
SA	Store Address
SA	Structured Analysis
SA	Sub-Assembly
SA	Subject to Approval
SA	System Administrator
SA	Systems Address|Analysis
SA	Systems Analyst
SA0	Stuck-At 0 (zero)
SA1	Stuck-At 1 (one)
SAA	Service Action Analysis
SAA	Slot Array Antenna
SAA	Standards Association of Australia

SAA Step Adjustable Antenna
SAA System(s) Application Architecture
SAAC Schedule Allocation And Control
SAAGS Specification, Anticipation, Acquisition, Generation, Specification
SAAL Signalling ATM Adaption Layer (ATM = Asynchronous Transfer Mode)
SAAM Simulation Analysis And Monitoring
SAAOC System of Analysis and Assignment of Operations according to Capacities
SAB Secondary Application Block
SAB Session Awareness Block
SAB Solid Assembly Block
SAB Stack Access Block
SAB Storage|Synchronization Address Bus
SAB System Advisory Board
SABE Society for Automation in Business Education
SABF Sub-Array Beam Former
SABIR Semi-Automatic Bibliographic Information Retrieval
SABLE System for the Analysis of the Behaviour of Logic Elements
SABM Set Asynchronous Balanced Mode
SABME Set Asynchronous Balanced Mode Extended
SABO Sense Amplifier Blocking Oscillator
SABOD Same As Basic Operations Directive
SABR Symbolic Assembler for Binary Relocatable programs
SABRE Semi-Automatic Business Research Environment
SAC Self-Adaptive Control
SAC Semi-Automatic Coding|Core
SAC Service Area Code
SAC Serving Area Concept
SAC Signalling Access Controller
SAC Simplified Access Control
SAC Single Address Code
SAC Single-Attachment Concentrator
SAC Special Area Code
SAC Standard Amplitude Calibrator
SAC Storage Access Channel|Control
SAC Storage Address Character
SAC Store And Clear
SAC System Alert Control
SACCS Semi-Automatic Computer Conversion System
SACCS Strategic Automated Command & Control System
SACE Semi-Automatic Checkout Equipment
SACF Single Association Control Function
SACI Secondary Address Code Indicator
SACK Selective Acknowledgment
SACMAP Selective Automatic Computational Matching And Positioning
SACNET Secure Automatic Communications Network

SACS	Scientific and Administrative Computing System
SACS	Simulation for the Analysis of Computer Systems
SACS	Synchronous Altitude Communications Satellite
SACSTOR	Sequential Access Storage
SACT	Semi-Automatic Cell Tester
SACTS	Semi-Automatic Cell Test Set
SAD	Situation Attention Display
SAD	Stand-Alone Device
SAD	Store Address Director
SA&D	Structured Analysis & Design
SAD	System Analysis Drawing
SADAP	Simplified Automatic Data Plotter
SADAR	Satellite Data Reduction
SADC	Sequential Analog-Digital Computer
SADG	System Application Design Guide
SADIC	Solid-state Analog to Digital Computer
SADL	Synchronous Data Link
SADP	Stand-Alone Data Processing
SADP	System Architecture Design Package
SADS	Single Application Data Sheet
SADS	Structured Analysis and Design System
SADSAC	Sampled Data Simulator And Computer
SADT	Structured Analysis and Design Technique
SADT	Surface Alloy Diffused-base Transistor
SADV	String Array Dope Vector
SAE	Shaft-Angle Encoder
SAE	Software Administration Environment
SAE	Stand Alone Executive
SAF	Segment Address Field
SAF	Store And Forward
SAF	Structural Adjustment Facility
SAF	Subnetwork Access Facility
SAF	Symmetry Adapter Functions
SAFE	Security And Freedom through Encryption
SAFECOMP	Computer Safety, reliability and security (international conference)
SAFENET	Survivable Adaptable Fiber-optic Embedded Network
SAFER	Spectral Application of Finite Element Representation
SAFER	Structural Analysis, Frailty Evaluation and Redesign
SAFF	Store And Forward Facsimile
SAFRAS	Self-Adaptive Format Retrieval And Storage system
SAG	Structured query language Access Group
SAG	Systems Analysis Group
SAGE	Semi-Automatic Ground Environment
SAGFET	Self-Aligned Gate Field-Effect Transistor
SAGMOS	Self Aligning Gate Metal-Oxide Semiconductor
SAH	Sample And Hold
SAI	Sense Amplifier Inhibit

SAI	Serving Area Interface
SAI	Sub-Architectural Interface
SAIC	Switch Action Interrupt Count
SAID	Semi-Automatic Integrated Documentation
SAIM	Systems Analysis and Integration Model
SAINT	Symbolic Automatic Integrator
SAINT	Symposium on Applications and the Internet
SAIV	Schedule As Independent Variable
SAK	Selective Acknowledgment
SAKDC	Swiss Army Knife Data Compression
SAL	Service Action Log
SAL	Shift Arithmetic Left
SAL	Structured|Symbolic Assembly Language
SAL	Systems Activity Log
SAL	Systems Assembly Language
SALE	Simple Algebraic Language for Engineers
SALI	Stand-alone Automatic Location Identification
SALINET	Satellite Library Information Network
SALK	Signs Alike
SALMON	SNA Application Monitor
SALMS	Systematic Asset Library Management System
SALS	Solid-state Acoustoelectric Light Scanner
SALT	Sequential Analyzer Logic Tester
SALT	Sequential Automatic Logic Test
SALT	Symbolic Algebraic Translator
SAM	Security and Administration Module
SAM	Selective Auto(matic) Monitoring
SAM	Semantic Association Model
SAM	Sensor Access Manager
SAM	Sequential Access Memory|Method
SAM	Serial Access Memory
SAM	Service Attitude Measurement
SAM	Simulation of Analog Methods
SAM	Simultaneous Access Memory
SAM	Single Address Message
SAM	Single Application Mode
SAM	Software Associative Memory
SAM	Solenoid Array Memory
SAM	Symantec Antivirus for Macintosh
SAM	System Application Manual
SAM	System for Accumulating Measurements
SAM	Systems Activity Monitor
SAM	Systems Adapter Module
SAM	System(s) Analysis Machine
SAM	Systems Architecture Methodology
SAMA	Step-by-step Automatic Message Accounting
SAMANTHA	System for the Automated Management of Text from a Hierarchical Arrangement

SAMF	System Activity Measurement Facility
SAMI	Systems Activity Measurement Instruction
SAMIS	Structural Analysis and Matrix Interpretation System
SAMM	Systematic Activity Modelling Method
SAMMS	Standard Automated Materiel Management System
SAMON	SNA Application Monitor
SAMOS	Silicon and Aluminum Metal Oxide Semiconductor
SAMPS	Subdivision And Map Plotting System
SAMS	Sampling Analog Memory System
SAMS	Satellite Automatic Monitoring System
SAMSC	Semi-Automatic Message Switching Center
SAMSO	Support Availability Multi-Systems Operations
SAMSON	Strategic Automatic Message Switching Operational Network
SAN	Small Area Network
SAN	Storage Area Network(ing)
SAND	Sorting and Assembly of New Data
SANDS	Structural Analysis Numerical Design System
SANE	Standard Apple Numeric Environment|Extension
SANR	Subject to Approval – No Risk
SANS	Simplified Account Numbering System
SANS	System Administration, Networking and Security (conference)
SAO	Select Address and Operate
SAO	Single Association Object
SAO	Systems Analysis Office
SAOS	Select Address Output Signal
SAP	Service Access Point
SAP	Service Advertising Protocol
SAP	Share Assembly Program
SAP	Structural Analysis Program
SAP	Subordinate Application Program
SAP	Survey Analysis Program
SAP	Symbolic Address|Assembly Program
SAP	System Access Point
SAP	Systems, Applications and Products (for data processing)
SAP	Systems Assurance Program
SAPCH	Semi-Automatic Program Checkout
SAPI	SCSI Application Programming Interface
SAPI	Service Access Point Identifier
SAPR	Semi-Annual Progress Report
SAR	Search And Rescue
SAR	Segment Address Register
SAR	Segmentation And Reassembly
SAR	Shift Arithmetic Right
SAR	Source Address Register
SAR	Special Apparatus Rack
SAR	Stack Address Register
SAR	Stor(ag)e Address Register
SAR	Successive Approximation(s) Register

SARA	Semi-Automatic Registration Analyzer
SARA	System Availability and Reliability Analysis
SARA	System for Activity Recording and Analysis
SARA	Systems Analysis and Resource Accounting
SARAH	Search And Rescue And Homing
SARAN	Satellite Range
SARC	Symantec Antivirus Research Center
SARF	Security Alarm Reporting Function
SARG	Self-Adapting Report Generator
SARM	Set Asynchronous Response Mode
SART	Synchronous/Asynchronous Receiver/Transmitter
SARTS	Switched Access Remote Test System
SARUMAN	Secure Access and Restricted Use of Images on the Net
SAS	Setter And Shaper
SAS	Single Attached Station
SAS	Single Audio System
SAS	Stand-Alone System
SAS	Statistical Analysis System
SAS	Statistically Assigned Sockets
SAS	Switch(ed) Access System
SASC	Semi-Automatic Switching Center
SASD	Structured Analysis, Structured Design
SASE	Statistical Analysis of Series of Events
SASFE	SEF/AIS Alarm Signal, Far End (SEF/AIS = Severely Errored Framing/Alarm Indication Signal)
SASI	Shugart Associates Systems Interface
SASIG	STEP Automotive Special Interest Group (STEP = Standard for The Exchange of Product model data)
SASPA	Sub-Assembly Staging Pre-Audit
SASR	Storage Address Select Register
SAT	Satellite
sat	saturation
SAT	Special Access Termination|Technique
SAT	System Analysis|Audit Table
SATAN	Satellite Automatic Tracking Antenna
SATAN	System Administrator's Tool for Analyzing Networks
SATCOM	Satellite Communications (network)
SATCOM	Satellite for Communicating television broadcasts
satd	saturated
SATF	Security Audit Trail Function
SATF	Shortest Access Time First
SATNET	Satellite Network
SATS	Selected Abstract Test Suite
SAU	Smallest Addressable Unit
SAUL	Seismic Application Users Language
sav	saved (file extension indicator)
SAVDM	Single Application Virtual DOS Machine (DOS = Disk Operating System)

SAVE	System for Automatic Value Exchange
SAVT	Save (area) Table
SAVT	Secondary Address Vector Table
SAW	Surface Acoustic Wave
SAWR	Surface Acoustic Wave Resonator
SAX	Simple API for XML (API = Application Programming Interface)
SB	Secondary Breakdown
SB	Semi-Balance (model)
SB	Sense Bytes
SB	Side-Band
SB	Solomon Islands
SB	Sound Board
SB	Stabilized Breakdown
SB	Stack Base
SB	Straight Binary
SB	Synchronous Bit
SBA	Scene Balance Algorithms
SBA	Sequential Boolean Analyzer
SBA	Set Buffer Address
SBA	Shared Batch Area
SBA	Small Business Administration
SBA	Standard Beam Approach
SBC	Single Board Computer
SBC	Small Binary|Business Computer
SBC	Standard Batch Control
SBC	Standard Buried Collector
SBCA	Sensor Based Control Adapter
SBCS	Single-Byte Character Set
SBCU	Sensor Based Control Unit
SBD	Schottky Barrier Diodes
SBD	Structured Block Diagram
SBDC	Small Business Development Center
SBDT	Surface-Barrier Diffused Transistor
SBE	System Based Education
SBH	Sequencing By Hybridization
SBI	Single Byte Interleaved
SBI	Sound Blaster Instrument
SBIC	Small Business Investment Corporation
SBIR	Storage Bus In Register
SBLC	Standard Base Level Computer
SBM	Small Business Machine(s)
SBM	Solution Based Modelling
SBM	Space Block Map
SBP	Semiconductor Bipolar Processor
SBQ	Standard Batch Quantity
SBR	Storage Buffer Register
SBS	Satellite Business System(s)

SBS	Sensor Based System
SBS	Shared Business System
SBS	Single Business Service
SBS	Small Business System
SBS	Subscript (character)
SBS	System Building System
SBSSC	Sensor Based Systems Support Center
SBT	Screen-Based Telephone
SBT	Selective Bottom-to-Top (parsing)
SBT	Sic Bit Transcode
SBT	Surface Barrier Transistor
SBT	System Backup Tape (drive)
SBTAM	Sensor Based Terminal Access Method
SBTS	Small Business Terminal System
SBU	Shipped But Uninstalled
SBU	Station Buffer Unit
SBV	Single Board Video
SC	Sample Clock
SC	Satellite Communication(s)|Computer
SC	Saturable Core
sc	scale
SC	Scanner Controller
SC	Script
SC	Secondary Channel
SC	Sectional Center
SC	Security Code
SC	Selector Channel
SC	Self Check
SC	Send Common
SC	Sequence Controller|Counter
SC	Session Control
SC	Set Clear
SC	Seychelles
SC	Shift Control
SC	Shoe Connector
SC	Short Circuit
SC	Signal Comparator
SC	Single Column|Contact
SC	Single Counter
sc	single crystal
SC	Small Card
SC	Solar Cell
SC	Solid-state Circuit
SC	Source Code
SC	South Carolina (US)
SC	Special Circuit
SC	Start Computer
SC	Statistical Control

SC	Stop Code
SC	Stop/Continue (register)
SC	Storage Capacity
SC	Store Character
SC	Stored Command
SC	Stress Compensated
SC	Structure Card
SC	Sub-Committee
SC	Subroutine Call
SC	Subscriber Computer|Connector
SC	Supervisory Control
SC	Suppressed Carrier
SC	System Control(ler)
SC4	Subcommittee 4 of ISO TC184 (ISO = International Organization for Standardization; TC = Technical Committee)
SCA	Secondary Communications Authorization
SCA	Selectivity Clear Accumulator
SCA	Short Code Address
SCA	Subsidiary Communications Authority
SCA	Sunlink Channel Adapter
SCA	Synchronous Communications Adapter
SCA	System Communication|Control Area
SCAB	Stochastic Computerized Activity Budgeting
SCAC	Syntax Controlled Acoustic Classifier
SCAD	Small Current Amplifying Device
SCAD	Subprogram Change Affect Diagram
SCADA	Supervisory Control And Data Acquisition
SCADC	Standard Central Air Data Computer
SCAI	Scientific Computing and Algorithms Institute (DE)
SCAI	Switch to Computer Application Interface
SCALD	Structural Computer Aided Logic Design
SCALP	Small Card Automated Layout Program
SCAM	Synchronous Communications Access Method
SCAME	Screen-oriented Anti-Misery Editor
SCAMP	Scientific Computer And Modular Processor
SCAMP	Single-Chip A-series Mainframe Processor
SCAMP	System Compiler Assembly Program
SCAN	Stock Control and Analysis
SCAN	Switched Circuit Automatic Network
SCANDI	Surveillance Control And Driver Information system
SCANNET	Scandinavian Network
SCAR	Sub-Cell Address Register
SCARA	Selective Compliance Assembly Robot Arm
SCARFU	Simultaneous Charge And Read Function Unit
SCARS	Status, Control, Alerting and Reporting System
SCAS	Scan String
SCAT	Schottky Cell Array Technology
SCAT	Share Compiler Assembler and Translator

SCAT Strip-Chip Architecture Technology
SCAT System Configuration Audit Table
SCATS Sequentially Controlled Automatic Transmitter Start
SCB Stack|Station Control Block
SCB String Control Byte
SCB Subscriber|Subsystem Control Block
SCB Supervisory Circuit Breaker
SCB System Control Block
SCBAR System Control Block Address Register
SCC Satellite Communications Concentrator|Control(ler)
SCC Satellite Control Center
SCC Semi-Conductor Component
SCC Sequential Control Counter
SCC Serial Communications Controller(s)
SCC Serial Controller Chip
SCC Single Channel Communications controller
SCC Source Code Control
SCC Spark Controlled Computer
SCC Specialized Common Carrier(s)
SCC Status Change Character
SCC Storage Connecting Circuit
SCC Strongly Connected Component
SCC Switching Control Center
SCC Synchronous Channel Check
SCC Synchronous Communications Controller
SCC System Control Center|Command
SCCB Software Configuration Control Board
SCCC Shared Contingency Computer Center
SCCFF Second Check Character Flip-Flop
SCCP Signalling Connection Control Part
SCCS Software Configuration Control System
SCCS Source Code Control System
SCCS Specialized Common Carrier Service
SCCS Switching Control Center System
SCCU Single Channel Control Unit
SCD Serial Cryptographic Device
SCD Service Computation Date
SCD Slow(ly) Changing Dimension
SCD Standard Colour Display
SCD System output Class Directory
SCD System(s) Contents Directory
SCDA Small Card Design Automation
SCDC System Control Distribution Computer
SC/DKI Serial Communication/Datakit Interface
SCDP Society of Certified Data Processors
SCDR Store|Subsystem Controller Definition Record
SCDSB Suppressed-Carrier Double Sideband
SCDT Special Committee on Data Transmission

SCE	Signal Conversion Equipment
SCE	Signal Correlation Effects
SCE	Single-Cycle Execute
SCE	Software Capability Evaluation
SCE	Standard Card Enclosure
SCE	Structure Chart Editor
SCEC	Small Card Engineering Change
SCED	SCSI/Ethernet/Diagnostics
SCERT	Systems and Computer Evaluation and Review Technique
SCEU	Selector Channel Emulation Unit
SCF	Secondary Control Field
SCF	Selective Call Forwarding
SCF	Self-Consistent Field
SCF	Switched Capacitor Filter
SCF	Synchronous Communications Feature
SCF	System Control Facility
SCFA	Self-Consistent Field Approximation
SCFM	Sub-Carrier Frequency Modulation
SCFP	Software Council Fellowship Program
SCFTS	Small Card Functional Test System
SCH	Schedule(r)
SCH	Special Character
SCH	Synchronization Channel
SCHED	Schedule
SCI	Scalable Coherent Interface
SCI	Science
SCI	Serial Communication Interface
SCI	Stacker Control Instruction
SCI	Strategic Computing Initiative
SCI	System Control Interface
SCIA	Smart Card Industry Association
SCIA	Switched Computer Interface Application
SCIFI	Science Fiction
SCIL	System Core Image Library
SCIM	Selected Categories In Microfiche
SCIM	Speech Communication Index Meter
SCIP	Seniors Computer Information Project
SCIP	System Control Interface Package
SCIS	Survivable Communications Integration System
SCL	Scale
SCL	Send Control Loader
SCL	Sequential Control Logic
SCL	Small Card Layout
SCL	Supervisory Control Language
SCL	Switch to Computer Link
SCL	System Communication Location
SCL	System Control Language
SCLA	Section Carry Look Ahead

SCLM	Software Configuration and Library Management
SCLP	Signalling Channel Link Protocol
SCM	Single Channel Monitoring
SCM	Single-Chip Microcomputer|Module
SCM	Single Column Model
SCM	Small Core Memory
SCM	Software Configuration|Contract Management
SCM	Source Code Management
SCM	Station Class Mark
SCM	Subscriber Carrier Module
SCM	Supply-Chain Management
SCM	Switch Core Matrix
SCM	System Control Manager
SCMO	Subsidiary Communication Multiplex Operation
SCMODS	State, County, and Municipal Offender Data System
SCMP	Simple Cost-effective Micro-Processor
SCMP	Single-Cycle Micro-Processor
SCMP	Software Configuration Management Plan
SCMS	Serial Copy Management System
SCMS	Service Contract Management System
SCN	Scan
SCN	Self Checking Number
SCNR	Sorry, Couldn`t Resist
SCO	Santa Cruz Operation
SCO	Stanford Computer Optics
SCO	Sub-Carrier Oscillator
SCOI	Standing Communication Operating Instruction
SCOM	System Communication
S-COMA	Simple Cache-Only Memory Access(ing)
SCOOP	Silicon-Compatible Optoelectronics (EU)
SCOOP	System for Computerization Of Office Processes
SCOOPS	Scheme Object Oriented Programming System
SCOPE	Screen Oriented Program Editor
SCOPE	Simple Communications Programming Environment
SCOPE	Software evaluation and Certification Programme Europe
SCOPE	Supervisory Control Of Program Execution
SCOPE	System for Control Program Evaluation
SCORE	Solar Cell Optical Reading Equipment
SCORE	Supplier Cost Reduction Effort
SCORE	System Cost and Operational Resource Evaluation
SCORPIO	Subject-Content Oriented Retriever for Processing Information On-line
SCOT	Stepper Central Office Tester
SCOTS	Surveillance and Control Of Transmission Systems
SCP	Save Cursor Position
SCP	Secondary Control Program
SCP	Service|Signal Control Point
SCP	Signal Conversion Point

SCP	Single-Chip Processor
SCP	Subsystem Control Port
SCP	Sunlink Communications Processor
SCP	Supervisory Control Program
SCP	System Control Program(ming)|Point
SCPC	Single Channel Per Carrier
SCPD	Scratch Pad
SCPD	Supplementary Central Pulse Distributor
SCPE	System Control Program Extended
SCPI	Standard Commands for Programmable Instruments
SCR	Scan Control Register
SCR	Screen(ing)
scr	script (file extension indicator)
SCR	Secure Conversion
SCR	Sequence Control Register
SCR	Serial Clock Receive
SCR	Signal-to-Clutter Ratio
SCR	Silicon Controlled Rectifier(s)
SCR	Single Character Recognition
SCR	Software Change Report
SCR	Split Cycle Random
SCR	Subtract Character Register
SCR	Sustainable Cell Rate
SCR	System Change Request
SCRA	Single Channel Radio Access
SCRAM	Static Column Random Access Memory
SCRAM	Strip Cylindrical Random Access Memory
SCRATSTOR	Scratchpad Storage
SCRIBE	System for Computerized Reporting of Information for Better Education
SCRID	Silicon-Controlled Rectifier Indicator Driver
SCRN	Screen (video display)
SCROLL	String and Character Recording Oriented Logogrammatic Language
SCROM	Scratchable Read-Only Memory
SCRS	Scalable Cluster of RISC Systems (RISC = Reduced Instruction Set Computer)
SCS	Satellite Communications Systems
SCS	Scanning and Conversion System
SCS	Schedule Control System
SCS	Secondary Clear to Send
SCS	Security Control System
SCS	Self Contained System
SCS	Silicon Controlled Switch(es)
SCS	Single Channel Simplex
SCS	Small Computer System
SCS	SNA Character String
SCS	Society for Computer Simulation

SCS	Standard Character Set
SCS	Supervisory Control System
SCS	Switch Central System
SCS	System Communication Services
SCS	System Conformance Statement
SCS	System Control Software
SCSA	Signal Computing System Architecture
SCSA	Sun Common SCSI Architecture
SCSC	Summer Computer Simulation Conference
SCSI	Small Computer System Interface
SCSTOR	Semi-Conductor Storage
SCT	Sector
SCT	Semi-Conductor Technology
SCT	Serial Clock Transmit
SCT	Service Counter Terminal
SCT	Special Characters Table
SCT	Step Control Table
SCT	Subroutine Call Table
SCT	Surface-Charge Transistor
SCT	System Communication Table
SCT	System Component Test
SCT	Systems (and) Computer Technology
SCTL	Short-Circuited Transmission Line
SCTO	Soft Carrier Turn-Off
SCTR	Sector
SCTR	Sector Register
SCTS	Secondary Clear To Send
SCU	Secondary|Selector Control Unit
SCU	Sequence|Shared Control Unit
SCU	Special|Station Control Unit
SCU	Storage|System Control Unit
SCU	Subscriber Channel|Connection Unit
SCU	Synchronous Control Unit
SCUBA	Small Card Unit Block Analysis
SCULL	Serial Communication Unit for Long Links
SCV	Scientific Computing and Visualization
SCWID	Spontaneous Call Waiting Identification
SCX	Selector Channel Executive
SCY	Successful Cycles
SD	Sample Data|Delay
SD	Schematic Diagram
SD	SCSI Disk
SD	Section Definition
SD	Send Data
SD	Sequential Disk
SD	Signal Distributor
SD	Single Density
SD	Sort-file Description

SD	South Dakota (US)
SD	Start Delimiter
SD	Structured Design
SD	Sudan
SD	Sum of Differences
SD	Switch Driver
S-D	Synchro to Digital
SD	System Description|Development
SD	System Directory|Design
SD	System-wide Data
SDA	Screen Design Aid
SDA	Security Domain Authority
SDA	Send Data with (immediate) Acknowledgment
SDA	Serial Data Adapter
SDA	Share Distribution Agency
SDA	Software Disk Array
SDA	Source Data Acquisition|Automation
SDA	Standard Data Adapter
SDA	Subcarrier Demodulator Assembly
SDA	Synchronous Data Adapter
SDA	System Display Architecture
SDAI	STEP Data Access Interface (STEP = Standard for the Exchange of Product model data)
SDAL	Switched Data Access Line
SDAP	Standard Document Application Profile
SDAS	Scientific Data Automation System
SDAT	Symbolic Device Address|Allocation Table
SDB	Segment Descriptor Block
SDB	Software Development Board
SDB	Storage Data Buffer|Bus
SDB	Sub Data Base
SDB	Symbolic Debugger (Unix)
SDBD	Software Data Base Document
SDBI	Storage Data Bus-In
SDBO	Storage Data Bus-Out
SDBS	Shared Data Bank Systems
SDC	Send Data Condition
SDC	Signal Data Converter
SDC	Software Distribution Center
SDC	Space Digital Computer
SDC	Synchro-to-Digital Converter
SDCD	Secondary Data Carrier Detect
SDCE	Software Development Capability Evaluation
SDD	Software Description Database
SDD	Software Design Description
SDD	Stored Data Description
SDD	System for Distributed Databases
SDDB	Somewhat Distributed Data Base

SDDF	Self-Defining Data Format
SDDI	Serial Digital Data Interface
SDDL	Stored Data Definition Language
SDE	Software Design Engineer
SDE	Software Development Environment
SDE	Source Data Entry
SDE	Spatial Database Engine
SDE	Storage Distribution Element
SDEP	Source Data Entry Package
SDF	Screen Definition Facility
SDF	Secondary Distribution Frame
SDF	Serial Data Field
SDF	Single Disk File
SDF	Software Development Facility|File
SDF	Space Delimited File|Format
sdf	standard data format (file extension indicator)
SDF	Sub-Distribution Frame
SDFD	System Data Flow Diagrams
SDFS	Standard Disk Filing System
SDFT	Schottky Diode Field effect Transistor logic
SDH	Synchronous Digital Hierarchy
SDH	System Digital Hierarchy
SDI	Scalable Data Interconnect
SDI	Selective Dissemination of Information
SDI	Single Document Interface
SDI	Software Development Interface
SDI	Source Data Information
SDI	Standard Data|Disk Interface
SDI	Storage Device Interconnect
SDI	Supplier Declaration of Inter-operation
SDI	Switch Discrete In
SDI	Switched Digital International
SDI	Systems Design and Installation
SDIF	Standard Document Interchange Format
SDILINE	Selective Dissemination of Information on-Line
SDIO	Serial Digital Input/Output
SDIS	Switched Digital Integrated Service
SDK	Software Development Kit
SDL	Software Design Language
SDL	Sonic Delay Line
SDL	Specification and Description|Design Language
SDL	Structured Design Language
SDL	System Description|Design Language
SDL	System Directory List
SDLA	Supply Distribution and Load Analysis
SDLC	Synchronous Data Link Control
SDLC	System Development Life Cycle
SDM	Schematic Data Model

SDM	Selective Dissemination on Microfiche
SDM	Semiconductor Disk Memory
SDM	Shared Data Memory
SDM	Space-Division Multiplexing
SDM	Sub-rate Data Multiplexer
SDM	Synchronous Digital Machine
SDM	System Design Methodology
SDM	System Development Management|Methodology
SDM	System Development Multitasking
SD&MA	Software Development & Maintenance Agreement
SDMA	Space-Division Multiple Access
SDMA	Storage Device Migration Aid
SDMAC	Shared Direct Memory Access Controller
SDMS	SCSI Device Management System
SDMSS	Software Development and Maintenance Support System
SDN	Software Defined Network
SDN	Symbolic Destination Name
SDN	Synchronous Digital (transmission) Network
SDN	System Development Notification
SDNS	Secure Data Network System
SDOC	Selective Dynamic Overload Control
SDOS	Scientific Disk Operating System
SDP	Scientific Data Processing
SDP	Sequential Disk Processor
SDP	Software Development Plan
SDP	Source Data Processing
SDP	Streaming Data Procedure
SDP	Structured Data Processing
SDP	Structured Design and Programming
SDP	Survey Data Processing
S&DP	Systems & Data Processing
SDR	Search Decision Rule
SDR	Sender
SDR	Signal Data Recorder
SDR	Signalling Data Rate
SDR	Software Delivery Report
SDR	Start Delimiter Received
SDR	Statistical Data Recording
SDR	Stor(ag)e Data Register
SDR	Streaming Data Request
SDR	System Definition Record
SDR	System Design Review
SDRAM	Synchronous Dynamic Random Access Memory
SDRC	Structural Dynamics Research Corporation
SDRL	Subcontract Data Requirements List
SDS	Schema Definition Set
SDS	Scientific Data Systems
SDS	Sequential Data Set

SDS	Shared Data Set
SDS	Shortest Distance to Station
SDS	Significant Digit Scanner
SDS	Single Disk Storage
SDS	Smart Distributed System
SDS	Software Development System
SDS	Software Distribution Services
SDS	Status Display Support
SDS	Summarized Demand Statement
SDS	Switched Data Service
SDS	Synchronous Data Set
SDS	Sysops Distribution System
SDS	System Data Set
SDS	System Development System
SDSAF	Switched Digital Services Applications Forum
SDSC	San Diego Super Computer
SDSC	San Diego Supercomputer Center
SDSC	Synchronous Data Set Controller
SDSD	Single Disk Storage Device
SDSF	Spool Display and Search Facility
SDSI	Shared Data Set Integrity
SDSL	Single-line Digital Subscriber Line
SDSL	Symmetric Digital Subscriber Line
SD-STB	Streaming Data Strobe
SDSU	Switched Data Service Unit
SDT	Signal Detection Theory
SDT	Special Dial Tone
SDT	Start Data Traffic
SDT	Start Descriptor Table
SDT	Storage Display Tube
SDT	Supply and Demand Table
SDT	Syntax Directed Translation
SDT	System Down Time
SDTDL	Saturating Drift Transistor-Diode Logic
SDTM	Screen Dialog Transaction Manager
SDTRL	Saturating Drift Transistor-Resistor Logic
SDTS	Spatial Data Transfer Standard
SDTS	Syntax Directed Translation Scheme
SDU	Signal Distribution Unit
SDU	Source Data Utility
SDU	Station Display Unit
SDU	Storage Distribution Unit
SDU	Subscriber Data and message Unit
SDV	String|Structure Dope Vector
SDV	Switched Digital Video
SDW	Segment Descriptor Word
SDW	System Development Workbench
SDX	Satellite Data Exchange

SE	Scheduling Entity
Se	Selenium
SE	Session Entity
SE	Sign Extend(ed)
SE	Single Element|End
SE	Single Entry
SE	Software Engineering
SE	Special Equipment
SE	Stack Empty
SE	Standard Error
SE	Storage Element
SE	Sweden
SE	System Element|Extension
SE	System(s) Engineer(ing)
SE	System-wide Exchange
SEA	Scanning, Editing, Archiving
SEA	Self Extracting Archive
SEA	Single Ended Amplifier
SEA	Software Engineering and Applications (international conference)
SEA	Static|Systems Error Analysis
SEAC	Standards Eastern Automatic Computer
SEAL	Screening External Access Link
SEAL	Segmentation and (re-)Assembly Layer (protocol)
SEAL	Simple and Efficient Adaptation Layer
SEAL	Simulated Evolution And Learning (international conference)
SEAL	Simulation, Evaluation and Analysis Language
SEAM	Software Engineering And Maintenance
SEAM	Systems Error Analysis Machine
SEAMS	Systems Effectiveness And Management System
SEAS	SHARE European Association
SEAS	Simulated Evolution approach for Analog circuit Synthesis
SEASAT	Sea Satellite
SEB	Scanning Electron Beam
SEB	Source Evaluation Board
SEC	Second(ary)
SEC	Secondary Electron Conduction
sec	section
SEC	Simple Electronic Computer
SEC	Single-Edge Contact
SEC	Single Engineering Control
SEC	Single Error Correction
SECAL	Selective Calling
SECAM	Système Electronique Couleur Avec Mémoire (FR)
SECAM	System Enhanced with Colours And Memory
SECDED	Single Error Correcting and Double Error Detecting
SECLEV	Second Level

SECMAN	Security Management
SECO	Sequential Control
SECORD	Secure voice Cord board
SECP	Seldom Ending Channel Program
SECPDED	Single Error Correcting and Partial Double Error Detecting
SECS	Small Engineering Computer System
SECSTOR	Secondary Storage
SECT	Section
SECU	Slave Emulator Control Unit
SED	Special Energy Distribution
SED	Stream Editor
SEDED	Single Error and Double Erasure Detecting
SEDIT	Source program Editor
SEDR	System Effective Data Rate
SEE	Software Engineering Environment
SEE	Standard Error of Estimate
SEE	Systems Effectiveness Engineering
SEEA	Software Error Effects Analysis
SEEC	Single Error and Erasure Correcting
SEED	Self-Explaining Extended DBMS
SEER	System for Event Evaluation and Review
SEF	Software Engineering Facility
SEF	Source Explicit Forwarding
SEF	Standard External File
SEG	Segment
SEG	Simulation Event Generator
SEG	Special Effects Generator
SEGTAB	Segment Table
SEH	Structure(d) Exception Handling
SEI	Software Engineering Institute (US)
SEID	Systems Engineering Integration Directorate
SEL	Select(or)
SEL	Self-Extensible (programming) Language
SEL	Software Engineering Laboratory
SELCAL	Selective Calling
SELCH	Selector Channel
SELDADS	Space Environmental Laboratory Data Acquisition and Display System
SELDAM	Selective Data Management system
SELEAC	Standard Elementary Abstract Computer
SELECT	Search, Extract/sort, List, Edit, Count and Total
SELREG	Select Register
SELV	Safety Extra-Low Voltage
SEM	Scanning Electron Microscope
SEM	Standard Electronic Module
SEM	Standard Error (of the) Mean
SEM	Systems Engineering Management
SEMATECH	Semiconductor Manufacturing Technology (consortium)

SEMBEGS	Simply Extended and Modified Batch Environmental Graphical System
SEMCOR	Semantic Correlation
SE/ME	Single Entry/Multiple Exit
SEMI	Semiconductor Equipment and Materials Institute
SEMICON	Semiconductor Conference
SEMMA	Sample, Explore, Modify, Model, Assess
SEMP	Standard Electronic Modules Program
SEN	Scanning Encoding
SEN	Scheduling Entity Number
SEN	Sense
SEN	Software Error Notification
SENET	Slotted Envelope Network
SENS	Sprint Enterprise Network Services
SENSEG	Sensitive Segment
SENWDG	Sense Winding
SE-ODP	Support Environment for Open Distributed Processing
SEON	Send Only
SEP	Separate
SEP	Simulator Event Processor
SEPC	Secure Electronic Payment Control
SEPG	Software Engineering Process Group
SEPOL	Soil Engineering Problem Oriented Language
SEPP	Secure Electronic Payment Protocol
SEQ	Sequential
SEQCH	Sequence Check
SEQ-IC	Sequencer – Iteration Control
SEQUEL	Structured English Query Language
SER	Sequential Events Recorder
SER	Serial
ser	series
SERADD	Serial Add
SERC	Software Engineering Research Center
SERCOS	Serial Real-time Communications System
SERDES	Serializer/Deserializer
SERE	Systems Error Record and Entry
SEREP	System Environmental Recording, Editing and Printing
SEREP	System Error Recording Editing Program
SERIM	Software Engineering Risk Management
SERLINE	Serials on-Line
SERNO	Serial Number
SERQL	Structural Entity Relationship Query Language
SERR	Systems Error Record and Retry
SES	Service Evaluation System
SES	Severely Errored Second
SES	Software Engineering Services
SES	System Evaluation System
SES	System External Storage

SES Systems Engineering Services
SESAR Systems Engineering Services Activity Record
SESDIP Structural Evaluation and Synthesis Design of distributed Industrial Processes (EU)
SE/SE Single Entry/Single Exit
SESFE Severely Errored Second, Far End
SESP Severely Errored Second, Path
SESR Segment Entry Save Register
SESR System Equipment Status Report
set (driver) set (file extension indicator)
set (image) settings (file extension indicator)
SET Secure Electronic Transaction
SET Shock Excited-Tones
SET Software Engineering Technology
SET Standard|Système d'Echange et de Transfert (FR)
SET Stepped Electrode Transistor
SETEXT Structure Enhanced Text
SETS Selected Executable Test Suite
SEU Smallest Executable Unit
SEU Source Entry Utility
SEVA System Evaluation
SEW Software Engineering Workbench
SEWOSA Software Engineering Workbench – Open Systems Architecture
SEWP Scientific & Engineering Workstation Procurement
SEX Software Exchange
SF Safety|Scale Factor
SF Select Frequency
SF Shift Forward
SF Short Format
SF Side Frequency
SF Signal Frequency
SF Sign Flag
SF Single Frequency
SF Skip Flag
SF Stability Factor
SF Stack Full
SF Standard Form|Frequency
SF Start Field
SF Step Function
S&F Store & Forward
SF Summarization Function
SFA Segment Frequency Algorithm
SFA Single-Frequency Amplifier
SFAR System Failure Analysis Report
SFBI Shared Frame Buffer Interconnect
SFC Sectored File Controller
SFC Selector File Channel
SFC Sequential Function Chart

SFC	Shop Floor Control
SFC	Single Feed Carriage
SFC	Space Fill Command
SFC	Switched Fabric Controller
SFC	System(s) Flow Chart
SFCM	Small Flat Collector Motor
SFCU	Serial File Control Unit
SFCU	State and Function Control Unit
SFD	Simple Formattable Document
SFD	Software Functional Description
SFD	Start of Frame Delimiter
SFD	Successive Feed
SFD	Switching Field Distribution
SFDI	Standard Format of Digital Images
SFDM	Statistical Frequency Division Multiplexing
SFE	Scriptable Forms Environment
SFE	Smart Front End
SFFT	Semi-Fast Fourier Transformation
SFG	Signal Flow Graph
SFI	Single Frequency Interface
SFI	Special Feature Index
SFL	Substrate Fed Logic
SFL	Symbolic Flowchart Language
SFM	Swept Frequency Modulation
SFMC	Satellite Facility Management Center
SFMR	Stepped Frequency Microwave Radiometer
SFO	Single-Frequency Oscillator
SFO	Specified Functional Operation
SFO	Standard-Frequency Oscillator
SFOG	Shape Feature Object Graph
SFORTRAN	Structured Fortran
SFP	Security Filter Processor
SFP	Soft Front Panel
SFP	Sum First Pass
SFPC	Shop Floor Planning and Control
SFQL	Structured Full-text Query Language
SFR	Single-Frequency Receiver
SFS	Sequential File Structure
SFS	Suomen Standardisoimisliitto (FI)
SFS	Symbolic File Support
SFS	System File Server
SFSU	Signal Frequency Signalling Unit
SFT	Source Fetch Trigger
SFT	System Fault Tolerance
SFTO	Shortest Flow Time at Operation
SFU	Special Front end Unit
SFU	Special Function Unit
SFU	Store and Forward Unit

SFX	Special Effects
SG	Sample Gate
SG	Scanning Gate
SG	Screen Grid
SG	Set Gate
SG	Signal Generator|Ground
SG	Silicon Graphics
SG	Singapore
SG	Specific Gravity
SG	Symbol Generator
SG	System Gain
SGA	Shareable Global Area
SGC	Selective Geometric Complex
SGCN	System Generated Change Number
SGCS	Silicon Gate-Controlled Switch
SGDF	Super-Group Distribution Frame
SGDT	Store Global Descriptor Table
SGEN	Signal|System Generator
SGI	Super Geometric Intelligence
SGJP	Satellite Graphic Job Processor
SGL	Signal
SGL	Single
SGM	Shaded Graphics Modeling
SGM	Statistical Gateway Module
SGML	Standard Generalized Markup Language
SGMP	Simple Gateway Management|Monitoring Protocol
SGN	Scan Gate Number
SGND	Signal Ground
SGNF	Significant
SGR	Set Graphics Rendition
SGRAM	Synchronous Graphics Random Access Memory
SGS	Simultaneous Graphics System
SGS	Status Group Select
SGT	Segment Table
S&H	Sample & Hold
SH	Screen|Session Handler
SH	Shared
SH	Shift
SH	Source Handshake
SH	St. Helena
SH	Super high-Frequency
SH	Switch Hook
SHA	Sample Hold Amplifier
SHA	Secure Hash Algorithm
SHAR	Shell Archive(r)
SHARE	Software Help in Applications, Research and Education
SHARP/SHNS	Self Healing Alternate Route Protection/Self Healing Network Service

S-HDSL	Single-pair High-bit-rate Digital Subscriber Line
SHED	Standard Hotspot Editor
SHF	Super-High Frequency
SHG	Second Harmonic Generation
SHG	Segmented Hyper-Graphics
SHIFT	Scalable Heterogeneous Integrated Facility Testbed
SHIOER	Statistical Historical Input/Output Error Rate
SHL	Shift Logical Left
SHL	System Header Label
SHLD	Shield
SHOC	Software/Hardware Operational Control
SHOCON	Shorts and Continuity (test)
SHOPTS	Shorts Open Tester
SHORAN	Short Range Aid to Navigation
SHP	Single Highest Peak
SHP	Standard Hardware Program
SHR	Share
SHR	Shift Logical Right
SHR	Subtract Halfword Register
SHR	Synchronous Hubbing Regeneration
SHRD	Shared
SHT	Short
SHT	Short Holding Time
SHTML	Special Hyper-Text Markup Language
SHTSK	Short Skip
SHTSTB	Short Stub
SHTTP	Secure Hyper-Text Transport Protocol
SHV	Safe High Voltage
SHV	Standard High Volume
SHY	Syllable Hyphen (character)
SI	Sample Interval
SI	Scheduled Interruption
SI	Screen-grid Input
SI	Selected Item
SI	Serial Input
SI	Shift-In (character)
SI	Signal Interface
SI	Signalling rate Indicator
S/I	Signal-to-Intermodulation ratio
Si	Silicon
SI	Simulator program
SI	Single Instruction
SI	Slovenia
SI	Source Index
SI	SPDU Identifier
SI	Special Instruction
SI	Status Indicator
SI	Step Index

S&I	Storage & Intermediate
SI	Système Internationale (de couleurs) (FR)
SI	System Incidents|Information
SI	System Integrator|Integrity
SIA	Semiconductor Industry Association
SIA	Service in Informatics and Analysis
SIA	Software Industry Association
SIA	Stable Implementation Agreements
SIA	Standard Interface Adapter
SIAM	Society for Industrial and Applied Mathematics
SIB	Screen Image Buffer
SIB	Serial Input/output Board
SIB	Serial Interface Board
SIB	Service Independent Building block
SIB	Session Information Block
SIB	Standard Information Base
SIB	System Interface Board|Bus
SIC	Semiconductor Integrated Circuit
SIC	Shift-In Character
SIC	Silicon Chip
SIC	Silicon Integrated Circuit
SIC	Simulated Input/output Control
SIC	Single Input Change
SIC	Specific Inductive Capacity
SIC	Statistical Inventory Control
SICB	Sub-Interrupt Control Block
SICICA	Symposium on Intelligent Components and Instruments for Control Applications
SICL	Short Interrupt Class List
SICL	Standard Instrument Control Library
SICM	Simultaneous Insertion of Circuit Modules
SICS	Service Implementation Conformance Statement
SICS	Swedish Institute of Computer Science
SID	Serial Input Data
SID	Silicon Imaging Device
SID	Society for Information Display
SID	Source Identifier
SID	Stroboscopic Indicator Display
SID	Structure Identification
SID	Sudden Ionospheric Disturbance
SID	SWIFT Interface Device
SID	Switch Interface Device
SID	Symbolic Interactive Debugger
SID	System Identification
SIDES	Source Input Data Edit System
SIDF	Standard Interchange Data Form
SIDF	System Independent Data Format
SIDS	Standard Interoperable Datalink System

SIDT	Store Interrupt Descriptor Table
SIE	Single Instruction Execute
SIED	Static Instruction Executed Device
SIEP	Static Instruction Executed Plotter
SIF	Standard Interchange Format
SIF	Storage Interface Facility
SIF	System's Information File
SiFA	Single Function Agent
SIFT	Software Implemented Fault Tolerance
SIFT	Stanford Information Filtering Tool
SIG	Signal
SIG	Significant
SIGACT	Special Interest Group on Automata and Computability Theory
SIGARCH	Special Interest Group on computer Architecture
SIGART	Special Interest Group on Artificial intelligence
SIGBDP	Special Interest Group on Business Data Processing
SIGBIO	Special Interest Group on Biomedical computing
SIGCAPH	Special Interest Group on Computers And the Physically Handicapped
SIGCAS	Special Interest Group on Computers And Society
SIGCAT	Special Interest Group on CD-ROM Applications and Technology
SIGCHI	Special Interest Group on Computer-Human Interaction
SIGCOMM	Special Interest Group on data Communications
SIGCOSIM	Special Interest Group on Computer System Installation Management
SIGCPR	Special Interest Group on Computer Personnel Research
SIGCSE	Special Interest Group on Computer Science Education
SIGCUE	Special Interest Group on Computer Use in Education
SIGDA	Special Interest Group on Design Automation
SIG/DAT	Signal Data
SIGEFT	Special Interest Group on Electronic Funds Transfer
SIGFET	Silicon Gate Field-Effect Transistor
SIGFIDET	Special Interest Group on File Description and Translation
SIGGRAPH	Special Interest Group in computer Graphics
SIGhyper	Special Interest Group on hypertext and multimedia
SIGI	System of Interactive Guidance and Information
SIGINT	Signal Intelligence
SIGIR	Special Interest Group on Information Retrieval
SIGLA	Sigma Language
SIGLASH	Special Interest Group on Language Analysis and Studies in the Humanities
SIGLE	System for Information on Gray Literature in Europe
SIGMAP	Special Interest Group on Mathematical Programming
SIGMET	Significant Meteorological information
SIGMICRO	Special Interest Group on Microprogramming
SIGMINI	Special Interest Group in Minicomputers
SIGMOD	Special Interest Group on Management Of Data

SIGNUM	Special Interest Group on Numerical Mathematics
SIGOA	Special Interest Group on Office Automation
SIGOPS	Special Interest Group on Operating Systems
SIGOUT	Signal Output
SIGPC	Special Interest Group on Personal Computing
SIGPLAN	Special Interest Group on Programming Languages
SIGSAM	Special Interest Group on Symbolic and Algebraic Manipulation
SIG/SDI	Special Interest Group for Selective Dissemination of Information
SIGSIM	Special Interest Group on Simulation
SIGSOC	Special Interest Group on Social and behavioral science in computing
SIGSOFT	Special Interest Group on Software engineering
SIGUCC	Special Interest Group on University Computing Centers
SIL	Scanner Input Language
SIL	Semiconductor Injection Laser
SIL	Speech Interference Level
SIL	Store Interface Link
SILS	Standard for Interoperable LAN Security
SILT	Stored Information Loss Tree
SIM	Selective Item Management
SIM	Set Initialization Mode
SIM	Set Interrupt Mask
SIM	Simulator
SIM	Society for Information Management (US)
SIM	Spatial Information Management
SIM	Stack Interface Module
SIM	Subscriber Identification|Interface Module
SIM	Synchronous Interface Module
SIM	System Information Management
SIM	System Integration Modulus
SIMAN	Simulation Modeller and Analyzer
SIMBAD	Set of Identifications, Measurements and Bibliography for Astronomical Data
SIMCOM	Simulation and Computer
SIMCOM	Simulator Compiler
SIMCON	Simulation Control
SIMD	Single Instruction/Multiple Data (stream)
SIMGEN	Simulation Generating system
SIML	Simulation Language
SIMM	Serial|Single In-line Memory Module
SIMM	Symbolic Integrated Maintenance Manual
SIMM	System Integrated Memory Module
SIMNET	Simulation Network
SIMON	Software Implementation Monitor
SIMOS	Stacked-gate Injection Metal Oxide Semiconductor
SIMOSC	System-Independent Method of Operator/System Communication

SIMP	Structured Implementation (language)
SIMPA	Simplified Parallel
SIMPAC	Simulation Package
SIMPARAG	Simultaneous Parallel Array Grammars
SIMPL	Simulation Implementing Machine Programming Language
SIMPL/1	Simulation Programming Language 1
SIMPLE	System for Integrated Maintenance and Program Language Extension
SIMPRO	Simulation Program
SIMPS	Simple and Mnemonic file Processing System
SIMS	Simultaneous(ly)
SIMS	Sun Internet Mail Server
SIMSER	Simple Serial
SIMSYS	Simulation System
SIMTEL	Simulation and Teleprocessing
SIMTRAN	Simple Translation
SIMULA	Simulation Language
SIMULACRE	Standard Image for Multi-Unit Laminar Circuit Representation
sin	sine
SIN	Symbolic Integrator
SINA	Static Integrated Network Access
SINAD	Signal-to-Noise-And-Distortion
SINCGARS	Single Channel Ground-Air Radio System
SINGAN	Singularity Analyzer
sinh	sinus hyperbolicus
SIO	Security Information Object
SIO	Serial Input/Output (communications driver)
SIO	Shortest Imminent Operation
SIO	Start Input/Output
SIOA	Serial Input/Output Adapter
SIOC	Serial Input/Output Channel|Control
SIOC	Serial Input/Output Converter
SIOC	Simulated Input/Output Control
SIOC	Synchronous Input/Output Control
SIOF	Start Input/Output Fast release
SIOM	Starting Input/Output Message
SIOP	Selector Input/Output Processor
SIOP	Single Integrated Operations Plan
SIOT	Step Input/Output Table
SIP	Self-Interpreting Program generator
SIP	Semi-direct Iterative Procedure(s)
SIP	Session Initiation Protocol
SIP	Short Irregular Pulses
SIP	Simulated Input Processor
SIP	Single In-line Package
SIP	SMDS Interface Protocol
SIP	Società Italiana per l'Esercizio delle Telecommunicazioni
SIP	System Integration Program

SIPE	System Internal Performance Evaluator
SIPO	Serial In, Parallel Out
SIPO	System Installation Productivity Option
SIPOGA	Storage Intensive Project Oriented Graphics Application
SIPP	Single In-line Pin Package
SIPRNET	Secret Internet Protocol Router Network
SIPROS	Simultaneous Processing Operating System
SIPS	Satellite Imagery Processing System
SIPS	Statistical Interactive Programming System
SIPSS	Science Informatics Professional Services Sector
SiQUIC	Silicon Quantum Integrated Circuits (EU)
SIR	Segment Identification Register
SIR	Selective|Semantic Information Retrieval
SIR	Serial Infra-Red
SIR	Service Interface Routine
SIR	Stratified Indexing and Retrieval (JP)
SIRB	System Interrupt Request Block
SIRDS	Single-Image Random Dot Stereogram
SIRE	Symbolic Information Retrieval
SIRS	Selective|Standards Information Retrieval System
SIRW	Stuffed Indirect Reference Word
SIS	Scheduling|Scientific Information System
SIS	Scientific Instruction Set
SIS	Shared Information System
SIS	Software Information Services
SIS	Software Integrated Schedule
SIS	Sort-Interval Scheduling
SIS	Standard Instruction Set
SIS	Standardiseringskommissionen I Sverige (SE)
SIS	Strategic Information System
SIS	Structured Information Store
SIS	Successor Instruction Set
SIS	System Interrupt Supervisor
SISAL	Streams and Iteration in a Single-Assignment Language
SISAM	Sorted Indexed Sequential Access Method
SISD	Single Instruction, Single Data
SISFO	Sorted Indexed Sequential File Organization
SIS-MDS	Single Instruction Stream, Multiple Data Stream
SISO	Serial In/Serial Out
SI/SO	Shift In/Shift Out
SISO	Single Input, Single Output
SIS-SDS	Single Instruction Stream, Single Data Stream
SIT	Special Information Tones
SIT	Stand-alone Intelligent Terminal
SIT	Static Induction Transistor
SIT	System Initialization Table
SITA	Society for International Telecommunications for Aeronautics
SITAR	System for Interactive Text-editing Analysis and Retrieval

SITD	Still In The Dark
SITOR	Simplex Teletype Over Radio
SITRAN	Simultaneous Transmission
SIU	System Identification Unit(s)
SIU	System Input Unit
SIU	System Interface Unit
SIVC	Simple Internet Version Control
SJ	Single Job
SJ	Source Jamming
SJ	Svalbard and Jan Mayen
SJD	Silicon Junction Diode
SJD	Sysout Job Directory
SJF	Shortest Job First
SJP	Serialized Job Processor
SJQ	Selected Job Queue
SK	Skip
SK	Slowakia
SKB	Skew Buffer
SKE	Structured Knowledge Engineering
SKIL	Scanner Keyed Input Language
SKIP	Secure Key Interchange Protocol
SKL	Skip Lister
SKM	Sine-cosine Multiplier
SKP	Skip line Printer
SKR	Skip Record
SkSP	Skip-lot Sampling Plan(s)
SL	Section List
SL	Session Layer
SL	Shift Left
SL	Sierra Leone
SL	Signal Level
SL	Simulation Language
SL	Sink Loss
SL	Standard Label
SL	Statement List
SL	Statistical List
SL	Stereo-Lithography
SL	Subroutine Library
SL	Synchronous Line
SL	Systems Language|Library
SLA	Service Level Agreement
SLA	Shared Line Adapter
SLA	Site Level Aggregator
SLA	Stereo-Lithography Apparatus
SLA	Stored Logic Array
SLA	Synchronous Line Adapter
SLAM	Simulation Language for Alternative Modelling
SLAM	Symbolic Language Adapted for Microcomputers

SLAMS Simplified Language for Abstract Mathematical Structures
SLANG Systems Language
SLAP Simulation|Symbolic Language Assembly Program
SLATE Structural Linguistic Analysis and Text Evaluation
SLAVE Solid Logic Automated Variable Evaluator
SLC Selector Channel
SLC Semiconductor Laser Configurations
SLC Set Location Counter
SLC Shared Language Component
SLC Shift Left and Count (instructions)
SLC Single Line Controller
SLC Straight Line Coding
SLC Subscriber Line Charge
SLC Subscriber Loop Carrier
SLC Super Low Cost
SLC Synchronous Line with Clock
SLC Systems Life Cycle
SLCC Store Level Communications Controller
SLCM Software Life Cycle Management
SLCU Synchronous Line Control Unit
SLD Second Logic Diagram
SLD Simplified Logic Diagram
SLD Solid
SLD Solid Logic Diagram
SLD Super-Luminescent Diode
SLD Synchronous Line Driver
SLDA Solid Logic Design Automation
SLDC Synchronous Data Link Control
SLDR System Loader
SLDT Store Local Descriptor Table
SLDTSS Single Language Dedicated Time-Sharing System
SLE Screening Line Editor
SLE Sequential Logic Element
SLED Single Large Expensive Disk
SLED Specification Language for Encoding and Decoding
SLED Synchronous Logic Element Display
SLF Straight Line Frequency
SLF System Library File
SLG Selecting
SLG Synchronous Line Group
SLI Socket List Interface
SLI String Language Interpreter
SLI Suppress Length Indicator
SLI Synchronous Line Interface
SLI System Load and Initialization
SLIB Source Library
SLIB Subsystem Library
SLIC Simulation Language for Integrated Circuitry

SLIC	Subscriber Line Interface Circuit(s)
SLIC	System Link and Interrupt Controller
SLICE	System Life Cycle Estimation
SLIDE	Source Library Image Delivery Expeditor
SLIH	Second Level Interrupt Handler
SLIM	Software Lifecycle Management
SLIM	Subscriber Line Interface Module
SLIMS	Supply Line Inventory Management System
SLIP	Serial Line Interface|Internet Protocol
SLIP	Serviceability Level Indicator Processing
SLIP	Symbolic List Processor
SLIP	Symmetric List Interpretative Program
SLIP	Symmetric List Processor
SLIS	School of Library and Informational Sciences
SLIS	Shared Laboratory Information System
SLL	Statically Linked Library
SLL	Synchronous Line (low) Load
SLM	Selective Level Meter
SLM	Solid Logic Module(s)
SLM	Spatial Light Modulator
SLM	Synchronous Line|Link Module
SLN	Selection
SLO	Segment Limits Origin
SLOA	Starting Location of Output Area
SL/OP	Shortest Slack per remaining Operation
SLO/SRI	Shift Left Out/Shift Right In
SLOT	Sequential Logic Tester
SLP	Second Level Package
SLP	Segmented Level Programming
SLP	Selective Line Printing
SLP	Self-Loading Program
SLP	Service Location Protocol
SLP	Super Long Play
SLPA	Solid Logic Process Automation
SLPC	Second Level Product Commitment
SLPQ	Sleep Queue
SLR	Service Level Report(er)
SLR	Single Lens Reflex
SLR	Storage Limits Register
SLRUM	Simple Least Recently Used stack Model
SLS	Selective Laser Sintering
SLS	Selective Listening Station
SLS	Sequential Logic Systems
SLS	Shop Load Scheduler
SLS	Side-Lobe Suppression
SLS	Single List Structure
SLS	Source Library System
SLS	Subscriber Logic Shelf

SLSI	Super Large Scale Integration
SLT	Select
SLT	Solid Logic Technology
SLTF	Shortest Latency Time First
SLTFES	Self-Learning and -Testing Fuzzy-Expert System
SLTU	Self-Loading Tape Unit
SLU	Secondary Logical Unit
SLU	Serial Line Unit
SLU	Source Library Update
SLU	Synchronous Link Unit
SLUR	Systems Logic Usage Recorder
SLV	Stipulated Loss Value
SLW	Straight Line Wavelength
SM	San Marino
SM	Scheduled Maintenance
SM	Segment Map
SM	Semiconductor Memory
SM	Sequence Monitor
SM	Sequential Machine(s)
S&M	Service & Maintenance
SM	Service Manual
SM	Service Model|Module
SM	Set Mode (character)
SM	Shared Memory
SM	Short Message
S/M	Signal to Mean
SM	Sort Merge
SM	Spatial Media
SM	Standard Matched
SM	Standby Monitor
SM	Status Modifier
SM	Storage Mask|Module
SM	Structured Macro-assembly
SM	Structure Memory
SM	Switching Module
SM	Synchronous Modem
SM	Systems Manager
SMA	Sequence Model Approximation
SMA	Software Maintenance Association
SMA	Structured Markov Algorithm
SMA	Sub-Miniature Assembly
SMA	Synchronous Mode Adapter
SMAC	Scene Matching Area Correlator
SMAC	Store Multiple Access Control
SMACS	Scheduled Machine Assembly Control System
SMAD	Self-Maintained Audio Device
SMAE	Systems Management Application Entity
SMAL	Structured Macro-Assembly Language

SMALGOL	Small computer Algorithmic Language
SMALL	Selenium Matrix Alloy Logic
SMAP	Systems Management Application Process
SMART	Scheduler Manager And Resource Translator
SMART	Scheduling Management and Allocating Resources Technique
SMART	Self-Monitoring Analysis and Reporting Technology
SMART	Sort/Merge All Records Technique
SMART	Systematic Master Receiver Tabulator
SMART	Systems Management Analysis, Research and Test
SMAS	Supplementary Main Store
SMAS	Switched Maintenance Access System
SMASE	Systems Management Application Service Element
SMASF	Supplementary Main Store Frame
SMASPU	Supplementary Main Store Power Unit
SMB	Server Message Block (protocol)
SMB	Service|System Message Block
SMB	System Monitor Board
SMC	Single Memory Cycle
SMC	Solaris Management Console
SMC	Station Monitor and Control
SMC	Storage Module Controller
SMC	System Monitor Controller
SMCA	System Management Control Area
SMCC	System Monitoring and Coordinating Center (US)
SMD	Service Mount Device
SMD	Shared Memory Data
SMD	Source Macro Definition
SMD	Storage Module Device|Drive(r)
SMD	Surface Mounted Device
SMD	System Macro Definition
SMD	Systems Maintenance Design
SMDF	Subscriber Main Distributing Frame
SMDI	Subscriber Message Desk Interface
SMDL	Standard Music Description Language
SMDR	Station Message Detail Recorder
SMDS	Switched Multimedia|Multi-megabit Data Service
SME	Semantic Modelling Extension
SME	Small and Medium size Enterprises
SME	Subject Matter Expert
SMEC	Single Module Engine Controller
SMF	Standard Message Format
SMF	Supplier Master File
SMF	System Management Facility|Function
SMF	System Message Field
SMF	System(s) Measurement Facilities
SMFA	Systems Management Functional Area
SMG	Screen Management Guidelines
SMG	Super Master-Group

SMG	System Management Group
SMI	Start-of-Message Indicator
SMI	Static Management|Memory Interface
SMI	Station Management Interface
SMI	Structure of Management Information
SMI	System Management Interrupt
SMI	System Memory Interface
SMIB	System Management Information Base
SMIL	Synchronized Multimedia Integration Language
SMILES	Semiconductor Micro-cavity Light Emitters (EU)
SMILES	Simplified Molecular Input Line Entry System
SMIMD	Switched Multiple Instruction, Multiple Data
S/MIME	Secure Multipurpose Internet Mail Extension
SMIP	Specific Management Information Protocol
SMIP	Structure Memory Information Processor
SMIS	Sequential Multilayer Interconnection System
SMIS	Society for Management Information Systems
SMIS	Specific Management Information Service
SMIS	Symbolic Matrix Interpretation System
SMIS	Systems Management Information System
SMISE	Specific Management Information Service Element
SMIT	System Management Interface Tool
SMK	Shared Management Knowledge
SMK	Software Migration Kit
SMK	System Monitor Kernel
SML	Source Module Library
SML	Spool Multi-Leaving
SML	Standard Meta|Modelling Language
SML	Symbolic Machine Language
SMM	Semiconductor Memory Module
SMM	Start of Manual Message
SMM	System Management Methodology|Mode
SMM	System Management Monitor
SMMC	Simultaneous Multi-indenture Multi-echelon Computations
SMMR	Scanning Multichannel Microwave Radiometer
SMO	System Management Overview
SMOC	Small Matter Of Commitment
SMOH	Since Major Overhaul
SMON	Switch Monitoring
SMOP	Small Matter Of Programming
SMP	Sample(r)
SMP	Session|Simple Management Protocol
SMP	Sort and Merge Program
SMP	Standby Monitor Present
SMP	Symbolic Manipulation Program
SMP	Symmetric(al) Multi-Processing
SMP	Symmetric Multi-Processor
SMP	System Maintenance|Modification Program

SMPC	Shared Memory Parallel Computer
SMPDU	Service Message Protocol Data Unit
SMPDU	System Management Protocol Data Unit
SMPL	Sample
SMPO	Software Migration Project Office
SMPS	Switched-Mode Power Supply
SMPT	Simple Mail Transport Protocol
SMPTE	Society of Motion Picture and Television Engineers
SMPX	System Maintenance Program Extensions
SMP/X	System Modification Program/Extended
SMQ	Save/restore Message Queues
SMR	Series Mode Rejection
SMR	Shift-out Modular Redundancy
SMRA	Stepwise Multiple Regression Analysis
SMRAM	System Management Random Access Memory
SMRT	Signal Message Rate Timing
SMS	Satellite Multiservice System
SMS	Service-Management System
SMS	Shared Mass Storage
SMS	Short Message Service
SMS	Standard Modular System
SMS	Storage Management Services|System
SMS	Structured Multiprocessor System
SMS	Systematic Management Services
SMS	System Measurement Software
SMS	Systems Management Server|Services
SMSM	Shared Main Storage Multiprocessing
SMSP	Storage Management Services Protocol
SMST	Switching Module System Test
SMSW	Store Machine Status Word
SMT	Scheduled Maintenance Time
SMT	Station Management
SMT	Surface Mounted Technology
SMT	Switch Maintenance Terminal
SMT	System Management Team
SMT	System(s) Maintenance Time
SMTE	Society of Motion picture and Television Engineers
SMTP	Simple Mail Transfer Protocol
SMU	Store Monitor Unit
SMU	System Management Utility
SMU	System Monitoring Unit
SMUX	Statistical Multiplexer
SMV	Symbolic Model Verifier
SMW	Sniff Master (for) Windows
SMX	Signalling Multiplex
SMX	System Matrix
SN	Sector Number
SN	Semiconductor|Shaping Network

SN	Senegal
SN	Sequence|Serial Number
SN	Serial Number
SN	Sign
SN	Signal Node
S/N	Signal to Noise ratio
SN	Subnetwork Number
SN	Switched|Systems Network
SNA	Systems|Synchronous Network Architecture
SNAC	Single Number Account Control
SNACP	Sub-Network Access Protocol
SNACS	Share News on Automatic Coding Systems
SNADS	Systems Network Architectural Distribution Services
SNADS	Systems Network Architecture Delivery System
SNAFU	Situation Normal, All Fouled Up
SNAP	Shipboard Non-tactical Automated data Processing program
SNAP	Significant News About Programming
SNAP	Simulation Network Analysis Program
SNAP	Stylized Natural Procedural (language)
SNAP	Sub-Network Access Protocol
SNAP	System and Network Administration Program
SNAPP	System Networking, Analysis, and Performance Pilot
SNAP/SHOT	System Network Analysis Program/Simulated Host Overview Technique
SNARE	Sub-Network Address-Resolution Entity
SNA/SDLC	Systems Network Architecture/Synchronous Data Link Control
SNBU	Switched Network Back-Up
SNCP	Single Node Control Point
SND	Send
SND	Sound
SNDCF	Sub-Network-Dependent Convergence Facility
SNDCP	Sub-Network-Dependent Convergence Protocol
SNES	Super Nintendo Entertainment System
SNEWS	Secure News Server
SNF	Sequence Number Field
SNF	Server Natural|Normal Format
SNGL	Single
SNGOPR	Single Operation
SNI	Selective Notification of Information
SNI	Sequence Number Indicator
SNI	SNA Network Interconnection
SNI	Subscriber Network Interface
SNI	System Network architecture Interconnection
SNI	System Network Integration|Interconnect
SNIA	Storage Network Industry Association
SNICF	Sub-Network-Independent Convergence Facility
SNICP	Sub-Network-Independent Convergence Protocol

SNIF	Signal-to-Noise ratio Improvement Factor
SNIP	Service Node Intelligent Peripheral
SNIT	Syndicat Nationale des Installateurs de Télécommunications (FR)
SNK	Sink
SNM	Sun-Net Manager
SNMP	Simple Network Management Protocol
SNMP	Small Network Management Packet
SNMPD	Simple Network Management Protocol Daemon
SNNS	Stuttgart Neural Network Simulator
SNO	Serial Number
SNOBOL	String Oriented Symbolic Language
SNORE	Signal-to-Noise Ratio Estimator
SNP	Sequence Number Protection
SNP	Serial Number/Password
SNP	Statistical|Synchronous Network Processor
SNPA	Sub-Network Point of Attachment
SNQ	Shared In Queue
SNR	Signal-to-Noise Ratio
SNRM	Set Normal Response Mode
SNS	Secondary Network Server
SNTSC	Super National Television Standard Code
SNZ	Sum Not Zero
SO	Send Only
SO	Serial Output
SO	Shift Out (character)
SO	Shut Off (sequence)
SO	Slow Operate
SO	Somalia
SO	Special Operation(s)
SO	Stop Order
SO	Substance Of
SO	Support Operations
SO	System Operation|Override
SO	Systems Orientation
SOA	Safe Operating Area
SOA	Start Of Address|Authority
SOA	State Of the Art
SOAP	Simple Object Access Protocol
SOAP	Stage Operation Analysis Project
SOAP	Symbolic Optimizing Assembler Program
SOAR	Safe Operating Area
SOAR	State, Operator, And Result
SOB	Start Of Block
SOC	Self-Organizing Control
SOC	Service Oversight Center
SOC	Shift-Out Character
SOC	Social issues (USENET newsgroup category)

SOC	Span Of Control
SOC	System On a Chip
SOCC	Satellite Operations Control Center (US)
SOCKS	Socket Secure (server)
SOCOCO	Symposium Of software for Computer Control
SOCS	Secondary Operator Control Station
SOD	Serial Output Data
SOD	System Operational Design
SoDA	Software Documentation Automation
SODA	System Optimization and Design Algorithm
SODAS	Satellite Operation planning and Data Analysis System
SOE	Server|Standard Operating Environment
SOF	Start Of Frame|Format
SOF	Structured Oriented Fortran
SOFAR	Sound Fixing And Ranging
SOFE	Stop On First Error
S/OFF	Sign Off
SOFLC	Self-Organizing Fuzzy Logic Controller
SOFM	Self-Organizing Feature Map
SOFPIN	Software Process Improvement Networks
SOFRECOM	Société Française d'études et de Réalisations d'Equipments de Communications
SOFT	Software
SOGITS	Senior Officials' Group for Information Technology Standardization
SOGT	Senior Officials' Group for Telecommunications
SOH	Section Overhead
SOH	Service Order History
SOH	Start Of Header (character)
SOHC	Single Overhead Cam(shaft)
SOHO	Small Office/Home Office
SOI	Signal Operating Instruction
SOI	Silicon-On-Insulator
SOI	Standard Operating Instructions
SOIC	Small Outline Integrated Circuit
SOIF	Summary Object Interchange Format
SOJ	Small Outline J-lead
SOJ	Stand Off Jamming
SOL	Simulation Oriented Language
sol	soluble
sol	solution
SOL	Speed Of Light
SOLA	Source Language
SOLACE	Sacred On-Line Active Communal Environment
SOLD	System On-Line Diagnosis
SOLDES	Self-Organizing Low-Dimensional Electronic Structures (EU)
SOLE	Society Of Logistics Engineers
SOLID	Specials Oriented Logic Interactive Design

SOLIS	STEP On Line Information Server (STEP = Standard for The Exchange of Product model data)
SOLIST	Source Listing
SOLN	Solution
SOLOMON	Simultaneous Operation Linked Ordinal Modular Network
SOM	Scripting Object Model
SOM	Self-Organizing Machine|Map
SOM	Small-Office Microfilm
SOM	Start Of Message (character)
SOM	System Object Model
SOMA	Semantic Object Modelling Approach
SOMADA	Self-Organizing Multiple-Access Discrete Address
SOMM	Stop On Micro-Match
SOMOD	Source Module
SOMS	System Optimization and Monitoring Services
S/ON	Sign On
SONAN	Sonic Noise Analyzer
SONAR	Sound Navigation And Ranging
SONDS	Small Office Network Data System
SONET	Synchronous Optical Network
SOON	Solar Observing Optical Network (US)
SOP	Standard Operating Procedure(s)
SOP	Sum Of Products
SOP	Symbolic Optimum Program
SOPO	Set Of Possible Occurrences
SOR	Specific Operation(al) Requirement
SOR	Start Of Record
SOR	Statement Of Requirements
SORC	Source
SORCES	Service Order Record Computer Entry System
SORDID	Summary Of Reported Defects, Incidents and Delays
SORM	Set Oriented Retrieval Module
SOS	Self-Organizing System
SOS	Share Operating System
SOS	Silicon On Sapphire
SOS	Sophisticated Operating System
SOS	Standards and Open Systems
SOS	Start Of Significance
SOS	Station Operator Support
SOS	Storage Operating System
SOSAS	Symposium for an Open System Architecture Standard
SOS_CMOS	Silicon On Sapphire Complementary Metal Oxide Semiconductor
SOSIC	Silicon On Sapphire Integrated Circuit
SOSIG	Social Science Information Gateway
SOSP	Symposium on Operating System Principles
SOST	Special Operator Service Traffic
SOSTEL	Solid-State Electronic Logic

SOSUS	Sound Surveillance System
SOT	Scanning Oscillator Technique
SOT	Send-Only Terminal
SOT	Short-Open Test
SOT	Sort Output Tape
SOT	Sound On Tape
SOT	Start Of Text
SOT	System Order Tape
SOT	Systems Operation Test
SOTUS	Sequentially Operated Teletypewriter Universal Selector
SOU	System Output Unit
SOUP	Simple Off-line Usenet Packet
SOUP	Software Updating Package
SOUT	Swap Out
SOW	Statement Of Work
SOX	Sound Exchange
SOYD	Sum Of the Years' Digits method
SP	Satellite Processor
SP	Scratch Pad
SP	Send Processor
SP	Sequence Processor
SP	Sequential Phase|Processing
S/P	Series to Parallel
SP	Service Pack
SP	Set Point
SP	Shift Pulse
SP	Short Play
SP	Signalling Point
SP	Signal Processor
SP	Single Phase
sp	single-pole
SP	Single Programmer|Purpose
SP	Space (character)
SP	Special Purpose
SP	Stack Pointer
SP	Standards Project
SP	Standing Procedure
SP	Start Permission
SP	Stick Printer
SP	Storage Protect
SP	Stored Program
SP	Structured Programming
SP	Sub-Program
SP	Subscriber Processes
SP	Summary Plotter|Punch
SP	Supervisor Process
SP	Switch Panel
SP	Symbolic (assembly) Program(s)

SP	System Package|Product
SP	System Parameter
SP	Systems Planning|Procedure
S&P	Systems & Procedures
SP	Systems Program(med)
SPA	Scratch Pad Area
SPA	Shared Peripheral Area
SPA	Single Parameter Analysis
SPA	Software Publishers Association
SPA	System Performance Analysis
SPA	Systems Programmed Application
SPACE	Self-Programming Automatic Circuit Evaluator
SPACE	Serial Programming by Associative Coordinate Execution
SPACE	Standard Packaged And Computer Enclosures
SPACE	Stored Program Accounting and Calculating Equipment
SPACE	System Precision Automatic Checkout Equipment
SPAD	Scratch Pad
SPADE	Single channel per carrier, Pulse code modulation, multiple Access, Demand assignment Equipment
SPAG	Standards Promotion and Application Group
SPAG	Strategic Planning Advisory Group
SPAM	Satellite Processor Access Method
SPAM	Send Phenomenal Amounts of Mail
SPAN	Space Physics Analysis Network
SPAN	Statistical Processing and Analysis
SPAN	System Performance Analysis
SPANS	Simple Protocol for ATM Network Signalling (ATM = Asynchronous Transfer Mode)
SPAR	Special Area
SPAR	Stellar Pattern Recognition
SPAR	Symbolic Program Assembly Routine
SPAR	System Performance Analysis and Reporting
SPAR	System Planning And Review
SPAR	Systems Performance Activity Recorder
SPARC	Scalable Performance Architecture
SPARC	Standards Planning And Review Committee
SPARC	Steel Processing Automation Research Center
SPAS	Stored Procedure Address Space
SPASM	Special Purpose Application Service Module
SPAT	Silicon Precision Alloy Transistor
SPATS	Semiconductor Process Area Test System
SPB	Stored Program Buffer
SPB	Surface Pigment Bonding
SPC	Setpoint Control
spc	silicon-point contact
SPC	Small Peripheral Control(ler)
SPC	Software Productivity Consortium
SPC	Software Program Control

SPC	Southern Pacific Communications
SPC	Special Purpose Computer
SPC	Speech Processor Chip
SPC	Statistical Process Control
SPC	Stored Program Computer|Collector
SPC	Stored Program Control(ler)
SPC	Summary Punch Control
SPC	Switching and Processing Center
SPCC	Staggered Phase Carrier Cancellation
SPCL	Self-Propagating Core Logic
SPCL	Special
SPCS	Storage and Processing Control System
SPCS	Stored Program Control Systems
SPC/SQC	Statistical Process Control/Statistical Quality Control
SPD	Sample Pulse Driver
SPD	Secondary Paging Device
SPD	Software Product Description
SPD	Structured Program Design
SPDE	Special Purpose Difference Equation
S/PDIF	Sony/Philips Digital Interface
SPDL	Standard Page Description Language
SPDM	Special Purpose Dexterous Manipulator
SPDM	Sub-Processor with Dynamic Microprogramming
SPDT	Single Pole, Double Throw
SPDU	Session Protocol Data Unit
SPE	Single Processing Element
SPE	Solid Phase Extraction
SPE	Special Purpose Equipment
SPE	Stored Program Element
SPE	Symbolic Programming Environment
SPE	Synchronous Payload Envelope
SPE	System(s) Performance Effectiveness
SPEAC	Special Purpose Electronic Area Correlator
SPEAL	Special Purpose Engineering Analysis Language
SPEC	Special Character
SPEC	Specification
SPEC	Standard Performance Evaluation Corporation
SPEC	System Performance Evaluation Cooperative
SPECDP	Specialized Data Processing
SPECLE	Specification Language
SPECMARK	System Performance Evaluation Cooperative Mark
SPECOL	Special Customer Oriented Language
specs	specifications
SPECS	Stored Program Electronic Circuit Switch
SPECT	Semiconductor Product Engineering Chip Tester
SPECTA	Structure Preserved Error Correcting Tree Automata
SPECTRE	System for Prediction and Evaluation of Circuits with Transient Radiation Effects

SPEDAC	Solid-state Parallel, Expandable, Differential-Analyzer Computer
SPEED	Self-Programmed Electronic Equation Delineator
SPEED	Systematic Plotting and Evaluation of Enumerated Data
SPEND	Spacing End Distortion
SPENT	Summary Punch Entry
SPEX	Summary Punch Exit
SPF	Shortest Path First
SPF	Software Productivity Facility
SPF	Static Permutation Flowshop
SPF	Structured|System Programming Facility
SPFW	Single-Phase, Full-Wave
SPG	System Programmers Guide
SPI	SCSI Parallel Interface
SPI	Serial Peripheral Interface
SPI	Service Provider Interface
SPI	Shared Peripheral Interface
SPI	Signal Point Identification
SPI	Signal Propagation on Interconnects (workshop)
SPI	Single Processor Interface
SPI	Single Program Initiator
SPI	Specific Productivity Index
SPI	Subsequent Protocol Identifier
SPICE	Simulation Program with an Integrated Circuit Emphasis
SPID	Service Provider Identification
SPIE	Scientifically Programmed Individualized Education
SPIF	Sequential Prime Implicant Form
SPIKE	Science Planning Intelligent Knowledge-based Environment
SPIN	Searchable Physics Information Notices
SPIN	Set Pin
SPIN	Software Process Interest
SPIN	Strategies and Policies in Informatics
SPINDEX	Selective Permutation Indexing
SPIRES	Stanford Public Information Retrieval System
SPIRS	Silver Platter Information Retrieval System
SPIRT	Special Program for Instant Registration Testing
SPIT	Selective Printing of Items from Tape
SPK	Storage Protection Key
SPKR	Speaker
SPL	Selected Product Line
SPL	Signal Processing Language
SPL	Simulation Programming Language
SPL	Sound Pressure Level
SPL	Source Program Library
SPL	Special Purpose Language
spl	spell (checker) (file extension indicator)
SPL	Spooler
SPL	Symbol Processing Language

SPL	System Program Loader
SPL	System(s) Programming Language
SPLC	Standard Point Location Code
SPLICE	Shorthand Programming Language In Cobol Environment
SPLIT	Space Programming Language Implementation Tool
SPLIT	Sundstrand Processing Language Internally Translated
SPLS	Storage Protect Local Store
SPM	Scanning Probe Microscopy
SPM	Scratch Pad Memory
SPM	Session Protocol Machine
SPM	Source Program Maintenance
SPM	Special Purpose Multiplexer
SPM	Storage Protect Memory
SPM	Structured Pattern Matching
SPM	System Performance Monitor
SPM	System Preventive Maintenance
SPMAR	Scratch Pad Memory Address Register
SPMD	Single Program Multiple Data
SPMF	Servo Play-Mode Function
SPMOL	Source Program Maintenance On-Line
SPMP	Software Project Management Plan
SPN	Shared|Signal Processing Network
SPN	Standard Power Network(s)
SPN	Stochastic Petri Net
SPN	Subscriber Premises Network
SPN	Switched Public Network
SPNS	Switched Private Network Service
SPO	Separate Partition Option
SPO	Storage Printout
SPO	Synchronized Power On
SPOA	Single Point Of Access
SPOOF	Structure and Parity Observing Output Function
SPOOL	Simultaneous Peripheral Operation Off-Line|On-Line
SPOR	Sub-Portion
SPOT	Shared Product Object Tree
SPP	Sequenced Packet Protocol
SPP	Service Provision Point
SPP	Signal Processing Peripheral
SPP	Simultaneous Print/Plot
SPP	Special Purpose Processor
SPP	Standard Pin Pad
SPP	Storage Parity Protection
SPPDG	Special Purpose Processor Development Group
SPPRG	Special Program
SPPS	Scalable Power Parallel System
SPPS	Subsystem Program Preparation Support
SPPU	Summary Punch Pickup
SPQ	Special Performance Quality

SPQE	Sub-Pool Queue Element
SPR	Software Problem Report
SPR	Special Purpose Register
SPR	Statistical Pattern Recognition
SPR	Storage Protection Register
SPR	Supervisory Printer Read
SPR	System Parameter Record
SPRACH	Speech Recognition Algorithms for Connectionist Hybrids (EU)
SPRB	Senior Performance Review Board
SPRINT	Selective Printing
SPRINT	Switched Private Network Telecommunications
SPRN	Suppression
SPRO	Space Robotics (international conference)
SPROM	Switched Programmable Read Only Memory
SPRT	Sequential Probability Ratio Test
SPS	Secondary Processing Sequence
SPS	Selective Photo Sensitizer
SPS	Sequential Partitional Scheduler
SPS	Sequential Partitioned System
SPS	Serial Parallel Serial
SPS	Special Processor System
SPS	Stand-by Power Supply|System
SPS	String Process System
SPS	Superscript (character)
SPS	Support|Symbolic Programming System
SPSELIM	Space Shift Elimination
SPSR	Single Product, Single Row
SPSS	Serial, Parallel, Serial, Solid
SPST	Single Pole, Single Throw
SPSW	Summary Punch Switch
spt	sectors per track
SPT	Shared Page Table
SPT	Shortened|Shortest Processing Time
SPT	Standard Penetration Test
SPT	Structured Programming Techniques
SPT	System Parameter Table
SPTF	Shortest Process Time First
SPTS	System Performance Test Set
SPU	Slave Processing Unit
SPU	System Processing Unit
SPUCDL	Serial Peripheral Unit Controller/Data Link
SPUD	Storage Pedestal Upgrade Disk|Drive
SPUFI	SQL Processor Using File Input
SPUN	Streamed Pipelined Unifier
SPUR	Source Program Utility Routine
SPUR	Systech Pluraxial Unplug Repeater
SPUSP	Special User Program

SPUTNIC	Synchronously Programmed User Terminal and Network Interface Control
SPX	Sequenced Packet Exchange
SPX	Simplex
sq	square
SQ	Squeezed (files)
SQA	Software Quality Assurance
SQA	System Queue Area
SQAP	Software Quality Assurance Plan|Procedure
SQC	Statistical Quality Control
SQD	Signal Quality Detection
SQE	Signal Quality Error
SQE	System Queue Element
SQIID	Simultaneous Quad Infrared Imaging Device
SQL	Structured Query Language
SQL/DS	Structured Query Language/Data System
SQL/J	Structured Query Language for Java
SQP	Sequential Quadratic Programming
SQP	Supplier Quality Program
SQPSK	Staggered Quadri-Phase Shift Keying
SQR	Square Root(er)
SQRT	Square Root (function)
SQS	System Queue Space
SQUARE	Specifying Queries As Relational Expressions
SQUARE	Statistical Quality Analysis Report
SR	Sales Representative
SR	Sample Rate
SR	Saturable Reactor
SR	Scientific Report|Research
SR	Secondary Runout
SR	Self-Rectifying
S/R	Send/Receive
SR	Senior
SR	Service Record
S/R	Set/Reset
SR	Shift Register|Reverse
SR	Silicon Rectifier|Rubber
SR	Sorter/Reader
SR	Source Routing
SR	Special Register|Report
SR	Specification Requirement
SR	Speech Recognition
SR	Starting Relay
SR	Status Register
SR	Storage Register|Ring
SR	Subtract Register
SR	Summary Report
SR	Suriname

SR	Switch Register
SR	System Reader|Residence
SRA	Secure RPC Authentication (RPC = Remote Procedure Call)
SRA	Security Research Alliance
SRA	Selective Routing Arrangement
SRA	Systems Requirements Analysis
SRAC	Supplier Relations Action Council
SRAM	Scratchpad Random Access Memory
SRAM	Semiconductor Random Access Memory
SRAM	Semi-Random Access Memory
SRAM	Shadow Random Access Memory
SRAM	Sort Reentrant Access Method
SRAM	Static|Structure Random Access Memory
SRB	Service Request Block
SRB	Sorter Reader Buffered
SRB	Source-Route Bridge
SRBI	Smooth and Rotate Base Index
SRC	Scan Request Channel
SRC	Source
SRC	Stored Response Chain
SRC	Synchronous Remote Control
SRCB	Sentence Record Control Byte
SRCCOM	Source Compare program
SRCH	Search
SRCR	System Run Control Record
SRCS	Smallest Remaining Capacity at Station
SRCU	Shared Remote Control Unit
SRD	Schedule Request Date
SRD	Screen Reader (system)
SRD	Secondary Receive Data
SRD	Software Requirements Document
SRD	Step Recovery Diode
SRDS	Symposium on Reliable Distributed Systems
SRE	Sending Reference Equivalent
SRE	Single Region Execution
SREJ	Selective Reject
SREM	Software Requirements Engineering Methodology
SREP	Software Requirements Engineering Program
SRETL	Screened Resistor Epitaxial|Etched Transistor Logic
SRETL	Screened Resistor Evaporated Transistor Logic
SRF	Self-Resonant Frequency
SRF	Software Recording|Recovering Facility
SRF	Sorter-Reader Flow
SRF	Specifically Routed Frame
SRF	System Recorder File
SRG	Short and Ring Ground (test)
SRGP	Simple Routing Gateway Protocol
SRI	Subtract Register Immediate (instruction)

SRIM	Selected Research In Microfiche
SRJE	SNA Remote Job Entry
SRL	Scheme Representation Language
SRL	Shift Register Latch
SRL	Systems Reference Library
SRM	Security Reference Monitor
SRM	Short Range Modem
SRM	Storage Resource Management
SRM	System Resources Manager
SRN	Selector Recorder Network
SRN	System Reference Number
SRO	Self-Regulatory Organization
SRO	Sharable and Read Only
SRP	Serial Reader Punch
SRP	Shared Resources Processing
SRP	Station Readout Pickup
SRP	System Routing Program
SRPI	Server Requester Programming Interface
SRPP	Serial Reader Parallel Punch
SRPT	Shortest Remaining Processing Time
SRPT	Systems Reliability Prediction Technique
SRQ	Service Request
SRR	Serially Reusable Resource
SRR	Software Requirements Review
SRR	System Requirements Review
SRS	Selective Recording System
SRS	Send/Receive Switch
SRS	Slave Register Set
SRS	Software Requirements Specification
SR/SK	Source/Sink
SRT	Segmentation Register Table
SRT	Set-Reset Trigger
SRT	Single Requesting Terminal
SRT	Sort(ing)
SRT	Source Route/Transparent (bridging)
SRT	System Resource Table
SRTM	Simplified Real-Time Monitor
SRTOS	Special Real-Time Operating System
SRTS	Secondary Ready To Send
SR-UAPDU	Status Report-User Agent Protocol Data Unit
SRVIFS	Service Installable File System
SR/W	Scatter Read/Write
SRWG	Software Review Working Group
ss	samples per second
SS	Satellite Switched|System
SS	Schedule Status
SS	Selective Signalling
SS	Select Standby

SS	Server-to-Server
SS	Session Service
SS	Shared Segment
SS	Signalling System
SS	Signal Selector
SS	Single Scan|Segment
SS	Single Shift|Sideband
SS	Single Sided|Step
SS	Single System
SS	Small Signal|System(s)
SS	Solid State
S/S	Source/Sink
SS	SPARC Station (SPARC = Scalable Performance Architecture)
S/S	Spooler/Scheduler
SS	Stack Segment
SS	Start of Special (sequence)
SS	Start/Stop (character)
S-S	Storage to Storage
SS	Sub-Scan
SS	Sub-Structure
SS	Sub-System
SS	Support System
SS	Switch Selector
SS	System Segment|Specifications
SS	System Service|Software
SS	System Status|Supervisor
SS	System Support
SSA	Segment Search Argument
SSA	Serial Storage|System Architecture
SSA	Static Single-Assignment
SSA	Status Save Area
SSA	Structured Systems Analysis
SSADM	Structured System(s) Analysis and Design Method
SSAM	Skip Sequential Access Method
SSAP	Session Service Access Point
SSAP	Source Service Access Point
SSAS	Station Signalling and Announcement Subsystem
SSB	Single Side-Band
SSBAM	Single Side-Band Amplitude Modulation
SSBM	Single Side Band Modulation
SSBS	Super Station Bar Strips
SSBSC	Single Side-Band Suppressed Carrier
SSC	Self-Scheduling Channel
SSC	Signalling and Supervisory Control
SSC	Software Support Center
SSC	Solid State Circuit
SSC	Station Selection Code
SSCA	Social Science Computing Association

SSCE	Support Services Conference and Exposition
SSCF	Secondary System Control Facility
SSCF	Service Specific Coordination Function
SSCL	Serial SCSI Command Language
SSCOP	Service Specific Connection Oriented Protocol
SSCP	Subsystem|System Services Control Point
SSCS	Service Specific Convergence Sublayer
SSCS	Step-by-Step Control System
SSD	Signal Strength Detector
SSD	Solid State Design|Disk
SSD	Structured System Design
SSD	System Software Disk
SSDA	Synchronous Serial Data Adapter
SS/DD	Single-Sided/Double-Density
SSDL	Storage Structure Definition Language
SSDM	Systematic Software Development and Maintenance
SSDU	Session Service Data Unit
SSE	Software Support Engineer
SSE	Solid State Element
SSE	Space-Shift Elimination
SSE	Special Support Equipment
SSEC	Selective Sequence Electronic Calculator
S-SEED	Symmetric Self-Electro-optic Effect Device
SSEL	Session (service) Selector
SSES	Software Specification and Evaluation System
SSF	Shared Shell Face
SSF	Skip Sequential File
SSF	Substitution Selector F
SSF	Supplier Structure File
SSF	Symmetrical Switching Function
SSFF	Showcase Software Factory of the Future
SSFM	Single sideband Subcarrier Frequency Modulation
SSG	Symbolic Stream Generator
SSGA	System Support Gate Array
SSH	Secure Shell
SSI	Server Side Include
SSI	Single System Image|Interface
SSI	Small-Scale Integration
SSI	Solid-State Imager
SSI	Standard System Interface
SSI	Sub-Systems Incidents
SSI	Subsystem Support Interface
SSI	Synchronous Systems Interface
SSI	System Status Index
SSIA	Sub-Schema Information Area
SSID	Software Service Identification
SSID	Sub-System Identification
SSII	Société de Service en Ingénierie Informatique (FR)

SSIO	Small System Input/Output
SSIT	Shared Segment Index Table
SSIT	Single-Source Interconnect Tree
SSK	Set Storage Key
SSK	Stereo Soft-copy Kit
SSL	Scientific Subroutine Library
SSL	Secure Socket Layer
SSL	Shift and Select
SSL	Single Stuck-Line
SSL	Software Specification|Synthesis Language
SSL	Source Statement Library
SSL	Storage Structure Language
SSL	Synthesizer Specification Language
SSL	System Specification Language
SSLC	Synchronous Single Line Controller
SSM	Semiconductor Storage Module
SSM	Set System Mask
SSM	Single Segment Message
SSM	Small Semiconductor Memory
SSM	Synchronous Subscriber Module
SSMA	Spread Spectrum Multiple Access
SSM/I	Special Sensor Microwave/Imager
SSN	Segment Stack Number
SSN	Switched Service Network
SSN	System Segment Name
SSNA	Steady-State (a-c) Network Analysis
SS/ND	Single-Sided/Normal Density
SSO	Satellite Switching Office
SSO	Single Sign-On
SSO	Software Service Organization
SSO	Standard Secondary Output
SSO	Systems and Service Organization
SSOP	Shrink Small-Outline Problem
SSP	Scientific Software|Subroutine Package
SSP	Scientific Statistical Program
SSP	Service|Signal Switching Point
SSP	Start/Stop Processing
SSP	Steady-State Pulse
SSP	System Service|Support Program
SSP	System Status Panel
SSPC	System Status Panel Controller
SSP-ICF	System Support Program – Interactive Communication Feature
SSPLIB	Scientific Subroutine Package Library
SSPRU	System Status Panel Relay Unit
SSR	Software Specification Review
SSR	Solid State Relay
SSR	Storage Select Register
SSR	Symposium on Software Reusability

SSR	System Status Report
SSRB	Sequential Scan Request Block
SSRP	Simple Server Redundancy Protocol
SSRR	Strict Source and Record Route
SSRS	Source Storage and Retrieval System
SSRT	Sub-Second Response Time
SSS	Selective Service System
SSS	Sequential Scheduling System
SSS	Server Session Socket
SSS	Simulation and Scheduling System
SSS	Smart Sensory Systems (EU)
SSS	Solid State Software
SSS	Standard Supply System
SSS	Step-by-Step Switch
SSS	Subsystem Support Services
SSS	Supply Support Systems
SSS	System Support Specialist
SSSC	Single Sideband Suppressed Carrier
SSSD	Single Sided Single Density
SST	Scanned Storage Tube
SST	Single Sideband Transmitter
SST	Standard Secondary Transactions
SST	Start/Stop Transmission
SST	Super-Sonic Transport
SST	Synchronous System Trap
SST	System Scheduler|Segment Table
SST	System Software Tape
SST	Systems Services and Technology
SST	Systems Switching Transients
SSTDMA	Satellite|Spacecraft Switched Time Division Multiple Access
SSTF	Shortest Search Time First
SSTOR	Start/Stop Storage
SSTV	Slow Scan Television
SSU	Subsequent Signal Unit
SSU	System Services Unit
SSW	Satellite Switched
SSW	Scene Switch
SSWI	Scene Switch Input
SSX	Small Systems Executive
SSX	Starter System Executive
SSYN	Slave Synchronization
ST	Sao Tome and Principe
ST	Segment Table|Type
ST	Self-Tapping
ST	Self-Test
ST	Sequence Timer
ST	Sort Time
ST	Special Text|Tool(s)

ST	Start (signal)
ST	Status
ST	Storage Tube
ST	Store
ST	Stored Time
ST	Straight Time
ST	System Test
STA	Spanning Tree Algorithm
sta	station(ary)
STA	Status
stab	stability
STAC	Software Timing And Control
STAC	Storage Access Cycle
STAC	System Test Adapter Complex
STACK	Start Acknowledge|Acknowledgment
STAD	Subprogram Table Affect Diagram
STAE	Specify Task Abnormal|Asynchronous Exit
STAF	Statistical Analysis of Files
STAG	Standard Application Generator
STAGE	Structured Text And Graphics Editor
STAGES	Structured Techniques for Analysis and Generation of Expert Systems
STAI	Sub-Task Abend Intercept
STAIRS	Storage And Information Retrieval System
STALL	Storage Allocation(s)
stalo	standardized oscillator
STAM	Shared Tape Allocation Manager
STAM	Single Terminal Access Method
STAP	Space-Time Adaptive Processing
STAR	Self-Testing And Repairing
STAR	Self-Tuning Architecture
STAR	Serial Titles Automated Records
STAR	Special Telecommunication Action for Regional development (EU)
STAR	Standard Routines
STAR	Storage Address Register
STAR	String and Array (processing, processor)
STAR	System Training Analysis Report
STARLAB	Space Telecommunications And Radioscience Laboratory (CA,US)
StarLAN	Star Local Area Network
STARS	Satellite Telemetry Automatic Reduction System
STARS	Software Technology for Adaptable, Reliable Systems
STARS	Standard Terminal Automation Replacement System
STARS	Standard Time And Rate Setting
START	Status And Reporting Technique
START	System for Analysis, Research, and Training
STAT	Statistical

STAT	Status
STAT	Status of Transfer
STATDSB	Status Disable
STATPAC	Statistical Package
STAVOSTOR	Static Volatile Storage
STB	Selective Top-to-Bottom (parsing)
STB	Set Top Box
STB	Simple Two-Band
STB	Software Technical Bulletin
STB	Start of Text Block
STB	Start To Build
STB	Strobe
STBY	Standby
STC	Sensitive Time Constant|Control
STC	Serving Test Center
STC	Set Carry (flag)
STC	Society for Technical Communication
STC	Standard Transmission Code
STC	Start Conversion
STC	Stop Transfer Code
STC	Store Character
STC	Switching and Testing Center
STC	Switching Technical Center
STC	System Test Coordinator(s)
STCB	Sub-interrupt Control Block
STCB	Sub-Task Control Block
STCK	Store Clock
STCL	Storage Control
STCR	System Test Configuration Requirements
STD	Secondary Transmit Data
STD	Set Direction (flag)
STD	Software Test Description
STD	Standard
STD	State Transition Diagram
STD	Stored
STD	Subscriber Trunk Dialling
STD	Superconductive Tunnelling Devices
STD	Symbolic Timing Diagram
STD	Synchronous Time Division
STDIN	Standard Input
STDIO.H	Standard Input/Output Header
STDM	Statistical|Synchronous Time Division Multiplexing
STDOUT	Standard Output
STDPRN	Standard Printer
STDS	Set-Theoretic Data Structure
STE	Segment Table Entry
STE	Segment Translation Exception
STE	Signal Terminal Equipment

STE	Spanning Tree Explorer
STE	Subscriber Terminal Equipment
STEDR	Staging Effective Data Rate
STEM	Strategic Telecommunications Evaluation Model
STEP	Simple Transition to Electronic Processing
STEP	Software Test and Evaluation Panel
STEP	Specification Technology Evaluation Program
STEP	Standard for The Exchange of Product (model) data
STEP	Supervisory Tape Executive Program
STEP-2DBS	Standard for The Exchange of Product data 2-Dimensional Building Subset
STEP/ENV	Draft European STEP standard (STEP = Standard for The Exchange of Product data)
STEPPS	Some Tools for Evaluating Parallel Programs
STEPS	Strategic Evaluator and Planning System
STEPWISE	STEP Web Integrated Supplier Exchange pilot (STEP = Standard for The Exchange of Product data)
STESD	Software Tool for Evaluating System Designs
STET	Società Finanziaria Telefonica (IT)
STF	Standard Transaction Format
STF	Statens Teleforvaltning (NO)
stf	structured file (file extension indicator)
STF	Supervisory Time Frame
STFT	Short-Term Fourier Transform
STG	Storage
STG	Subtree Transformational Grammar
STH	Satellite To Host
STH	Store Halfword
STI	Scientific and Technical Information
STI	Set Interrupt flag
STI	Single Tuned Interstage
STI	Special Test Instruction
STI	Standard Tape Interface
STIC	Support Tools for Installation Control
STIC	System Tailoring and Installation Component
STIF	Standard Interface
STIL	Standard Identification Label
STIL	Statistical Interpretive Language
STIR	STEP TDP Interoperability Readiness (project) (STEP = Standard for The Exchange of Product model data; TDP = Technical Data Package)
STIR	Storage Tag In Register
STIRE	Service Time Requirement
S/TK	Sectors Per Track
STK	Stack
STKR	Stacker
STL	Schottky Transistor Logic
STL	Selective Tape Listing

STL	Send-Transmit Leader
STL	Space To Letters
STLS	Stream Load and Store
STM	Short Term Memory
STM	Statement
STM	Store Multiple
STM	Switching Time Meter
STM	Synchronous Transfer Mode|Module
STMF	State Management Function
STM-i	Synchronous Transport Module i
STMIS	Systems Test Management Information System
STML	STEP Template Markup Language (STEP = Standard for The Exchange of Product data)
STMT	Statement
STN	State-Task Network
STN	Station
STN	Super-Twist(ed) Nematic
STN	Switched Telecommunication Network
STND	Standard
STNR	Signal-To-Noise Ratio
STO	Segment Table Origin
STO	Stor(ag)e
STOC	Stock Transfer On-line Control
STOC	Symposium on Theory Of Computing
STOH	Since Top Overhaul
STOIC	Stack Oriented Interactive Compiler
STOL	Systems Test and Operation Language
SToMP	Software Teaching of Modular Physics
STONE	Structured and Open Environment
STOPF	Simple Transmission Of Program Files
STOQ	Storage Queue
STOQUE	Storage Queue
STOR	Segment Table Origin Register
STOR	Storage Tag Out Register
STORET	Storage and Retrieval
STORM	Statistically Oriented Matrix (language)
STOS	Store String
STP	Science and Technology Policy
STP	Secure Transfer Protocol
STP	Shielded Twisted Pair
STP	Signal Transfer Point
STP	Software Test Plan
STP	Software Through Pictures
STP	Spanning Tree Protocol
STP	Standard Program
STP	Stop (character)
STP	Synchronized Transaction Processing
STPG	Stepping

STR	Segment Table Register
STR	Self-Turning Regulator
STR	Serial Transmit-Receive
STR	Status Register
STR	Store
STR	Store Task Register
STR	Synchronous Transmit(ter) Receive(r)
STRADIS	Structured Analysis, Design and implementation of Information Systems
STRAIN	Structural Analytical Interpreter
STRAM	Synchronous Transmit and Receive Access Method
STRAPP	Structural Analysis Program Package
STRB	Strobe
STRC	Synchronous Transmit and Receive Code
STRESS	Structural Engineering System Solver
STRICOM	Simulation Training and Instrumentation Command
STRIDE	Science and Technology for Regional Innovation and Development in Europe (EU)
STRIPS	Stanford Research Institute Problem Solver
STRL	Schottky Transistor-Resistor Logic
STRLD	Start Load
STROBES	Shared Time Repair Of Big Electronic Systems
STRUBAL	Structured Basic Language
STRUC	Structure
STRUDL	Structural Design Language
STRUM	Structured Microprogramming
STS	Scientific Terminal System
S/TS	Simulator/Test Set
STS	Slow Time Scale
STS	Space-Time-Space network
STS	Space Transportation System
STS	Static Test Stand
STS	Statistics and Traffic measurement Subsystem
STS	Status
STS	Support for Time-Sharing
STS	Switch Time Sensitivity
STS	Synchronous Transport Signal
STSN	Set and Test Sequence Number
STST	System Task Set Table
STSTB	Status Strobe
STT	Secure Transaction Technology
STT	Seek Time per Track
STT	Sequence Test Trigger
STT	Seven Track Tape
STT	Single Transition Time
STT	Small Tactical Terminal
STT	State Transition Table
STTL	Schottky Transistor-Transistor Logic

STTR	Small-business Technology Transfer Program
STU	Segment Time Unit
STU	Set Top Unit
STU	Storage (control) Unit
STU	Subscriber Terminal Unit
STUFF	System To Uncover Facts Fast
STUR	Standard Utility Routine
STX	Start of Text (character)
sty	style (file extension indicator)
SU	Selectable Unit
SU	Service User
SU	Set Up
SU	Set User-id
SU	Signal(ing) Unit
SU	Signs Unlike
SU	Speed Up
SU	Status of Uses
SU	Stor(ag)e Unit
SU	Sub-Unit
SU	Super User
SU	Switching Unit
SU	Systems Usage
SUA	Stored Upstream Address
SUABORT	Session User Abort
SUB	Subaddressing
SUB	Subroutine
SUB	Subscriber
SUB	Substitute (character)
SUB	Subtract
SUBCY	Sub-Cycle
SUBR	Sub-Routine
SUBST	Substitute
SUBT	Subtract
SUC	Successive
SUCCESS	Sun Corporate Catalyst Electronic Support Service
SUCS	Staffordshire University Computing Society
SUD	Structural Unit Descriptor
SUD	System Utility Device
SUDM	Single User Drive Module
SUE	System User Engineered
SUFF	Suffix
SUG	Sun User's Group
SUI	Standard Universal Identification number
SUID	Set User Identification
SUL	Standard and User Label
SULK	Signs Unlike
SUM	Symantec Utilities for Macintosh
SUM	System User's Manual

SUM System Utilization Monitor
SUMIT Standard Utility Means for Information Transformation
SUMM Summarize
SUMT Sequential Unconstrained Minimization Technique
SUN Single-User Network
SUNOS Sun Operating System
SUP Software Update Protocol
SUP Subtract Units Position
SUP Supervisor(y)
sup supplemental (dictionary) (file extension indicator)
SUP Suppress(ion)
SUPER Systems Usage and Performance Environment Report
SUPER System Used for Prediction and Evaluation of Reliability
SUPERJANET Super Joint Academic Network
SUPP Support
SUPPAK Support Package
SUPV Supervisory
SUR Start Up Rate
SURANET Southeastern Universities Research Association Network
SURF Selective Unit Record File
SURF Sequent Users Resource Forum
SURF Support of User Records and Files
SURF Systems and Units Requirements Forecast
SURF System Unit Requirements Forecast
SURGE Sorting, Updating, Report Generating
SURP Service and Unit Record Processor
SUS Silicon Unilateral Switch
SUSFD Successive Feed
SUSI Simple Unified Search Index
SUSP System Use Sharing Protocol
SUT Socket|System Under Test
S-UTP Screened Unshielded Twisted Pair
SUTY System Utility
SUW Synchronized Unit of Work
SV El Salvador
SV Scientific Visualization
SV Shared Variable
SV Single Value
SV Status Valid
SVA Service point and Variable interval Allocation
SVA Shared Virtual Area
SVAVISCA Space Variant Visual Sensor with Colour Acquisition (EU)
SVC Semi-permanent Virtual Circuits
svc service
SVC Supervisory Call
SVC Switched Virtual Call|Channel
SVC Switched Virtual|Voice Circuit
SVC Systems Validation Center

SVCI	Supervisory Call Interrupt
SVD	Simultaneous Voice/Data
SVD	System Validation Diagram
SVDF	Segmented Virtual Display File
SVF	Simple Vector File|Format
SVF	Software Validation Facility
SVGA	Super Video Graphics Array
SVGL	Sub-Voice Grade Lines
S-VHS	Super Video Home System
SVI	Standard Virtual Interface
SVID	System V Interface Definition
SVM	Single Value Modulation
SVM	Supervisor Virtual Machine
SVM	System Virtual Machine|Memory
SvMs	Service Methods
SVP	Schematic Verification Program
SVP	Service Processor
SVP	Shared Variable Processor
SVR	Server
SVRB	Supervisor Request Block
SVS	Single Virtual Storage|System
SVS	Structural Verification Simulator
SVS	Switched Virtual|Voice System
SVSPT	Single Virtual Storage Performance Tool
SVSS	Shared Virtual Storage System
SVSTOR	Single Virtual Storage
SVT	System Validation Test(ing)
SVT	System Variable Table
SVVP	Software Verification and Validation Plan
SVVS	System V Verification Suite
SW	Short Wave
SW	Single Weight
SW	Software
SW	Status Word
SW	Structured Walkthrough
SW	Switch
SWA	Scheduler|System Work Area
SWADS	Scheduler Work Area Data Set
SWAIS	Simple Wide Area Information Server
SWAMI	Software Aided Multi-font Input
SWAN	Satellite Wide-Area Network
SWAN	Sun Wide Area Network
SWAP	Shared Wireless Access Protocol
SWAP	Standard Wafer Array Programming
SWC	Serving Wire Center
SWCEPP	Software Engineering Code of Ethics and Professional Practice
SWD	Smaller Word
SWDT	Synchronization Word Detect Time

SWDW	Shared Windowed Digital World
SWE	Status Word Enable
SWEAT	SoftWare Engineers Automation Tool
SWEBOK	Software Engineering Body Of Knowledge
SWECC	Software Engineering Coordinating Committee
SWEEP	Software Engineering Education Project
SWF	Shortwave Fading
SWF	Simulation|Standard Work File
SWI	Software Interrupt
SWIF	Software Information File
SWIFT	Society for Worldwide Interbanking Financial Telecommunications
SWIM	See What I Mean
SWIPNET	Swedish Internet Provider Network
SWIR	Short Wavelength Infra-Red
SWISH	Simple Web Indexing System for Humans
SWITCH	Swiss Telecommunication for Higher education and research
SWL	Safe Working Load
SWL	Short Wave Listener
SWL	Software Writer's Language
SWM	Software Monitor
SWM	Solbourne Window Manager
SWOP	Specifications for Web-Offset Publications
SWOP	Standard Web Offset Printing
SWORD	Software Optimization for the Retrieval of Data
SWORDS	Software Order and Distribution Support
SWORDS	Standard Work Order Recording and Data System
SWOT	Strength Weakness Opportunity Threat (analysis)
SWP	Short Wavelength Pass (filter)
swp	swap (file extension indicator)
SWR	Spin Wave Resonance
SWR	Standing Wave Ratio
SWR	Status Word Register
SWS	Scientific Workstation Support
SWS	Selective Work Schedule
SWS	Single Wire Seal
SWS	Software Services
SWS	System Work Sheet
SWT	Sorting Work Tape
SWT	Structured Walk-Through
SWTL	Surface Wave Transmission Line
SWU	Switching Unit
SX	Secondary Instruction
SX	Simplex (signalling)
SXN	Section
SXS	Step-by-Step Switch(ing)
SY	Synchronized
SY	Syrian Arab Republic

SYC	System Control
SYCOM	System Communication Module
SYFA	System For Access
SYL	Syllable|Syllabus
SYLCU	Synchronous Line Control Unit
SYLK	Symbolic Link
SYM	Symbol(ic)
sym	symbols (file extension indicator)
SYM	Symmetrical
SYMAN	Symbol Manipulation
SYMAP	Synagraphic Mapping (system)
SYMBAL	Symbolic Algebra
SYMBUG	Symbolic Debugger
SYMP	Symposium
SYMPAC	Symbolic Program for Automatic Control
SYN	Synchronize (flag)
SYN	Synchronous (idle character)
syn	synonym (file extension indicator)
SYNC	Synchronous
SYNCOM	Synchronous Communications
SYNGLISH	Synthetic English
SYNIC	Synchronous Idle Character
SYNTRAN	Synchronous Transmission
SYNTRAN	Syntax Translation
SYRIUS	Symbolic Representations for Image Understanding System
SYROCO	Symposium on Robot Control
SYS	System
sys	system (configuration|device driver) (file extension indicator)
SYSADMIN	System Administrator
SYSAN	System Analyst
SYSCAP	System of Circuit Analysis Programs
SYSCOM	System Communication
SYSCON	System Configuration|Control
SYSCR	System Control Register
SYSDA	System Design Analysis
SYSEX	System Exclusive|Executive
SYSG	System tape Generator
SYSGEN	System Generation|Generator
SYSID	System Identification (conference)
SYSIN	System Input stream
SYSIPT	System Input (unit)
SYSJOB	System Job
SYSL	System description Language
SYSLIB	System (subroutine) Library
SYSLIST	System List
SYSLNK	System Link
SYSLOG	System Log

SYSMOD	System Modification
SYSMON	System Monitor
SYSOP	System Operator
SYSOPO	System Oriented Programmed Operator
SYSOUT	System Output stream
SYSPCH	System Punch
SYSPEC	System Specification
SYSRDR	System Reader
SYSREC	System Recorder
SYSREQ	System Request (key)
SYSRES	System Residence
SYSTRAN	Systems (analysis) Translator
SyTOS	Sytron Tape Operating System
SYTRAN	Synchronous Transmission
SYU	Synchronization Unit
SZ	Send Zmodem (Unix)
SZ	Size
SZ	Swaziland

T

t	target
T	Telecommunication(s)
T	Temperature
T	Tera (prefix)
T	Terminate
T	Time
T	Top
T	Transfer
T	Transformer (symbol for)
T	Transmit(ter)
T	True
T1	North American standard transmission speed
T1FE	T1-carrier Front End
T1OS	T1-carrier Out of State
T2L	Transistor-Transistor Logic
T2V	Timing To Value
T3	44.736 Mbits/sec digital carrier facility
T&A	Taken & Accepted
TA	Tape|Terminal Adapter
TA	Technical Applications
TA	Telegraphic Address
TA	Telex (network) Adapter
TA	Terminal Address
T&A	Time & Attendance
TA	Track Address
TA	Transaction Analysis
TA	Transfer Address|Agent

TA	Transfer Allowed
TAA	Tactical Asset Allocation
TAA	Track Average Amplitude
TAAS	Trunk Answer from Any Station
TAB	Tabulate
TAB	Tape Assembly Bonding
TABOU	The Alternative Bytemovers Of the Underground
TABR	Tabulator
TABSIM	Tabulating Simulator
TABSOL	Tabular System Oriented Language
TAC	Terminal Access Controller
TAC	Terrain Avoidance Computer
TAC	TRANSAC Assembler Compiler
TAC	Transformer Analog Computer
TAC	Transistorized Automatic Computer|Control
TACACS	Terminal Access Controller Access Control System
TACADS	Tactical Automated Data processing System
TACINTEL	Tactical Intelligence
TACOS	Tool for Automatic Conversion of Operational Software
TACPOL	Tactical Procedure Oriented Language
TACS	Time and Attendance Computation System
TACS	Total Access Communications System
TACT	Terminal Activated Channel Test
TACT	Transient Area Control Table
TACT	Transistor And Component Tester
TAD	Telephone Answering Device
TAD	Terminal Address Designator
TAD	Transaction Applications Driver
TAD	Transient Area Descriptor
TADAC	Tracking Analog-to-Digital And Comparator
TADIL	Tactical Digital Information Link
TADP	Terminal Area Distribution Processing
TADP	Tests and Analyses of Data Protocols
TADS	Tactical Automatic Digital Switch(ing)
TADS	Teletypewriter Automatic Dispatch System
TADSS	Tactical Automatic Digital Switching System
TAEE	Transverse Acousto-Electric Effect
TAEX	Typewriter Address exit
TAF	Terminal Access Facility
TAF	Time And Frequency
TAF	Transaction Facility
TAFIM	Technical Architecture Framework (for) Information Management
TAG	Time Automated Grid
TAG	Transfer Agent
TAGOS	Tech-Tran Annotated Graphics Operating System
TAHA	Tapered Aperture Horn Antenna
TAIS	Telecommunications & Automated Information Systems

TAL	Target Analyze List
TAL	Technical Administration List
TAL	Terminal|Transaction Application Language
TALAB	Tape Label
TALC	Terminal And Line Configuration
TALES	Test Analyzer Logic Evaluation System
TALK	Tradeoff Analysis based on Line cast and Know-how
TAM	Tag-Addressed Memory
TAM	Task Analysis Method
TAM	Telecommunications Access Method
TAM	Telephone Answering Machine
TAM	Terminal Access Method
TAM	Threaded Abstract Machine
TAMALAN	Table Manipulation Language
TAMOS	Terminal Automatic Monitoring System
TAMPFETS	Technology for Advanced Microwave Power Field-Effect Transistor Structures (EU)
TAMPR	Transformation Aided Multiple Program Realization
TANJ	There Ain't No Justice
TANJUG	Telegrafska Agencija Nova Juglavia
TAP	Telephone Administration Package
TAP	Telephone Assistance Plan
TAP	Telocator Alphanumeric Protocol
TAP	Term Availability Plan
TAP	Terminal Access Processor
TAP	Terminal Assistance Package
TAP	Test Access Port
TAP	Test Aids Package
TAP	Time-sharing Assembly Program
TAP	Timing Analysis Program
TAP	Total Action Program
TAP	Trace|Transistor Analysis Program
TAPAC	Tape Automatic Positioning And Control
TAPCIS	The Access Program for the CompuServe Information Service
TAPE	Tape Automatic Preparation Equipment
TAPGEN	Terminal Applications Program Generator
TAPI	Telephony Applications Program(ming) Interface
TAPS	Terminal Applications Processing System
TAPSTOR	Tape Storage
TAR	Tape Archive(r)
TAR	Temporary Accumulator
TAR	Temporary|Terminal Address Register
TAR	Tilt Arrest time
TAR	Total Assets Reporting
TAR	Track Address Register
TAR	Transfer Address Register
TARAN	Test And Repair|Replace As Necessary
TARE	Telegraph Automatic Relay Equipment

TARE	Telemetry Automatic Reduction Equipment
TARE	Transistor Analysis Recording Equipment
TARFU	Things Are Really Fouled Up
TARLAN	Target Language
TARP	Test And Repair Processor
TARR	Test Action Request Receiver
TAR.Z	Tape Archived files compressed (Unix)
TAS	Telephone Access Server
TAS	Telephone Answering Service
TAS	Terminal Address Selector
TAS	Terminal Automation System
TAS	Test And Set
TAS	Total Application Solution
TAS	TotalNet Advanced Server
TASC	Tabulate-Assemble-Sort-Collate
TASC	Telecommunications Alarm Surveillance and Control system
TASC	Terminal Area Sequence and Control
TASC	Translator Auto-Scaler and Coder
TASCON	Television Automatic Sequence Control
TASD	Technical and Administrative Support Division
TASI	Time Assignment Speech Interpolation
TASK	Temporary Assembled Skeleton
TASM	Turbo Assembler
TASO	Television Allocations Study Organization
TAT	Tape Auto-Testing
TAT	Time and Attendance Terminal
TAT	Trans-Atlantic Telecom
TAT	Trans-Atlantic Telephone (cable)
TAT	Turn-Around Time
TAT-8	Trans-Atlantic Telephone cable (fiber optic)
TATG	Tuned-Anode Tuned-Grid
TATS	Tactical Transmission System
TAU	Tape Adapter Unit
TAU	Telematic Access Unit
TAU	Trunk Access Unit
TAXI	Transparent Asynchronous Transceiver/Receiver Interface
TAXIR	Taxonomic Information Retrieval
Tb	Terabit
TB	Terabyte
TB	Terminal Block|Board
TB	Time Base
T/B	Top and Bottom
TB	Translation Buffer
TB	Transmitter Blocker
TB	Transparent Bridging
TB	Type Bar
TBA	Table Base Address
TBA	To Be Activated|Advised

TBA	To Be Announced|Arranged
TBA	To Be Assigned
TBB	Transnational Broadband Backbone
TBBS	The Bread Board System
TBC	Time Base Correction
TBC	Token Bus Controller
TBD	To Be Determined
TBE	Time Base Error
TBE	Tube
TBEM	Terminal Based Electronic Mail
TBGA	Tape Ball Grid Array
TBITS	Treasury Board Information Technology Standard
tbk	tool book (file extension indicator)
TBK	Tool Builder Kit
TBL	Table
TBM	Tone Burst Modulation
TBMT	Transmitter Buffer Empty
Tbps	Terabits per second
TBps	Terabytes per second
TBR	Table Base Register
TBS	Terminal Business System
TBSP	Table Spoon (quantity)
TBTF	Taste Bits from the Technology Front
TBU	Tape Backup Unit
TBW	To Be Written
TC	Tabulating Card
TC	Tape Core
TC	Tape to Card
TC	Technical Change|Committee
TC	Telecommunication(s)
TC	Terminal Computer|Concentrator
TC	Terminal Control(ler)
TC	Test Control
TC	Thermocouple
TC	Time Clock
TC	Timed Closing
TC	Time to Computation
TC	Timing Counter
TC	Total Colour
TC	Transaction Capabilities
TC	Transfer Count
TC	Transmission Check|Control
TC	Transmission Convergence (sublayer)
TC	Transmitter Clock
TC	Transport Connection
TC	Trunk Control
TC	Turks and Caicos Islands
TC	Type category

TC184	Technical Committee 184 (industrial standards) of ISO (= International Organization for Standardization)
TCA	Task Communication|Control Area
TCA	Telecommunications Association
TCA	Temporary Control Amplifier
TCA	Terminal Communication Adapter
TCA	Terminal Control Area
TCA	Terminal Controlled Airspace
TCA	Ternary Coded Asynchronous
TCA	Time Cost Analyzer
TCAM	Telecommunications Access Method
TCAME	Telecommunications Access Method Extended
TCAP	Transaction Capability Application Part
TCAS	T-Carrier Administration System
TCAS	Terminal Control Access|Address Space
TCB	Task|Terminal Control Block
TCB	Thread|Transmission Control Block
TCB	Transaction|Transfer Control Block
TCB	Trusted Computer Base
TCC	Terminal Chromaticity Coordinates
TCC	Tessera Compliant Chip
TCC	Transmission Control Character(s)
TCC	Trunk Class Code
TCCA	Technical Committee on Computer Architecture
TCCC	Technical Committee for Computer Communications
TCCO	Temperature-Compensated Crystal Oscillator
TCDE	Technical Committee on Data Engineering
TCDS	Test Case Development System
TCE	Top Computer Executive
TCE	Total Composite Error
TCE	Transaction Cost Estimator
TCE	Transmission Control Elements
TCF	Terminal Configuration Facility
TCF	Transparent Computing Facility
TCFIMS	Tightly Coupled Federated Information Management System
TCG	Test Call Generation
TCH	Test Channel
TCI	Technical Committee on Internet
TCI	Telecommunication Interface
TCIP	Telecommunications Interface Protocol
TCIS	Telex Computer Inquiry Service
TCL	Terminal Command|Control Language
TCL	Test Case Library
TCL	Time Clock at Limit
TCL	Tool Command Language
TCL	Transaction Control Language
TCL	Transistor Coupled Logic
TCL	Tymeshare Conversational Language

TCLP	Time-Constraint Loop Pipelining
TCL/TK	Tool Command Language/Toolkit
TCM	Telecommunications Manager|Monitor
TCM	Terminal, Computer, and Multiplexer
TCM	Terminal Control Mode
TCM	Terminal-to-Computer Multiplexer
TCM	Test Control Module
TCM	Thermal Conduction Module
TCM	Time Compression Multiplexing
TCM	Transmission Control Mode
TCM	Trellis Coded Modulation
TCMS	Telecommunications Management System
TCMS	Transistorized Computer Machine Switch
TCN	Telecommunications Networks
TCN	Train Communication Network
TCNS	Thomas-Conrad Network System
TCO	Telenet Central Office
TCO	Trunk Cut-Off
TCOP	Teleprocessing Control Program
TCP	Tape Conversion Program
TCP	Task Control Portfolio|Program
TCP	Telecommunication Carrier Products
TCP	Telemetry and Command Processor
TCP	Terminal Control Program
TCP	Time, Cost, and Performance
TCP	Transmission Control Program|Protocol
TCP	Transmitter Clock Pulse
TCPC	Telephone Cable Process Controller
TCP/IP	Transmission Control Protocol/Internet Protocol
TCR	Tape Cartridge Reader
TCR	Tape Cassette Recorder
TCR	Temperature Coefficient of Resistance
TCR	Total Closeness Rating
TCR	Transient Call Record
TCS	Tape Control System
TCS	Technical and Computing Services
TCS	Telecommunication Satellite
TCS	Telecommunications|Terminal Control System
TCS	Telecommunication System
TCS	Testing Computer Software (international conference)
TCS	Text Composition System
TCS	Thermal Control System
TCS	Tightly Coupled System
TCS	Transaction Control System
TCS	Transmission Controlled Speed
TCS	Two-Channel Switch
TCS-1	Trans-Caribbean System-1
TCS-AF	Telecommunications Control System – Advanced Function

TCSEC	Trusted Computer System Evaluation Criteria
TCSP	Technical Committee on Security and Privacy
TCSP	Telecommunications Special Product
TCSS	Terminal Control and Screen handling System
TCT	TCAM Control Task
TCT	Terminal Control Table
TCT	Timing Control Table
TCT	To Challenge Tomorrow
TCT	Toll Connecting Trunk
TCT	Total Cycle Time
TCT	Transaction Code|Control Table
TCTAP	Transistor Circuit Transient Analysis Program
TCTS	Trans-Canada Telephone System
TCTTE	Terminal Control Table Terminal Entry
TCU	Tape Control Unit
TCU	Telecommunications Control Unit
TCU	Teletypewriter Control Unit
TCU	Terminal|Timing Control Unit
TCU	Transmission Control Unit
TCU	Transport Conditioning Unit
TCU	Trunk Coupling Unit
TCUI	Transmission Control Unit Interface
TCVG	Technical Committee on Visualization and Graphics
TCW	Time Code Word
TCWG	Telecommunications Working Group
TCXO	Temperature Controlled Crystal Oscillator
TCZD	Temperature-Compensated Zener Diode
TD	Chad
TD	Tape Drive
TD	Tape to Disk
TD	Tapped Delay
TD	Telemetry Data
TD	Test Data
TD	Time Delay|Division
TD	Time Distribution
TD	Top Down
TD	Track Data
TD	Transfer Date
TD	Transmission Distributor
TD	Transmit(ted) Data
TD	Transmitter/Distributor
TD	Tunnel Diode
TD	Typed Data
TDA	Tape Deck Assembly
TDA	Top-Down Approach
TDA	Transistorized Drum Assembly
TDA	Tunnel Diode Amplifier
TDAF	TCAM Destination Address Field

TDAS	Tactical Data Automation System
TDAS	Traffic Data Administration System
TDBM	Test Data Base Manipulator
TDC	Tabular Data Control
TDC	Tape Data Controller
TDC	Telemetry Data Center
TDC	Terrestrial Data Circuit
TDC	Test Data Control system
TDCC	Transportation Data Coordinating Committee
TDCS	Time Division Circuit Switching
TDD	Telecommunications Device for the Deaf
TDD	Time Division Demultiplexing|Duplexing
TDD	Two Double Diode
TDDI	Twisted-pair Distributed Data Interface
TDDL	Time-Division Data Link
TDE	Terminal Display Editor
TDE	Time Delay Estimation
TDE	Total Data Entry
TDE	Transactional Data Entry
TDE	Transition Diagram Editor
TDEM	Test Data Effectiveness Measurement
TDF	Test Data File
tdf	trace definition file (file extension indicator)
TDF	Transborder Data Flow(s)
TDF	Transnational Data Flow(s)
TDF	Trunk Distribution Frame
tdf	typeface definition file (file extension indicator)
TDG	Test Data Generator
TDGL	Test Data Generating Language
TDI	Technical Data Interface
TDI	Telecommunications Data Interface
TDI	Transit Delay Indication
TDI	Transport Device Interface
TDI	Trusted Database Interpretation
TDI	Two-wire Direct Interface
TDIA	Transient Data Input Area
TDID	Trade Data Interchange Directory
T-disk	Temporary disk
TDL	Task Description Library
TDL	Terminal Display Language
TDL	Test Description Language
TDL	Total Dictionary Language
TDL	Transaction|Transformation Definition Language
TDL	Transistor-Diode Logic
TDM	Time-Division Multiplexing
TDM	Time Driven Monitor
TDMA	Time-Division Memory|Multiple Access
TDMA	Time Domain Multiple Access

TDMAM	Time-Division Multiple Access Modem
TDMC	Time-Division Multiplexed Channel
TDMS	Teleprocessing Data Management System
TDMS	Terminal Display Management System
TDMS	Text Data Management System
TDMS	Time-shared Data Management System
TDOA	Transient Data Output Area
TDOS	Tape/Disk Operating System
TDOS	Test Data Output System
TDP	Task Dispatcher
TDP	Technical Data Package
TDP	Tele-Data Processing
TDP	Telelocator Data Protocol
TDP	Test Data Package
TDP	Top-Down Programming
TDP	Traffic Data Processor
TDP	Transactional Data Processing
TDPL	Top Down Parsing Language
TDPS	Tactical Data Processing System
TDR	Tape Data Register
TDR	Test Data Reduction
TDR	Test Discrepancy Report
TDR	Time Delay Relay
TDR	Tone Dial Receiver
TDR	Track Description Record
TDR	Transmit Data Register
TDRS	Tracking (and) Data Relay Satellite
TDRSS	Tracking (and) Data Relay Satellite System
TDS	Tape Data Selector
TDS	Telecommunication and Data Systems
TDS	Television Display System
TDS	Temporary Data Set
TDS	Test Data System
TDS	Time Distribution Schedule|System
TDS	Time-Division Switching
TDS	Top-Down Structure(d)
TDS	Transaction Data Set
TDS	Transaction Distribution|Driven System
TDS	Transient Data Structure
TDS	Transistor Display System
TDS	Transit Delay Selection
TDSAI	Transit Delay Selection And Indication
TDSP	Top Down Structured Programming
TDSR	Transmitter Data Service Request
TDT	Telephone Data Transmission
TDTG	True Date-Time Group
TDTG	Tuned Drain, Tuned Gate
TDTL	Tunnel-Diode Transistor Logic

TDtoDP	Tablet Coordinates to Display Coordinates
TDU	Thumbwheel Dial Unit
TDU	Topographical Display Unit
tdu	total defects per unit
TDV	The Digital Village
TDWG	Taxonomic Databases Working Group
TDX	Time Division Exchange
TDY	Task Dictionary
TE	Task Element
TE	Terminal Equipment
T&E	Test & Evaluation
TE	Text Editor
TE	Threshold Elements
TE	Time Emitter|Equipment
TE	Trailing Edge
TE	Transverse Electric (wave)
TE1	Terminal Equipment Type 1 of ISDN (= Integrated Services Data Network)
TE2	Terminal Equipment Type 2 (not ISDN)
TEA	TCL Extension Architecture (TCL = Tool Command Language)
TEAM	Teleterminals Expandable Added Memory
TEAM	Together Everyone Accomplished More
TEAMS	Test Evaluation And Monitoring System
TEB	Tape Error Block
TEB	Task Entry Block
TEB	Thread Environment Block
TEBOL	Terminal Business Oriented Language
TEC	Triple Erasure Correction
TECH	Technical
TECHEVAL	Technical Evaluation
TECHEX	Technical Exchange
TECHMEMO	Technical Memorandum
TECHREPT	Technical Report
TECHTRACS	Technology Tracking System
TECMT	Test Engineering Circuit Master Tape
TECO	Tape|Text Editor and Corrector
TED	Text Editor
TED	Tiered Electronic Distribution
TEDIS	Trade Electronic Data Interchange System
TEHO	Tail End Hop Off
TEI	Terminal Endpoint Identifier
TEI	Text Encoding Initiative
TEI	Thermal Error Index
TEIKADE	The Environment Informally Known As Dejava
TEL	Task Execution Language
tel	telegram
tel	telegraph
tel	telephone

TELCO	Telephone Company
Telecom	Telecommunication(s)
TELECON	Telephone Conference
Telefónica	Telefónica de España S.A.
TELEMUX	Telegraph Multiplexer
TELENETE	Telephone Network Evaluation
Telepac	The Portuguese packet-switched network
Telepoint	Public access cordless telephony
telex	telegraphy exchange
TELEX	Teletype(printer) Exchange
TELEX	Teletype(writer) Exchange
TELINT	Telemetry Intelligence
TELNET	Telecommunications Network
telocator	tele-locator
TELOPS	Telemetry On-line Processing System
TELRY	Telegraph Reply
TELSAM	Telephone Service Attitude Measurement
TELSCOM	Telemetry Surveillance Communications
TELSET	Telephone Set
TEM	Transmission Electron Microscope
TEM	Transverse Electromagnetic Mode
TEM	Transverse Electro-Magnetic (wave)
TEMA	Telecommunication Engineering and Manufacturing Association
TEML	Turbo Editor Macro Language
TEMOD	Terminal Environment Module
TEMP	Temperature
TEMP	Temporary (register)
TEMPEST	Transient Electro-Magnetic Pulse Emanation Standard
TEMPo	Tenor Element Management Platform
TEMPOS	Timed Environment Multi-Partitioned Operating System
TEMPY	Temporary
TEMWAVE	Transverse Electro-Magnetic Wave
TEN	Telephone Equipment Network
TENET	Texas Education Network
TEP	Terminal Error Program
TEP	Thermographic Evaluation Program
TEPE	Time-sharing Event Performance Evaluator
TEPOS	Test Program Operating System
TER	Translation Error Rate
tera	trillion
TERC	Telecommunications Equipment Re-marketing Council
TERCOM	Terrain Contour Mapping
TERENA	Trans-European Research and Education Networking Association
TERM	Terminal
TERM	Termination
TERMA	Terminator Assignment

TERMPWR	Terminator Power
TERNR	Territory Number
TES	Text Editing System
TEST	Teleprocessing Environmental Simulator Tester
TEST	Time-Space-Time network
TEST	Transaction Step Task
TESTRAN	Test Translator
TET	Test Environment Toolkit
TETFT	Test Engineering Translate Functions Tape
TEU	Twenty foot Equivalent Unit
TEX	Teletype Exchange
TEX	Telex
tex	TeX files extension (file extension indicator)
TEXEL	Texture Element
TEXTIR	Text Indexing and Retrieval
TF	French Southern Territories
TF	Tabulating Form
TF	Tape Feed
TF	Telecom Finland
TF	Terminal Frame
TF	Time Frame
TF	Total Float
TF	Transfer Function
TF	Transmit Filter
TF	Transmitter Frequency
TFA	Test Form Analyzer
TFA	Transaction Flow Auditing
TFA	Transparent File Access
TFC	Total Full Colour
TFD	Transferred
TFDD	Text File Device Driver
TFEL	Thin-Film Electro-Luminescent
TFF	Try For Fit
TFFET	Thin Film Field-Effect Transistor
TFLAP	T-carrier Fault-Locating Applications Program
tflops	trillion floating point operations per second
tfm	tagged font metric (file extension indicator)
TFM	Tamed Frequency Modulation
TFM	Tape File Management
tfm	TeX font metric (file) (file extension indicator)
TFM	Time-Frequency Modulation
TFM	Trusted Facilities Manual
TFMR	Transformer
TFMS	Text and File Management System
TFN	Tribe Flood Network
TFO	Test For Overflow
TFP	TOPS Filing Protocol (TOPS = Terminal Office Processing System)

TFR Transaction Formatting Routines
TFR Transfer
TFR Trouble Failure Report
TFS Tape File Supervisor
TFS Task Form Specification
TFS Temporary File System
TFS Thin Film Store
TFS Translucent File System
TFS Trunk Forecasting System
TFSF Time to First System Failure
TFT Thin-Film filed Triode
TFT Thin-Film Technology|Transistor
TFT Translate Functions Tape
TFTP Trivial File Transfer Protocol
TFU Test File Update
TFZ Transfer Zone
TG Task Group
TG Terminal Guidance
TG Terminator Group
TG Togo
tga targa image format (file extension indicator)
TGAL Think Globally, Act Locally
TGB Trunk Group Busy
TGC Terminal|Terminator Group Controller
TGE Tape Generator and Editor
TGID Transmission Group Identifier
TGL Toggle
TGN Trunk Group Number
TGR Trigger
TGS Transaction Generating System
TGT Transformational Grammar Tester
TGTP Tuned Grid, Tuned Plate
TGW Trunk Group Warning
tgz tarred and g-zipped (file extension indicator)
TH Temporary Hold
TH Thailand
TH Thermal
TH Transmission Header
TH Trouble History
TH True Heading
THA Technology Hackers Association
THAID Theta Automatic Interaction Detector
THC Temperature and Humidity Chamber
THD Third
thd thread (file extension indicator)
THD Total Harmonic Distortion
THEMIS Three-Hole Element Memory with Integrated Selection
THEOR Theoretical

thermistor	thermally sensitive resistor
THF	Tremendously High Frequency
THI	Temperature Humidity Index
THIC	Tape Head Interface Committee
THIS	Total Health Information System
THK	Thick
THLD	Threshold
THOMAS	The House (of representatives) Open Multimedia Access System (US)
THOMIS	Total Hospital Operating and Medical Information System
THOR	Tandy High-performance Optical Recording
THOR	Tape Handling Option Routines
THOR	Thermal Optical Recording
THP	Terminal Handling Processor
THP	Terminal Holding Power
THPLAS	Thermoplastic
THR	Transmit(ter) Holding Register
THRE	Transmit(ter) Holding Register Empty
THREAD	Three-dimensional Reconstruction And Display
ths	thesaurus (file extension indicator)
THT	Token Holding Time(r)
thy	thyratron
THz	Terahertz
TI	Tape Indication|Inverter
TI	Technical Integration
TI	Temporary Instructions
TI	Terminal Instruction
TI	Terminal Interchange|Interface
T&I	Test & Integration
TI	Time In
TI	Tone Index
TI	Top Inputs
TI	Transformational Implementation
TI	Transmission Identification
TIA	Task Item Authorization
TIA	Technique for Information Analysis
TIA	Telecommunications Industry Association (US)
TIA	Telecommunications Information Administration
TIA	Telematic Internetworking Application
TIA	Telephone Information Access
TIA	Thanks In Advance
TIA	The Internet Adapter
TIAS	Telematic Internetworking Abstract Service
TIB	Terminal Interchange Buffer
TIB	The Information Bus
TIB	Transparent Interleaved Bipolar
TIC	Tape Inter-system Connection
TIC	Task Interrupt Control

TIC	Technical Information Center
TIC	Terminal's Identification Code
TIC	Term-Info Compiler
TIC	Time Interval Counter
TIC	Time Issue Control
TIC	Token-ring Interface Coupler
TIC	Tongue In Cheek
TIC	Transducer Information Center
TIC	Transfer In Channel
TICCIT	Time-shared Interactive Computer Controlled Information Television
TICS	Telecommunication Information Control System
TID	Target Identification
TID	Telephone Inquiry Device
TID	Transaction Identification
TID	Tuple Identifier
TIDE	Tele-Immersive Data Explorer
TIDIC	Time Interval Distribution Computer
TIDMA	Tape Interface Direct Memory Access
TIE	Technical Information Exchange
TIE	Terminal Interface Equipment
TIE	Track In Error
TIES	Time Independent Escape Sequence
TIES	Transmission and Information Exchange System
tif	tagged image file (file extension indicator)
TIF	Tape Inventory File
TIF	Telephone Interference Factor
TIF	Terminal Independent Format
TIF	The Information Facility
TIFF	Tagged Image|Information File Format
TIGA	Texas Instruments Graphics Architecture
TIGER	Topologically Integrated Geographic Encoding and Referencing
TIGRE	Time-Invariant Gray Radiance Equation
TIGS	Terminal Independent Graphics System
TIH	Trunk Interface Handler
TIIAP	Telecommunications and Information Infrastructure Assistance Program
TIIF	Time-Initiated Input Facility
TIM	Table Input to Memory
TIM	The Inventory Machine
TIM	Transient Inter-Modulation (distortion)
TIME	Telecommunication Information Management Executive
TIMES	Time Interval Measuring System
TIMM	Thermionic Integrated Micro Modules
TIMM	Toshiba Integrated Multimedia Monitor
TIMOTHY	Timely Information and Messaging On-line To Help You
TIMS	Technical Information Management System
TIMS	Transmission Impairment Measuring Sets

TIN	Triangulated Irregular Network
TINA	Telecommunications Information Networking Architecture
TINA-C	Telecommunications Information Networking Architecture Consortium
TI-NET	Transparent Intelligent Network
TINWIS	That Is Not What I Said
TIO	Test Input/Output
TIO	Time Interval Optimization
TIOA	Terminal Input/Output Area
TIOB	Terminal Input/Output Block
TIOB	Test Input/Output and Branch
TIOC	Terminal Input/Output Coordinator
TIOM	Terminal Input/Output Manager
TIOT	Task Input/Output Table
TIOT	Terminal Input/Output Task
TIOWQ	Terminal Input/Output Wait Queue
TIP	Team Improvement Program
TIP	Telemedicine Instrumentation Pack
TIP	Terminal Interface Package|Processor
TIP	Transaction Interface Processor
TIPP	Timed Processes and Performance evaluation
TIPS	Technology Integration Panel Study
TIPS	Technology|Text Information Processing System
TIPSI	Transport Independent Printer System Interface
TIPTOP	Tape Input/Tape Output
TIQ	Task Input Queue
TIR	Target Instruction Register
TIRB	Task Input/output Request Block
TIRIS	Texas Instruments Registration and Information System
TIRKS	Trunks Integrated Records Keeping System
TIROS	Television and Infra-Red Observation Satellite
TIRPC	Transport Independent Remote Procedure Call
TIRS	The Integrated Reasoning System
TIS	Technical Information System
TIS	Telecommunication Information System
TIS	Telephone Information Service
TIS	Test|Total Information System
TISA	Technical Infrastructure (and) System Architecture
TISN	Tokyo International Science Network (JP)
TITOS	Technological Interface To Our Senses
TITS	The Interactive Task Switcher
TIU	Telematic Internetworking Unit
TIU	Terminal|Trusted Interface Unit
TJ	Tajikistan
TJB	Time-sharing Job control Block
TJID	Terminal Job Identification
TK	Teletype Keyboard
TK	Ten Keyboard

TK	Tokelau
TK	Track
TKS	Turn-Key Set
TK/TK	Track to Track
TL	Target Language
TL	Test Loop
TL	Tie Line
TL	Time Limit
TL	Tool
TL	Transaction Language|Listing
TL	Transit Lock
TL	Transmission Level|Line
TL	Transport Layer
TL	Tuple Length
TL-10	Toyota robot Language-10
TLA	Time Line Analysis
TLA	Top Level Aggregator
TLA	Transmission Line Adapter|Assembly
TLAB	Tape Label
TLAP	Token-talk Link Access Protocol
TLB	Table Look-aside Buffer
TLB	Translation Look-aside Buffer
TLB	Type Library
TLC	Task Level Checkpoint|Controller
TLC	The Learning Channel
TLC	Thin Layer Chromatography
TLCAP	Transmission Line Circuit Analysis Program
TLCS	Tape Library Control System
TLCSC	Top-Level Computer Software Component
TLCT	Total Life Cycle Time
TLD	Top Level Domain
TLD	Transmission Line Driver
TLF	Trunk Line Frame
TLGCD	Telegraphic Code
TLI	Transport Layer|Level Interface
TLI	Transport Library Interface
TLK	Talk
TLK	Test Link
TLM	Trouble Locating Manual
TLMA	Telematic Agent
TLMAU	Telematic Access Unit
TLMS	Tape Library Management System
TLN	Trunk Line Network
TLNCE	Tele-Learning Network of Centres of Excellence
TLOG	Transaction Log
TLP	Telefonas de Lisboa e Porto (PT)
TLP	Telegraph Line Pair|Patch
TLP	Transmission Level Point

TLR	Toll Line Release
TLR	Transmission Line Receiver
TLS	Tape Librarian System
TLS	Thesaurus and Linguistic (integrated) System
TLS	Transport Layer Security
TLSA	Transparent Line-Sharing Adapter
TLSPP	Transport Layer Sequenced Packet Protocol
TLT	Transmission Line Terminator
TLTP	Trunk Line and Test Panel
TLU	Table Look-Up
TLU	Threshold Logic Unit(s)
TLUE	Table Look-Up Equals
TLUEH	Table Look-Up Equals or High
TLUL	Table Look-Up Low
TLV	Type-Length-Value
tlx	telex (file extension indicator)
TLXAU	Telex Access Unit
TLZ	Transfer on Less than Zero
TM	Tape Mark|Module
TM	Telemetry
TM	Temperature Modulator
TM	Terminal Management|Monitor
TM	Time Monitor
TM	Training Manual
TM	Transaction Manager
TM	Transverse Magnetic (wave)
TM	Turing Machine
TM	Turkmenistan
TM	Type Mismatch
T/M	Type Model
TM/1	Tables Manager/1
TMA	Telecommunications Management
TMC	Tape Management Catalog
TMC	Task Management Component
TMC	Temporary Microcode Change
TMC	Traffic Message Channel
TMDF	Trunk Main Distributing Frame
TMDL	Total Maximum Daily Load
TME	Tivoli Management Environment
TMG	Timing
TMGP	Test Matrix Generating Program
TMI	Too Much Information
TMIS	Technician Maintenance Information System
TMIS	Telecommunications Management Information System
TMM	Transmission Mode Message
TMMS	Telephone Message Management System
TMN	Telecommunications Management Network
TMN	Transmission

TMO	Time-triggered Message-triggered Object
TMOS	Telecommunications Management Operations Support
tmp	temporary (file extension indicator)
TMP	Terminal Message Program
TMP	The Morrow Project
TMP	Transaction Monitor Program
TMPDU	Test Management Protocol Data Unit
TMPL	Template
TMPP	Test Message Processing Program
TMR	Timer
TMR	Transient Memory Record
TMR	Transmitter
TMR	Triple Modular Redundancy
TMRS	Traffic Measurement and Recording System
TMRS	Traffic Metering Remote System
TMS	Table|Tape Management System
TMS	Telecommunications Message Switcher
TMS	Telephone|Terminal Management System
TMS	Text Management System
TMS	Time-Multiplexed Switching
TMS	Transaction Monitoring Software
TMS	Transmission Measuring Set
TMS	Truth Maintenance System
TMSC	Tape Mass Storage Control
TMSF	Time, Minutes, Seconds and Frames
TMT	Transmit
TMTC	Through-Mode Tape Convertor
TMTR	Transmitter
TMU	Test Maintenance Unit
TMU	Time Measurement Unit
TMU	Transmission Message Unit
TMX	Transaction Management Executive
TN	Tennessee (US)
TN	Terminal Node
TN	Test Number
TN	Transaction Number
TN	Transferable Note
TN	Transition|Transport Network
TN	Tunisia
TN	Twisted Nematic
TNA	Transient Network Analyzer
TNC	Terminal Node Controller
TNC	The Networking Center
TNC	Threaded Navy|Nut Connector
TNC	Threaded Neill-Concelnan connector
TNC	Transport Network Controller
TND	The Neon Dragon
TNDS	Total Network Data System

TNET	Terminal (and computer) Network
TNF	Third Normal Form
TNIC	Transit Network Identification Code
TNMS	Telephony Network Management System(s)
TNN	Trunk Network Number
TNOP	Total Network Operation Plan
TNP	Transportation
TNPC	Traffic Network Planning Center
TNPP	Telocator Network Paging Protocol
TNS	Transaction Network Service
TNT	Terminal Name Table
TNT	The News Toolkit
TNT	Turner Network Television
TNX	Thanks
TNZ	Transfer on Non-Zero
TO	Table of Organization
TO	Telecommunications Organization
TO	Telegraph Office
TO	Test Object|Operation
TO	Tonga
TO	Transmitter Order
TO	Turn Over
TOA	Type Of Allocation
TOADS	Terminal Oriented Administrative Data System
TOAF	TCAM Origin Address Field
TOAKE	Tape-Oriented Advanced Key Entry
TOALS	Time Of Arrival Location System
TOB	Type Of Business
TOC	Table Of Contents
toc	table of contents file (file extension indicator)
TOC	Theory Of Constraints
TOC	Time Out Circuit
TOC	Transfer Of Control
TOC/DBR	Theory Of Constraints/Drum-Buffer-Rope
TOCOTOX	Too Complicated To Explain
TOCS	Terminal Oriented Computer System
TOCTTOU	Time Of Check To Time Of Use
TOD	Time Of Day
TODC	Time Of Day Clock
TODS	Test Oriented Disk System
TODS	Transactions On Database Systems
TOE	Tape Overlap Emulator
TOF	Top Of File|Form
TOF	Total Overflow
TOGAF	Technical Open Group Architectural Framework
TOL	Test Oriented Language
TOL	Tolerance
TOLAR	Terminal On-Line Availability Reporting

TOLT	Total On-Line Testing
TOLTEP	Teleprocessing On-Line Test Executive Program
TOLTS	Total On-Line Testing System
TOM	Type Of Maintenance
TOMS	Transactions On Mathematical Software
TONIC	The On-line Netskills Interactive Course
TONICS	Telecommunications Operations Integrated Network Control System
TOO	Target Of Opportunity
TOOIS	Transactions On Office Information Systems
TOOL	Test Oriented Operator Language
TOOLS	Technology of Object-Oriented Languages and Systems (international conference)
TOP	Task Oriented Practice
TOP	Technical Office Protocol
TOP	Thematic Organization Packet
TOP	Transaction Oriented Package
TOPC	Test and Operational Program Console
TOPFIT	Tailored Oligomers and Polymers For Information Technology (EU)
TOPICS	Terminal Oriented Project Information and Control System
TOPICS	Total On-line Program and Information Control System
TOPLAS	Transactions On Programming Languages And Systems
TOPMS	Telemarketing Operations Management System
TOPO	Topographic
TOPS	Terminal Office Processing System
TOPS	Testing and Operating System
TOPS	The Operating Planning System
TOPS	Time-sharing Operating System
TOPS	Total Operations Processing System
TOPS	Traffic Operator Position System
TOPS	Trajectory Optimization System
TOPS	Transcendental Operating System
TOPSMP	Traffic Operator Position System Multipurpose
TOR	Telephone Order Register
TORES	Text Oriented Editing System
TOROS	Tantalum Oxide Read-Only Store
TORTOS	Terminal Oriented Real-Time Operating System
TOS	Tape|Team Operating System
TOS	Technical and Office Systems
TOS	Time Operation System
TOS	Top Of Stack
TOS	Tramiel Operating System
ToS	Type of Service
TOSCW	Top Of Stack Control Word
TOSL	Terminal Oriented Service Language
TOSM	Through, Open, Short, Match
TOSP	Top Of Stack Pointer

TOSS	Terminal Oriented Support System
TOT	Time On Target
tot	total
TOT	Transfer Overhead Time
TOTE	Teleprocessing On-line Test Executive
TOX	Time Of Expiration
TOXLINE	Toxicology information on-Line
TP	East Timor
TP	Tape to Printer
TP	Telecommunication Program
TP	Telecom Portugal
TP	Telenet Processor
TP	Teleprinter
TP	Tele-Processing
TP	Teletype Printer
TP	Terminal Point|Portability
TP	Terminal Post|Printer
TP	Terminal Protocol
TP	Test Plan|Procedure
TP	Thermoplastic
TP	Time Pass
TP	Toll Point
TP	Transaction Processing
TP	Transition Period
TP	Transport Protocol
TP	Triple Play
TP	Twisted Pair
TP	Type
TPA	Thermal Plasma Analyzer
TPA	Transient Program Area
TPACS	Total Product Auto Control System
TPAD	Tele-Processing Analysis and Design
TPAD	Terminal Packet Assembler/Disassembler
TPAM	Tele-Processing Access Method
tpaML	trading partner agreement Markup Language
TPAP	Transaction Processing Application Program
TPAS	Traffic Profile Analysis System
TP-ASE	Transaction Processing Application Service Element
TPB	Technical Programming Bulletin
TPB	Telecommunications Policy Bureau (J)
TPB	Teletype Printer Buffer
TPBVP	Two-Point Boundary Value Problem
TPC	Test Program Control
TPC	Text Processing Center
TPC	Total Print Control
TPC	Transaction Processing performance Council
TPC	Trans-Pacific Cable
TPCB	Tens Position C Bit

TPD	Tape Packing Density
TPD	Technical Product Documentation
TPD	Thermo-plastic Photoconductor Device
TPD	Time Pulse Distribution
TPD	Transaction Processing Description
TPDDI	Twisted Pair Distributed Data Interface
TPDRV	Tape Drive
TPDT	Tree-walking Push Down Transducer
TPDU	Transport Protocol Data Unit
TPE	Transaction Processing Executive
TPE	Transmission Parity Error
TPE	Twisted Pair Ethernet
TPF	Transaction Processing Facility
TPF	Two Photon Fluorescence
TPFDD	Time Phased Force Deployment (and) Data
TPFI	Terminal Pin Fault Insertion
TPFP	Transaction Processing Function Processor
TPFS	Teleprocessing Performance Forecasting System
TPG	Telecommunication Program Generator
TPG	Test Pattern Generator
TPG/C	Test Pattern Generator/Comparator
tpi	tracks per inch
TPI	Transport Protocol|Provider Interface
TPINDR	Tape Indicator
TPIS	Telecommunications Products Information retrieval and Simulation
TPL	Table Producing Language
TPL	Terminal Processing|Programming Language
TPL	Test Procedure Language
TPL	Transaction Processing Language
TPL	Triple
TPLAB	Tape Label
TPL/F	Test Procedure Language/Fortran
TPLIB	Transient Program Library
TPLT	Tape Left
TPM	Tape Preventive Maintenance
TPM	Tape Processing Machine
TPM	Technical Performance Measures
TPM	Tele-Processing Monitor
tpm	transactions per minute
TPM	Two-Phase Modulation
TPMF	Tele-Processing Multiplex Feature
TPMK	Tape Mark
TPMM	Tele-Processing Multiplexer Module
TPMP	Total network data system Performance Measurement Plan
TPMR	Trunk Private Mobile Radio
TPMS	Tele-Processing Monitor System
TPN	Timed Petri Net

TPNS	Tele-Processing Network Simulator
TPOA	Telecommunication Private Operating Agency
TPOP	Time Phased Order Point
TPORT	Twisted pair Port Transceiver
TPP	Technical Programming Practice
TPPMD	Twisted Pair, Physical Media Dependent
TPPN	Trans-Pacific Profiler Network
TPR	Tape Programmed Raw data
TPR	Tele-Printer
TPR	Tele-Processing Region
TPR	Telescopic Photograph Recorder
TPR	Thermal Print (mechanism)
TPR	Thermo-Plastic Recording
TPR	Transaction Processing Routine
TPR	Transmission Performance Rating
TPRDR	Tape Reader
TPRT	Tape Right
TPS	Tape Processing|Programming System
TPS	Tape Program Search
TPS	Tape to Print System
TPS	Task Performance Services
TPS	Technical Programming Standard
TPS	Technical Publishing Software
TPS	Technology Program(s) Support
TPS	Telemetry Processing Station
TPS	Tele-Processing Services|System
TPS	Tele-Processing Supervisor
TPS	Terminal Polling|Programming System
TPS	Terminals Per Station
TPS	Thermal Protection System
TPS	Transaction Processing System
tps	transactions per second
TPS	Transportation Programming System
TPSE	Transaction Processing Service Element
TPST	Two Processor Switch
TPSU	Transaction Processing Service Unit
TPT	Throughput Time
TPT	Total Processing Time
TPT	Transient Program Table
TPT	Twisted Pair Transceiver
TPTB	The Powers That Be
TPTC	Tele-Processing Test Center
TPTG	Tuned-Plate, Tuned-Grid
TPU	Tape Preparation Unit
TPU	Tape Unit
TPU	Telecommunications Processing Unit
TPU	Terminal Processing Unit
TPVM	Tele-Processing Virtual Machine

TPW	Turbo Pascal for Windows
TPWR	Typewriter
TPx	Transport Protocol, class x (x= 0, 1, 2, 3, or 4)
TQA	Transformational Question Answering
TQE	Time(r) Queue Element
TQF	Terminal Query Facility
TQFP	Thin Quad Flat Pack
TQM	Task Queue Management
TQRD/C	Technology, Quality, Responsiveness, Delivery/Cost reduction
TQS	Transaction Query Subroutine
TQTBL	Task Queue Table
TR	Tape Reader|Recorder
TR	Tape Resident
TR	Temporary Routing
TR	Terminal Ready
TR	Test Register|Request
TR	Test Responder|Run
TR	Throughput Ratio
TR	Token Ring
TR	Track
TR	Transaction Recorder
TR	Transfer Register|Reset
TR	Transformer/Rectifier
TR	Translation Register
TR	Transmitter
TR	Transmit(ter) and Receive(r)
TR	Triangular
TR	Trigger
TR	Trouble Report
TR	Turkey
TRA	Tape Read Alpha
TRA	Telephone Recording Attachment
TRAC	Text Reckoning And Compiling
TRAC	Total Risk Analysis Calculation
TRAC	Tracking Reporting Analysis and Control
TRACE	Time-shared Routines for Analysis, Classification and Evaluation
TRACE	Total Remote Access Center
TRADACOMS	Trading Data Communications Standards
TRADIC	Transistorized Airborne Digital Computer
TRAFFIC	Transaction Routing And Form Formatting In Cobol
TRAFO	Transformer
TRAIN	Tele-Rail Automated Information Network
TRAM	Test Reliability And Maintenance
TRAM	Trademark Reporting And Monitoring system
TRAN	Transmit
TRAN-PRO	Transaction Processing
trans	transform(er)

TRANS	Translation
trans	transmit(ter)
trans	transverse
TRANSAC	Transistor Automatic Computer
transceiver	transmitter/receiver
TRANSDUMP	Transparent Dump
TRANSFOR	Translator for Structured Fortran
transistor	transfer and resistor
TRANSLIS	Transforming our Libraries and Information Services
transponder	transmitter/responder
TRANSPOTEL	International Electronic information system for the Transportation industry
transputer	transmission/computer
TRAP	Trajectory Analysis Program
TRASS	Test Results Analysis Standard System
TRAWL	Tape Read And Write Library
TRAY	Translation Array
TRC	Table Reference Character
TRC	Timed Readout Control
TRC	Trace
TRC	Transmit/Receive Control unit
TRC	Transverse Redundancy Check
TRC	Tuple Relational Calculus
TRC-AS	Transmit/Receive Control unit – Asynchronous Start/stop
TRC-SC	Transmit/Receive Control unit – Synchronous Character
TRC-SF	Transmit/Receive Control unit – Synchronous Framing
TRD	Tape Read
TRDL	Translator Rule Description Language
TRDMC	Tears Running Down My Cheeks
TRDPS	Topographic Retrieval and Data Presentation System
TRDPS	Transistor Radar Data Presentation Sets
TRE	Timer Request Element
TREAT	Trouble Report Evaluation Analysis Tool
TREC	Text Retrieval Conference
TREE	Transient Radiation Effects on Electronics
TREM	Tape Reader Emulator Module
TRENDS	Transmission Environmental Digital Studies
TRF	Transfer
TRF	Tuned Radio Frequency
TRG	Tip and Ring Ground
TRG	Trigger
TRIAL	Technique for Retrieving Information from Abstracts of Literature
TRIB	Transfer Rate of Information Bits
TRIB	Transmission Rate In Bits
TRI-BITS	Three Bits
TRICLOPS	The Real-time Intelligently Controlled Optical Positioning System

TRIDAC	Three-Dimensional Analog Computer
TRIL	Token Ring Interoperability Laboratory
TRIM	Tailored Retrieval and Information Management
TRIM	Terminal Reader In Magnetics
TRIP	Text Retrieval and Indexing Product
TRIP	Transcontinental ISDN Project (ISDN = Integrated Services Data Network)
TRIPM	Trip Master
TRIPS	Teleconferencing Remote Interactive Presentation System
TRIQ	Time-controlled Reasonable Issue Quantity
TRIS-On-LINE	Transportation Research Information Services On-Line
TRK	Track
TRL	Trail Label
TRL	Transistor-Resistor Logic
TRLR	Trailer
TRM	Technical Reference Model
trm	terminal (file extension indicator)
TRM	Terminal Response Monitor
TRM	Test Register under Mask (instruction)
TRM	Test Request Message
TRML	Terminal
TRMS	Terminal Report Management System
TRMS	Transmission Resource Management System
TRMTR	Transmitter
TRN	Threaded Read News
TRN	Token Ring Network
TRN	Transfer
TRN	Transmission Repeat Number
TRNDT	Transaction Date
TRNSP	Transpose
TRNT	Transient
TRO	Timer Run Out
TROFF	Text Run-Off
TROFF	Tracer Off
TROLL	Time-shared Reactive On-Line Laboratory
TROM	Transformer Read-Only Memory
TRON	The Real-time Operating system Nucleus
TRON	Tracer On
TROS	Tape Read-Only Storage
TROS	Transducer|Transformer Read-Only Storage
TRP	Transaction Processor
TRQ	Task Ready Queue
TRR	Tape Read Register
TRR	Test Readiness Review
TRR	Tip-Ring Reverse
TRR	Token Ring Repeater
TRR	Transaction Routing Routines
TRS	Tandy Radio Shack

TRS	Terminal Receive Side
TRS	Tip Ring Sleeve
TRS	Transmit/Receive Switch
TRSB	Transcribe
TRSFR	Transfer
TRSL	Translate
TRSN	Transaction
TRSOTW	The Right Side Of The Web
TRSP	Transport
TRST	Transit
TRT	Token Rotation Time(r)
TRT	Trace Table
TRT	Translate and Test
TRU	Transmit/Receive Unit
TRUPACT	Trans-Uranic Package Transporter
TRUSIX	Trusted Unix
TRV	True Resistance Voltage
TRW	The Real World
TRX	Transaction
TS	Technical Specification|Support
TS	Telecommunication System
TS	Tele Sonderjylland (DK)
TS	Tensile Strength
TS	Time Sharing|Switch
TS	Traffic Sensitive
TS	Trail Stock
TS	Transaction Services
TS	Transition System
TS	Transit Storage
TS	Transmission Service(s)|System
TS	Transmission Subsystem
TS	Transport Service|Station
TS	Tri-State
TS	True Space
TS	Type Specification
TSA	Target Service Agent
TSA	Time Series Analysis
TSA	Time Slot Access
TSA	Transport Service Access (point)
TSA	Tributary Signal Adaptation
TSAM	Time Series Analysis and Modelling
TSAP	Transport Service Access Point
TSAPI	Telephony Server Application Program Interface
TSAR	Task Switch Analysis Routine
TSAS	Time-Sharing Accounting System
TSAU	Time Slot Access Unit
TSB	Task Scheduling Block
TSB	Terminal|Termination Status Block

TSBP	Time Sharing Business Package
TSC	Technical Sub-Committee
TSC	Temporary Specification Change
TSC	Thermally Stimulated Conductivity
TSC	Three State Control
TSC	Time-Sharing Control
TSC	Totally Self-Checking
TSC	Transit Switching Center
TSC	Transmitter Start Code
TSC	Transportation Systems Center
TSCB	Task Set Control Block
TSCE	Time-Slice Control Element
TSCL	Time-Sharing Command Language
TSCP	Time-Sharing Control Program
TSCPF	Time Switch and Call Processor Frame
TSD	Time-Sharing Driver
TSD	Total System Design
TSD	Tree Structure Diagram
TSD	Tri-State Device
TS/DMS	Time-Shared Data Management System
TSDS	Time-Sharing Disk Supervisor
TSDU	Transport-layer Service Data Unit
TSE	Terminal Source Editor
TSE	Terminal Switching Exchange
TSE	Time-Sharing Executive
TSE	Time-Slice End
TSE	Transaction Set Editor
TSEE	Technical Software Engineering Environment
TSEL	Tape no-op Select
TSEL	Transport (service) Selector
TSF	Time to System Failure
TSFP	Time to System Failure Period
TSGAS	Time-Shared General Accounting System
TSI	Task Status Index
TSI	Test Structure Input
TSI	Time Slot Interchange(r)
TSID	Track Sector Identification
TSIO	Time Shared Input/Output
TSIOA	Temporary Storage Input/Output Area
TSIU	Telephone System Interface Unit
TSIU	Time Slot Interchange Unit
TSK	Task
TSL	Test Source Library
TSL	Thermally Stimulated Luminescence
TSL	Time Series Language
TSL	Time-Sharing Library
TSL	Tri-State Logic
TSLT	Translate

TSM	Time-Sharing Monitor
TSM	Total Storage|Systems Management
TSM	Transformation Server Module
TSMIT	Transmit
TSMS	Time Series Modeling System
TSO	Terminating Service Office
TSO	Time-Sharing Option
TSODB	Time Series Oriented Data Base
TSO/E	Time-Sharing Option/Extensions
TSOP	Thin Small Outline Package
TSORT	Transmission System Optimum Relief Tool
TSOS	Time-Sharing Operating System
TSO/VTAM	Time-Sharing Option for the Virtual Telecommunications Access Method
TSP	Teleprocessing Services Program
TSP	Terminal Service Profile
TSP	Time Series Processor
TSP	Time Synchronization Protocol
TSP	Touch Screen Program
TSPL	Telephone System Programming Language
TSPS	Telecommunications Support Processor System
TSPS	Traffic Service Position System
TSR	Temporary Storage Register
TSR	Terminate and Stay Resident
TSR	Test Summary Report
TSR	Translation State Register
TSRS	Transistorized Switching Regulators
TSRT	Task Set Reference Table
TSS	Task State Segment
TSS	Technical System Specification
TSS	Terminal Send Side
TSS	Time-Sharing Service|System
TSS	Transaction Security System
TSS	Trunk Servicing System
TSSDU	Typed data Session Service Data Unit
TS/SI	Top Secret/Sensitive Information
TSSS	Time-Sharing Support System
TSST	Time-Space-Space-Time (network)
TSS&TP	Test Suite Structure & Test Purposes
TST	Temporary Storage Table
tst	test (file extension indicator)
TST	Time-Sharing Terminal
TST	Time-Space-Time (network)
TST	Transaction Status Table
TST	Transaction Step Task
TSTEQ	Test Equipment
TSTN	Triple Super-Twisted Nematic
TSTR	Tester

TSTR	Transistor
TSTS	Time-Space-Time-Space (network)
TSU	Technical Support Unit
TSU	Time-Sharing User
TSU	Time Speed Up
TSU	Transmission Service Unit
TSV	Tab Separated Values
TSVQ	Tree-Structured Vector Quantization
TSX	Time-Sharing Executive
TT	Tape to Tape
TT	Teletype
TT	Teletypewriter
TT	Teller Terminal
TT	Transaction Telephone|Terminal
TT	Transmitting Typewriter
TT	Transmit Trailer
TT	Trinidad and Tobago
TT	Trunk Type
TT	Typewriter Text
TTA	Telecommunication Technology Association (US)
TTA	Trigger Trace Analyzer
TTAE	Trunk Traffic Analyzer Equipment
TTAP	Trust Technology Assessment Program
TTBL	Task Table
TTBOMK	To The Best Of My Knowledge
TTC	Telecommunications Technology Council (J)
TT&C	Telemetry, Telecommand & Control
TT&C	Telemetry, Tracking & Command
TTC	Terminating Toll Center
TTC	Timing, Trigger and Control
TTCN	Tree and Tabular Combined Notation
TTCN.GR	Tree and Tabular Combined Notation, Graphical Representation
TTCN.MP	Tree and Tabular Combined Notation, Machine Processable
TTCVI	Timing, Trigger and Control VME-bus Interface (VME = Virtual Memory Environment)
TTD	Tape To Disk
TTD	Target Transit Delay
TTD	Temporary Text Delay
TTD	Text (to) Telephone|Teletype device for the Deaf
TTD	Things To Do
TTDL	Terminal Transparent Display Language
TTE	Terminal-Table Entry
TTE	Transient Tape|Test Error(s)
TTF	Tabular Test Format
TTF	Terminal Transaction Facility
TTF	Time to Time Failure
ttf	true type font (file extension indicator)
TTFN	Ta-Ta For Now

TTHG	Two-Term Henyey-Greenstein (model)
TTK	Tie Trunk
TTL	Through The Lens
TTL	Transistor-Transistor Logic
TTL/LS	Transistor-Transistor Logic – Low power Schottky
TTLS	Transistor-Transistor Logic Schottky
TTMA	Tennessee Tech Microcomputer Association
TTN	Tandem Tie-line Network
TTP	Tape To Printer
TTP	Telephone Twisted Pair
TTP	Thermal-Transfer Printing
TTP	Touch Trigger Probe
TTP	Trunk Test Panel
TTPN	Timed Transition Petri Net
TTR	Terminal-to-Terminal Router
TTRT	Target Token Rotation Time
TTS	Teletype Setting
TTS	Temporary Threshold Shift
TTS	Text-To-Speech
TTS	Transaction Tracking System
TTS	Transmission Test Set
TTS	Trunk Time Switch
TTSPN	Two Terminal Series Parallel Networks
TTT	Tape To Tape
TTTC	Test Technology Technical Council
TTTN	Tandem Tie-Trunk Network
TTTP	Transmitting Typewriter with Tape Punch
TTW	Teletypewriter
TTX	Teletex
TTX	Torn Tape Exchange
TTXAU	Teletex Access Unit
TTY	Telephone-Teletypewriter
TTY	Teletype(writer)
TTYC	TTY Controller (TTY = Teletype)
TTYCD	Teletype Code
TTYL	Talk To You Later
TTYRS	Talk To You Real Soon
TU	Tape|Terminal Unit
TU	Time|Transfer Unit
TU	Transmission|Transport Unit
TUA	Telecommunications Users' Association (GB)
TUC	Technical Usage Code
TUCC	Triangle University Computing Center
TUCOWS	The Ultimate Collection Of Winsock Software
TUF	Time of Useful Function
TUF	Transmitter Underflow
TUFD	The User File Died
TUFF	Tape Updater for Formatted Files

TUFT	Tape Unit Functional Test
TUG	Tape Updater and Generator
T&UG	Telephone & Utilities Group
TUG	TeX User's Group
TUI	Text-based User Interface
TULIPS	Tsukuba University Library Information Processing System (JP)
TUMS	Table Update and Management System
TUP	Telephone User Part
TUR	Traffic Usage Recorder
TU-R	Transceiver Unit, Remote
TUR	Trunk Utilization Report
tut	tutorial (file extension indicator)
TUTPOC	The Urge To Print Or Copy
TV	Television
T/V	Temperature to Voltage
TV	Terminal Velocity
TV	Transfer Vector
TV	Tuvalu
TVC	Television Camera
TVC	Trunk Verification by Customer
TVD	Total Virus Defense
TVE	Temperature Variation Error
TVF	Table of contents Verbosely from File (UNIX)
TVG	Time Varying Gain
TVI	Television Interference
TVL	Television Listener
TVL	Two-Valued Logic
TVM	Thanks Very Much
TVM	Time-Varying Media
TVM	Transistor Voltmeter
TVO	Transistor-Volt-Ohmmeter
TVOL	Television On-Line
TVOM	Transistor-Volt-Ohm Meter
TVRO	Television Receive Only
TVS	Transient Voltage Suppressor
TVS	Trunk Verification by Station
TV-SAT	Television Satellite
TVT	Television Terminal
TVT	Television Typewriter
TW	Taiwan, Province of China
TW	Terawatt
TW	Time Word
TW	Travelling Wave
TW	Typewriter
TWA	Terminal|Transaction Work(ing) Area
TWA	Travelling Wave Amplifier
TWA	Two-Way Alternate
TWA	Typewriter Adapter

TWAIT	Terminal Wait
TWAVE	Transverse (electric) Wave
TWB	Tailor Welded Blank
TWB	Terrestrial Wide-Band
TWC	Two Wire Circuit
TWIG	Technical Wizard Interest Group
TWIMC	To Whom It May Concern
TWIP	Twentieth of a Point
TWIT$	The Wired Interactive Technology Fund
TWITS	Two-Way Image Transfer System
TWM	Travelling Wave Master
TWOM	Travelling Wave Optical Maser
TWP	The Wacko Programmer
TWP	Twin Wire Press
TWP	Twisted Wire Pair
TWR	Tape Writer Register
TWS	Translator Writing System
TWS	Two-Way Simultaneous
TWSD	Task Work Stack Descriptor
TWT	Teletype-Writer Terminal
TWT	Travelling Wave Tube
TWT	Type-Writer Terminal
TWTA	Travelling Wave Tube Amplifier
TWX	Teletypewriter Exchange (service)
TX	Telex
TX	Terminal Executive
TX	Texas (US)
TX	Transmit(ting)
TXA	Terminal Exchange Area
TXC	Transaction Code
TXD	Transmit(ted) Data
TXK	Telephone Exchange Crossbar
txr	texture plus (file extension indicator)
TXS	Telephone Exchange Strowger
TXT	Text
TXT2STF	Text To Structured File
TXTM	Text Maintenance
TYDAC	Typical Digital Automatic Computer
TYP	Type
TYP	Typewriter
TYPOUT	Typewriter Output (routine)
TYPWRTR	Typewriter
TYSH	Typewriter Shift
TYVM	Thank You Very Much
TZ	Tanzania

U

U	Uncoupled
U	Underflow
U	Union (of sets)
U	Unit
U	Universal (set)
U	Unprotected
U	Upper
UA	Ukraine
UA	Unauthorized Absence
UA	Unit|Universal Address
UA	Unnumbered Acknowledgment
UA	User Account
UA	User Agent|Area
UAA	Universally Administered Address
UAB	Unix Appletalk Bridge
UAC	Uniform Article Code
UAC	Uninterrupted Automatic Control
UACN	Unified Automated Communication Network
UACTE	Universal Automatic Control and Test Equipment
UADP	Uniform Automatic Data Processing
UADPS	Uniform Automatic Data Processing System
UADS	User Attribute Data Set
UAE	Uninterruptable Application Error
UAE	Unrecoverable Application Error
UAE	User Agent Entity
UAF	Unit Authorization File
UAI	Use As Is
UAID	User Automated Interface Data
UAL	Unit Authorization List
UAL	User Agent Layer
UAM	User Authentication Method
UAOS	User Alliance for Open Systems
UAP	User Area Profile
UAPDU	User Agent Protocol Data Unit
UAR	User Action Routine
UART	Universal Asynchronous Receiver/Transmitter
UAS	Unavailable Second
UASFE	Unavailable Second, Far End
UASL	User Agent Sub-Layer
UASTOR	Uniformly Accessible Storage
UAT	User Accounting Table
UB	Upper Bound
UB	User Board
UBA	Uni-Bus Adapter
UBC	Universal Block Channel

UBC	Universal Buffer Controller
UBCIM	Universal Bibliographic Control/International MARC (= Machine Readable Cataloguing)
UBD	Utility Binary Dump
UBE	Upper Band Edge
UBHR	User Block Handling Routine
UBITA	Upper Bound of Information Translation Amount
UBLK	Unblank
UBLK	Universal Block
UBR	Unspecified Bit Rate
UBS	Unit Backspace (character)
UC	Uncertainty
UC	Under Current
UC	Uni-Channel
UC	Unit Cell
UC	Up Converter
UC	Upper Case|Control
UC	Usable Control
UC	Utilization Control
UCA	Under Colour Addition
UCA	Upper Control Area
UCAID	University Corporation for Advanced Internet Development (US)
UCB	Unit Control Block
UCB	Universal Character Buffer
UCBTAB	User Control Block Table
UCC	Unallowable Code Check
UCC	Unit Communications Control
UCD	Uniform Call Distribution
UCD	Use Case Diagram
UCD	User Code Document
UCDP	Uncorrected Data Processor
UCE	Unit Checkout Equipment
UCF	Utility Control Facility
UCI	Upper Control|Constant Impulse
UCI	Utility Card Input
UCIS	Up-range Computer Input System
UCITA	Uniform Computer Information Transactions Act
UCK	Unit Check
UCL	Universal Communications Language
UCL	Utility Control Language
UCLAN	User Cluster Language
UCLX	Upper Control Limit Range
UCM	Universal Communications Monitor
UCN	Uniform Control Number
UCNDL	Unconditional
UCP	Uninterruptable Computer Power
UCP	Units Construction Practice

UCP	Users Control Program
UCPMT	Uncomplemented
UCR	Under Colour Removal
UCRL	Upper Control Limit
UCS	Unconditional Stimulus
UCS	Unicode Conversion Support
UCS	Uniform Chromaticity Scale
UCS	Uniform Communications Standard|System
UCS	Universal Call Sequence
UCS	Universal Character Set
UCS	Universal Classification Code
UCS	Universal Component System
UCS	User Control Store
UCS	User Coordinate System
UCS	Users Control System
UCSAR	Universal Character Set Address Register
UCSB	Universal Character Set Buffer
UCSTR	Universal Code Synchronous Transmitter/Receiver
UCT	Universal Coordinated Time
UCU	Universal Control Unit
UCW	Unit Control Word
UD	Ultra Density
UD	Undetected Defect
UD	Unit Data|Diagnostics
UD	Usage Data
UDA	Universal Data Access
UDAC	Universal Digital-Analog Controller
UDAC	User Digital-Analog Controller
UDAS	Unified Direct Access Standard(s)
UDB	Unified|Universal Data Base
UDC	Unit Distance Code
UDC	Universal Decimal Classification
UDC	User Defined Commands
UDE	Universal Data Entry|Exchange
UDEC	Unitized Digital Electronic Calculator
UDED	User Data Entry Dialog
UDF	Unit Development Folder
UDF	User Defined Function
U-DID	Unique Data Item Description
UDIF	Universal Data Interchange Format
UDL	Uniform Data Language
UDL	User Defined Logic
UDLC	Universal Data Link Control
UDOP	Ultra-high frequency Doppler system
UDP	Usenet Death Penalty
UDP	User Datagram Protocol
UDR	Universal Document Reader
UDS	United Data Set

UDS	Utility Definition Specifications
UDT	Uniform Data Transfer
UDT	Universal Document Transport
UDT	Update
UDT	User Defined Type(s)
UDTS	Universal Data Transfer Service
UDTV	Ultra High Definition Television
UDU	Update Data Units
UE	Unit Equipment|Exception
UE	Unprintable Error
UE	User Element|Equipment
UEC	User Environment Component
UEJ	Unchanged Eject
UEML	Unified Enterprise Modelling Language
UEP	Underwater Electric Potential
UEPB	Universal Electronic Program Board
UER	Uninstalled Equipment Report
UET	Universal Emulating Terminal
UETS	Universal Emulating Terminal System
UEX	Unit Exception
UF	Used For
UF	Utility File
UFAM	Universal File Access Method
UFAS	Universal File Access System
UFC	Universal Frequency Counter
UFD	User File Directory
UFET	Unipolar Field-Effect Transistor
UFF	Universal Flip-Flop
UFI	Usage Frequency Indicator
UFI	User Friendly Interface
UFLO	Underflow
UFM	Universal Function Module
UFM	User-to-File Manager
UFN	User Friendly Naming
UFO	Users File On-line
UFP	Utility Facilities Program
UFRZ	Unfreeze
UFS	Uniform Filing System
UFS	Universal|Unix File System
UFT	Unified File Transfer
UG	Uganda
UG	User Group
UGA	Unity Gain Amplifier
UGLI	Universal Gate for Logic Implementation
UGT	User Group Table
UH	Unit Head
UH	Upper Half
UHA	Ultra-High Aperture

UHF	Ultra-High Frequency
UHL	User Header Label
UHM	Universal Host Machine
uhr	ultrahigh resistance
UI	Unit Interval
UI	Unix International
UI	Unnumbered Information|Interrupt
UI	Unscheduled Interruption
UI	Useful Information
UI	User Interface
UIA	User Interface Agents
UIC	User Identification Code
UICP	Uniform Inventory Control Program
UID	Unique|User Identifier
UIDL	Unique-Identifier Listing
UIG	User Instruction Group
UIL	User Interface Language
UIM	Ultra-Intelligent Machine
UIMS	User Interface Management System
UIMX	User Interface Management systems for X windows
UIN	Universal Internet Number
UIO	Units In Operation
UIO	Universal Input/Output
UIOD	User Input/Output Device(s)
UIP	Usage Input
UIP	User Interface Program
UIR	User Instruction Register
UIS	Uncertain(ty) Inference System
UIS	Unit Identification System
UIS	Universal Information Services
UIST	User Interface Software and Technology
UIT	Union Internationale des Télécommunications
UITP	Universal Information Transport Plan
UJCL	Universal Job Control Language
UJT	Uni-Junction Transistor
UK	Ukraine
UK	United Kingdom
UKB	Universal Keyboard
UKIS	Universal Knowledge-based Imaging System
UKITO	United Kingdom Information Technology Organization
UKOLN	United Kingdom Office for Library Networking
UKRA	United Kingdom Registration Authority
UKSG	United Kingdom Serials Group
UL	Undefined Line
U/L	Universal/Local
UL	Unordered List
UL	Upload
UL	Upper Limit

U/L	Upper/Lower
UL	User Language|Location
ULA	Uncommitted Logic Array
ULA	Upper Layer Architecture
ULAN	Universal Local Area Network
ULANA	Unified Local-Area Network Architecture
ULANG	Update Language
ULB	Universal Logic Block
ULC	Uniform Loop Clock
ULC	Universal Logic Circuit
U&LC	Upper & Lower Case
ULC	Upper/Lower Case
ULCC	Ultra Large Crude Carrier
ULCC	University of London Computing Centre
ULCE	Unified Life Cycle Engineering
ULCT	Upper Layer Conformance Testing
ULD	Ultra-Low Distortion
ULD	Unit Load|Logic Device
ULD	Universal Language Description
ULE	Unit Location Equipment
ULF	Ultra-Low Frequency
ULL	Under Low Limit
ULL	Unix Linkable Library
ULM	Ultrasonic Light Modulator
ULM	Universal Line Multiplexer
ULM	Universal Logic Module
ULN	Universal Link Negotiation
ULP	Upper Layer Process|Protocol
ULPAA	Upper Layer Protocols, Architectures and Applications (conference)
ULS	Unit Level Switchboard
ULSI	Ultra Large Scale Integration
ULT	Ultimate(ly)
UM	Unified Messaging
UM	United States Minor Outlying Islands
UM	Unit of Measure(ment)
UM	Unscheduled Maintenance
UM	User Mode
UM	User's Manual
UMA	Unified Memory Architecture
UMA	Upper Memory Area
UMB	Upper Memory Block
UMC	Unibus Micro-Channel
UMCS	Unattended Multipoint Communications Station
UME	UNI Management Entity
UMF	Ultra Micro-Fiche
UMFC	Unit of Measure Family Code
UML	Unified Modelling Language

UML	Universal Markup|Modelling Language
UMLC	Universal Multi-Line Controller
UMOD	User Module
UMP	Universal Macro Processor
UMPDU	User Message Protocol Data Unit
UMS	Universal Maintenance Standards
UMS	Universal Message System
UMS	Universal Multiprogramming System
UMTS	Universal Mobile Telephone Service
UN	Unit
UN	United Nations
UNA	Unattended Answering Accessory
UNA	Universal Network Architecture
UNA	Upstream Neighbours Address
UNALC	User Network Access Link Control
UNAMACE	Universal Automatic Map Compilation Equipment
UNC	Unconditional
UNC	Universal Naming Convention
UNC	Uuencoded Netnews Collator (Unix)
UNCATLG	Uncataloging
UNCID	Uniform rules of Conduct for Interchange of trade Data by teletransmission
UNCJIN	United Nations Criminal Justice Information Network
UNCL	Unified Numerical Control Language
UNCLOS	United Nations Convention on the Law Of the Sea
UNCM	User Network Control Machine
UNCOL	Universal Computer Oriented Language
UNCSD	United Nations Center for Science and technology for Development
UNCT	Undercut
UNCTAD	United Nations Conference on Trade And Development
UNDEF	Undefined
UNEJ	Unchanged Eject
UNESCO	United Nations Educational, Scientific (and) Cultural Organization
UNGTDI	United Nations Guidelines for Trade Data Interchange
uni	single
UNI	User Network Interface
UNIBUS	Universal Bus
UNICC	United Nations International Computing Centre
UNICOM	Universal Integrated Communication (system)
UNICOMP	Universal Compiler (Fortran-compatible)
UNICOS	Universal Compiler System (Fortran compatible)
UNICS	Uniplex Information Computer Services
UNIFET	Unipolar Field-Effect Transistor
UNII	Unlicensed National Information Infrastructure
UNIMOD	Universal Module
UNIPOL	Universal Procedure Oriented Language

UNIPRO	Universal Processor
UNIQUE	Uniform Inquiry Update and Edit
UNIQUE	Uniform Inquiry Update Element
UNISIST	Universal System for Information in Science and Technology
UNISTAR	Universal Single call Telecommunications Answering & Repair
UNISTAR	User Network for Information Storage, Transfer, Acquisition and Retrieval
UNISWEP	Unified Switching Equipment Practice
UNISYS	United Information Systems
UNIV	Universal
Unix	Uniform Executive
UNIX	Universal Interactive Executive
UNJEDI	United Nations Joint Electronic Data Interchange
UNK	Unlinked
UNL	Unlatch(ed)
UNLD	Unload(ing)
UNMA	Unified Network Management Architecture
UNMSR	Unit of Measure(ment)
UNN	Unnormalized
UNPK	Unpack(ed)
UNRJ	Unit (not busy) Reject
uns	unstable
uns	unsymmetrical
UNSM	United Nations Standard Messages
UNSOL	Unsolicited
UNSZ	Undersize
UNTDED	United Nations Trade Data Elements Directory
UNTDI	United Nations Trade Data Interchange
UNU/IIST	United Nations University – International Institute for Software Development (UN)
UOB	Unit Of Behaviour
UODDL	User Oriented Data Display Language
UOF	Unit Of Functionality
UOLAN	User Oriented Language
UOV	Units Of Variance
UP	Uncovered Position
UP	Uni-Processor
UP	Upper
UPA	Ultra-sparc Port Architecture (sparc = scalable performance architecture)
UPACS	Universal Performance Assessment and Control System
UPB	Upper Bound
UPC	Uniform Product Code
UPC	United and Philips Communications (NL)
UPC	United Pan-Europe Communications
UPC	Unit of Processing Capability
UPC	Universal Peripheral Control
UPC	Usage Parameter Control

UPCB	Units Position C Bit
upconv	up converter
UPCS	Upper Case
UPD	Unaxial Plastic Deformation
UPD	Update
UPDS	Uninterruptable Power Distribution System
UPDS	User Profile Data Set
UPE	User Portable Extension
UPG	Unified Parametric Geometry
UPG	Upgrade
UPI	Universal Peripheral Interface
UPIC	Universal Personal Identification Code
UPL	Universal Programming Language
UPM	Unix Programmer's Manual
UPM	User Profile Management
UPOS	Utility Program Operating System
UPP	Universal Post-Processor
UPP	Universal PROM Programmer
UPS	Uninterruptable Power Supply|System
UPS	Uninterrupted Power Supply
UPS	Universal Processing System
UPSG	Unique Phrase Structure Grammar
UPSI	User Program Sense|Switch Indicator
UPSP	Upstroke Space
UPT	Universal Personal Telecommunications
UPT	User Process Table
UPVF	User Program Verification Facility
UQBT	University of Queensland Binary Translator
UQT	User Queue Table
UQUM	Use Quick Update Methods
UR	Unit Record|Register
UR	Units Restore
UR	User Record
UR	Utility Register
URA	User Requirements Analysis
URA	Utilization Review Agency
URC	Uniform Resource Characteristics|Citation
URC	Unit Record Control
URD	User Record Description|Dictionary
URD	User Requirements Definition
URDS	User Requirements Database System
URE	Unit Record Equipment
UREP	Unix RSCS Emulation Protocol (protocol) (RSCS = Remote Spooling and Control Subsystem)
URG	Urgent (flag)
URI	Uniform Resource Identifier
URISA	Urban and Regional Information Systems Association
URL	Uniform Resource Locator

URL	User Requirements Language
URLS	Uniform Resource Locations
URMP	United Records Maintenance Program
URN	Uniform Resource Name|Number
URP	Unit Record Processor
URP	Users Review Panel
URS	Uniformly Reflexive Structure
URS	Uniform Reporting System
URS	Unit Record System
URS	User Requirements Specification
URSI	Union Radio-Scientifique Internationale
URT	User Registration Tool
US	United States
US	Unit (record) Signal
US	Unit Separator (character)
US	Unit Specify|String
USA	Undedicated Switch Access
USA	Unique Search Argument
USA	United States of America
USAC	United States Answer Center
USACII	USA Code for Information Interchange
USACM-Net	United States Association for Computational Mechanics Network
USAM	Unique Sequential Access Method
USAN	University Satellite Network
USART	Universal Synchronous/Asynchronous Receiver/Transmitter
USASCII	USA Standard Character set for Information Interchange
USASCSOCR	USA Standard Character Set for Optical Characters
USASI	USA Standards Institute
USAT	Ultra Small Aperture Terminal
USB	Unified S-Band
USB	Universal Serial Bus
USB	Upper Side-Band
USBN	Universal Standard Book Number
USC	Under Supervisory Control
USC	User Service Center
USD	Uniform Symbol Description
USD	Universal Standard Data
USDA/CRIS	United States Department of Agriculture/Current Research Information System
USE	User System Emulator|Executive
usec	microsecond
USENET	User's Network
USER	User System Evaluator
USERID	User Identification
USF	Universal Source File
USG	Unix Support Group
USG	Usage

USI	Unique Secondary Index
USI	User System Interface
USIA	United States Information Agency
USIO	Unlimited Sequential Input/Output
USIS	United States Information Service(s)
USIT	Unit Share Investment Trust
USITA	United States Independent Telephone Association
USITS	Usenix Symposium on Internet Technologies & Systems
USL	Uncommitted Storage List
USL	Unix System Laboratories
USLIB	User Library
USLP	Unix System Laboratories Pacific
USM	Unbound Storage Manager
USM	Unifying Semantic Model
USM	Unsharp Masking
USMARC	United States Machine Readable Cataloguing
USMTF	United States Message Text Format
USN	Update Sequence Number
USO	Universal Service Order
USO	Universal Services Obligation utilization system
USO	Unix Software Operation
USOA	Uniform System Of Account
USP	Uninterrupted Server Page
USP	Universal Sampling Plan
USP	Upstroke Space
USP	User Program
USQ	Unsqueezed (files)
USR	United States Robotics (corporation)
USR	Universal Series Regulator
USR	User Service(s) Routine
USRT	Universal Synchronous Receiver/Transmitter
USS	Unformatted System Services
USS	Uniform Symbol(ogy) Specification
USS	User Support System
USSA	User Supported Software Association (GB)
USSC	Upper Sideband, Suppressed Carrier
USSD	Unstructured Supplementary Service Data
USSS	User Services and Systems Support
UST	Usage Structure Tape
UST	User Symbol Table
USTA	United States Telephone Association
USV	User Services
USW	Und So Weiter (DE)
UT	Uniform Taper
UT	Units Tens
UT	Unit Tester
UT	Universal Time
UT	Upper Tester

UT	User Terminal
UT	Utah (US)
UT	Utility (program)
UTA	User Transfer Address
UTC	Universal Telex Controller
UTC	Universal Time Coordinates
UTC	Unstable Trigger Circuit
UTC	Utilities Telecommunications Council
UTD	Universal Transfer Device
UTE	Unit Test Equipment
UTEMP	Under Temperature
UTF	Universal Text Format
UTI	Universal Text Interchange|Interface
UTIL	Utility
UTIL	Utilization
UTL	Unit Transmission Loss
UTL	User Trailer Label
UTLD	United Telephone Long Distance
UTOL	Universal Translator Oriented Language
UTP	Unshielded Twisted Pair
UTR	Unprogrammed Transfer Register
UTS	Ultimate Tensile Strength
UTS	Unattended Terminal Syndrome
UTS	Unbound Task Set
UTS	Unified Test Systems
UTS	Universal Terminal|Time-sharing System
UTT	Universal Teleprocessing Tester
UTTC	Universal Tape-to-Tape Converter
UTTP	Unshielded Telephone Twisted Pair
UTYPE	Unit Type
UU	Ultimate User
UU	Unix-to-Unix
UU	Uuencode/Uudecode
UUC	Undernet Users Committee
UUCICO	Unix to Unix Copy Incoming Copy Outgoing
UUCP	Unix to Unix Copy Program
UUD	Unix-to-Unix Decoding
Uudecode	Unix-to-unix decoding
UUE	Unix-to-Unix Encoding
Uuencode	Unix to unix encoding
UUG	Unix User Group
UUI	User-to-User Information
UUNET	Unix-to-Unix Network
UUS	User to User Signalling
UUT	Unit Under Test
UV	Ultra-Violet (light)
UV	Under Voltage
UVE	Ultra-Violet Erasing

UVM	Universal Vendor Marking
UVM	Universal Virtual Machine
UVPROM	Ultra-Violet Programmable Read Only Memory
UVROM	Ultra-Violet light erasable Read Only Memory
UVV	Usenet Volunteer Votetakers
UW	Underwriter
UW	Used With
UWA	User Work(ing) Area
UWB	Ultra-Wide Band
UWCSA	University of Washington Computer Science Affiliates
UWE	Unified Workstation Environment
UWG	User Working Group
UWS	Ultrix Workstation Software
UY	Uruguay
UZ	Uzbekistan

V

V	Potential
v	vector
v	velocity
V	Version
V	Vertical
V	Volt
V	Volume
V.110	Specification of DTE with synchronous/asynchronous serial interface on an ISDN network (DTE = Data Terminal Equipment; ISDN = Integrated Services Data Network)
V.120	Specification of DTE with synchronous/asynchronous serial interface on an ISDN network using protocols
V.13	Simulating carrier control on full-duplex modems
V.14	Specification of synchronous modems to carry asynchronous data
V.17	Fax standard using TCM modulation at 12000 and 14400 bps (TCM = Trellis Coded Modulation)
V.21	300 bps, 2-wire, dial/lease lines, full duplex, asynchronous
V.22	600/1200 bps, 2-wire, dial/lease lines, full duplex, (a)synchronous
V.22bis	1200/2400 bps, 2-wire, dial/lease lines, full duplex, (a)synchronous
V.23	1200/75 bps, 2-wire, dial/lease lines, half duplex, (a)synchronous
V.24	Definitions for interchange between DTE and DCE (DCE = Data Circuit Equipment)
V.25	Automatic calling and answering equipment with parallel interface
V.25bis	Automatic calling and answering equipment with serial interface

V.26	1200/2400 bps, 4-wire, lease lines, full duplex, synchronous
V.26bis	1200/2400 bps, 2/4-wire, dial/lease lines, full/half duplex, synchronous
V.27bis	2400/4800 bps, 2/4-wire, lease lines, full/half duplex, synchronous
V.27ter	2400/4800 bps, 2/4-wire, dial lines, full/half duplex, synchronous
V.28	Electrical characteristics of the V.24 interface
V.29	9600/7200/4800 bps, 4-wire, lease lines, full/half duplex, synchronous
V.32	9600/4800 bps, 2-wire, dial/lease lines, full/half duplex, (a)synchronous
V.33	14400/9600/4800 bps, 4-wire, lease lines, full duplex, synchronous
V.35	Group band modems that combine the bandwidth of several telephone circuits (like high-speed RS-32 interface)
V.42	Error correction and data compression for modems
V.42bis	Modem data compression
V.52	Error rate and distortion test for modems
V.54	Loop test devices for modems
V.56	Method of testing modems
VA	Video Amplifier
VA	Virginia (US)
VA	Virtual Address
VA	Voltage Amplifier
VA	Volt-Ampere(s)
VAA	Virtual Address Area
VAA	Voice Access Arrangement
VAB	Voice Answer-Back
vac	vacuum
VAC	Value-Added Carrier
VAC	Vector Analog Computer
VAC	Vibration Adaptive Control
VAC	Video Amplifier Chain
VAC	Virtual Address Cache
VAC	Voltage Alternating Current
VAC	Volt-Ampere Characteristics
VAC	Volts of Alternating Current
VACC	Value-Added Common Carrier
VAD	Velocity-Azimuth Display
VAD	Voltmeter Analog-to-Digital converter
VADAC	Voice Analyzer Data Converter
VADC	Video Analog-to-Digital Converter
VADD	Value Added Disk Driver
VADE	Versatile Automatic Data Exchange
VADS	Value-Added Data Services
VADS	Verdix Ada Development System
VAG	Vertex Adjacency Graph
VAG	VRML Advisory Group

VAIO	Video Audio Integrated Operation
val	validity (checks) (file extension indicator)
VAL	Value
VAL	Voice Application Language
VALDEFD	Value Defined
VAM	Virtual Access Method
VAM	Vogel's Approximation Method
VAMP	Vector Arithmetic Multi-Processor
VAMS	Visual Analog Mood Scales
VAN	Value Added Network
VAN	Variable|Vehicle Area Network
VANDAT	Vancouver Data
VANDL	Vancouver Data Language
VANS	Value Added Network Services
VAP	Value Added Process(es)
VAP	Value Additivity Principle
VAP	Vertex Allocation Program
VAP	Video(tex) Access Point
VAP	Vision As Process (EU)
VAPS	Volume, Article, Paragraph, Sentence
VAPT	Variable Automatic Pulse Tester
VAR	Value Added Remark(et)er
var	variable
VAR	Variance
VAR	Vector Auto-Regressive model
VAR	VHF Aural Range
VAR	Visual Aural Range
VAR	Volt-Amperes Reactive
VARBLK	Variable Blocked
VARBT	Variable Block-size and data-Transfer-time
varco	variable condenser
VARIAC	Variable Capacitor
varicap	variable capacitance
VARUNB	Variable Unblocked
VAS	Value Added Statement|Service(s)
VAS	Vector Addition System
VAS	Virtual Address Space
VASCAR	Visual Average Speed Computer And Recorder
VAST	Variable Array Storage Technology
VAST	Vector and Array Syntax Translator
VAST	Virtual Archive Storage Technology
VAT	Value Added Translator
VAT	Variable Auto Transformer
VAT	Virtual Address Translation
VAT	Voice Actuated Typewriter
VATE	Versatile Automatic Test Equipment
VAU	Vertical Arithmetic Unit
VAX	Virtual Access Executive|Extended

VAX	Virtual Architecture Extended
VAXBI	Virtual Address Extension Bus Interconnect
VAXELN	Virtual (memory) Architectural Extension Executive (for) Local (area) Network
VAX/VMS	Virtual Address Extension/Virtual Memory System
VB	Variable Block
VB	Visual Basic
VB	Voice Band
VBA	Visual Basic for Applications
VBBS	Virtual Backbone Service
VBE	VESA BIOS Extender (BIOS = Basic Input/Output System)
VBE	VGA BIOS Extensions (VGA = Video Graphics Adaptor)
VBE/AI	VESA BIOS Extension/Audio Interface
VBF	Variable Block Format
VBI	Vertical Blanking Interval
VBLC	Virtual Building Life-Cycle
VBNS	Very high Bandwidth Network Service
VBNS	Very high performance Backbone Network Service
vBNS	very high speed Backbone Network Service
VBOMP	Virtual Base Organization and Maintenance Processor
VBP	Valid Built-in-test Processor
VBP	Virtual Block Processor
VBR	Variable Bit Rate
VBRUN	Visual Basic Runtime
VBV	Video Buffer Verifier
VBX	Visual Basic custom Controls
VBX	Visual Basic Extension
vbx	visual basic (file extension indicator)
VC	Saint Vincent and the Grenadines
VC	Validity Check
VC	Vector Control
VC	Verification Condition
VC	Video Conferencing|Correlator
VC	Virtual Call|Channel
VC	Virtual Circuit|Computer
VC	Virtual Container
VC	Voice Call|Channel
VC	Voltage Comparator
VCA	Video Capture Adapter
VCA	Voice Connecting Arrangement
VCA	Voltage Controlled Amplifier
VCB	Variable|Volume Control Block
VCBA	Variable Control Block Area
VCC	Video Compact Cassette
VCC	Video Control Chip
VCC	Virtual Channel Connection
VCC	Void Compensation Circuit
VCCO	Voltage-Controlled Crystal Oscillator

VCD	Variable Capacitance Diode
VCD	Virtual Communications Driver
VCDFS	Virtual CD-rom File Specification (CD = Compact Disk)
VCE	Virtual Collaborative Environment
VCE	Visual Colour Efficiency
VCF	Verified Circulation Figure
VCF	Voltage Controlled Filter
VCF	Voltage Control of Frequency
VCG	Verification Condition Generator
VCG	Voltage Controlled Generator
VCGEN	Verification Condition Generator
VCGL	Versatec Computer Graphic Language
VCI	Virtual Channel|Circuit Identifier
VCL	Visual Component|Control Library
VCM	Videotex Communication Monitor
VCM	Virtual Call Mode
VCM	Voice Coil Motor
VCN	Visitor and Community Network
VCNA	VTAM Communication Network Application (VTAM = Virtual Terminal Access Method)
VCO	Variable Cycles Operation
VCO	Voltage Controlled Oscillator
VCORE	Variable Core
VCOS	Visual Caching Operating System
VCP	Videotex Control Program
VCPI	Virtual Control Program Interface
VCR	Video Cassette Recorder
VCS	Validation Control System
VCS	Video Communications System
VCS	Virtual Circuit Switch|System
VCS	Virtual Computing System
VCSEL	Vertical Cavity Surface Emitting Laser
VCSR	Voltage-Controlled Shift Register
VCT	Voice-Coded Translation|Translator
VCTCA	Virtual Channel To Channel Adapter
VCTD	Vendor Contract Technical Data
VCV	Variable Compression Vector
VCXO	Voltage-Controlled Crystal Oscillator
VD	Voltage Drop
VD	Volume Deleted
VDA	Variable Data Area
VDA	Velocity De-aliasing Algorithm
VDA	Visual Data Analysis
VDAM	Virtual Data Access Method
VDB	Vector Data Buffer
VDB	Video Display Board
VDB	Virtual Data Base
VDC	Vendor Data Control

VDC	Video Display Controller
VDC	Voltage Direct Current
VDC	Voltage Doubler Circuit
VDC	Volts of Direct Current
vdcw	voltage direct current working
VDD	Version Description Document
VDD	Virtual Device Driver
VDD	Visual Display Data
VDDL	Virtual Data Description Language
VDDM	Virtual Device Driver Manager
VDDP	Video Digital Data Processing
VDE	Variable Display Equipment
VDE	Video Display Editor
VDE	Visual Development Environment
VDE	Volume Data Entry
VDETS	Voice Data Entry Terminal System
VDF	Video Disk File
VDFM	Virtual Disk File Manager
VDG	Value Dependence Graph
VDG	Video Display Generator
VDH	Variable-length Divide or Halt
VDI	Video Display Input|Interface
VDI	Virtual Device Interface
VDI	Visual Display Input
VDISK	Virtual Disk
VDL	Virtual Database Level
VDLIB	Virtual Disk Library
VDM	Video Display Module
VDM	Virtual Device Metafile
VDM	Virtual DOS Machine (DOS = Disk Operating System)
VDM	Visual Display Module
VDM	Voice Data Multiplexer
VDMAD	Virtual Direct Memory Access Device
VDP	Variable-length Divide or Proceed
VDP	Vertical Data Processing
VDP	Video Data Processor
VDP	Video|Visual Display Processor
VDPS	Voice Data Processing System
VDR	Video-Disk Recorder
VDR	Voice Digitization Rate
VDR	Voltage-Dependent Resistor
VDS	Video Direct Slot
VDS	Virtual DMA Services (DMA = Direct Memory Access)
VDS	Visionary Design System
VDS	Voice/Data Switch
VDS	Volume Data Set
VDSL	Very high bit-rate Digital Subscriber Line
VDSL	Very high data Digital Subscriber Line

VDT	Validate
VDT	Video Data|Display Terminal
VDT	Video Dial Tone
VDT	Visual Display Terminal
VDU	Video|Visual Display Unit
VDUC	VDA Data Utilization Center (VDA = Velocity De-aliasing Algorithm)
VDWQT	Vertical Deferred Write Queue Threshold
VE	Value Engineering
VE	Venezuela
VE	Virtual Environment
VEA	Value Engineering Audit
VEC	Value Engineering Change
VEC	Vector analog Computer
VEC	Videotext Enquiry Center
VEC	Volume Expansion Coefficient
VECP	Value Engineering Change Proposal
VECT	Vector
VEF	Videotex Editor Facility
VEFCA	Value Engineering Functional Cost Analysis
VEGA	Video-7 Enhanced Graphics Adapter
vel	velocity
VEM	Value Engineering Model
VEMM	Virtual Expanded Memory Manager
VEMMI	Videotex Enhanced Man-Machine Interface
VENUS	Valuable and Efficient Network Utility Service
VEOT	Virtual End Of Tape
VER	Valid Exclusive Reference
VER	Verification
VER	Version
VERA	Verify Read Access
VERNET	Virginia Research Network
VERONICA	Very Easy Rodent-Oriented Net-wide Index to Computer Archives
VERT	Venture Evaluation and Review Technique
vert	vertical
VERTRDN	Vertical Redundancy
VERW	Verify Write (access)
VES	Value Exchange System
VESA	Video Equipment Standards Association
VEST	Volunteer Executive Service Team
VETAF	Virtual Emergency Task Force
VETC	Virtual Environment Technology Center
VETLA	Very Enhanced Transmission Line Adapter
VEU	Volume End User
VEV	Voice-Excited Vocoder
VEX	Video Extension for X window
VF	Variable Factor|Frequency

VF	Vector Facility
VF	Vibratory Feeder
VF	Video|Voice Frequency
VF	Virtual Floppy
VF	Visual(ly) Filed
V/F	Voltage-to-Frequency (converter)
VFAT	Virtual File Allocation Table
VFB	Vertical Format Buffer
VFC	Variable File Channel
VFC	Variable Frequency Clock
VFC	Vector Function Chainer
V.FC	Version First Class (communications standard)
VFC	Vertical Form(at) Control
VFC	Very Fast Class (modems)
VFC	Voltage Frequency Channel
VFCT	Voice Frequency Carrier Telegraph
VFD	Vacuum Fluorescent Display
VFEA	VMEbus Futurebus Extended Architecture
VFFT	Voice Frequency Facility Terminal
VFG	Variable Function Generator
VFL	Variable Field Length
VFM	Variable Format Message(s)
VFMED	Variable Format Message Entry Device
VFO	Variable Frequency Oscillator
VFP	Variable Floating Point
VFP	Virtual File Protocol
VFRP	VINES Fragmentation Protocol (VINES = Virtual Networking System)
VFS	VINES File System
VFS	Virtual File Store|System
VFT	Voice Frequency Telegraph(y)
VFTG	Voice Frequency Telegraph
VFU	Vertical Format Unit
VFU	Vocabulary File Utility
V-F-V	Voltage to Frequency to Voltage
VfW	Video for Windows
VFY	Verify
VG	Vector Generator
VG	Very Good
VG	Virgin Islands (GB)
VG	Voice Grade
VGA	Variable Gain Amplifier
VGA	Video Graphics Accelerator|Adapter
VGA	Video Graphics Array
VGA_HC	Video Graphics Array High Colour
VGAM	Vector Graphics Access Method
VGC	Video Graphics Controller
VGCA	Voice Gate Circuit Adapters

VGF	Voice Grade Facility
VGI	Virtual Graphics Interface
VGL	Voice Grade Lines
VGM	Variational Geometry Modeller
VGS	VAX Gateway Server (VAX = Virtual Address Executive)
V-H	Vertical-Horizontal
VHD	Very|Video High Density
VHDL	VHSIC Hardware Definition Language
VHF	Very High Frequency
VHFO	Very High Frequency Oscillator
VHFR	Very High Frequency Receiver
VHLL	Very High Level Language
VHM	Virtual Hardware Monitor
VHN	Vassal Half Name
VHO	Very High Output
VHOL	Very High Order Language
VHP	Very High Performance
VHPCC	Very High Performance Computing and Communication
VHR	Very High Resistance
VHRVM	Very High Resistance Volt-Meter
VHS	Very High Speed
VHS	Video Home System
VHS	Virtual Host Storage
VHSI	Very High Speed Integration
VHSIC	Very High Speed Integrated Circuit
VI	Variable Interval
VI	Video Input
VI	Virgin Islands (US)
VI	Visual Interactive (editor)
VI	Visual Interface
VI	Volume Indicator
VIA	Versatile Interface Adapter
VIA	Virtual Interface Architecture
VIABLE	Vertical Installation Automation Base-Line
VIAS	Voice Interference Analysis Set
vib	vibrate
VIBA	Virtual-Instruction-Buffer Address
VIBGYOR	Violet, Indigo, Blue, Green, Yellow, Orange, and Red
VIBROT	Vibrational/Rotational
VIC	V.35 Interface Cable
VIC	Variable Instruction Computer
VIC	Virtual Interaction Controller
VICAM	Virtual Integrated Communications Access Method
VICAR	Video Image Communication And Retrieval
VICC	Visual Information Control Console
VICNET	Victoria's Network (AU)
VICP	VINES Internet Control Protocol (VINES = Virtual Networking System)

VICS	Vehicle Information Communication System
VID	Video
VIDAT	Visual Data Acquisition
VIDC	Video Controller
VIDCA	Video Cassette
VIDEO	Visual Data Entry On-line
VIDEOS	Visual Data Entry On-line System
VIE	Virtual Information Environment
VIEW	Visible, Informative, Emotionally appealing, Workable
VIEWS	Virtual Environment Workstation
VIF	Virtual Interface
VIF	Virtual Interrupt Flag
VIFRED	Visual Forms Editor
VIG	Video Integrating Group
VIL	Vertically Integrated Liquid
VILE	Visual editor Like Emacs
VIM	Video Interface Module
VIMCOS	Vehicle for Investigation of Maintenance Control Systems
VIMTPG	Virtual Interactive Machine Test Program Generator
VINE	Vine Is Not Emacs
VINES	Virtual Networking System
VINES	Virtual Network Software
VINT	Virtual Inter-Network Test bed
VIO	Very Important Object
VIO	Video Input/Output
VIO	Virtual Input/Output
VIP	Value In Performance
VIP	Variable Information Processing
VIP	Vector Instruction Processor
VIP	Verifying Interpreting Punch
VIP	Verify, Interpret, Punch
VIP	Versatile Information Processing
VIP	Video Information Provider
VIP	Video Input Processor
VIP	Video Programming
VIP	VINES Internet Protocol (VINES = Virtual Networking Systems)
VIP	Virtual Interrupt Pending
VIP	Virtual Plane
VIP	Visual Image Processor
VIP	Visual Information Processing|Projection
VIP	Visualized Input
ViP	Visual Programming
VIPC	VINES Inter-Process Communications (VINES = Virtual Networking Systems)
VIPER	Verifiable Integrated Processor for Enhanced Reliability
VIPP	Variable Information Processing Package
VIPS	Variable Item Processing System

VIPS	Voice Information Processing System
VIPS	Voice Interruption Priority System
VI/QCAV	Vision Interface & Quality Control by Artificial Vision (conference)
VIR	Vertical Interval Reference
VIR	Virtual
VIRGIL	Virtual Gnomic Information Library
VIS	Vector Instruction Set
VIS	Verification Information System
VIS	Video Information System
VIS	Viewable Image Size
VIS	Visual Inquiry Station
VIS	Visual Instruction Set
VIS	Visual Instrumentation System
VIS	Voice Identification System
VIS	Voice Information Services
VISA	Virtual Instrument Software Architecture
VISAM	Variable-length Indexed Sequential Access Method
VISAM	Virtual Indexed Sequential Access Method
VISCA	Video System Control Architecture
VISDA	Visual Information System Development Association (US)
VisiCalc	Visible Calculator
Visiscan	Visitor scanning
VISP	Videotext Information Service Provider
VISPA	Virtual Storage Productivity Aid
VISSR	Visible and Infrared Spin-Scan Radiometer
VITAL	Virtually Integrated Technical Architecture Lifecycle
VITC	Vertical Interval Time Code
VITS	Vertical Interval Test Signal
VIU	Voiceband Interface Unit
VIURAM	Video Interface Unit Random Access Memory
VIVA	Viewpoint-Invariant Visual Acquisition (EU)
VIVID	Video, Voice, Image, Data
VJ++	Visual Java plus-plus
VJHC	Van Jacobson Header Compression
VL	Vector Length
VL	Virtual Link
VL	Visual Languages (symposium)
VLA	Very Large Array
VLA	Virtual Learning Agent
VLAM	Variable Level Access Method
VLAN	Very Large Area Network
VLB	VESA Local Bus
VLBA	Very Long Baseline Array
VLBI	Very Long Baseline Interferometry
VL-BUS	VESA Local-Bus
VLC	Variable Length Code
VLC	Videotext Library Center

VLCBX	Very Large Computerized Branch Exchange
VLCM	Very Large Capacity Memory
VLD	Valid
VLD	Variable Length Decoder
VLDB	Very Large Data Base
VLDL	Very Low Density Lipoprotein
VLF	Variable Length Field
VLF	Very Low Frequency
VLISP	Visual LISP (= List Processor)
VLIW	Very Large|Long Instruction Word
VLLC	Very Low Level Compatibility
VLM	Variable-Length Multiply
VLM	Very Large Memory
VLM	Virtual Loadable Module
VLP	Video Long Player
VLR	Variable Length Record
VLR	Very Long Range
VLR	Very Low Resistance
VLR	Visitor Location Register
VLS	Very Long Shot
VLS	Virtual Linkage Subsystem
VLSI	Very Large Scale Integration
VLSIPS	Very Large Scale Immobilized Polymer Synthesis
VLSW	Virtual Line Switch
VLT	Variable List Table
VLT	Video Layout Terminal
VLT	Video Lookup Table
VLTP	Variable Length Text Processor
VLW	Visible Language Workshop
VM	Vertical Merger
VM	Virtual Machine|Memory
VM	Voice Mail|Message
VM	Voice Messaging
VM	Voltage Memory
VMA	Valid Memory Address
VMA	Virtual Machine Assist
VMA	Virtual Memory Address|Allocation
VMA	Virtual Memory Area
VMAPS	Virtual Memory Array Processing System
VM/AS	Virtual Machine/Application System
VMB	Virtual Machine Boot
VMBLOK	Virtual Machine control Block
VM/BSE	Virtual Machine/Basic System Extension
VMC	Vertical Machining Center
VMC	Vertical Micro-Code
VMC	Virtual Memory Computer
VMCB	Virtual Machine Control Block
VMCC	Videotext Management and Control Center

VMCF	Virtual Machine Communication Facility
VM/CMS	Virtual Machine/Conversation(al) Monitor System
VMCP	Virtual Machine Control Program
VMD	Vector Memory Display
VMD	Virtual Manufacturing Device
VMD	Virtual Model Display
VME	Versabus Module Europe
VME	Versa Modular Europa
VME	Virtual Memory Environment
VMEbus	Versa Modular European bus
VMF	Variable Message Format
VMF	Virtual Machine Facility
VM/HPO	Virtual Machine/High Performance Option
VMID	Virtual Machine Identification
VM/IFS	Virtual Machine/Interactive File Sharing
VML	Virtual Markup Language
VML	Virtual Memory Level
VMM	Variable Mission Machining
VMM	Virtual Machine Manager|Monitor
VMM	Virtual Memory Manager
VMMAP	Virtual Machine Monitor Analysis Program
VM-MS	Virtual Machine – Memory System
VMOS	Vertical|V-groove Metal Oxide Silicon
VMOS	Virtual Memory Operating System
VMOSFET	Vertical Metal-Oxide Semiconductor Field-Effect Transistor
VMP	Virtual Multi-Processor
VMPE	Virtual Memory Performance Enhancement
VMPPF	Virtual Machine Performance Planning Facility
VMR	Vertical Market Reveler
VMR	Violation Monitor and Remover
VMRS	Voice Message Relay System
VM/RSP	Virtual Machine/Remote System Programming
VM/RTM	Virtual Machine/Real-Time Monitor
VMS	Variable Mesh Simulator
VMS	Velocity Measurement System
VMS	Vertical Motion Simulator
VMS	Video Motion Sampler
VMS	Virtual Mass Storage
VMS	Virtual Memory System
VMS	Voice Mail|Message System
VMS	Voice Management System
VMSD	Visualization and Media Systems Design
VM/SE	Virtual Machine/System Extension
VM/SP	Virtual Machine/System Product
VMS/XA	Virtual Memory System with External Addresses
VMT	Variable Micro-cycle Timing
VMT	Virtual Memory Technique
VMTL	Voltage Mode Transistor Logic

VMTP	Versatile|Virtual Message Transaction Protocol
VM/TSO	Virtual Machine/Time-Sharing Option
VMUIF	Voice Mail User Interface
VM/VSE	Virtual Machine/Virtual Storage Extended
VM/XA	Virtual Machine/Extended Architecture
VN	Viet Nam
VNA	Very New Amplifiers
VNA	Virtual Network Architecture
VNC	Voice Numerical Control
VNC	VTAM Network Controller (VTAM = Virtual Terminal Access Method)
VNCA	VTAM Node Control Application
VNET	Virtual Network
VNF	Virtual Network Feature
VNM	Virtual Network Monitor
VNS	Virtual Network System
VNUG	Vancouver Netware Users' Group
VOA	Volt-Ohm-Ammeter
VoATM	Voice over Asynchronous Transfer Method
VOB	Vacuum Optical Bench
VOBANC	Voice Band Compression
VOC	Variable Output Circle|Circuit
voc	voice (files) (file extension indicator)
VOCAB	Vocabulary
VOCAL	Vocabulary Language
vocoder	voice coder
VOCOM	Voice Communications
VOD	Vertical Obstruction Detector
VOD	Video On Demand
VODACOM	Voice Data Communications
VODAS	Voice Operated Device Anti-Singing
VODAS	Voice Over Data Access Station
VODAT	Voice Operated Device for Automatic Transmission
VODER	Voice Coder
VODIS	Voice Operated Database Inquiry System
VODSL	Voice Over Digital Subscriber Line
VOF	Variable Operating Frequency
VOFDM	Vector Orthogonal Frequency Division Multiplexing
VOFR	Voice Over Frame Relay
VOGAD	Voice Operated Gain-Adjusting Device
VOGAD	Voice Operated Gain(ing) Adjustment Device
VOI	Value Of Identification
VOICE	Validating OSA in Industrial CIM Environments (OSA = Open System Architecture; CIM = Computer Integrated Manufacturing)
VOICE_II	Validating OSA in Industrial CIM Environments by Integration and Implementation (EU)
VOIN	Vehicle-Oriented Inertial Navigation

VOIP	Voice Over Internet Protocol
VOIS	Voice-Operated Information System
vol	volume
VOLASTOR	Volatile Storage
VOLCAL	Volume Calculator
VOLID	Volume Identifier
VOLLAB	Volume Label
VOLSER	Volume/Serial
volt	voltage
VOLT	Volume Table
VOM	Volt-Ohm Meter
VOM	Volt-Ohm-Milliammeter
VOMM	Volt-Ohm-Millimeter
VON	Voice On the Net
VOP	Velocity Of Propagation
VOP	Visualized Output
VOR	Very high frequency Omni-Range
VOR	Voice Operated Relay
VORT	Very Ordinary Rendering Toolkit
VORT	Voltage Regulator Tester
VOS	Verbal|Virtual Operating System
VOS	Voice Operated Switch
VOSIM	Voice Simulation
VOSL	Variable Operating and Safety Level
VOX	Voice Operated Circuit|Control
VOX	Voice Operated Keying
VOX	Voice Operated Transmission
voxel	volume element
VP	Vector Processor
VP	Verification Program (branch)
VP	Verifying Punch
VP	Vertical Parity
VP	View Processor
VP	Virtual Path|Prototyping
VP	Virtual Processing|Processor
VPAM	Virtual Partitioned Access Method
VPC	Vertical Parity Check
VPC	Virtual Path Connection
VPC	Virtual Processor Complex
VPC	Voice Processing Computer
VPC	Voltage to Probability Converter
VPCB	Virtual Processor Control Block
VPD	Virtual Printer Device
VPD	Vital Product Data
VPDM	Virtual Product Development Management
VPDN	Virtual Private Data Network
VPDS	Virtual Private Data Service
VPE	Vector Processing Element

VPE	Visual Programming Environment
VPF	Vector Parameter File
VPF	Vector Product Format
VPF	Vertical Processing Facility
VPI	Virtual Path Identifier
VPIB	Virtual Processor Identification Block
VPID	Virtual Processor Identification
VPIM	Volume Product Inventory Management
VPL	Virtual Programming Language
VPLS	Volume Product Lead-time Scheduling
VPM	Versatile Packaging Machine
vpm	vibrations per minute
VPM	Virtual Processor Monitor
VPM	Virtual Product Model
VPN	Vickers Pyramid Number
VPN	Virtual Page Number
VPN	Virtual Private Network
VPOE	Volume Product Order Entry
VPP	Velocity Prediction Program
VPR	Variable Point Representation
VPS	Vector Processing System
vps	vibrations per second
VPS	Video Programming System
VPS	Voice Processing System
VPSC	Vault, Process, Structure, Configuration
VPSC	Volume Product Stock Control
VPSS	Vector Processing Subsystem Support
VPSW	Virtual Program Status Word
VPT	Virtual Path Terminator
VPT	Virtual Print(er) Technology
VPTR	Value Pointer
VPU	Video Programming Unit
VPU	Virtual Processing Unit
VPW	Virtual Paper Writer
VPZ	Virtual Processing Zero
VQ	Vector Quantization
VR	Variety Reduction
V=R	Virtual Equals Real
VR	Virtual Reality|Route
VR	Visible Record
VR	Voltage Regulator
VRA	Value Received Analysis
VRAD	Very Rapid Application Development
VRAIS	Virtual Reality Annual International Symposium
VRAM	Variable Random Access Memory
VRAM	Variable Rate Adaptive Multiplexing
VRAM	Video Random Access Memory
VRAP	Vertical Registration Analysis Program

VRC	Vector-to-Raster Converter
VRC	Vertical Redundancy Character|Check(ing)
VRC	Video Recorder
VRC	Visible Record Computer
VRCC	Vertical Redundancy Check Character
VR-DIS	Virtual Reality – Design Information System
VRE	Visual Resource Editor
VRE	Voltage Regulated Extended
VREF	Voltage Reference
VRFN	Verification
VRID	Virtual Route Identifier
VRM	Virtual Resource Manager
VRM	Visitor Relationship Management
VRM	Voice Recognition Module(s)
VRM	Voltage Regulator Module
VRML	Virtual Reality Markup|Modelling Language
VRMS	Volt Root Mean Square
VROOMM	Virtual Real-time Object Oriented Memory Manager
VRP	Video RISC Processor (RISC = Reduced Instruction Set Computer)
VRP	Visual Record Printer
VRP	Vital Record Protection
VRR	Visual Radio Range
VRS	Voice Recognition|Recording System
VRS	Voice Response System
VRT	Visual Reaction Time
VRT	Voice Response Terminal
VRT	Voltage Regulation Technology
VRTX	Versatile|Virtual Real-Time Executive
VRU	Voice Response Unit
VRUP	VINES Routing Update Protocol (VINES = Virtual Networking System)
VRX	Virtual Resource Executive
VS	VAX-Station (VAX = Virtual Address Extension)
VS	Virtual Storage|System
VS	Visual Surveillance (international workshop)
VS	Vocal Synthesis
VSA	Value Systems Analysis
VSA	Visual Scene Analysis
VSA	Voice-Stress Analyzer
VSA	Voltage Sensitive Amplifier
VSAG	VHDL Standards Acceptance Group
VSAM	Virtual Sequential Access Method
VSAM	Virtual Storage|System Access Method
VSAT	Very Small Aperture (satellite) Terminal
VSB	Vestigial Side-Band
VSBS	Very Small Business System
VSC	Video Scan Converter

VSC	Virtual Subscriber Computer
VSC	Vision System Computer
VSCF	Variable-Speed Constant Frequency
VSD	Variable Slope Delta
VSD	Vertical Situation Display
VSD	Very Small Device
VSD	Virtual Shared-Disk
VSDM	Variable Slope Delta Modulation
VSDT	Very Small Device Technology
VSE	Virtual Storage Extended
VSE/PT	Virtual Storage Extended/Performance Tool
VSE/SP	Virtual Storage Extended/System Package
VSF	Vertical Scanning Frequency
VSF	Vestigial Sideband Filter
VSF	Virtual Software Factory
VSF	Voice Store-and-Forward (system)
VSI	Virtual Socket Initiative
VSI	Visual Simulator Interface
VSIO	Virtual Serial Input/Output
VSL	Voltage Sense Level
VSM	Video Switching Matrix
VSM	Virtual Shared Memory
VSM	Virtual Storage Management
VSM	Visual System Management
VSOP	Very Special Old Product
VSOS	Virtual Storage Operating System
VSP	Vector/Scalar Processor
VSPC	Virtual Storage Personal Computing
VSPP	VINES Sequenced Packet Protocol (VINES = Virtual Networking System)
VSPT	Virtual Storage Performance Tool
VSR	Validation Summary Report
VSR	Virtual Service Representative
VSR	Voice Storage and Retrieval
VSS	Video Storage System
VSS	Virtual Storage|Support System
VSS	Voice Signalling|Storage System
VSS	Voltage for Substrate and Sources
VSSP	Voice Switch Signalling Point
VST	Video Scroller Terminal
VSW	Very Short Wave
VSWR	Voltage/Standing Wave Ratio
VSX	Verification Suite for X/open
VSX	Videotext Storage and Exchange system
VSYNC	Vertical Synchronization
VT	Vacuum Tube
VT	Validation Test
VT	Variable Time

VT	Vermont (US)
VT	Vertical Tab(ulation)
VT	Video|Virtual Terminal
VT100	Video Terminal 100
VT220	Virtual Terminal 220
VTA	Variable Transfer Address
VTA	Virtual Terminal Agent
VTAB	Vertical Tabulation (character)
VTAC	Video Timing And Controller
VTAM	Virtual Telecommunication|Terminal Access Method
VTAM	Vortex Telecommunications Access Method
VTAME	Virtual Telecommunication Access Method Entry
VTB	Video Terminal Board
VTC	Vertical Tabulation Character
VTC	Video Tele-Conference
VTC	Videotext Terminal Concentrator
VTC	Viewdata Terminal Controller
VTC	Virtual Terminal Control
VTC	Volume Texture Compression
VTCP	VTAM Terminal Control Program (VTAM = Virtual Terminal Access Method)
VTD	Variable Time Delay
VTD	Vertical Tape Display
VTDI	Variable Threshold Digital Input
VTE	Visual Task Evaluation|Evaluator
VTEP	Virtual Terminal Emulation Package
VTF	Vertical Tracking Force
VTF	Voltage Transfer Function
VTI	Video Terminal Interface
VTI	Virtual Terminal Interface
VTI	Voluntary Termination Incentive
VTIGRE	Vacuum Time-Invariant Gray Radiance Equation
VTIOC	VTAM Terminal Input/Output Coordinator (VTAM = Virtual Terminal Access Method)
VTL	Variable Threshold Logic
VTL	Vendor Transistor Logic
VTLC	Virtual Terminal Line Controller
VTM	Vocal Tract Models
VTMS	Voice/Text Message System
VTO	Voltage-Tuned Oscillator
VTOA	Voice and Telephony Over ATM (= Asynchronous Transfer Method)
VTOC	Volume Table Of Contents
VTP	Verification Test Plan
VTP	Viewdata Terminal Programming
VTP	Virtual Terminal Program|Protocol
VTR	Video Tape Recorder
VTRS	Video Tape Recording System

VTRU	Variable Threshold Recently Used
VTS	Video Teleconferencing System
VTS	Virtual Teaming System
VTS	Virtual Terminal Service
V+TU	Voice plus Teleprinter Unit
VTVM	Vacuum Tube Volt-Meter
VTX	Videotext
VU	Vanuatu
VU	Vertical arithmetic Unit
VU	Voice|Volume Unit
VUE	Visual User Environment
VUI	Video User Interface
VUIT	Visual User Interface Tool
VUV	Vacuum Ultra-Violet
V-V	Velocity-Volume
V&V	Verification & Validation
V=V	Virtual Equals Virtual
VV	Volume in Volume
VV&C	Verification, Validation & Certification
VVC	Voltage-Variable Capacitor
VVCD	Voltage-Variable Capacitor Diode
VVLSI	Very Very Large Scale Integration
VVM	Valve Voltmeter
VVR	Voltage-Variable Resistor
VVV	test signal
VW	Volts Working
VWB	Visual Work Bench
VWC	Virtual Workstation Concept
VWS	Variable Word Size
VWS	VAX Workstation Software (VAX = Virtual Address Extension)
VXD	Virtual Extended (device) Driver
VxD	Virtual extended Driver
VXD	Virtual X Driver
VxDS	Virtual extended device Drivers
VxFS	Veritas File System
VXI	VMEbus Extension for Instrumentation (VMEbus = Versa Modular Europa bus)
VXO	Variable Crystal Oscillator
VY	Very
VZ	Virtual Zero

W

W	Wait (time)
W	Watt
w	week(s)
w	weight
W	Width

W	Wolfram (tungsten)
W	Word
W	Work
W	Write
W2k	Windows 2000
W3	World Wide Web
W3A	World Wide Web Applets
W3C	World Wide Web Consortium
W4	What-Works-With-What
WA	Washington (US)
WA	Wave
WA	Weighted Average
WA	Word Added|Address
WA	Write Audit
WAA	Wide-Angle Acquisition
WAAS	Wide Area Augmentation System
WAB	Was Authorized By
WABI	Windows Application Binary Interface
WAC	Wide Area Collimated
WAC	Worked All Continents
WAC	Working Address Counter
WAC	Write Address Counter
WACK	Wait (before transmit positive) Acknowledgment
WACK	Wait for Acknowledgment
WACS	Wire Automated Check System
WADS	Wide Area Data Service
WAE	Wireless Application Environment
WAF	With All Faults
WAF	Word Address Format
WAFL	Write Anywhere File Layout
WAH	Working At Home
WAI	Wait for Interrupt
WAI	Work Area Information
WAIS	Wide Area Information Server|Service
WAITS	Wide Area Information Transfer System
WAK	Wait Acknowledge
WAK	Write Access Key
WAKPAT	Walking Pattern
WALT	Write Ahead Log Tape
WAM	Wave Model
wam	words a minute
WAM	Work Analysis and Measurement
WAN	Wide Area Network
WANKER	Wide Area Network Key Engineering Resource
WAP	Wireless Access|Application Protocol
WAPI	Windows Application Programming Interfaces
WAR	With All Risks
WARC	World Administrative Radio Conference

WARE	Wafer Alignment for Resist Exposure
WARES	Workload And Resources Evaluation System
WARF	Weekly Audit Report File
WARP	Wavelength Routing Protocol
WARP	Widely Adaptive & Responsive Display
WARP	Worldwide Automatic digital information network Restoration Plan
WAS	Wax In Shell
WAS	Worked All States
WASAR	Wide Application System Adapter
WASP	Work Activity Sampling Plan
WASP	Workshop Analysis and Scheduling Programming
WASU	Wireless Access Subscriber Unit
WATCC	World Administrative Telephone and Telegraph Conferences
WATER	Wonderful And Total Extinguishing Resource
WATFOR	Waterloo Fortran
WATS	Wide Angle Transmission Service
WATS	Wide Area Telecommunications|Telephone Service
WATTC	World Administrative Telegraph and Telephone Conference
WAU	Wavelet Audio Unit
WAUCS	Wiring Alternate Unit Codes
wav	waveform audio/video (file extension indicator)
WAVES	Wireless Audio Visual Emergency System
WAW	Waiter-Actor-Webmaster
WAWOT	What A Waste Of Time
WAZ	Worked All Zones
WB	Weber
WB	Welcome Back
WB	Workbench
WB	Write Buffer
WBA	Wire Bundle Assembly
WBAT	Wide-Band Adapter Transformer
WBC	Wide-Band Channel|Coupler
WBCT	Wide-Band Current Transformer
WBDX	Wide-Band Data Switch
WBEM	Web-Based Enterprise Management
WBIS	Wide-Band Information System
WBL	Web-Based Learning
WBL	Weird Blinking Lights
WBL	Wide-Band Lines
WBL	Work-Based Learning
WBM	Web-Based Management
WBR	Word Buffer Register
WBS	Webchat Broadcasting System
WBS	Work Breakdown Structure
WBTS	Wide-Band Transmission System
WC	Wire Center|Contact
WC	Wired Control

WC	Without Charge
WC	Word Count
WC	Workstation Cluster
W&C	Write & Compute
WC	Write Control
WCB	Way Control Block
WCB	Will Call Back
WCC	Wire Common Carrier
WCC	Word Count Cycle
WCC	Work Center Code
WCC	Work Control Center
WCC	World Congress on Computing
WCC	Write Control Channel|Character
WCCE	World Conference on Computers in Education
WCDB	Work Control Data Base
WCDMA	Wideband Code Division Multiple Access
WCET	Worst Case Execution Time
WCF	Workload Control File
WCGM	Writable Character Generation Memory
WCK	Write Clock
WCL	Word Control Logic
WCM	Wired Core Matrix|Memory
WCM	Word Combine and Multiplexer
WCM	Writable Control Memory
WCO	Wired Control Operator
WCP	Unix-to-Unix Copy Program
WCPC	Wire Center Planning Center
WCR	Wire Contact Relay
WCR	Word Control|Count Register
WCS	Work Control System
WCS	World Coordinate System
WCS	Writable Control Stor(ag)e
WCSSS	Workshop on Compiler Support for Systems Software
WC/WS	Work Center/Work Station
WD	When Distributed
WD	Width
WD	Wiring Diagram
WD	Word
WD	Working Dimension(s)
WD	Working Document|Draft
WD	Write Data
WDA	Write Data
WDB	Word Driver Bit
WDB	Working Data Base
WDC	Washington, District of Columbia
WDC	Wideband Directional Coupler
WDC	Working Device Code
WDC	World Data Center

WDC	Write Data Channel
WDCS	Writable Diagnostic Control Store
WDDX	Web Distributed Data Exchange
WDEF	Window Definition
WDG	Winding
WDIR	Working Directory
WDL	Windows Driver Library
WDL	Wireless Data Link
WDM	Wave(length-)Division Multiplexing
WDMK	Word Mark
WDRAM	Windows Dynamic Random Access Memory
WDT	Warning Display Terminal
WDT	Watch Dog Timer
WDT	Width
WDV	Written Down Value
WDYMBT	What Do You Mean By That?
WE	With Equipment
WE	Write Enable
WebCT	World wide web Course Tool
WebDAV	Web Distributed Authoring and Versioning
WebNS	Web Network Service(s)
WebQoS	Web Quality of Service
WEBSOM	Web Self-Organizing Map
WECA	Wireless Ethernet Compatibility Alliance
WEDM	Wire Electrical Discharge Machine
WEEB	Western European EDIFACT Board
WELL	Whole Earth 'Lectronic Link
WEM	Welded Encapsulated Module
WEM	Write Enable Mask
WEN	Waive Exchange if Necessary
WENUS	Weekly Estimate (of) Net Usage System
WEP	Well-known Entry Point
WEP	Wired Equivalent Privacy
WER	Write Enable Ring
WESS	Workshop on Empirical Studies of Software maintenance
WESTAR	Western Union Satellite
WETARFAC	Work Element Timer And Recorder For Automatic Coupling
WETICE	Workshop on Enabling Technologies: Infrastructure for Collaborative Enterprises
WEU	Western European Union
WF	Wallis and Futuna Islands
WF	Web Farming
WF	Write Fault|Forward
WF	Wrong Font
WFAD	Work Force Adjustment Directive
WFC	Windows Foundation Classes
WFD	Working Form Database
WFF	Well-Formed Formula

WFFS	Waiting For First completed Station
WFL	Work Flow Language
WfM	Wired for Management
WFM	Work Flow Management
WFMC	Work Flow Management Coalition
WFMS	Work Flow Management System
WFPC	Wide Field/Planetary Camera
WFQ	Weighted Fair Queuing
WFW	Windows For Workgroups
WfWG	Windows for Work Groups
WG	Weight Guaranteed
WG	Wire Gauge
WG	Working Group
WG	Write Gate
WG-ARC	Working Group on Architectures
WGC	Working Group on CASE (= Computer Aided Software Engineering)
WGD	Working Group on Data
WGIR	Working Group on Interface Registries
WGS	Work-Group System
WH	Watt-Hour
WHAM	Waveform Hold And Modify
WHC	Workstation Host Connection
WHP	Water Horse-Power
WHSE	Warehouse
WI	When Issued
WI	Wisconsin (US)
WI	Word Intelligibility
WIAPP	Workshop on Internet Applications
WIB	When Interrupt Block
WIBFD	Will Be Forwarded
WIBIS	Will Be Issued
WIBNI	Would It Be Nice If
WIC	WAN Interface Card
WID	Window Identifier
WIDL	Web Interface Definition Language
WID_RAM	Window Identifier Random Access Memory
WIE	Wrong Input Error
WiFi	Wireless Fidelity
WILCO	Will Comply
WILLOW	Washington Information Looker-upper Layered Over Windows
WILMA	Wissensbasiertes LAN Management (DE)
WIM	Write Inhibit Mask
WIMC	Whom It May Concern
WIMP	Windows, Icons, Mice and Pointers|Programs
WIMP	Windows, Icons, Mice and Pull-down menu
WIMP	Windows/Icons/Mouse Programming
WIMPS	Windows, Icons, Mice and Pointers

WIMPS	Windows, Icons, Mice, Pointers, Scroll bars
WIN	Wissenschaftsnetz (DE)
WINC	Wideband Infrared Communication
WINCIM	Windows Compuserve Information Manager
WINCS	WWMCCS Intercomputer Network Communication Subsystem (WWMCCS = Worldwide Military Command and Control System)
WinDD	Windows Distributed Desktop
WINDO	Wide Information Network for Data On-line
Windows_DNA	Windows Distributed inter-Net applications Architecture
WINE	Windows Emulator
WINForum	Wireless Information Networks Forum
WinHEC	Windows Hardware Engineering Conference
WINS	Warehouse Information Network Standard
WINS	Windows-Internet Naming Service
WINS	Wollongong Integrated Networking Solutions
WINSOCK	Windows Sockets
WINWORD	Word for Windows
WIP	Work In Process|Progress
WIP	Wrist Interface Plate
WIPO	World Intellectual Property Organization (UN)
WIR	Wiring
WIR	Write Inhibit(ing) Ring
WIRDES	Wire Description
WISDM	World-Class Information Strategy for Data-driven Marketing
WISE	WordPerfect Information System Environment
WISE	World Information Systems Exchange
WISP	Waves In Space Plasma
WIT	Web Interactive Talk
WIT	Weekly Information Tape
WITS	Washington Interagency Telecommunications System
WITS	Windows Integrated Test Suite
WITSA	World Information and Technology and Services Alliance
WK	Week
WK	Work
wk1	worksheet (file extension indicator)
wkb	workbook (file extension indicator)
wke	worksheet (file extension indicator)
WKG	Working (storage)
wkq	spreadsheet (file extension indicator)
WKS	Well Known Services
wks	worksheet (file extension indicator)
wkz	wordperfect compressed spreadsheet (file extension indicator)
WL	Wavelength
WL	Wiring List
WL	Word Line
WLAN	Wireless Local Area Network
WLAS	Waiting at Last Arrival Station

WLL	Wireless Local Loop
WLM	Wire Line Modems
WLN	Western Library Network
WLN	Wiswesser Line Notation
WLOG	Without Loss Of Generality
WLR	Wrong Length Record
WLT	Wire Laying and Termination
WM	Wattmeter
WM	Word Mark
wm	words per minute
WM	Work Manager
WM	Work Measurement
WMATM	Wireless Mobile Asynchronous Transfer Mode
WMC	Workflow Management Coalition
wmf	windows metafile (file extension indicator)
WMF	Windows Metafile Format
WMF	Workload Monitoring Function
WML	Wireless Markup Language
WMRM	Write Multiple, Read Multiple
WMS	Windows Media Services
WMS	Wire Matrix Switch
WMS	Work Measurement System
WMSCR	Weather Message Switching Center Replacement
WMT	Windows Media Technology
WNG	Wrong
WNIC	Wide-area Network Interface Coprocessor
WNIM	Wide-area Network Interface Module
WNL	Within Normal Limits
WNP	Will Not Process
WNT	Windows New Technology
WO	Wait Order
WO	Wipe(d) Out
W/O	Without
WO	Work Order
WO	Write Off|Out
WO	Write Only
WOAD	World Offshore Accident Databank
WOC	Wafer Order Control
WOCS	Work Order Control System
WOE	Without Equipment
WOFACS	Workshop On Formal and Applied Computer Science
WOLAP	Workplace Optimization and Layout Planning
WOM	Write Only Memory
WOMBAT	Waste Of Money, Brains, And Time
WOMS	Works On My System
WON	Waiver Of Notice
WOOD	Write Once Optical Disk
WOPAST	Work Plan Analysis and Scheduling Technique

WOPE	Without Personnel and Equipment
WORAM	Word Oriented Random Access Memory
WORDS	Workshop on Object-oriented Real-time Dependable Systems
WORKSTOR	Working Storage
WORLDCOM	World Communications
WORM	Write Once/Read Many (memory)
WORM	Write Only (once) Read Many (times)
WORM	Write Only Read Multiple
WORP	Word Processing
WOS	Word Organized Storage
WOS	Workstation Operating System
WOSA	Windows Open Services|System Architecture
WOT	Word Overlap Trigger
WOTAN	Workstations der Technischen Am Netz (DE)
WOTC	Wizards Of The Coast
WOTW	Women Of The Web
WOUDE	Wait On User Defined Event
WOW	Way Of (the) Web
WP	White Pages
WP	WordPerfect
WP	Word Processing|Processor
WP	Working Paper|Party
WP	Workspace Pointer
WP	Write Protected|Protection
WPA	With Particular Average
WPA	Works Progress Administration
WP/AS	Word Processing/Administrative Support
WPB	Write Printer Binary
WPBX	Wireless Private Branch Exchange
WPC	Wired Program Computer
WPC	Word Processing Center
WPCF	Word Processing Control Function
wpd	word processing document (file extension indicator)
WPD	Write Printer Decimal
WPDA	Writing Push-Down Acceptor
WPDN	Wind Profiler Demonstration Network
wpg	wordperfect graphics (file extension indicator)
WPHD	Write-Protect Hard Disk
WPI	World Patent Index
wpk	wordperfect keyboard (file extension indicator)
WPL	Wire and Plug Lists
WPLOT	Wiring Plot
wpm	wordperfect macro (file extension indicator)
wpm	words per minute
WPM	Work Package Management
WPM	Write Program|Protect Memory
WPMA	Windows and Presentation Manager Association
WPMA	Workshop on Perception for Mobile Agents

WPOE	Word Processing and Office Equipment
WPOS	Work Place Operating System
WPR	Word Processor
WPR	Write Permit(ting) Ring
WPS	Windows Printing System
WPS	Word Processing Software|System
wps	words per second
WPS	Workplace Shell
wpt	word processing template (file extension indicator)
WPU	Write Punch
WPVM	Windows Parallel Virtual Machine
wq!	wordperfect compressed spreadsheet (file extension indicator)
wq1	spreadsheet (file extension indicator)
WR	Wire Recorder
WR	Word Request
WR	Working Register
WR	Write
W/R	Write/Read
WRAIS	Wide Range Analog Input Subsystem
WRAM	Windows Random Access Memory
WRAP	Weighter Record Analysis Program
WRBND	Wire Bound
WRC	Workstation Resource Center
WRCALL	Write Call
WRE	Write Enable
WRG	Within Range
wri	write file (file extension indicator)
WRIPS	Wave Rider Information Processing System
WRIU	Write Interface Unit
WRK	Work
WRM	World Reference Model
WRPT	Write Protect(ion)
WRS	Word Recognition System
WRS	Working Reference System
WRSTAT	Write Status
WRT	With Regard(s)|Respect To
WRT	Write
WRTC	Write Control
WRTH	World Radio Television Handbook
WRU	Who Are You?
WS	Samoa
WS	Weakest Source
WS	Word Select|Station
WS	WordStar
WS	Word Synchronization
WS	Work(ing) Space
WS	Working Storage
WS	Work Station

WSC	Work Station Control
WSD	Work Station Division
WSDCU	Wideband Satellite Delay Compensation Unit
WSDL	Wire Sonic Delay Lines
WSE	Work Station Element
WSF	Work Station Function
WSF	Workstation Search Facility
WSI	Wafer Scale Integration
WSN	Wirth Syntax Notation
WSP	Wireless Session Protocol
WSS	Work Summarization System
WSSE	Working Set Size Estimate
WST	Word Study
WST	Word Synchronizing Track
WST	Work Station Terminal
WSU	Work Station Utility
WT	Wait(ing) Time
WT	Walk Through
WT	Waveguide Transmission
WT	Weight
WT	Wireless Telegraphy
WT	Wire Tap
WT	Without Thinking
WT	Word Terminal
WT	Word Type
WT	Write Through
WTB	Wanted To Buy
WTBD	Work To Be Done
WTC	Wire Through Connection
WTD	World Telecommunications Directory
WTG	Way To Go
WTLS	Wireless Transport Layer Security
WTM	Write Tape Mark
WTO	Write To Operator
WTOR	Write To Operator with Reply
WTP	Wireless Transaction Protocol
WTP	Write To Programmer
WTR	Writer
WTS	Word Terminal, Synchronous
WTT	Working Time-Table
WTTM	Without Thinking Too Much
WU	Where Used
WU	Write-Up
WUBR	Weighted Unspecified Bit Rate
WUC	Where Used Code
WUDB	Work Unit Data Bank
WUF	Where Used File
WUGNET	Windows Users Group Network

WUPS	Windows UDP Port Scanner (UDP = User Datagram Protocol)
WUS	Where Used Suppress
WUS	Word Underscore (character)
WV	West Virginia (US)
WV	Working Voltage
WVdc	Working Voltage for direct currents
WVPN	Worldwide Virtual Private Network
WVR	Within Visual Range
WW	Wagner-Whitin
WW	Wire Wrap
WW	With Warrants
WWC	Wire Wrap Connection
WWF	World Wide Fulfillment
W-WIRE	Word Wire
WWIS	World Wide Information System
WWMI	World-Wide Master Index
WWOPS	World-Wide Operations
WWR	Write While Read
WWW	World Wide Web
WWWW	World Wide Web Worm
WXTRN	Weak External (reference)
WY	Wyoming (US)
WYGIWYS	What You Get Is What You See
WYNIWYG	What You Need Is What You Get
WYNN	What You Need Now
WYSAWYG	What You See Ain't What You Get
WYSBYGI	What You See Before You Get It
WYSIAWYG	What You See Is Almost What You Get
WYSIAYG	What You See Is All You're Getting
WYSISWYG	What You See Is Sort of What You Get
WYSIWIS	What You See Is What I See
WYSIWYG	What You See Is What You Get
WYSIWYM	What You See Is What You Mean
WYSYHYG	What You See You Hope You Get

X

X	Abscissa
X	Execute
X	Executive
X	Extended
X	Horizontal deflection on Cathode Ray Tube
X	Index(ed)
X	Index register
x	multiplication
x	number of carriers
X	Reactance
x	unknown quantity

X.11R5	X window system version 11 Revision 5
X.12	Electronic Data Interchange standards committee
X.121	International numbering plan for PDN's (PDN = Public Data Network)
X.20	DTE/DCE interface in PDN (DTE/DCE = Data Terminal Equipment/Data Circuit Equipment)
X.20bis	DTE/V-series modems interface in PDN
X.21	DTE/DCE interface for synchronous operation in PDN
X.21bis	DTE/V-series synchronous modems interface in PDN
X.24	Definitions for DTE/DCE interchange in PDN
X.25	DTE/DCE interface with packet mode PDN's
X.28	DCE/DTE interface: start/stop – PAD (PAD = Public Access Device)
X.29	Procedures for PAD/packet-mode DTE or PAD/PAD communications
X2B	Hexadecimal to Binary
X2C	Hexadecimal to Character
X2D	Hexadecimal to Decimal
X.3	PAD in PDN (PAD = Public Access Device; PDN = Public Data Network)
X.32	Packet mode DTE/packet switched DCE interface (DTE = Data Terminal Equipment; DCE = Data Circuit Equipment)
X3D	Extensible 3-Dimensional
X.400	Protocol Standards for electronic mail interexchange
X.500	Mail and messaging protocol standard
X.75	Procedures for communications between packet switched networks
XA	Auxiliary Amplifier
XA	Cross-Assembler
XA	Extended Architecture|Attribute
XA	Transmission Adapter
XADA	X-caiver's Archaic Digital Anarchy
XALC	Extended Assembler Language Coding
XALS	Extended Application Layer Structure
XAM	External Address Modifier
XAND	Exclusive logical AND-function
XAPIA	X.400 Application Program Interface Association
XASM	Cross-Assembler
XAW	Extended Athena Widget
XB	Crossbar
XBC	External Block Controller
XBF	Experimental Boundary File
XBM	Extended Basic Mode
XBM	Extended Bit Map format
XBM	Extended Buffer Manager
XBM	Extended Burst Mode
XBM	X window Bit Map
XC	Xerox Copy

XCD	Exceed
XCF	Experimental Computing Facility
XCH	Exchange
XCHG	Exchange
XCL	Exclusive
XCMD	External Command
XCMDS	External Commands
XCONN	Cross Connection
XCR	Exclusive Character Register
XCT	X-band Communications Transponder
XCTL	Transfer Control
XCU	Transmission Control Unit
XCVR	Transceiver
XDAP	Execute Direct Access Program
XDF	Extended Density Format
XDFLD	secondary Index Field
XDI	External Data Interface
XDK	Xcert Developer Kit
XDMCP	Extended Display Manager Control Protocol
XDMS	Experimental Data Management System
XDR	Extended Data Representation
XDR	Transducer
xDSL	particular version of a Digital Subscriber Line
XDT	Executive Debugging Tool
Xducer	Transducer
XDUP	Extended Disk Utilities Program
XEC	Execute
XEC	Extended Emulator Control
XED	Extension to Event Detection
XEDIT	Extended Editor
XEQ	Execute
XETB	Transparent End-of-Text Block
XETX	Transparent End-of-Text
XFC	Extended Function Code
XFC	External Function Call
XFC	Transferred charge Call
XFCN	External Function
XFE	X-Front End
XFER	Transfer
XFM	X-band Ferrite Modulator
xfmr	transformer
XFN	Generated Flag Negative
xformer	transformer
XFR	Transfer
XG	Extended Graphics
XGA	Extended Graphics Adapter|Array
XGL	Extended Graphics Language
XGM	X.25 Gateway Module

XGMIDI	Extended General Musical Instrument Digital Interface specification
XHR	Exclusive Halfword Register
XHTML	Extensible Hyper-Text Markup Language
XIC	Transmission Interface Converter
XID	Exchange Identification|Identifier
XIE	Extended Image Extension
XIM	Extended Input Method
XIM	Extended I/O Monitor
XIO	Execut(iv)e Input/Output
XIO	Transfer Input/Output
XIOS	Extended Input/Output System
XIP	Execute In Place
XIS	Cross Interface Switch
xistor	transistor
XIT	Extra Input Terminal
XITB	Transparent Intermediate Block
XL	Cross-reference List
XL	Execution Language
XL	Extend List
XL	Extra Large
XLAT	Translate
Xlib	Extended library
XLL	Extensible Link Language
XLM	Excel Macro language
XLPE	Cross Linked Poly-Ethylene (cable)
XLR	Extra Long Run
XLSI	Extremely Large Scale Integration
XM	Expanded Memory
XMD	Extensible Module Designer
XMH	Extended Mail Handler
XMI	Extended Memory Interconnect
XMI	XML Meta data Interchange
xmission	transmission
xmit	transmit
xmitter	transmitter
XML	Extended|Extensible Markup Language
XML-AS	Extended Markup Language – Application Server
XML-DTD	Extended Markup Language – Data Type Definition
XMM	Extended Memory Manager
XMP	Experimental Mathematical Programming system
XMS	Extended Memory Specification
XMS	Extended Multiprocessor operating System
XMSN	Transmission
XMT	Transmit(ter)
XMTLMT	Transmission Limit
XMTR	Transmitter
XMT-REC	Transmit/Receive

XMTR-REC	Transmitter/Receiver
XMUX	X.25 Multiplexer
XN	Execution Node
XNA	Xerox Network Architecture
XNOS	Experimental Network Operating System
XNS	Xerox Network Services
XNS/ITP	Xerox Network Services/Internet Transport
XOFF	Transmitter Off
XOI	Exclusive-OR Inverter
XON	Transmitter On
XOP	Extended Operation
XOR	Exclusive logical OR-function
XORL	Exclusive OR Latch
XOT	Extra Output Terminal
xover	crossover
xovr	cross over
XP	Extreme Programming
XPD	Cross Polarization Discrimination
XPD	Expedite
XPDR	Transponder
XPG	X/open Portability Guide
XPL	Explanation
XPM	Extended Pixel Map
XPNF	Expedited Negative Flag
xponder	transponder
XPORT	Transport
XPSDU	Expedited Presentation Service Data Unit
XPSW	External Processor Status Word
XPT	External Page Table
XPU	Extended Pick-Up
XPYB	X Position Y Bit
XQ	Cross Question
XQL	Structured Query Language
XR	Exclusive|Extended Register
XR	External Reset
XR	Index Register
XRC	Extended Remote Copy
XRDY	External Ready
XREF	Cross-Reference
XRF	Extended Recovery Facility
XRF	X-Ray Fluorescence (technique)
XRM	Extended Relational Memory
XRM	Extended Resource Manager
XRN	X/open Reader of News
XRP	X window Record (and) Playback
XRT	Extensions for Real-Time
XRYB	X Register Y Bit
XS	Cross Section

XS–Y

XS	Transform Services
XS4ALL	Access For All
XSECT	Cross Section
X-section	Cross section
XSI	X/open System Interface
XSL	Extensible Style(sheet) Language
XSLT	Extensible Stylesheet Language Transformation
XSMD	Extended Storage Module Drive (interface)
XSP	Extended Set Processor
XSPT	External Shared Page Table
XSSDU	Expedited Session Service Data Unit
XSSO	Cross Systems Sign On
XST	External Segment Table
XSTR	Transistor
XT	Extended Technology
XT	X.11 Toolkit
XTA	X-band Tracking Antenna
XTAL	Crystal
xtalk	cross-talk
XTC	External Transmit Clock
XTCLK	External Transmit Clock
XTENT	Extent
XTERM	Extended Terminal
XTI	X/open Transport(-layer) Interface
XTP	Express Transfer Protocol
XTRAN	Experimental Translation
XTSI	Extended Task Status Index
XU	Extended Unit
XUI	Extended User Interface
XUI	X window User Interface
XUL	Extensible User interface Language
XUV	Extreme Ultra-Violet
XVAR	External Variable
XVERS	Transverse
XVGA	Extended Video Graphics Array
XVIEW	X window system-based Visual/Integrated Environment for Workstations
XVT	Extended Virtual Toolkit
XVTR	Transverter
XWSDS	X Window System Display Station
XX	Double Excellent
XYAT	X-Y Axis Table
XYP	X-Y Plotter

Y

Y	Admittance
Y	Ordinate

Y	Vertical deflection on Cathode Ray Tube
Y	Yellow
Y	Young's modulus
Y2K	Year 2000
YA	Yet Another
YAC	Yeast Artificial Chromosome
YACC	Yet Another Compiler-Compiler
YAFIYGI	You Asked For It, You Got It
YAHOO	You Always Have Other Options
YAM	Yet Another Modem
YAOTM	Yet Another Off-Topic Message
YAP	Yield Analysis Pattern
YATI	Yet Another Trek Inconsistency
YCCAA	Youth Committee of Chinese Association of Automation
YDC	Yaw Damper Computer
YE	Yemen, Republic of
YEC	Youngest Empty Cell
YES	Y2K Expert Service
YES	Yorktown Expert System
YG	Yield Grade
YGWYPF	You Get What You Pay For
YIPES	Young's Internet Protocol Ethernet Service
YIPP	Yield Improvement Prediction Program
YK	Yoke
YMC	Yellow, Magenta and Cyan
YMCK	Yellow, Magenta, Cyan and Black
YMS	Yield Measurement System
YMS	Yorktown Monitor System
YMU	Y-net Management Unit
YN	Yes-No
YOE	Year Of Entry
yon	yonder
YP	Yield Point
yr	year
YR	Your(s)
YRS	Ysgarth Rules System
YS	Yield Strength
YSF	Yield Safety Factor
YSM	Yourdon Structured Method
yT	y-matrix of Transistor
YTC	Yield To Call
YTD	Year To Date
YU	Yugoslavia
YUV	Intensity, Hue and Value
yV	y-matrix of Vacuum tube
YWSYLS	You Win Some, You Lose Some
YY	Year

Z

z	electrochemical equivalent
Z	Impedance symbol
Z	Zero
z	zone
ZA	South Africa
ZA	Zero Adder
ZA	Zero and Add
ZA	Zone Analyzer|Auditor
ZAB	Zinc Air Battery
ZAC	Zero Administration Client
ZAI	Zero Address Instruction
ZAW	Zero Administration for Windows
ZB	Zero Base|Bit
ZBA	Zero-Bracket Amount
ZBR	Zone Bit Recording
ZBTSI	Zero Byte Time Slot Interchange
ZC	Zone Common
ZCO	Zero Cross-Over
ZCR	Zero Crossing Rate
ZD	Zero Defect
ZDBOp	Ziff-Davis Benchmark Operation
ZDL	Zero Delay Lockout
ZDV	Z-Direction Vector(s)
ZE	Zero (balance) Entry
ZEDAT	Zentral Einrichtung für Datenverarbeitung (DE)
zel	z-element
ZF	Zero Frequency
ZFC	Zero Failure Criteria
ZFS	Zenworks For Servers
ZFS	Zone Field Selection
ZFT	Zero Fill Trigger
ZG	Zero Gravity
ZGDV	Zentrum der Graphischen Daten-Verarbeitung (DE)
ZGS	Zero Gradient Synchrotron
ZI	Zero Input
ZIB	Zuse-zentrum für Informationstechnik Berlin (DE)
ZIF	Zero-Insertion Force (socket)
zip	compressed file (file extension indicator)
ZIP	Zigzag In-line Package
ZIP	Zone Improvement Plan|Protocol
ZIS	Zone Information Socket
ZIT	Zone Information Table
ZLA	Zero Level Address
ZLB	Zero Length Buffer
ZM	Zambia

ZMA Zone Multicast Access
ZO Zero Output
ZOD Zero Order Detector
ZOE Zero Energy
ZOH Zero Order Hold
ZOI Zero Order Interpolator
zoo compressed file (file extension indicator)
ZOP Zero Order Predictor
ZPC Zero Print Control
ZPM Zoned Pressure Molding
ZPV Zoomed Port Video
ZR Zaire
ZRE Zero Rate Error
ZRR Zero Remainder Remembered
ZS Zero Shift
ZS Zone Setting
ZSE Zero Suppress End
ZSL Zero Slot LAN
ZSUP Zone Suppression
ZTS Zero To Space
ZW Zimbabwe
ZWC Zero Wind Computer
ZWC Zero Word Count
ZZC Zero-Zero Condition
ZZF Zentralamt für Zulasungen im Fernmeldwesen (DE)

#

0K Zero Kilobytes
0TLP Zero Transmission Level Point
10BASE2 IEEE standard for 10Mb/s thin-wire ethernet with 200 m. segments
10BASE5 IEEE standard for 10Mb/s baseband ethernet with 500 m. segments
10BASEF IEEE standard for 10Mb/s ethernet over fiber optic cable
10BASET IEEE standard for 10Mb/s ethernet over twisted pair cable
1BASE5 IEEE standard for 1Mb/s baseband ethernet with 500 m. segments
1-D One-Dimensional
1f lower operating frequency
1NF First Normal Form
1TR6 Technische Richtlinien ISDN d-kanal protocol (DE) (ISDN = Integrated Services Data Network)
2.5D Two-and-a-half Dimensional
2B1Q Two Binary, One Quaternary
2CS Two Carrier System
2D Double Density
2D Two-Dimensional

2f	higher operating frequency
2NF	Second Normal Form
2PC	Two Phase Commit
3ACC	3A Central Control
3D	Three-Dimensional
3#D	Triple Diffused
3DMF	Three-Dimensional Meta-File
3DP	Three-Dimensional Plotting|Printing
3DTV	Three-Dimensional Terminal Viewer
3GL	Third Generation Language
3M	Models, Methodologies, Metadata
3PM	Three-Position Modulation
3PRO	Three PRO's: Product, Process, Project
3-S	Three Schema architecture
4CITE	For a Competitive Information and Technology Economy
4DDA	Four-Dimensional Data Assimilation (scheme)
4GL	Fourth Generation Language
5A	Assess, Access, Analyze, Act, Automate
5XBCOER	Five X-Bar Central Office Equipment Reports
650	IBM's first major computer (1954)
6DOF	Six Degrees Of Freedom
802.3	IEEE CSMA/CD bus
802.4	IEEE token passing bus
802.5	IEEE token passing ring